Current Topics in Behavioral Neurosciences

Volume 64

Current Topics in Behavioral Neurosciences provides critical and comprehensive discussions of the most significant areas of behavioral neuroscience research, written by leading international authorities. Each volume in the series represents the most informative and contemporary account of its subject available, making it an unrivalled reference source. Each volume will be made available in both print and electronic form.

With the development of new methodologies for brain imaging, genetic and genomic analyses, molecular engineering of mutant animals, novel routes for drug delivery, and sophisticated cross-species behavioral assessments, it is now possible to study behavior relevant to psychiatric and neurological diseases and disorders on the physiological level. The Behavioral Neurosciences series focuses on translational medicine and cutting-edge technologies. Preclinical and clinical trials for the development of new diagnostics and therapeutics as well as prevention efforts are covered whenever possible. Special attention is also drawn on epigenetical aspects, especially in psychiatric disorders.

CTBN series is indexed in PubMed and Scopus.

Mohammed R. Milad • Seth D. Norrholm
Editors

Fear Extinction

From Basic Neuroscience to Clinical Implications

Springer

Editors
Mohammed R. Milad
Department of Psychiatry and Behavioral Sciences
UTHealth Houston
Houston, TX, USA

Seth D. Norrholm
Psychiatry and Behav Neurosciences
Wayne State University
Detroit, MI, USA

ISSN 1866-3370 ISSN 1866-3389 (electronic)
Current Topics in Behavioral Neurosciences
ISBN 978-3-031-43004-6 ISBN 978-3-031-43005-3 (eBook)
https://doi.org/10.1007/978-3-031-43005-3

This Springer imprint is published by the registered company Springer Nature Switzerland AG
The registered company address is: Gewerbestrasse 11, 6330 Cham, Switzerland

Paper in this product is recyclable.

Preface

What Is This Book About?

Fear is a feeling that we all experience at one point or another in our lifetime. It is highly adaptive and it is fundamental to our survival. Expressing fear within inappropriate contexts or for extended periods of time can become maladaptive and unhealthy. In this book, we aim to provide the reader a neuroscientific understanding pertaining to a very simple question: how do we learn not to fear? Exploring answers to this question is very important for two reasons. First, learning about the brain mechanisms underlying fear and its extinction is of relevance to everyone's navigation through everyday life – it is such a basic, yet intriguing question that we are all curious about. Understanding brain mechanisms of fear and its regulation is therefore essential from a basic neuroscience point of view. Second, excessive fear and the inability to regulate its expression is one of the hallmarks of psychopathologies such as anxiety-, fear-, trauma-, and stressor-related disorders. And, as such, learning about how fear is acquired, stored, expressed, and regulated could help advance our understanding of the etiology of psychopathology, the maintenance of symptoms pertaining to a failure to regulate fear, and may help us develop novel therapeutics to better equip patients to quell their fears.

Chapters in this book were written by experts in the fields of basic and clinical neuroscience, experimental and clinical psychology, and neuropsychology and neuropsychiatry. The contributions are organized into four parts: Part 1 provides the reader with some theoretical and experimental concepts related to fear extinction including a discussion of basic cellular and neural circuits underlying fear extinction across species. Part 2 of the contributions is focused on data and theoretical concepts related to factors that contribute to individual differences amongst us regarding our capacity to acquire and regulate conditioned threat and fear responses. Factors impacting individual differences discussed in these chapters include sex and hormonal variations, sleep, stress, and exercise. In Part 3, we shift to the role of extinction learning processes to psychopathology and its treatment. Chapters in this part explore the relevance of fear extinction to the etiology and maintenance of psychiatric symptoms, with a high focus on post-traumatic stress disorder (PTSD) and anxiety disorders, at the developmental stage and during adulthood. A description of how dysfunction of fear extinction neural networks might contribute to psychopathology and how current treatments might restore fear homeostasis in the human brain will be reviewed in this part. In the last part of the book, chapters will focus on approaches to strengthen the functioning of the neural circuits of fear extinction, and the therapeutic implications of such enhancement. State-of-the-art approaches included novel pharmacotherapeutics, reconsolidation blockade, and technological advances including virtual reality and device-based approaches. Our aim is that contributions included in this volume will hopefully trigger and invite increased discussion into how to best generate the next sets of questions and hypotheses for the coming decade that aim to further our understanding of fear and its extinction. Domains that we expect will be of great interest to the future of this area of research include more focused studies on enhancing fear extinction with the aim of improving clinical care and treatment outcomes for fear- and anxiety-based disorders. Additionally, a growing area of research that we hope would also see major expansion soon is the application of novel computational and artificial intelligence approaches to large datasets with the aim of enhancing our current understanding of whole-brain mechanisms involved in the emotion–cognition interface pertaining to emotion regulation. Lastly, a third area of research would be in improving device-based neuromodulation tools to enhance fear extinction.

We end with an expression of gratidude and appreciation to all contributing authors for their commitment and support to this project and for conducting great science that helped build the knowldege base for this book. We also appreciate the many exceptional and talented scientists that we were not able to include in this book or that they were unable to contribute – we are very grateful to your contributions to the field of fear extinction specifically but to the larger field of emotional learning and memory broadly speaking.

Houston, TX, USA Mohammed R. Milad
Detroit, MI, USA Seth D. Norrholm

Contents

Part III Psychopathology

Part IV Therapeutics and Neuromodulations of Fear Extinction

Part I
To Fear or Not

Fear Extinction as a Psychologist Views It

Bram Vervliet

Essentially only one thing in life interests us: our psychical constitution, the mechanism of which was and is wrapped in darkness. (Pavlov 1904)

Contents

Abstract Fear extinction is a topic of central importance in translational neuroscience. It integrates knowledge from various disciplines, including clinical psychology, experimental psychology, psychiatry, cellular and systems neuroscience, and pharmacology. The experimental phenomenon of extinction was first discovered by Ivan P. Pavlov more than 100 years ago and still forms the basis for investigating the psychological and physiological mechanisms that drive extinction of fear. Here, I present old and new ways to think about fear conditioning and extinction from a psychologist's point of view. Extinction is a simple phenomenon with a complex machinery. Enhancing the behavioral analysis of extinction is necessary to advance research in neighboring disciplines as well and to increase our chances to develop extinction enhancers that might further improve efficacy of extinction-based therapies to treat dysfunctional fears. For that purpose, I address a number of fundamental questions in this chapter to clarify psychological viewpoints on the process of fear extinction. What is extinction? What is an association? What is fear? What can we

B. Vervliet (✉)
Brain and Cognition, KU Leuven, Leuven, Belgium

Leuven Brain Institute, Leuven, Belgium
e-mail: bram.vervliet@kuleuven.be

Curr Topics Behav Neurosci (2023) 64: 3–18
https://doi.org/10.1007/7854_2023_433
Published Online: 28 July 2023

learn from fear extinction? My goal is to reinforce critical thinking about basic assumptions underlying fear extinction and to open up new avenues for further research.

Keywords Associative learning · Conditioning · Extinction · Fear · Pavlov

1 What Is Extinction?

A good way to introduce the experimental phenomenon of extinction is to trace its historical roots. Extinction was first discovered, examined, and conceptualized by physiologist Ivan P. Pavlov. Following his groundbreaking work on the digestive system, which earned him a Nobel Prize in 1904, he set out to investigate the physiological underpinnings of human consciousness. He proposed, in line with a long philosophical tradition, that our stream of consciousness is not controlled by rational reasoning processes, but by arbitrary associations that connect independent ideas with each other (Pavlov 1904; Eelen and Vervliet 2006). If we want to understand human consciousness, he argued, we need to know how these associations are formed and regulated. In his eyes, the digestive system provided the perfect tool for answering this question (Pavlov 1906).

The iconic picture of Pavlov's conditioning experiment is a dog that salivates to a bell, after the bell has been paired with food. Yet, this does not characterize the typical experiment in his lab (Todes 2014). He preferred a metronome (easier to control than a ringing bell) and acid powder (elicits more saliva than food). Also, since there is no natural connection between the sound of a metronome and acid powder, Pavlov was confident that the dog had formed a new association when salivating to the metronome (rather than expressing some inborn physiological reflex). Thus, by counting drops of saliva during the sound of the metronome, Pavlov had a window on the associative principles that form the basis of consciousness. In dogs, for sure, but to him, dogs are little humans and humans are big dogs.

Pavlov developed the theory that a novel association is formed between cortical representations of the CS and US in the brain of the dog (Pavlov 1927). This association activates the representation of the acid powder (US, unconditional stimulus) during the metronome (CS, conditional stimulus), which thereby elicits drops of saliva to prepare for the US (he calls the CR a "purposeful reflex"). Then, he took his research one step further and investigated how associations are updated. That was when he discovered the phenomenon that he coined "extinction" of the previously established CR (conditional response). The procedure of extinction consisted of systematic repetitions of the metronome (CS) without the acid powder (US). The result he observed was a gradual decrease of saliva drops during the metronome (Pavlov 1927). This is how Pavlov defined extinction – which remains to this date as the definition of extinction within the field: a reduction in CR caused by presentations of the CS alone.

After having discovered extinction, the next question was how it works. What is the process that links its procedure (CS-alone presentations) to its effect

(CR decrease)? This came to be a cardinal question in his work. Most importantly, Pavlov observed that the effect of extinction is only temporary (Pavlov 1927). If, a few days after extinction, he sounded the metronome to the same dog, he saw drops of saliva falling into the observation tube again. This "spontaneous recovery" showed that extinction is a fragile phenomenon. Pavlov reasoned that the original CS–US association must survive extinction (how else could the CR return?) and as such is not erased. But if the association remains intact, the challenge is to explain the decrease of CR during extinction itself. The solution offered by Pavlov was to assume that the CS-alone presentations form a new, inhibitory CS–US association that suppresses the US representation in the neocortex of the dog and thereby counteracts the previous CS–US association. As a result, the CR decreases. His final argument was that inhibition must be weaker than excitation, and so the activity of the new, inhibitory association is more easily disrupted by the passage of time (spontaneous recovery) or by a completely novel stimulus when it appears together with the CS (disinhibition). The outcome is a return of the extinguished CR.

This view is still the backbone of extinction theory today. But now we know why inhibitory associations are more easily disturbed: they are more easily forgotten (Bouton 1994a). Retrieving inhibitory associations from memory is more difficult and depends on additional retrieval cues (other than the CS itself). The passage of time decreases the chance that cues from the original extinction learning remain retrievable, and so the CR recovers in the absence of cues that could otherwise retrieve the inhibitory association and suppress the CR. This idea was formally tested by Mark Bouton in the late 1970s (Bouton and Bolles 1979a). He manipulated cues in the environment between extinction learning and delayed testing. He found that the extinction effect (low CR) was intact only when the cues overlapped. He called the ensemble of cues the "context" in which learning and testing about specific stimuli (CS and US) take place. Whether an extinguished CR recovers depends on the context in which the CS is encountered during test.

One way to understand the role of context is to think of how contexts sometimes disambiguate the meaning of words (Bouton 1994b). The word "fire!" has a different meaning in the context of a gun versus the context of a building. The point is that, just like the word "fire," a conditioned-and-extinguished CS has two opposing meanings that are captured by its excitatory and inhibitory associations: dangerous and safe. The context in which the CS is later encountered disambiguates this double meaning: if the context is the same as during extinction learning, the meaning of safety prevails. In any other context, the danger meaning prevails. It is not that the extinction context itself has become safe, it only disambiguates the meaning of the CS. In associative learning terms, this is known as occasion-setting ("the context sets the occasion for the inhibitory association of the CS"). The extinction context forms an association with the inhibitory CS–US association and henceforth controls this association, and thereby the CR (Bouton et al. 2021).

2 What Is an Association?

Pavlovian conditioning is a cornerstone for translational research, and, as such, so is extinction. Researchers from diverse fields that range from genetics to neuroscience and experimental psychology converge on this simple but elegant paradigm to investigate how we learn and remember emotional material. This is in large part due to the concepts that Pavlov created and that allow to flexibly adjust the paradigm to different needs: CS, US, and CR can signify any stimulus or response, as long as they are bound by an association in the experiment. But what do we mean by association?

This question is more difficult than it seems. The simplest answer is that an association is the little line that connects a CS with a US when we write down what happens during Pavlovian conditioning (Eelen, personal communication). In that case, the association does little more than depicting the spatio-temporal contingency that exists between the occurrences of the CS and the US: the time between CS and US is shorter than the time between the US and the next CS (intertrial-interval). This is what constitutes a conditioning experiment.

For learning psychologists, the association is often thought of as a copper wire that transmits energy from one node to another (see Vervliet and Boddez 2020a). According to this "spreading of activation" view, perceiving a CS will activate the mental representation of this CS and, via the association that was formed during conditioning, the mental representation of the US as well. Activating the US representation then retrieves action schemas that are linked to the US or its anticipation. Because the CS typically precedes the US, the association captures something of an expectation. Indeed, psychologists often interpret "association" as an implicit or explicit expectation of the US (Boddez et al. 2013).

But associations do not necessarily invoke expectations. Memory researchers, for example, interpret Pavlovian associations as retrieval links (Bouton 1994a). From this perspective, a CS is but a retrieval cue that activates the memory of a US. Obviously, we can think of things without expecting them to happen. Receiving a holiday postcard from Turkey can retrieve memories of your own holidays in that wonderful country, yet without eliciting expectations that you will be there soon. Memory retrieval does not imply expectation, although the two are often mixed in the extinction literature. Yes, a return of CR after extinction reflects retrieval of the conditioning memory, but retrieving a conditioning memory is in itself not sufficient to produce a given CR.

Human research on extinction has revealed a distinction between what has been called expectation learning and referential learning. A CS that is paired with an aversive US typically acquires a negative *valence* consistent with the US and starts eliciting *fear* reactions in anticipation of the US. These are two classes of CRs that result from the same CS–US conditioning experiences. Remarkably, extinction decreases the fear CR, but not the negative valence (or only to a lesser degree). Thus, a conditioned-and-extinguished CS no longer elicits skin conductance reactions (related to fear), but continues to show negative valence effects in an implicit reaction time task (affective priming; Vansteenwegen et al. 2006).

One way to explain this dissociation is that fear is related to an expectation of the US, while acquired valence merely refers to the US. Hence, CS-alone presentations during extinction contradict fearful expectations that the US will arrive here and now, but do not contradict CS-valence based on mere reference to the US (Baeyens and De Houwer 1995; Baeyens et al. 2001). Both fear and valence depend on US memory retrieval, but the mere act of retrieving a memory is never 'right' or 'wrong'. Only expectations (here and now) can be violated. Thus, the reason why anticipatory fear decreases over extinction is because it is a behavioral correlate of an expectation of the US, which is contradicted by CS-alone presentations. Acquired valence, on the other hand, does not imply an expectation of the US and is therefore not contradicted by CS-alone presentations during extinction. This is why an extinguished CS elicits little fear, but remains a negative stimulus. And lingering negative valence has been shown to be a risk factor for a return of fear after extinction (Hermans et al. 2005; Zbozinek et al. 2015; but see van Dis et al. 2019).

In this light, it is good to realize that the term "fear" is a summary term for a broad array of CRs (Beckers et al. 2013), which may at times dissociate and extinguish at different speeds (Vansteenwegen et al. 2006). For example, fear does not only encompass the defensive reactions to danger, but also the subjective feeling of "being afraid." This subjective feeling requires a conscious interpretation of the current situation as dangerous and one's own feelings as fearful, whereas the defensive reactions in brain and body are thought to be regulated more automatically (LeDoux 2014). Because defensive reactions of "threat" and conscious appraisals of "fear" are subserved by specialized, separate neural circuits, they may also follow different extinction trajectories (LeDoux and Pine 2016).

Some psychologists reject the idea that automatic CS–US associations are the basis of conditioning. Instead, they argue that all conditioning relies on conscious and verbal learning, like a flash of insight (Mitchell et al. 2009). Rather than meaningless associations, they claim that propositions are acquired. Propositions are verbal hypotheses that can be right or wrong, like "the CS causes the US." In extinction, such proposition is violated and consequently updated ("the CS does not cause the US"). Conditional responses are thought to follow these propositions. The idea behind this approach is that our mental world is fundamentally verbal in nature, supported by the fact that there is little or no evidence for non-verbal, truly unaware conditioning in humans (Mertens and Engelhard 2020). This approach highlights a trivial but important fact: participants try to make sense of the situation they are in when they participate in an experiment (Norrholm et al. 2008; Vervliet and Boddez 2020b). They verbalize internally what is happening to them and summarize the experiences into propositions like "if CS, then US." But the implications of this approach are not trivial. *If* verbal propositions are indeed the sole cause of CRs, then the emphasis of conditioning research should be on investigating the environmental conditions that lead participants to form certain propositions, rather than trying to probe hypothetical and meaningless associations (De Houwer 2020). The debate is not over yet.

When clinical psychologists use the term association, they refer to something like a meaning generator. Through CS–US conditioning experiences, the CS acquires a new meaning of danger: it becomes a danger signal. Conversely, a CS that is never followed by the US transforms into a safety signal (Lissek et al. 2014). It is the association with the occurrence or non-occurrence of the US that constitutes the meaning of the CS: danger or safety. The goal of extinction, or exposure therapy, is to turn a (false) signal of danger into a signal of safety (Craske et al. 2022). As mentioned before, this means that the original excitatory association between CS and US is supplemented by an inhibitory association: An association that suppresses the expectation of danger and thereby transforms the CS into a safety signal. But how do inhibitory associations work?

There are two possibilities. The first one was proposed by Pavlov: an inhibitory association raises the activation *threshold* of the representation of the US, so that it requires more excitation to be activated (Pavlov 1927). For an extinguished CS, the inhibition equals the excitation, which means that the activation threshold is raised high enough for the excitation not to have any effect any more (which Pavlov coined "internal inhibition"). The other possibility is that the CS becomes associated with a representation of the *absence* of the US. This was first proposed by Konorski (1948), who reasoned that the omission of an expected US produces contrast effects in the brain (much like you see green spots after having stared at a piece of red paper). These contrast effects are captured in a "noUS" representation that becomes associated to the CS during extinction. When the CS activates the "noUS" representation, it counters the impact of the "US" representation and decreases the fear CR (see also Pearce and Hall 1980).

Finally, for neuroscientists, the term "association" takes on a physical dimension, in line with Pavlov's original intention, who forcefully disagreed with psychologists of his day who used terms like expectation and meaning to interpret associations. For him, associations could only have explanatory value if they are used in a physiological sense, namely, the formation of neural connections. He thought that neural representations of CS and US reside in the neocortex and that associations are formed through cortical synapses between the neurons that make up these representations. Why the cortex? Because he believed that associations are the building blocks of consciousness, which, according to many, resides in the cortex. He provided evidence for this hypothesis by lesioning parts of the cortex, which erased the CR in his dogs. Some of his contemporaries disagreed and pointed to subcortical areas as loci of association formation. We now know that they were, in part, right. At least in rats, aversive CS–US associations are formed in subcortical regions of the brain; specifically, the various nuclei of the amygdala as well as other brain regions like the thalamus (Myers and Davis 2002; Bouton et al. 2021). And, an association is not a single synapse between neurons representing the CS and the US, but an ensemble of many synapses that are formed between networks of neurons whose integrated activities represent the CS and the US (Pape and Pare 2010). Thus, neuroscientists use "association" as an abstract summary term for the complex physiological mechanisms that take place inside and between neurons during conditioning. Subsequent chapters in this book will provide detailed molecular and cellular neuronal mechanisms related to associative learning in the brain.

In conclusion, there is less agreement than you might think about what an "association" is. But this is a strength, not a weakness. The term "association" serves as a canvas onto which researchers from many different disciplines can paint their own theoretical interpretation and also provide neuroscientific evidence supporting its existence at the synaptic level in the brain. This runs the risk of losing each other in translation, for sure, but it also brings branches of research together that would otherwise remain distal to each other (Vervliet and Hermans 2013). Maybe the simplest answer was the most correct after all: an (inhibitory) association is the little line that we draw between the letters CS and US (or noUS).

3 What Is Fear?

Even if we don't really know *what* an association is, the concept still serves an important heuristic function. By depicting the memory trace of conditioning experiences as a CS–US association, it becomes clear that fear memories consist of three important components: cues from the environment (the CS), an aversive experience (the US), and the association that binds them (Davey 1989; Haesen and Vervliet 2015; Jasnow et al. 2017). Why is this important? It suggests that changes in any of these three components will have an impact on the fear elicited by new confrontations with the cues. And, it also illustrates the different processes by which CS-alone presentations might reduce fear in extinction.

First, the strength of the association represents the acquired property of the CS to activate the memory of the US, elicit US-expectancy, produce anticipatory reactions of fear. The stronger the association, the stronger the fear reaction. Extinction might entail a weakening of the association itself, or the formation of a novel inhibitory association that raises the threshold of the US representation or activates a "noUS" representation, as mentioned above. Thus, the effect of extinction might reflect changes at the level of the association. This is the dominant view. But it is not the only way to conceptualize extinction.

It is also possible that the CS–US association remains intact, but that the representation of the US weakens (Rescorla and Heth 1975). Memories of an intensely aversive experience arguably elicit stronger fear than memories of a mild experience. Thus, if the US memory, for some reason, weakens, the fear elicited by the CS should also weaken. This principle was established by showing in rats and humans that if, after a CS–US conditioning episode, the US is separately administered in decreasing (or increasing) intensity, later tests with the CS alone elicit less (or more) fear (Rescorla 1973; Haesen and Vervliet 2015; Hosoba et al. 2001; Leer et al. 2018). Also, the phenomenon of reinstatement is in line with this idea: administering the US in its original intensity reinstates fear of the CS by strengthening its representation again (Rescorla and Heth 1975). Indeed, extinction could weaken the US representation by repeatedly activating this representation in the absence of further US experiences. Although later research has discarded this explanation as a viable explanation of standard extinction (Bouton and Bolles 1979b), it has recently stimulated research for using cognitive and behavioral deflation techniques to weaken the memory of the US as an alternative route toward fear reduction (Dibbets et al. 2012; Woelk et al. 2021). The hope is that the deflation effect is less fragile than standard extinction, with clinical implications.

The CS representation is also part of the fear memory. When encountering the CS again, it activates its own representation and, by virtue of the association, also the US representation that leads to fear. Thus, an important part of the fear dynamics is that the physical CS activates the mental representation of this event. Memories fade over time, and so it is in principle possible that a fear memory loses details of the CS representation. The result will be that many more stimuli will resemble this blurred representation of the CS, activate the associated US, and produce fear. Indeed, fear generalization has been shown to depend on perceived similarity between a test stimulus and the CS, which is influenced by various psychological processes (perceptual attention, learning, memory; Zaman et al. 2021). Conversely, with less perceived similarity, generalization decreases and less fear is elicited by the test stimulus. Could these perceptual processes have an influence on the decrease of fear during extinction as well?

The gradual decrease of the fear CR over the course of extinction learning resembles the typical generalization gradient that is obtained by testing stimuli of increasing dissimilarity with the original CS. This correspondence suggests that the extinction effect could result from a changing perception of the CS over consecutive CS-alone trials (Bouton et al. 2021). With a changing perception, the difference between the currently perceived CS and the representation of the original CS in the fear memory will increase with each trial and lead to less generalization of the fear CR. The extinction CS will progressively lose its ability to activate the fear memory. But why would the perception of the CS change during extinction?

Rats are known to memorize series of trials (Capaldi 1966; Capaldi and Martins 2010). A reinforced trial that occurs after a nonreinforced trial is memorized differently than two reinforced trials in a row. This is the basis of the partial reinforcement extinction effect: the decrease of the fear CR is slowed down when the CS was only sometimes followed by the US during conditioning. The fact that the fear CR is stronger after partial reinforcement than after continuous reinforcement is paradoxical, because it means that the CR is reinforced more when given less reinforcement. It's a violation of the law of effect (Thorndike 1927) and it cannot be explained by most error-correction learning theories (e.g., Rescorla and Wagner 1972). The only way out is to assume that rats also memorize what happened on the previous trials and that they take that information into account when predicting the outcome on the next trial. After partial reinforcement, the rat continues to expect the shock even after several nonreinforced trials.

It is of course unclear whether the rat reasons about this. The more parsimonious stance is to assume that the experience of a reinforced trial is slightly different when it follows a nonreinforced trial or a reinforced trial. If, as Konorski proposed, the omission of a US leaves a separate memory trace ("noUS"), this lingering memory trace could influence the perception of the CS on the following trial. Much like the music note *sol* has a different character when it is played shortly after a *do* or a *re*. It now becomes clear how the typical extinction curve might result from decreasing generalization: As experiences of nonreinforcement cumulate over trials, the perception of the CS changes gradually, until it is no longer perceived as similar to the original acquisition CS and fear is no longer elicited (Capaldi 1966). The original fear memory remains intact, but the extinction CS is no longer able to activate it. Manipulations that could restore the perception of the CS, like a change of context (renewal), a new presentation of the US (reinstatement, reacquisition), or maybe the lapse of time (spontaneous recovery) would lead to a recovery of the intact fear response (Capaldi and Martins 2010).

The prototypical memory of a conditioning experience (CS – US) reminds us that three aspects are important for the elicitation of fear: a memory of the traumatic event (US), a memory of surrounding cues (CS), and a memory of how these cues preceded the aversive event (association). But one question remains unanswered: why would a reactivation of a memory *produce* fear? Is mere reactivation of a memory sufficient, or are there more conditions to be met? If we want to know how fear decreases over the course of extinction learning, we need to understand why we react with fear in the first place.

Although emotions are typically felt within the body, contemporary emotion theory emphasizes that emotions do not originate in bodily perceptions alone (Scherer and Moors 2019). Rather, emotions are complex interactions between appraisal of the situation, physiological arousal, felt affect (positive/negative), behavioral tendencies, and activated goals (Moors and Fischer 2019). The fear that you experience when you encounter a lion is the result of cognitive appraisals of threat (How dangerous is a lion? Am I carrying a gun? Do lions climb trees?), bodily symptoms of arousal (pounding heart, sweaty hands, tunneled attention), your experience of the moment as positive or negative, a tendency to run away, and a prioritized goal of safety that overshadows any other goal you might have pursued moments before (Moors et al. 2017).

The typical human fear conditioning study measures skin conductance as an index of fear (Lonsdorf et al. 2017). This tracks sympathetic nervous system activity, which is known to subserve physiological arousal in the body. It makes intuitive sense to think that when a re-encounter with a CS retrieves the aversive memory of a US, physiological arousal is triggered that reflects an emotional state of fear. Yet, some have argued that skin conductance is related to a defensive mechanism that is automatically triggered by threat and driven by subcortical structures of the brain, while the conscious feeling of fear additionally requires cortical elaboration that brings in cognitive appraisals and such (LeDoux and Pine 2016). This view is in line with the idea that emotions are complex interactions between cognitive and non-cognitive processes. But is skin conductance really an unconscious, automatic reaction to threat?

Some psychologists argue that even bodily reactions of fear are ultimately goal-driven.

Why does sympathetic arousal, as picked up by skin conductance, increase during moments of threat? Because it prepares the body for action: the action tendencies elicited by threat (fight/flight) require a momentary increase of energy to skeletal muscles, which is provided by the sympathetic nervous system (Kolb et al. 2019). But how you react to threat depends on the current goals that you have and an evaluation of how different actions might lead to those goals (Boddez et al. 2020). By comparing the expected utility of each action (an estimation of the extent to which it will lead to the desired goal), the action with highest utility is selected to maximize chances to reach the goal. If this goal is to remain harm-free, then fear-related actions will be activated in order to prepare for fight or flight. Of note, some goals serve basic needs, such as survival, and may be more innate and less flexible than other goals. This makes it more difficult, though not impossible, to override survival goals during moments of perceived threat. After all, this is what an anxiety patient needs to do: counteracting survival goals by approaching situations of perceived threat.

What other goals could one have during threat than staying harm-free? Wanting to contribute to science, for example. The mere fact that participants in human fear conditioning experiments do not rip off the electrodes and storm out of the laboratory shows how different goals compete during moments of threat. They probably would like to storm out, hence the increased skin conductance, but they don't. Why not? Because they understand that science is important (and they want to collect their participation fee at the end). These are competing goals to staying harm-free. Or consider the example of a terminally ill patient who requested euthanasia and welcomes, not fears, the syringe with lethal substance (Boddez et al. 2020). The goal of being pain-free can overcome the goal of staying harm-free.

From a psychological perspective, it is important to know which beliefs and goals participants entertain when they participate in a human fear conditioning and extinction experiment. It is well-known, for example, that anxiety patients display slower fear extinction learning: skin conductance or US-expectancy remains higher over a series of CS-alone trials in extinction (Duits et al. 2015). Maybe this is related to difficulties in building up or retrieving inhibitory CS – US memories that indicate safety (Milad and Quirk 2012). It is, however, equally possible that patients have intact memories of what happened on the previous trials, but do not believe that these prior experiences are informative for future confrontations with the CS. A crucial test would be to reinforce this belief in anxiety patients via instructions, and see whether it alleviates the safety deficit. If so, the anxiety problem lies in beliefs about the world, rather than in building or retrieving extinction memories.

Likewise, goals could be regulated through instructions, for example, by offering a vignette story that devalues the importance of living a harm-free life (Boddez et al. 2020). If this decreases the goal of staying harm-free in the experiment, it might reduce action tendencies to flee from harm and thereby lower skin conductance. Another possibility would be to instruct the participant that they could opt for an alternative experiment, but it involves more painful stimulations. In this case, staying within the fear conditioning experiment aligns with the goal of staying harm-free and should no longer activate action tendencies and skin conductance.

By critically reflecting on what fear is and how it relates to CS–US memories, we can approach extinction from more angles, including manipulations that influence CS processing or US processing, or even manipulations that sever the maybe simplistic link between fear and memory.

4 What Can We Learn from Fear Extinction?

It was acknowledged more than 100 years ago that fear extinction experiments are relevant for improving clinical treatment of dysfunctional fears. John B. Watson and Rosalie Rayner, in their infamous study of the conditioning of fear in Little Albert (Watson and Rayner 1920), proposed a suite of extinction-like interventions to remove such fears. Inspired by discoveries in Pavlov's lab, they mentioned standard extinction, counterconditioning, approach training, and vicarious extinction learning. These were and still are great ideas (Fullana et al. 2020). But how do we know whether findings from the fear extinction laboratory bear any relevance to clinical treatment?

The question at hand is whether human fear extinction has external validity with regard to exposure treatment of dysfunctional fears (Scheveneels et al. 2016). Several criteria of validity have been developed that can guide us to assess the relevance of fear extinction for treatment (Vervliet and Raes 2013). Face validity refers to a superficial similarity in terms of procedure and effect (if it quacks like a duck, walks like a duck, it is a duck). This criterion seems fulfilled: the core events in fear extinction and exposure treatment are repeated confrontations to a fear-eliciting stimulus (CS) in a safe situation (noUS) until the fear declines (extinction).

Construct validity relates to the degree of similarity between the theory of extinction and the theory of how exposure works. While older theories of exposure emphasized a process of habituation, more recent theories emphasize inhibitory learning via expectancy violation as the driver of fear decreases (Craske et al. 2014, 2022). This is in accordance with the central emphasis on the role of the prediction error in extinction learning (Bouton et al. 2021). There is a strong connection between extinction and exposure in terms of associative learning theory.

And then there are two empirical criteria: prospective and predictive validity. Prospective validity is tested in patients and increases when individual differences in fear extinction relate to individual differences in exposure outcome. Thus, it boils down to inviting anxiety patients to a fear extinction experiment in the lab, before they take part in exposure therapy (Scheveneels et al. 2016). Correlations between extinction and exposure increases the likelihood that fear extinction captures processes that are involved in exposure (Lange et al. 2020; Maples-Keller et al. 2022). Indirect tests of prospective validity consist of examining whether certain psychological traits influence *both* fear extinction and exposure outcome. Examples are trait anxiety, self-efficacy, intolerance of uncertainty and sense of control (Beckers et al. 2023).

Predictive validity, finally, is the degree to which fear extinction is able to pick out interventions that enhance exposure therapy outcome. Thus, it addresses the question whether extinction enhancers that reduce fear in the lab also boost the decline of fear in exposure therapy. For example, both extinction and exposure effects are enhanced when feared stimuli are encountered in multiple contexts or when multiple variations of the feared stimulus itself are used (see Craske et al. 2022). Likewise, training interventions to induce a positive mood have been shown to enhance both extinction and exposure (Zbozinek and Craske 2017). The same is true for training interventions that increase self-efficacy (Zlomuzica et al. 2020).

Fear extinction passes these criteria of validity relatively well (Beckers et al. 2023). This means that we can be relatively confident that new findings in fear extinction experiments will translate, to some extent, to exposure treatment. Why is this important? Randomized clinical controlled trials (RCCTs) are time- and effort-consuming; they can easily go on for years. Pre-clinical testing of candidate enhancers for exposure therapy represents a cost-efficient tool for making the best available choices when conducting RCCTs. And, no less important, insight in the working mechanisms of therapies provides clinicians with better handles to apply structured protocols in a personalized manner (Huibers et al. 2021).

One example is the emphasis on *prediction error* as the motor of fear extinction learning. The prediction error is the mismatch between the predicted occurrence of the US (based on the CS–US association) and the actual absence of the US. The size of the prediction error is thought to be the teaching signal that triggers novel CS–noUS learning to match the absence of the US. Formalized theories of associative learning have specified the dynamics of the prediction error signal over the course of extinction learning. These simulations show that the prediction error peaks on early trials of fear extinction, when the mismatch is strongest, where it triggers fast learning by building a competing CS–noUS association. On later extinction trials, when the CS–noUS association cancels the CS–US association, the absence of the US no longer elicits a prediction error and no further learning will occur. Most learning in extinction occurs when the expectation of the US is still high.

How does this translate to psychotherapy? In exposure treatment, patients are guided toward feared situations that trigger their expectation of a dreaded outcome. Contemporary exposure theory states that this constitutes a prediction error, as the dreaded outcome of unrealistic fears stays away. Hence, the stronger the prediction error, the more extinction learning will take place that can decrease the unwanted fear. From this perspective, exposure exercises should be designed to maximize the prediction error, or in cognitive terms, maximize the violation of the fearful expectations. It is becoming routine practice to ask patients about their concrete expectations when they enter the exposure situation and to ask how much their expectation was violated when the exposure is over. More violation means more extinction learning (Craske et al. 2022).

An open question in this new approach is how to measure prediction error processing during exposure exercises. Asking about expectations and violations is one way. In my own lab, we are investigating the emotion of relief as a possible index of prediction error in fear extinction (Vervliet et al. 2017; Papalini et al. 2021).

Relief is the pleasant surprise that is felt when an expected aversive event remains absent (Deutsch et al. 2015). We have shown in healthy individuals that subjective relief during omissions of an aversive US follows the temporal dynamics of a prediction error signal over the course of fear extinction (high during early trials and lower during later trials). We believe that subjective relief is an emotional correlate of the prediction error during fear extinction and might be helpful for tracking successful induction of extinction learning during the course of exposure treatment (Willems and Vervliet 2021).

5 Conclusion

Extinction is a simple phenomenon with a complex machinery. Fear extinction is a nexus of trans-disciplinary research that is advancing our understanding on multiple levels of analysis. In this chapter, I have put together what I believe is a psychologist's view on fear extinction. It invites us to critically reflect central concepts that we use to describe the process of fear extinction and to expand the types of intervention that are usually envisioned to enhance extinction learning and memory. We have arrived at surprising possibilities, such as manipulating participants' beliefs about the world when they are about to enter a fear extinction experiment.

Asking what fear extinction is immediately brings up questions about what fear is, what an association is, and how memories of aversive conditioning experiences produce fear. Pavlov purportedly claimed that "If you want new ideas, read old books." I have traced historical interpretations and reviewed alternative ways of thinking about fear extinction, in order to shed light on various psychological theories of extinction learning and memory. This is important not only for psychology itself. It has been argued before that neuroscience can only advance as far as a clear understanding of behavioral processes allows it to go (Krakauer et al. 2017). The neuroscience of fear extinction depends on a clear view on the behavioral processes that occur during fear extinction. My hope is that this contribution may help to critically reflect on some of the basic assumptions in fear extinction research and create opportunities for new discoveries.

References

Baeyens F, De Houwer J (1995) Evaluative conditioning is a qualitatively distinct form of classical conditioning: a reply to Davey (1994). Behav Res Ther 33:825–831

Baeyens F, Vansteenwegen D, Hermans D, Eelen P (2001) Chilled white wine, when all of a sudden the doorbell rings: mere reference and evaluation versus expectancy and preparation in human Pavlovian learning. In: Columbus F (ed) Advances in psychology research, vol 4. Nova Science Publishers, pp 241–277

Beckers T, Krypotos AM, Boddez Y, Effting M, Kindt M (2013) What's wrong with fear conditioning? Biol Psychol 92:90–96

Beckers T, Hermans D, Lange I, Luyten L, Scheveneels S, Vervliet B (2023) Understanding clinical fear and anxiety through the lens of human fear conditioning. Nat Rev Psychol
Boddez Y, Baeyens F, Luyten L, Vansteenwegen D, Hermans D, Beckers T (2013) Rating data are underrated: validity of US expectancy in human fear conditioning. J Behav Ther Exp Psychiatry 44:201–206
Boddez Y, Moors A, Mertens G, De Houwer J (2020) Tackling fear: beyond associative memory activation as the only determinant of fear responding. Neurosci Biobehav Rev 112:410–419
Bouton ME (1994a) Conditioning, remembering, and forgetting. J Exp Psychol Anim Behav Process 20:219–231
Bouton ME (1994b) Context, ambiguity, and classical conditioning. Curr Dir Psychol Sci 3:49–53
Bouton ME, Bolles RC (1979a) Contextual control of the extinction of conditioned fear. Learn Motiv 10:445–466
Bouton ME, Bolles RC (1979b) Role of conditioned contextual stimuli in reinstatement of extinguished fear. J Exp Psychol Anim Behav Process 5(4):368
Bouton ME, Maren S, McNally GP (2021) Behavioral and neurobiological mechanisms of Pavlovian and instrumental extinction learning. Physiol Rev 101:611–681
Capaldi EJ (1966) Partial reinforcement: a hypothesis of sequential effects. Psychol Rev 73:459
Capaldi EJ, Martins AP (2010) Applying memories of reinforcement outcomes mainly to Pavlovian conditioning. Learn Motiv 41:187–201
Craske MG, Treanor M, Conway CC, Zbozinek T, Vervliet B (2014) Maximizing exposure therapy: an inhibitory learning approach. Behav Res Ther 58:10–23
Craske MG, Treanor M, Zbozinek TD, Vervliet B (2022) Optimizing exposure therapy with an inhibitory retrieval approach and the OptEx Nexus. Behav Res Ther 152:104069
Davey GC (1989) UCS revaluation and conditioning models of acquired fears. Behav Res Ther 27: 521–528
De Houwer J (2020) Revisiting classical conditioning as a model for anxiety disorders: a conceptual analysis and brief review. Behav Res Ther 127:103558
Deutsch R, Smith KJ, Kordts-Freudinger R, Reichardt R (2015) How absent negativity relates to affect and motivation: an integrative relief model. Front Psychol 6:152
Dibbets P, Poort H, Arntz A (2012) Adding imagery rescripting during extinction leads to less ABA renewal. J Behav Ther Exp Psychiatry 43:614–624
Duits P, Cath DC, Lissek S, Hox JJ, Hamm AO, Engelhard IM et al (2015) Updated meta-analysis of classical fear conditioning in the anxiety disorders. Depress Anxiety 32:239–253
Eelen P, Vervliet B (2006) Fear conditioning and clinical implications: what can we learn from the past? In: Craske MG, Hermans D, Vansteenwegen D (eds) Fear and learning: from basic processes to clinical implications. American Psychological Association, pp 17–35
Fullana MA, Dunsmoor JE, Schruers KRJ, Savage HS, Bach DR, Harrison BJ (2020) Human fear conditioning: from neuroscience to the clinic. Behav Res Ther 124:103528
Haesen K, Vervliet B (2015) Beyond extinction: habituation eliminates conditioned skin conductance across contexts. Int J Psychophysiol 98:529–534
Hermans D, Dirikx T, Vansteenwegenin D, Baeyens F, Van den Bergh O, Eelen P (2005) Reinstatement of fear responses in human aversive conditioning. Behav Res Ther 43:533–551
Hosoba T, Iwanaga M, Seiwa H (2001) The effect of UCS inflation and deflation procedures on 'fear'conditioning. Behav Res Ther 39:465–475
Huibers MJ, Lorenzo-Luaces L, Cuijpers P, Kazantzis N (2021) On the road to personalized psychotherapy: a research agenda based on cognitive behavior therapy for depression. Front Psych 11:1551
Jasnow AM, Lynch JF III, Gilman TL, Riccio DC (2017) Perspectives on fear generalization and its implications for emotional disorders. J Neurosci Res 95:821–835
Kolb B, Whishaw IQ, Teskey GC (2019) An introduction to brain and behavior, 6th edn. Macmillan Learning
Konorski J (1948) Conditioned reflexes and neuron organization. Cambridge University Press

Krakauer JW, Ghazanfar AA, Gomez-Marin A, MacIver MA, Poeppel D (2017) Neuroscience needs behavior: correcting a reductionist bias. Neuron 93:480–490

Lange I, Goossens L, Michielse S, Bakker J, Vervliet B, Marcelis M et al (2020) Neural responses during extinction learning predict exposure therapy outcome in phobia: results from a randomized-controlled trial. Neuropsychopharmacology 45:534–541

LeDoux JE (2014) Coming to terms with fear. Proc Natl Acad Sci 111:2871–2878

LeDoux JE, Pine DS (2016) Using neuroscience to help understand fear and anxiety: a two-system framework. Am J Psychiatr 173:1083–1093

Leer A, Haesen K, Vervliet B (2018) Beyond extinction: prolonged conditioning and repeated threat exposure abolish contextual renewal of fear-potentiated startle discrimination but leave expectancy ratings intact. Front Psych 9:117

Lissek S, Bradford DE, Alvarez RP, Burton P, Espensen-Sturges T, Reynolds RC, Grillon C (2014) Neural substrates of classically conditioned fear-generalization in humans: a parametric fMRI study. Soc Cogn Affect Neurosci 9:1134–1142

Lonsdorf TB, Menz MM, Andreatta M, Fullana MA, Golkar A, Haaker J et al (2017) Don't fear 'fear conditioning': methodological considerations for the design and analysis of studies on human fear acquisition, extinction, and return of fear. Neurosci Biobehav Rev 77:247–285

Maples-Keller J, Watkins LE, Nylocks KM, Yasinski C, Coghlan C, Black K et al (2022) Acquisition, extinction, and return of fear in veterans in intensive outpatient prolonged exposure therapy: a fear-potentiated startle study. Behav Res Ther 154:104124

Mertens G, Engelhard IM (2020) A systematic review and meta-analysis of the evidence for unaware fear conditioning. Neurosci Biobehav Rev 108:254–268

Milad MR, Quirk GJ (2012) Fear extinction as a model for translational neuroscience: ten years of progress. Annu Rev Psychol 63:129–151

Mitchell CJ, De Houwer J, Lovibond PF (2009) The propositional nature of human associative learning. Behav Brain Sci 32:183–198

Moors A, Fischer M (2019) Demystifying the role of emotion in behaviour: toward a goal-directed account. Cognit Emot 33:94–100

Moors A, Boddez Y, De Houwer J (2017) The power of goal-directed processes in the causation of emotional and other actions. Emot Rev 9:310–318

Myers KM, Davis M (2002) Behavioral and neural analysis of extinction. Neuron 36:567–584

Norrholm SD, Vervliet B, Jovanovic T, Boshoven W, Myers KM, Davis M, Rothbaum B, Duncan EJ (2008) Timing of extinction relative to acquisition: a parametric analysis of fear extinction in humans. Behav Neurosci 122(5):1016–1030. https://doi.org/10.1037/a0012604

Papalini S, Ashoori M, Zaman J, Beckers T, Vervliet B (2021) The role of context in persistent avoidance and the predictive value of relief. Behav Res Ther 138:103816

Pape HC, Pare D (2010) Plastic synaptic networks of the amygdala for the acquisition, expression, and extinction of conditioned fear. Physiol Rev 90:419–463

Pavlov IP (1904) The physiology of digestion. https://www.nobelprize.org/prizes/medicine/1904/pavlov/lecture. Accessed 25 Mar 2022

Pavlov IP (1906) The scientific investigation of the psychical faculties or processes in the higher animals. Science 24:613–619

Pavlov IP (1927) Conditioned reflexes: an investigation of the physiological activity of the cerebral cortex. Oxford University Press, London

Pearce JM, Hall G (1980) A model for Pavlovian learning: variations in the effectiveness of conditioned but not of unconditioned stimuli. Psychol Rev 87:532–552

Rescorla RA (1973) Effects of US habituation following conditioning. J Comp Physiol Psychol 82(1):137

Rescorla RA, Heth CD (1975) Reinstatement of fear to an extinguished conditioned stimulus. J Exp Psychol Anim Behav Process 1:88–96

Rescorla RA, Wagner AR (1972) A theory of Pavlovian conditioning: variations in the effectiveness of reinforcement and nonreinforcement. In: Black AH, Prokasy WF (eds) Classical conditioning II: current research and theory. Appleton-Century-Crofts, New York, pp 64–99

Scherer KR, Moors A (2019) The emotion process: event appraisal and component differentiation. Annu Rev Psychol 70:719–745

Scheveneels S, Boddez Y, Vervliet B, Hermans D (2016) The validity of laboratory-based treatment research: bridging the gap between fear extinction and exposure treatment. Behav Res Ther 86: 87–94

Thorndike EL (1927) The law of effect. Am J Psychol 39:212–222

Todes DP (2014) Ivan Pavlov: a Russian life in science. Oxford University Press

van Dis EA, Hagenaars MA, Bockting CL, Engelhard IM (2019) Reducing negative stimulus valence does not attenuate the return of fear: two counterconditioning experiments. Behav Res Ther 120:103416

Vansteenwegen D, Francken G, Vervliet B, De Clercq A, Eelen P (2006) Resistance to extinction in evaluative conditioning. J Exp Psychol Anim Behav Process 32:71–79

Vervliet B, Boddez Y (2020a) Memories of 100 years of human fear conditioning research and expectations for its future. Behav Res Ther 135:103732

Vervliet B, Boddez Y (2020b) Aversive stimulus pairings are an unnecessary and insufficient cause of pathological anxiety. Biol Psychiatry 87:870–871

Vervliet B, Hermans D (2013) Translation: that's the question. Biol Psychol 92:1

Vervliet B, Raes F (2013) Criteria of validity in experimental psychopathology: application to models of anxiety and depression. Psychol Med 43:2241–2244

Vervliet B, Lange I, Milad MR (2017) Temporal dynamics of relief in avoidance conditioning and fear extinction: experimental validation and clinical relevance. Behav Res Ther 96:66–78

Watson JB, Rayner R (1920) Conditioned emotional reactions. J Exp Psychol 3:1–14

Willems AL, Vervliet B (2021) When nothing matters: assessing markers of expectancy violation during omissions of threat. Behav Res Ther 136:103764

Woelk M, Krans J, Raes F, Vervliet B, Hagenaars MA (2021) Imagery rescripting versus extinction: distinct and combined effects on expectancy and revaluation learning. Clin Psychol Sci:21677026211055169

Zaman J, Chalkia A, Zenses AK, Bilgin AS, Beckers T, Vervliet B, Boddez Y (2021) Perceptual variability: implications for learning and generalization. Psychon Bull Rev 28:1–19

Zbozinek TD, Craske MG (2017) The role of positive affect in enhancing extinction learning and exposure therapy for anxiety disorders. J Exp Psychopathol 8:13–39

Zbozinek TD, Hermans D, Prenoveau JM, Liao B, Craske MG (2015) Post-extinction conditional stimulus valence predicts reinstatement fear: relevance for long-term outcomes of exposure therapy. Cognit Emot 29:654–667

Zlomuzica A, Schneider S, Konrad C, Merz CJ, Wolf OT, Raeder F, Margraf J (2020) Clinical implications of fear extinction in anxiety disorders. Neuroforum 26(3):143–149

Beyond Fear, Extinction, and Freezing: Strategies for Improving the Translational Value of Animal Conditioning Research

Christopher K. Cain

Contents

Abstract Translational neuroscience for anxiety has had limited success despite great progress in understanding the neurobiology of Pavlovian fear conditioning and extinction. This chapter explores the idea that conditioning paradigms have had a modest impact on translation because studies in animals and humans are misaligned in important ways. For instance, animal conditioning studies typically use imminent threats to assess short-duration fear states with single behavioral measures (e.g., freezing), whereas human studies typically assess weaker or more prolonged anxiety

C. K. Cain (✉)
Department of Child and Adolescent Psychiatry, NYU Langone Health, New York, NY, USA

Emotional Brain Institute, Nathan Kline Institute for Psychiatric Research, Orangeburg, NY, USA
e-mail: Christopher.Cain@nki.rfmh.org

Curr Topics Behav Neurosci (2023) 64: 19–58
https://doi.org/10.1007/7854_2023_434
Published Online: 3 August 2023

states with physiological (e.g., skin conductance) and self-report measures. A path forward may be more animal research on conditioned anxiety phenomena measuring dynamic behavioral and physiological responses in more complex environments. Exploring transitions between defensive brain states during extinction, looming threats, and post-threat recovery may be particularly informative. If care is taken to align paradigms, threat levels, and measures, this strategy may reveal stable patterns of non-conscious defense in animals and humans that correlate better with conscious anxiety. This shift in focus is also warranted because anxiety is a bigger problem than fear, even in disorders defined by dysfunctional fear or panic reactions.

Keywords Fear · Anxiety · Translation · Threat · Extinction · Freezing

Translational neuroscience for anxiety has fallen short of expectations (Hyman 2013), and much of the blame has been placed on animal studies used to predict treatment efficacy in the clinic (Haller et al. 2013; Kaffman et al. 2019). This is particularly disappointing for researchers studying Pavlovian fear conditioning (FC), otherwise considered one of the most successful paradigms in modern neuroscience. Detailed models of the psychological and neural processes mediating FC are supported by converging evidence from many labs over about 50 years (Johansen et al. 2011; Fadok et al. 2018; Josselyn and Tonegawa 2020). Research on human subjects using brain imaging largely confirms that conditioning phenomena are associated with activity in conserved threat processing circuits identified in animals (Delgado et al. 2008; Milad and Quirk 2012). And since people reacting to conditioned threats in the laboratory also report feelings of distress (Mobbs et al. 2007), there has been much hope for translating knowledge about conditioning mechanisms into better treatments.

Although it is true that progress in understanding the neurobiology of conditioned fear occurred in parallel with translational failures, the paradigm has played only a minor role in private drug discovery efforts. Academic animal researchers have used the task more frequently to study the neural mechanisms of memory (Bouton et al. 2001), perhaps growing complacent about the relevance of this work to emotion. Practical and ethical considerations in human research have also led to a situation where cross-species studies induce and measure "fear" in different ways. Indeed, an open secret in conditioning research is that animal studies typically use intense threats whereas conditioned humans "know they are participating in kind of a make-believe situation that never is allowed to become dangerous" (Ohman and Mineka 2001). Given this disconnect, it remains unclear if the wealth of information obtained from FC research can be exploited to improve the treatment of human anxiety.

A great deal has been written about this translational crisis, and I refer readers to a small sample of excellent reviews (Markou et al. 2009; Stewart et al. 2015; Beckers et al. 2013; Dunsmoor et al. 2022). Here I will suggest a path forward using animal aversive conditioning paradigms to induce lower-level defensive states. This simple shift in strategy leverages knowledge gained from FC studies while also helping to align the degree of threat in animal and human research. It will also develop basic knowledge about an understudied mode of defense that appears dysfunctional in human anxiety disorders. A renewed focus on conditioned anxiety is also warranted because anxiety may be the primary problem, even for disorders defined by fear and panic (Bouton et al. 2001). Studies using more intense threats in humans would complement this strategy but will not be covered here as this work is more difficult and less relevant to prolonged states of distress. I will also join others (Stewart et al. 2015; Adolphs and Anderson 2018; Fanselow and Pennington 2018; Haaker et al. 2019; Constantinou et al. 2021; LeDoux 2012; Shackman and Fox 2016; Pine and LeDoux 2017) in arguing that, to the extent possible, we must use identical paradigms and dependent measures in cross-species studies to improve translation and understand its limits.

1 Rodent Assays and Translational Strategies

Functional behavior systems approaches, like Predatory Imminence Theory (Fanselow and Lester 1988), provide a useful framework for considering response topography across varying levels of threat. Defensive survival circuits evolved primarily under predatory pressure to detect threats, prevent harm, and enable a return to preferred activities (e.g., foraging, mating, sleeping). Defensive modes are organized hierarchically and selected based on perceived threat imminence (Moscarello and Penzo 2022). Distant, diffuse, or uncertain threats activate the circuits causing minor deviations from preferred activity over long time frames. This "pre-encounter" mode is the most flexible stage of defense, characterized by risk-assessment, goal-directed behavior, and subtle changes in exploration and feeding (Blanchard and Blanchard 1989; Helmstetter and Fanselow 1993). As threats become more likely, better defined, and closer, different circuit elements are activated and behavior is restricted to species-specific defensive responses (SSDRs). These "post-encounter" (e.g., freezing) and "circa-strike" (e.g., flight) behaviors are short-duration, hard-wired, reflexive reactions that evolved to cope with immediate danger (Bolles 1960). Cognitive and autonomic responses also occur along this continuum to support defense and prime circuits for rapid reactions should threats escalate (Roelofs and Dayan 2022). Pre-encounter, post-encounter, and circa-strike defense modes have been related to emotional states of anxiety, fear, and panic in humans (Perusini and Fanselow 2015) (Fig. 1). Note, however, that in Predatory Imminence Theory and other "central state" models, these terms refer to non-subjective brain states indicated by specific response patterns to distinct levels of threat. To the extent that an organism is capable of conscious feelings, these are assumed to co-occur as part of a coordinated defensive response. Since feelings are unobservable, they are not a focus of study. Instead, observable behavior is the focus because this is the primary way an organism interacts with the environment to thwart threats to survival (Fanselow 2018a). To avoid confusion, I will refer to verbal

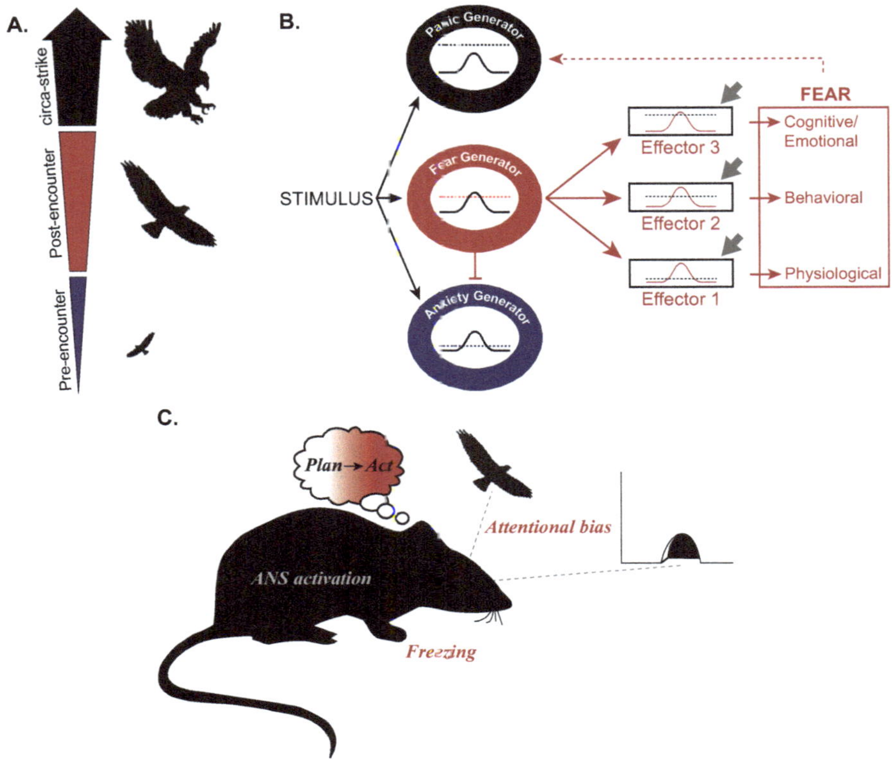

Fig. 1 Threat imminence, defensive modes, and coordinated fear responding in a central state framework. (**a**) Pre-encounter, post-encounter, or circa-strike response modes are engaged by different levels of threat, such as predators that are distant, near, or attacking. (**b**) Brain states of anxiety, fear, and panic are mediated by partially overlapping brain circuits. Distinct elements of these circuits that detect threats and orchestrate threat-appropriate responding are called generators. In the example, a stimulus is received by all three generators but fails to exceed the threshold for panic. Response modes are arranged hierarchically, so once the fear threshold is reached, the anxiety system is suppressed. Specific responses are initiated by downstream effector regions that have their own thresholds of activation. Effectors also have additional functions and unique regulatory influences (gray arrows). Thus, directional correlations between responses are typically high whereas response magnitude correlations at any snapshot in time are lower (Roberts and Young 1971). A central fear state is indicated by the coordinated threat-specific response. Fear influences multiple systems throughout the brain and body. For instance, one function of fear is to prime other circuits for rapid and strong responses should threats escalate. An example is potentiation of circa-strike reflexes like startle, represented here by the dashed line to the panic generator. Central state models are sometimes called "one-system" frameworks because all responses depend on a common input, though this does not mean that the common input has equal influence on all responses. Note that discordant responding can be explained by factors like effector thresholds or differential regulation of effectors. Treatments that affect generators should affect all responses in that mode, but selective effects are also possible by targeting effectors. (**c**) Illustration of a coordinated rat fear response to imminent threat. Cognitive responses include attentional biases to threats and escape routes (safety). The dominant post-encounter behavioral response is freezing. Physiological responses include bradycardia, reduced respiratory rate, release of stress hormones,

reports or other data based on introspection in humans as "conscious," "subjective," or "feelings of" fear.

More than 100 different assays have been used to trigger defense and study survival circuits in animals, mainly rodents (Haller et al. 2013). Historically, anxiolytic drug development relied on spontaneous anxiety assays that measure innate aversion to bright, open spaces against the motivation to explore (e.g., elevated plus maze; EPM). Punishment paradigms were also common (Jean-Richard-Dit-Bressel et al. 2018). The popularity of these classical assays is largely due to their sensitivity to known anxiolytics—especially benzodiazepines (Gray and McNaughton 2000). Though these tests evoke a range of defensive behavior, the procedures vary considerably between tasks and the underlying psychological mechanisms are often unclear, leading some to refer to them as "black box" assays that select for pharmacological isomorphism (Hyman 2013; Perusini and Fanselow 2015; Jean-Richard-Dit-Bressel et al. 2018; Cain et al. 2013). This coupled with the diffuse nature of eliciting stimuli (e.g., brightness/openness) and dependent measures (e.g., minutes in open area) makes it difficult to study underlying brain circuits with precision and hampers mechanism-based design of novel anxiolytics (Markou et al. 2009; Griebel and Holmes 2013; Geyer and Markou 2002). Finally, spontaneous anxiety tasks are less useful for discovery of agents that may facilitate therapy-related learning, as extinction tasks are meant to model.

Pavlovian conditioning protocols are popular rodent assays for activating survival circuits and studying neurobiological mechanisms of learning and post-encounter defense. These involve pairing a neutral conditioned stimulus (CS) with an innately aversive unconditioned stimulus (US). After conditioning, CS-alone presentations trigger a rapid, coordinated response in anticipation of the US (LeDoux and Daw 2018). Unlike the tasks above, researchers have a high degree of control over conditioning stimuli which allows standardization of threat levels and tracking of information flow in the brain from sensation to responding. Take acoustic startle in rats as an example (reviewed in Cain et al. 2013). Startle initiation begins 8 ms after sound onset. Lesion work identified five brain relays between the ear and hind leg muscles required for this response. Stimulation studies established the order of processing in this circuit from an early auditory relay (ventral cochlear nucleus) to the spinal cord. Other studies demonstrated that a projection of the central amygdala (CeA) to the brainstem was required for potentiation of startle by conditioned threats. Similar work established the basolateral amygdala (BLA) as an early site of CS–US convergence necessary for FC plasticity. Sensory pathways to BLA were identified for auditory, visual, and contextual CSs, as well as shock USs. Again, latency data helped establish the direction of information flow in this circuit. Other

Fig. 1 (continued) and increased skin conductance. Though volitional action is generally suppressed in high fear states (Fanselow 2018b), goal-directed planning may still occur so that directed action is prepared once threat imminence wanes (Roelofs and Dayan 2022; LeDoux and Daw 2018)

work demonstrated that CeA output pathways to different effector regions orchestrate behavioral and physiological responses to threat. This strategy for identifying a functional circuit proved invaluable for advancing our understanding of mammalian associative learning and post-encounter defense. Hundreds of subsequent studies anchored to this circuit identified microcircuits, cells, molecules, and genes contributing to acquisition, consolidation, retrieval, reconsolidation, and extinction of FC (Johansen et al. 2011; Josselyn and Tonegawa 2020; Cain et al. 2013; Sotres-Bayon and Quirk 2010; Janak and Tye 2015; Maren 2005; Tovote et al. 2015). Work in humans largely confirms animal work (Milad and Quirk 2012; Delgado et al. 2006; Fullana et al. 2020; Mobbs et al. 2015; Peng et al. 2022).

Despite this progress, the FC paradigm has not been used in a systematic way to identify and test novel treatments. Academic labs occasionally identified novel treatments with animal conditioning paradigms that were then evaluated in humans via verbal report, sometimes adding a direct test of the FC effect identified in rodents. Private drug discovery efforts seem to have used FC mainly as a secondary confirmation of agents identified with classical anxiety assays (Griebel and Holmes 2013). Promising candidates were routinely evaluated in humans via verbal reports assessing anxiety disorder symptoms—leaving unanswered the question of whether the FC effects identified in animals also occur in humans. Briefly reviewing two prominent translational efforts illustrates the current predicament.

1.1 Propranolol in Post-Traumatic Stress Disorder (PTSD)

The beta blocker propranolol has been evaluated in multiple clinical trials for its potential to prevent PTSD when given post-trauma (Stein et al. 2007; Vaiva et al. 2003; Reist et al. 2001; Pitman et al. 2002; Nugent et al. 2010; Hoge et al. 2012; McGhee et al. 2009; Rosenberg et al. 2018). These were inspired by studies showing dysregulated norepinephrine signaling in PTSD (Southwick et al. 1993) and impaired consolidation of inhibitory avoidance (IA) with post-training propranolol in rodents (Liang et al. 1986). IA is related to FC, but measures the reluctance to enter a context where shock occurred, rather than SSDRs directly elicited by a Pavlovian CS. Only one study found a significant reduction in PTSD as assessed by verbal report (Vaiva et al. 2003), though propranolol reduced some physiological responses to trauma reminders (Pitman et al. 2002; Hoge et al. 2012). There are many potential explanations for this apparent failure. The interval between trauma and drug administration may have been too long to block consolidation. Or propranolol may have blocked consolidation of hippocampal memories important for IA but left amygdala-dependent FC memories underpinning PTSD intact (Debiec and LeDoux 2004; Grillon et al. 2004). Another possibility is that propranolol does not block consolidation of IA in humans. Or maybe IA behavior correlates poorly with subjective PTSD symptoms. Unfortunately, it is difficult to differentiate between these and other possibilities because propranolol's ability to block IA consolidation in humans was never evaluated.

A second wave of interest in propranolol to treat established PTSD came after rodent studies demonstrated that protein synthesis inhibitors or propranolol weaken reconsolidation of Pavlovian freezing after a reminder of conditioning (CS presentation that reactivates and destabilizes the CS–US memory, requiring another round of consolidation (Giustino et al. 2016; Raut et al. 2022)). However, subsequent studies found that these effects are less robust than originally thought. "Boundary conditions" related to strength of memory, age of memory, and degree of prediction error during reactivation limit the effectiveness of treatments targeting reconsolidation (Kida 2020; Gerlicher et al. 2022). Perhaps more troubling, there is now evidence that memories reactivated in the same way may or may not destabilize for unknown reasons (Luyten et al. 2021; Cox et al. 2022; Stemerding et al. 2022). Not surprisingly (in retrospect), numerous clinical studies of propranolol in reconsolidation using different methodologies produced mixed results (Raut et al. 2022; Pigeon et al. 2022).

1.2 CRF1 Antagonists in Generalized Anxiety Disorder (GAD) and PTSD

Corticotropin releasing factor plays an important role in coordinating the autonomic, endocrine, and behavioral response to stress, primarily via activation of type 1 (CRF1) receptors (Bale and Vale 2004). Numerous reports of CRF1 antagonists reducing fear-, anxiety-, and stress-related responses in animals generated enthusiasm for the notion that these could represent a class of novel-mechanism, next-generation anxiolytics (Kehne and Cain 2010). CRF1 antagonists identified as potential anxiolytics in classical assays also showed potential in some aversive conditioning tasks. For instance, they impaired context-freezing as well as light-, CRF-, and context-potentiated startle. However, startle potentiated by short-duration CSs was unaffected or enhanced by CRF1 antagonists (Walker et al. 2009). This profile was taken as evidence for involvement of CRF1 receptors in anxiety since context conditioning, light-, and CRF-enhanced startle all depend on the bed nucleus of the stria terminalis (BST), a region implicated in anxiety, whereas conditioning with short-duration cues signaling imminent harm depends on BLA, a region more closely tied to fear (Davis et al. 2010; Waddell et al. 2006) Subsequent clinical trials assessing stress and FC phenomena in healthy controls produced a confusing pattern of results. In a CO_2 inhalation task, R317573 reduced anxious feelings but not cardiovascular or respiratory defensive responses (DRs) (Bailey et al. 2011). In a social stress test, NBI-34041 reduced hormonal responses but not anxious feelings (Ising et al. 2007). In a startle study, GSK561679 *increased* fear-potentiated startle, but had no effect on anxiety-potentiated startle or anxious feelings (Grillon et al. 2015). Unfortunately, large trials run by pharmaceutical companies relying mostly on verbal report found no efficacy for CRF1 blockers in treating GAD or PTSD (Coric et al. 2010; Dunlop et al. 2017). In the wake of these results, some have called for a pause in clinical trials for CRF1 antagonists until better methods for translation can be discovered (Murrough and Charney 2017).

Although translational neuroscience for anxiety has not lived up to expectations, it would be a mistake to lose sight of methods that have worked better than others. It is clear that we must abandon, for now at least, the practice of identifying candidate anxiolytics in animal behavioral tasks and heading straight to clinical trials that test efficacy only with verbal report. It also seems clear that preclinical FC tests have provided more accurate predictions of treatment success than classical anxiety assays—though there is certainly room for improvement. In the above examples, Pavlovian protocols predicted that propranolol may not prevent consolidation of amygdala-dependent traumatic memories. FC studies also suggest that propranolol has utility as a reconsolidation blocker, but not in all cases. And potentiated startle studies predicted that CRF1 blockers might facilitate fear and have equivocal effects on anxiety. Unfortunately, many of these "predictions" came after clinical trials began, highlighting the need to improve incentives for publishing null effects (Luyten et al. 2021). Finally, because FC is embedded in a strong theoretical and neurobiological framework, even null or disappointing results help refine models of behavior, brain function, and emotion that may improve translation. For instance, the finding that CRF1 antagonists selectively facilitate startle to short-duration CSs lends support to the hypothesis that BST promotes anxiety over fear via inhibition of CeA (Grillon et al. 2015) (but see Shackman and Fox 2016). A recent fear-potentiated startle study found that the CRF1 antagonist GSK561679 selectively reversed a fear inhibition deficit in PTSD patients (Jovanovic et al. 2020). Although this same treatment did not reduce PTSD symptoms on its own (Dunlop et al. 2017), this suggests that CRF1 antagonists may promote inhibitory learning processes that are critical for exposure-based therapies and are deficient in PTSD (Maren 2022). GSK561679 could be tested for its ability to make therapy more effective in PTSD. A positive result would be consistent with animal work in this area (Abiri et al. 2014). This illustrates how translational work using similar protocols and measures can lead to clues about anxiety pathology and new treatments, even when initial results are disappointing. It also suggests that confirmation of key animal findings in humans should occur early in the translation process before expensive and time-consuming clinical trials based on verbally reported symptoms.

2 A Challenge to Translational Research: The Two-System Framework for Fear

Joseph LeDoux and collaborators have recently questioned the practice of using non-conscious DRs to predict effects on conscious human emotions like fear (LeDoux 2019, 2012, 2015; Pine and LeDoux 2017; LeDoux and Daw 2018; Taschereau-Dumouchel et al. 2022; Mobbs et al. 2019; Brown and LeDoux 2020; LeDoux and Brown 2017). Their concerns stem from (1) disappointing translational results for anxiety, (2) weak correlations between physiological, behavioral, and conscious measures of fear in humans (response discordance), and (3) rare cases where patients with bilateral amygdala damage report conscious distress. To account for these findings, LeDoux and Pine (LeDoux and Pine 2016) propose a new "two-system" framework where a subcortical, amygdala-centered survival circuit detects and responds to threats with non-conscious DRs (system 1) and a cortical circuit responsible for working memory and consciousness mediates feelings of fear (system 2) (Fig. 2). Conscious fear is indirectly modulated by activity in survival circuits (and feedback from the responses they generate), but these are not required. Working memory, using autobiographical fear schema stored in long-term memory, can generate full conscious feelings via pattern completion (Brown and LeDoux 2020; LeDoux and Brown 2017). This assembled fear state may function to guide

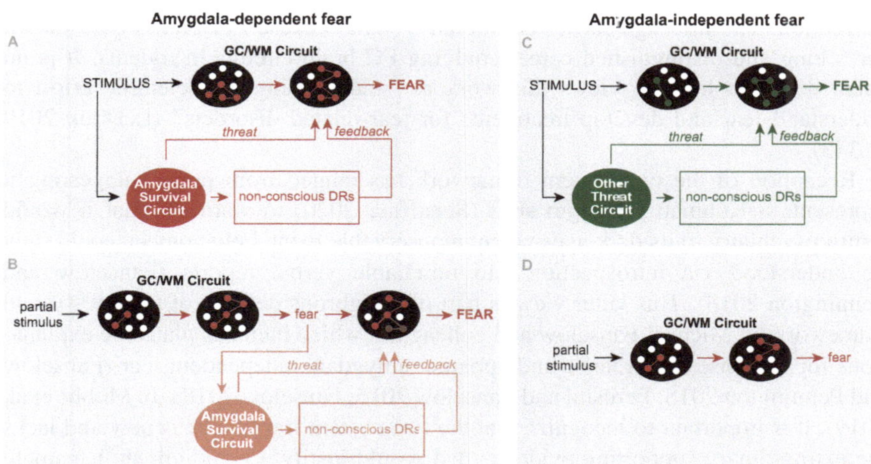

Fig. 2 Weak vs. strong two-system scenarios. (**a**) In a "weak" two-system scenario, a stimulus simultaneously activates separate circuits mediating conscious vs. non-conscious DRs. Conscious fear is constructed via pattern completion of fear schema in cortical circuits, facilitated in real time by amygdala inputs relaying threat information and/or feedback from amygdala-dependent DRs. (**b**) Alternatively, a partial threat (or feedback from a defense-like response; e.g. ANS arousal) motivates weak conscious fear via pattern completion of fear schema. Weak fear then recruits non-conscious responding by activating amygdala-dependent survival circuits, ultimately producing strong fear. In both scenarios (A & B), conscious and non-conscious responses should correlate and treatments that reduce amygdala threat processing should reduce fear. (**c**) In "strong" scenarios that may justify a two-system approach, stimuli or partial stimuli produce conscious fear via pattern completion in the absence of amygdala inputs or feedback from amygdala-dependent DRs. This may occur whether (*C*) or not (*D*) non-conscious responses are also produced via activation of different subcortical threat circuits. Thus, there is no expectation that amygdala-dependent DRs and fear will correlate and treatments that reduce amygdala threat processing are unlikely to reduce fear. Note, however, that non-conscious DRs should still correlate with fear in scenario C, and "fear" unaccompanied by behavioral or physiological DRs may be less destressing in scenario D. *ANS* autonomic nervous system, *GC/WM* general cognition/working memory networks (LeDoux and Brown 2017)

decisions by predicting future emotions (forecasting) using fear schema to plan goal-directed actions (avoidance), especially in dangerous situations that are novel. An important theme in this framework is that human cortical circuits mediating consciousness and emotion are not devoted to fear of imminent bodily harm based on inputs from an antipredator survival circuit. Fear can arise from schema based on different survival circuits (e.g., fear of starvation) or schema based on psychological harms (e.g., fear of embarrassment).

Though others have proposed multiple-system frameworks before (Ohman and Mineka 2001; Grillon 2009; Hugdahl 1981), LeDoux and Pine's two-system approach is unique in its emphasis on conscious fear over other responses. They correctly note that subjective distress is the main reason people seek treatment and the main way treatments are evaluated. In their model, non-conscious threat processing is not required for conscious fear and therefore non-conscious DRs should not be relied on to predict fear. This challenges the basic rationale for translation. The message carries additional weight because LeDoux is best known for a long and distinguished career studying FC brain circuits in rodents. It is no small deal that he now views this work as "undermining the scientific effort to understand fear and develop treatments for fear-related disorders" (LeDoux 2019 p. 343).

Reception of the two-system framework has ranged from praise suggesting it represents a Kuhnian paradigm shift (Schaffner 2020) to warnings that it would return psychiatry to its dark ages when unobservable mental phenomena could only be understood via introspection and unreliable verbal reports (Fanselow and Pennington 2018). This latter view is part of a vigorous defense of the one-system framework by Michael Fanselow and colleagues, which includes plausible explanations for response discordance and apparent amygdala-independent fear (Fanselow and Pennington 2018; Perusini and Fanselow 2015; Fanselow 2018a, b; Mobbs et al. 2019). It is important to recognize that the two-system framework is new and lacks the extraordinary supporting evidence that would justify a paradigm shift, namely clear demonstrations that strong fear routinely occurs in the absence of defense. Subcortical inputs and feedback from non-conscious DRs appear to make important contributions to two-system fear in most scenarios (Fig. 2). Fear without defense also suggests that fear is a functionless epiphenomenon, though that is clearly not what LeDoux has in mind (LeDoux and Daw 2018; LeDoux 2019). Thus, the two-system model needs more testing and clarification. Fortunately, resolving the one- vs. two-system debate and addressing the translational crisis require the same thing: more data on the coordination (or lack thereof) of multiple DRs at different levels of perceived danger. This includes the need for more work on situations that reliably induce conscious distress in humans. If care is taken to distinguish between different types of distress (e.g., panic, fear, anxiety) and to report response profiles for individuals, these data will also inform idiographic approaches to understanding emotion (Siegel et al. 2018) as well as process-based approaches to tailored treatment (Hofmann and Hayes 2019).

Considered with Ledoux's larger body of recent work (LeDoux and Daw 2018; LeDoux 2015; LeDoux and Brown 2017; LeDoux et al. 2017), the two-system framework can also be viewed as a blueprint for addressing vulnerabilities in the one-system model that reduce its value as a translational method. The standard one-system framework is a central state model that treats fear as a non-subjective psychological construct. Conscious fear is assumed to depend on this construct, but it is unclear how this happens and feelings are excluded. Instrumental avoidance is also excluded (but see Cain et al. 2013; Cain 2019; Moscarello and Hartley 2017), mainly because the field has been hyperfocused on post-encounter "antipredator" fear that seems incompatible with instrumental learning (Fanselow 1997). Finally, although one-system frameworks can account for panic, anxiety, and discordant responding (Fig. 1), research on circa-strike and pre-encounter defense has been neglected in favor of amygdala-dependent fear measured with behavior (e.g., freezing) in contexts with few response options. These deficiencies leave the standard one-system framework in a weak position to explain the major complaint in human anxiety (conscious distress), a major anxiety symptom (avoidance), a major impediment to translation (response discordance) or states of distress that do not depend on the amygdala. Going forward it will be important to determine if a revised one-system framework that addresses these issues can improve translation. The Survival Optimization System that Mobbs and colleagues have been developing in recent years seems like a positive step in this direction (Mobbs et al. 2015, 2020; Mobbs 2018). LeDoux's ideas about pattern completion, cortically-initiated fear, deliberative avoidance, and consciousness may even be a valuable addition to a revised one-system framework.

Note that authors of several recent publications conclude that their findings support the two-system framework. One study found widespread CS-specific functional connectivity changes in 30% of the brain during fear extinction. These changes extended well beyond the traditional "fear circuit" to include regions involved in attention, perception, working memory, and consciousness (Wen et al. 2021). Two studies using machine-learning with fMRI data found patterns of amygdala and prefrontal cortex activity that distinguish DRs from subjective fear (Taschereau-Dumouchel et al. 2020; Zhou et al. 2021). Another study found that while elements of a "fear circuit" identified in rodents were recruited in humans during FC, activation levels did not correlate with trait anxiety or subjective distress (Young et al. 2021). Finally, one study found that therapies targeting rapid, implicit threat reactions vs. slower, cognitive responses had additive benefits. Further, patients with dysfunctional amygdala–insula connectivity showed the greatest benefit from therapy targeting rapid reactions (White et al. 2017). Although these findings are consistent with the two-system framework, they also appear consistent with the one-system framework. This is especially true for studies that found significant correlations between conscious and non-conscious threat responses (Taschereau-Dumouchel et al. 2020) and amygdala activation patterns suggesting a role "in initiating or integrating a coordinated fear and threat response" (Zhou et al. 2021). The current one-system framework does not predict how conscious fear is represented in the brain, only that it requires input from subcortical circuits processing threats or DRs. Treatment effects on specific responses are also possible in a one-system framework via changes in effector regions downstream of the fear generator (Fig. 1).

Discordance can be explained by independent brain circuits or a host of factors that differentially moderate responses controlled by a common circuit (Hollenstein and Lanteigne 2014; Zinbarg 1998). However, LeDoux and colleagues spend comparatively little time discussing the possibility that study designs could be improved to enhance concordance and improve translation. Similarly, Fanselow and colleagues mount a strong defense of the one-system framework and acknowledge that translational work is difficult, but say little about how the situation can be improved. Response discordance is a tricky problem that may be insurmountable (Barrett and Satpute 2019), but it is worth considering new strategies in this vein given the high stakes and likelihood that consciousness will not be solved soon (Adolphs and Anderson 2018; Schaffner 2020).

3 Beyond Fear

For the remainder of this chapter, I will pursue the following related ideas: animal conditioning studies have focused on panic- and especially fear-related responses (circa-strike and post-encounter defense) while clinical studies have assessed (mainly) anxiety and pre-encounter responses. This has contributed to translational problems, as panic, fear, and anxiety are separate emotions/constructs, mediated by unique neurocircuits, supporting different response patterns. As the Blanchard's pointed out 20 years ago, adjusting human studies to evoke panic and fear "would involve a range of legal and ethical problems" (Blanchard et al. 2003). Thus, animal researchers may need to adjust their tasks to evoke pre-encounter defense modes more suitable for studying anxiety-relevant brain processes. This adjustment is also warranted because anxiety may be a bigger problem than fear, even for disorders defined by stronger and more frequent fear episodes (Fig. 3). Classical assays selected based on benzodiazepine sensitivity can help (Calhoon and Tye 2015), but these are less amenable to neurocircuit analyses (Cain et al. 2013) and have an uncertain relationship to fear vs anxiety (Perusini and Fanselow 2015; Headley et al. 2019). A more productive path may be to explore aversive conditioning phenomena using tightly controlled low-imminence threats in contexts with more response options (Boddez et al. 2020; Moscarello and Maren 2018; Lissek et al. 2006). This leverages the strong theoretical grounding developed by FC research and anchors neurobiological studies of anxiety to a partially elaborated survival circuit. It also aligns with calls for more stimulus-specific measures of conscious human distress that will help differentiate fear from anxiety (Craske et al. 2009). Problems related to response concordance and translation may be addressed by incorporating multiple measures in animal and human studies (Zinbarg 1998) and confirming animal findings in humans using similar procedures, measures, and threats (Grillon and Ernst 2020).

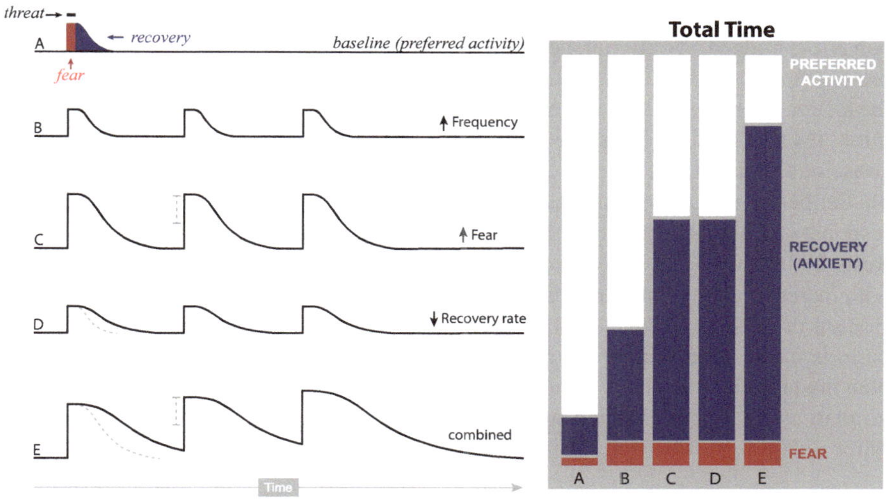

Fig. 3 Anxiety occupies more time than fear. (**a**) Illustration of a single fear episode (red) triggered by a brief threat (black rectangle) and followed by a more prolonged anxiety response in the post-threat recovery period (blue). (**b**) Increasing the frequency of fear episodes leads to much more anxiety than fear if time is the metric. (**c**) If fear is defined as a response to an immediate and present danger, increasing the magnitude of fear responses has no effect on total fear time but increases total anxiety time. (**d**) Reducing the rate of post-threat recovery can also increase total anxiety time. (**e**) Combining these three factors dramatically increases total anxiety time with minimal effect on total fear time

3.1 *Fear vs. Anxiety*

Fear is an intense, unpleasant, and relatively inflexible response to signals of imminent and significant harm. Fortunately, for most of us, these episodes are brief and rare. Anxiety is a close cousin to fear that operates on much longer timescales. Clear definitions that distinguish the two have been elusive (LeDoux 2015; Mobbs 2018; Barlow 2002). For instance, DSM-5 diagnostic criteria for various anxiety disorders describe fear and anxiety in terms of each other (Perusini and Fanselow 2015). Drugs used to treat fear and anxiety disorders are called anxiolytic or anti-anxiety but never anti-fear (Cain et al. 2013). No clear consensus is available in the animal literature either (Gray and McNaughton 2000; Davis et al. 2010; Waddell et al. 2006; Blanchard and Blanchard 2008). Perusini and Fanselow (2015) argue that if fear and anxiety are distinct, they should have different causes, manifest as different responses, and have dissociable underlying brain mechanisms. A defensive framework organized around perceived threat imminence captures much of the nuance in other definitions and relates panic, fear, and anxiety to modes of defense that meet these criteria. Importantly, tying emotion states to threat imminence also defines response modes by *duration* of activation. Threats signaling imminent harm require automatic, but short-lived responses. Low-imminence threats signaling uncertain or distant harm produce sustained states that allow for volitional behavior. Indeed, volitional action under threat may be seen as diagnostic of pre-encounter defense and anxiety (though not a requirement). As Mobbs suggests, the difference between fear and anxiety relates to how much time the organism has to think (Mobbs 2018). This conception fits well with the presumed function of conscious or cognitive threat responses: evading danger via execution of a flexible/deliberative action plan based on knowledge of the context and available responses (Roelofs and Dayan 2022; LeDoux and Daw 2018; Mobbs et al. 2020; Rolls 2008). It may also help explain why threat-related feelings get so unpleasant with increasing threat imminence and loss of control (Mobbs et al. 2007, 2009, 2010; Bouton et al. 2001; White et al. 2006). Anxiety is less intense because immediate harm is unlikely and effective, goal-directed actions are possible. Fear is an action plan thwarted by SSDRs as response options become fewer and riskier. And panic is no plan at all, where all good options are closed off and survival depends on pure reflex.

3.2 Is Fear the Problem?

If fear and anxiety are different, then which is the bigger problem? Fear is undoubtedly a problem in many disorders, but anxiety occupies more of a patient's time (Fig. 3). This is reflected in how fear vs. anxiety symptoms are captured by clinical data. Because fear episodes are brief (seconds to minutes) (Barlow 2002), fear is usually measured as increased frequency of events compared to controls (e.g., panic reactions, intrusions, nightmares). These frequencies can vary quite a bit from person to person. For instance, in studies of PTSD, intrusive symptoms range from about 0.5 to 10 events per day (Priebe et al. 2013). Current diagnostic criteria for panic disorder (PD) only require a single episode, but past criteria defined PD by a rate of panic attacks greater than one per week. In contrast, diagnostic criteria routinely reference the duration of persistent anxiety symptoms (DSM-5-TR). A PD diagnosis requires that an attack is followed by a month of "fear" of another attack. In PTSD, there is no minimum number of episodic symptoms, but specific categories of symptoms must persist at least 1 month. For a diagnosis of specific phobia, fear and anxiety are interchangeable but symptoms must persist for 6 months. In obsessive-compulsive disorder (OCD), obsessions and compulsions are time-consuming, occupying more than 1 h per day. Remarkably, most GAD patients report feeling tense, anxious, or worried more than 50% of their waking lives (Sanderson and Barlow 1990). Although fear can be a problem, anxiety-after-fear and in anticipation of the next fear event (so-called "fear of fear" or "fear of anxiety") takes up a greater proportion of a patient's time and deserves more attention from animal researchers (Bouton et al. 2001; Barlow 2002).

Clinical assessment tools that rely on introspection and verbal report often place more weight on anxiety symptoms, even for disorders where the primary problem is thought to be panic or fear processing. For instance, the *Panic Disorder Severity*

Scale has seven items rated on a five-point scale (Shear et al. 1997). Two of the items probe the frequency and intensity of panic attacks. The remaining five probe anxiety and avoidance, and avoidance is likely an anxiety response (Cain 2019). Similarly, the *Severity Measure for Specific Phobia* has two items devoted to fear symptoms and eight for anxiety/avoidance symptoms (Craske et al. 2013). These and other methods of assessment ask patients to rate emotion severity, mentally average ratings of multiple events, and integrate the impact of symptoms over time (Robinson and Clore 2002). If fear and anxiety are distinct states and anxiety occupies more time, it is likely that anxiety is contributing more to the final scores.

Finally, it is usually not practical or ethical to study human fear in typical laboratory settings. Even in FC experiments, there are many indicators that threats to survival are not imminent and escape is always possible. Conditioning protocols often use partial CS–US contingencies, weak USs, or instructed threats, and measure autonomic reactions that are good indices of learning but not necessarily level of distress (Ohman and Mineka 2001; LeDoux 2015; Craske et al. 2009; Risbrough 2010). Subjective reports likely skew toward anxiety for the reasons stated above. If anxiety is the bigger problem, easier to study in humans, and a major contributor to reports of distress, animal researchers should give more attention to pre-encounter defense. Once various conditioned anxiety responses are understood in healthy controls, abnormal responding can be studied as fear reactions to situations that are typically anxiety-provoking (among other possibilities) (Hugdahl 1981; Grillon and Ernst 2020; Fanselow 2022). This should also align animal work with a pattern of conditioning-related phenomena observed in anxious populations. Enhanced fear to stimuli that predict the US is occasionally observed (Lissek et al. 2005; Jovanovic et al. 2011), but this is often to background or contextual cues that are weaker predictors of the US than discrete cues (Grillon et al. 1998) A more robust finding in PTSD is impaired fear inhibition to safety signals or extinguished threats (Jovanovic et al. 2010, 2020; Milad et al. 2009). This sort of inhibition is a major way organisms use safety and context learning to reduce perceptions of threat imminence and transition to modes of flexible defense (Moscarello and Maren 2018; Cain and LeDoux 2007).

4 Beyond Extinction

Standard FC protocols in rodents typically use few trials where brief CSs co-terminate with painful, inescapable USs. Every CS is paired with the US and both training and testing occur in small chambers with few response options. These conditions are ideal for studying post-encounter DRs like freezing and potentiated startle. There are many procedural variations that can reduce threat imminence to study low-fear or anxiety states, even in these small chambers. These include things like reducing CS–US contingency, increasing CS duration, decreasing shock intensity, or studying behavioral phenomena observed when compound CSs are employed. These manipulations usually manifest as reduced fear DRs compared to control groups and are covered elsewhere (Beckers et al. 2013; Dunsmoor et al. 2022; Cain et al. 2013; Waddell et al. 2006; Lissek et al. 2006). Here I will focus on procedures that track transitions between response modes over time and/or induce persistent pre-encounter defensive states. Note that instrumental avoidance is clearly relevant to the current discussion but is omitted here due to space constraints. Interested readers are referred to recent reviews discussing its relevance to pre-encounter defense and conscious anxiety (Moscarello and Penzo 2022; LeDoux and Daw 2018; Cain 2019).

The discussion below reflects guidelines from other authors that seem especially useful for distinguishing between different modes of defense and for studying the brain states specific to these modes in animals (Adolphs and Anderson 2018; Perusini and Fanselow 2015; Mobbs et al. 2015, 2020; Hollenstein and Lanteigne 2014; Boddez et al. 2020; Moscarello and Maren 2018; Blanchard and Blanchard 2008). First, different emotion states reflect "scalability" and have different antecedents, meaning here that pre-encounter responses should be elicited by lower levels of threat imminence than post-encounter responses. Second, categorically different responses will be more useful for distinguishing between pre- and post-encounter defense than different degrees of a common behavior. Third, transitions between states, and especially abrupt transitions, are ideal for identifying distinct brain correlates of emotion states. Fourth, concordance of emotion response components (e.g., behavioral, physiological, and cognitive-emotional responses) is typically best when an emotion state is maximally activated. Fifth, unlike reflexes, emotion states persist beyond their eliciting stimuli to affect behavior on longer timescales. Finally, contextual information and response options can powerfully influence threat perception and the selection of different response modes.

4.1 Fear Extinction

Extinction training involves repeated presentations of threatening stimuli in the absence of an aversive outcome. Therapies with a strong extinction component remain among the most efficacious treatments for anxiety in use today (Craske et al. 2018; Foa and McLean 2016). Extinction can reduce all three components of the "fear" response in humans (Graham and Milad 2011), though some of these findings likely reflect anxiety extinction.

In rodents, CS-freezing is high early in extinction training, consistent with perceptions of high threat imminence and strong activation of circuits mediating post-encounter defense. With enough training, CS-freezing declines to zero, suggesting a drastic change in perception of threat imminence and perhaps a transition to pre-encounter defense. However, since most animal extinction sessions occur in small boxes, positive evidence for this is lacking. Extinction depends mostly on the formation of a new inhibitory CS–noUS memory that outcompetes the original CS–US memory for control of behavior. This is based mainly on recovery phenomena, where CS-fear returns with context changes, time, or exposure to stress.

Degradation (erasure) of the original CS–US association may also contribute. Extinction depends on plasticity in BLA, hippocampus, infralimbic cortex (IL), and intercalated cells (ITCs) that gate information flow between BLA and CeA. After training, extinction recall is achieved via suppression of amygdala-mediated threat responses by IL and ITCs. Hippocampus works in concert with IL help to disambiguate the meaning of the CS after extinction and ensure DRs are still elicited in non-extinction contexts. Recent reviews summarize research on the psychological and neural mechanisms of extinction (e.g., Bouton et al. 2021).

4.1.1 Extinction as a Translational Tool

The fear extinction paradigm is usually regarded as a success story for translational psychiatry (Milad and Quirk 2012; Ressler 2020; Anderson and Insel 2006), though certain methods of translation have been more successful than others. Extinction protocols developed in rodents to study memory phases have helped clinical researchers identify specific deficits in anxious patients linked to dysfunction in the threat processing circuitry. For instance, studies showing PTSD deficits in safety learning or extinction due to enhanced "fear load" were inspired by rodent studies of FC, extinction, and conditioned inhibition (Jovanovic et al. 2009; Norrholm et al. 2015). Similarly, studies showing that dysfunctional threat processing in ventromedial PFC is associated with extinction recall deficits in PTSD were inspired by rodent studies demonstrating a role for IL in extinction recall (Milad and Quirk 2012). Extinction is also a popular rodent paradigm for exploring potential adjuncts to therapies that involve exposure. For example, yohimbine (Cain et al. 2004; Morris and Bouton 2007; Hefner et al. 2008) and d-cycloserine (DCS; (Walker et al. 2002; Woods and Bouton 2006)) were shown to facilitate fear extinction in rodents before being tested in clinical trials. Yohimbine was then shown to enhance exposure therapy for social anxiety disorder (Smits et al. 2014), claustrophobia (Powers et al. 2009), and PTSD (Tuerk et al. 2018). One study found no effect when yohimbine was combined with virtual reality exposure for fear of flying (Meyerbroeker et al. 2012), possibly because of a strong effect of the therapy alone. DCS has been tested as an adjunct to therapy in clinical populations and in healthy subjects undergoing fear extinction in the laboratory (Bowers and Ressler 2015; Hertenstein et al. 2021). Although results have been mixed, DCS has shown promise in the treatment of social anxiety, PD, OCD, and PTSD. Considered with additional animal studies it appears that factors like genetic background, response to exposure, severity of disease, and competing responses modulate the effectiveness of both drugs (Craske et al. 2018; Bowers and Ressler 2015; Holmes and Quirk 2010). Although these and other studies of potential adjuncts paint a complex picture, the interplay between animal and human research continually refines models of extinction and suggests new options for tailored anxiety treatments. This is possible because extinction research in animals and humans provides strong theoretical and neurobiological grounding for translation. Studies employing similar protocols and measures in both species are especially informative.

Though extinction research is important, we should be cautious about relying too heavily on this paradigm to discover new treatments. First, extinction is highly context-dependent, even when combined with adjuncts like yohimbine or DCS, limiting its utility. Treatments that reduce context-dependency may prove useful, but this remains an open question (Sewart et al. 2021). Second, it is unclear if it will be possible to overcome the extinction deficits that define some anxiety disorders (Markowitz and Fanselow 2020). It may be wise to expand the evidence base for workarounds like active coping based on instrumental avoidance (Collins et al. 2014). It is also unclear how large a role extinction plays in normal recovery processes that contribute to resilience. To date, clinical trials have not found that animal extinction enhancers reduce subjective symptoms in the absence of extinction-based therapy (e.g., Stein et al. 2021).

4.1.2 Using Extinction to Study Pre-encounter Defense

Extinction may be a good place to begin studying conditioning-related pre-encounter defense as threat imminence is reduced and post-encounter SSDRs decline (Fig. 4a). Positive indicators of pre-encounter defense are preferred to determine when this mode is engaged (compared to negative indicators, like declining freezing or emergence of appetitive behavior). Rodent 22 kHz ultrasonic vocalizations (USVs) may be informative as these are emitted in dangerous situations but not at the highest levels of threat imminence (i.e., during CS or cat presentations) (Jelen et al. 2003).

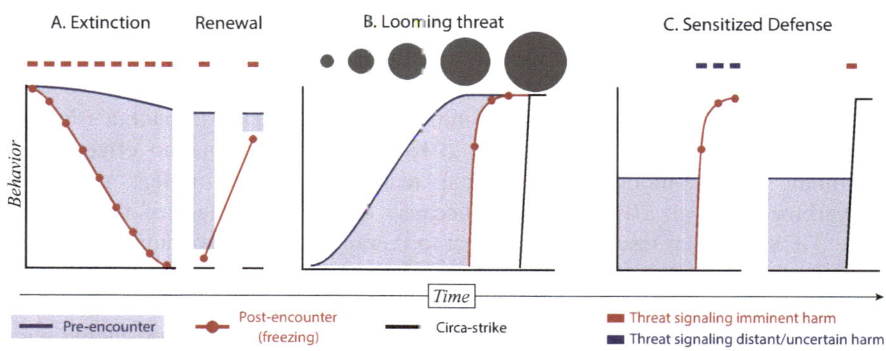

Fig. 4 Using aversive conditioning to study pre-encounter defense. (**a**) Pre-encounter DRs likely emerge during extinction as threat-elicited post-encounter DRs decline. Abrupt transitions between pre- and post-encounter modes can be induced by switching between extinction and non-extinction contexts (renewal). Stronger anxiety-like responses should occur earlier vs. later in extinction training. (**b**) A range of DRs may be elicited by looming stimuli that are innately aversive or transformed into threats by conditioning. Pre-encounter DRs should emerge before post-encounter DRs if the imminence progression is not too rapid. (**c**) Exposure to aversive stimuli can also produce long-lasting changes in defensive behavior thresholds (non-associative sensitization). This can produce inappropriate responding (pre-encounter DRs at baseline, post-encounter DRs to weak threats, and circa-strike DRs to imminent threats)

USVs emerge in the intertrial intervals during FC and partially return when safety signals suppress CS-freezing (Frysztak and Neafsey 1991). There is some indication that USVs re-emerge toward the end of fear extinction sessions in rats (Sakamoto 2020). Risk-assessment behaviors triggered by extinguished CSs are also possibilities (e.g., scanning or stretch-approach), although these have yet to be evaluated after extinction. In support of this general idea, drugs that reduce post-encounter SSDRs in fear-inducing situations lead to increases in risk-assessment behaviors, suggesting that pre-encounter defense is disinhibited as post-encounter defense wanes (Blanchard et al. 2003; Gozzi et al. 2010). To promote the observation of risk-assessment behaviors it may be helpful to test in environments with safe compartments connected to a more exposed open area (e.g., visible burrow system (Blanchard and Blanchard 1989)). Extinction has also been shown to reduce SSDRs and enable the learning and performance of instrumental actions (Cain and LeDoux 2007). The context-dependence of extinction may also be exploited to distinguish between brain states unique to fear vs. anxiety. Comparing reactions to the same threat in extinction- vs. non-extinction contexts (renewal) can create the abrupt transitions that are helpful for neurobiological studies.

4.2 Dynamic Responding to Changing Threat Imminence

Animal researchers interested in weak vs. strong anxiety typically use variations of spontaneous anxiety tasks to activate different degrees of pre-encounter defense (Haller et al. 2013; Kehne and Cain 2010; Calhoon and Tye 2015; Hoffman et al. 2022). As mentioned earlier, precision here is difficult given the diffuse nature of the anxiety-inducing stimuli used in these tasks (e.g., bright, open space). A better approach may be to examine dynamic defensive patterns as the perceived proximity of well-controlled threats changes with time.

4.2.1 Approaching Threats

Rapidly approaching or “looming” stimuli elicit innate DRs in many species including humans. Sensory systems likely evolved to preferentially process looming stimuli due to their association with dangerous predatory attacks or collisions. Studies of overhead looming stimuli in rodents have focused on stimulus properties and environmental conditions promoting different modes of defense. For instance, looming stimuli can evoke freezing or flight depending on distance to shelter and the speed/trajectory of the approaching stimulus (Yilmaz and Meister 2013; De Franceschi et al. 2016). The neural circuit controlling looming DRs includes the dorsal raphe (DRN), thalamus, BLA, hypothalamus, and dorsal periaqueductal gray (PAG) (Huang et al. 2017; Shang et al. 2015; Liu et al. 2022). A DR pattern triggered by visual looming stimuli can be elicited by optogenetic stimulation of a collicular–thalamic–BLA circuit and is blocked by lesions of BLA (Wei et al. 2015). Looming stimuli have not been thoroughly investigated in less-intense anxiety-inducing scenarios that directly elicit pre-encounter behavior, but some data suggest this may be worthwhile. In the above studies, activation of the looming defense circuit led to thigmotaxis or hiding that persisted long after stimulation—consistent with an anxiety-like state. Other studies investigating foraging behavior found that rats initially escape a simulated surging predator (Choi and Kim 2010). After the encounter, rats display risk-assessment behaviors before resuming foraging, presumably due to context conditioning. Humans also perceive looming sound sources (increasing volume) as closer than control sounds (Neuhoff 2018). This looming bias appears to be exaggerated in anxiety and has been referred to as a "looming vulnerability" that may indicate anxiety susceptibility (Riskind et al. 2014). Studies of looming in humans indicate that animate, threatening stimuli engage amygdala and PAG to a greater degree as threat imminence increases (Mobbs et al. 2010; Coker-Appiah et al. 2013). Thus, looming stimuli that begin as low-imminence threats may prove useful for evaluating dynamic DRs and transitions between all three defense modes as threats escalate (Fig. 4b).

4.2.2 Post-Threat Recovery

Anxiety disorders are characterized by exaggerated, inappropriate, or prolonged DRs to threats (Cain et al. 2013). Exaggerated and/or inappropriate responses have received the most attention in neurobiological studies (Jovanovic et al. 2010; Dymond et al. 2015). However, the neural processes controlling the duration of defensive episodes are rarely studied in animals, especially behaviorally (but see Quinn et al. 2002; Lee et al. 2017; Furlong and Carrive 2007). This is surprising, since prolonged anxiety reactions may hold the greatest potential to disrupt normal functioning in humans (Davidson 1998; Corchs and Schiller 2019). Interventions that could speed recovery duration but preserve initial adaptive threat reactions would have great therapeutic potential (Cain et al. 2012).

One simple strategy to study the range of pre-encounter defense is to strongly activate survival circuits with Pavlovian fear stimuli and then measure post-threat recovery from fear (RFF) as animals progress, in reverse, down the defensive behavior continuum and back to preferred activities. This should allow for the assessment of many different degrees of anxiety in one test while also anchoring neurobiological studies to the known FC circuit. As with extinction, this would also allow for comparisons between high and low threat imminence states as fear recedes in time. Recovery indices may track cessation of post-encounter SSDRs, emergence and then cessation of pre-encounter DRs, and finally emergence of non-defensive behaviors. As suggested by studies of predator encounters in a visible burrow system (VBS), this sort of recovery operates on much longer timescales than are typically evaluated in studies of survival circuits and could be particularly sensitive to anxiogenic and anxiolytic manipulations (Blanchard and Blanchard 1989). Recovery duration likely reflects at least three neurobiological processes that could be studied with the RFF procedure: passive decay of network activity after threat cessation, negative feedback (e.g., hormones), and active processes that delay or accelerate recovery based on information gathering and risk assessment. Large individual differences are possible in recovery rate independent of initial fear reactions, similar to what is observed for extinction in individuals that initially react similarly to threats (Bush et al. 2007).

Although speculative, several unpublished behavioral findings from my own work suggest that this could be fertile ground for studying dynamic anxiety-related processes. First, in a published study evaluating part of the extended amygdala (medial amygdala; MeA) in fear and avoidance conditioning, we found that lesions had no effect on Pavlovian freezing or overtrained shuttlebox avoidance (McCue et al. 2014). MeA lesions strongly impaired 22 kHz USVs and Pavlovian-instrumental transfer, consistent with a role for MeA in anxiety-like responding. To gain some insight into whether MeA plays a role in post-threat recovery, we subsequently analyzed freezing during the intertrial intervals of a Pavlovian cue fear test conducted in a neutral context. MeA lesions reduced freezing after the CS and sped the return of exploratory rearing, consistent with a role for MeA in maintaining defensive behavior beyond the threat (Fig. 5a). This pattern is reminiscent of effects observed after medial prefrontal cortex lesions in rats (Frysztak and Neafsey 1991). In a different pilot experiment evaluating stress effects on FC, we found that 1-h of restraint stress had no effect on the learning or performance of Pavlovian freezing to a discrete CS. However, animals stressed prior to conditioning (but not after) took almost 15 times longer to cease freezing after the test CS (Fig. 5b). This is reminiscent of effects in rodent PTSD models where exposure to predator odors or shock stress leads to abnormally strong FC (Rau and Fanselow 2009; Goswami et al. 2010). Our finding suggests that stress can have profound effects on post-threat recovery independent of extinction and even when FC itself is unaffected. Finally, in a pilot study to examine how non-defensive behaviors emerge after a threat ceases, we found that a single 30 s Pavlovian threat affected behavior in a familiar environment for at least 60 min (Fig. 5c). We initially predicted that non-defensive behaviors observed during habituation would be right-shifted in time by the CS presentation. Instead, we observed reductions in specific behaviors like rearing and especially wheel running. Wheel running is a rewarding activity for rats that has been shown to induce a place preference and reinforce lever-pressing (Belke and Wagner 2005). Thus, a protocol like this may be useful for studying brain states relevant to anxiety and pre-encounter behavior while extending previous work. Rewarding behavior is reduced as in conditioned suppression of bar-pressing for food, but this occurs well after the threat is removed—consistent with a persistent emotional state like anxiety. Similar to meal pattern reorganization in closed-economy experiments (Helmstetter and Fanselow 1993), threat induces a long-term change in topography of behavior, but without the potentially confounding effects of presenting shocks in the test context. And like defensive behavior to cats in the VBS (Blanchard and Blanchard 1989), recovery to a pre-threat baseline can be tracked over many hours, but with better control over threat levels initially experienced by all subjects.

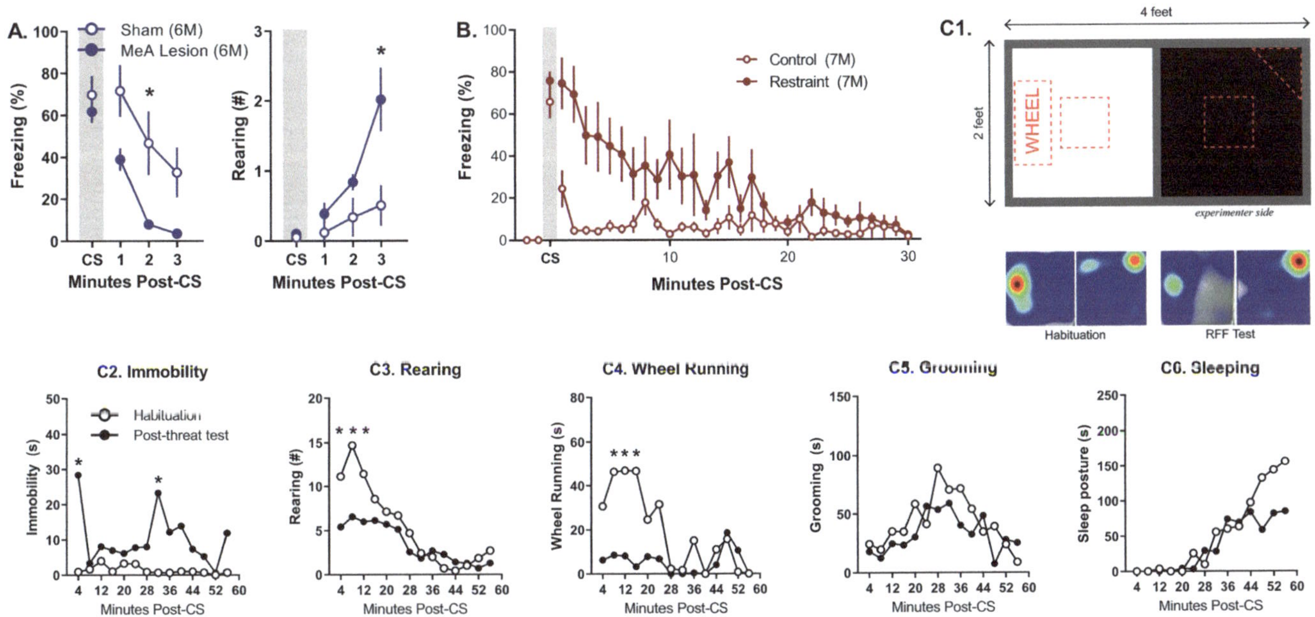

Fig. 5 Dynamic behavior patterns during post-threat recovery. (**a**) Medial amygdala (MeA) was bilaterally lesioned prior to Pavlovian FC and a CS-only test in a novel context 1 day later. MeA lesions had no effect on freezing (*left*) or rearing (*right*) during the CS, but reduced freezing and sped the re-emergence of rearing after the CS (freezing: main effect for Lesion $F_{(1,10)} = 6.7$, $p = 0.03$; rearing: time × lesion interaction: $F_{(2,20)} = 4.4$, $p = 0.03$). Data represent new analyses of intertrial interval behavior from Fig. 6b of McCue et al. 2014. (**b**) Adult, male Sprague-Dawley rats (Hilltop Lab Animals Inc.) received 1 h of restraint stress 1 day before or after Pavlovian FC (3 CS–US pairings; CS: 30 s, 5 kHz, 75 dB tone; US: 0.7 mA × 1 s shock, co-terminating), followed by a CS-only test in a novel context 2 or 1 day later (common training to test interval). Pre-conditioning restraint had no effect on CS-elicited freezing but significantly slowed the decline in freezing after the CS (stress × time interaction: $F_{(29,319)} = 3.5$, $p < 0.01$). Post-conditioning restraint had no effect on freezing during the test (F values for stress and stress × time < 0.9; not shown). Data were collected during a post-doctoral fellowship by C. Cain in Joseph Ledoux's laboratory at NYU (with approval from the NYU IACUC) (**c**) Seven adult, male, Long-Evans rats (bred at NKI with NKI IACUC approval) received Pavlovian

Fig. 5 (continued) FC (3 CS–US pairings, CS: 30 s, 5 kHz, 80 dB tone; US: 1 mA × 1 s shock, co-terminating), followed a day later by 5 daily 1 h habituation sessions to the large light/dark arena. Arena sides were separated by a divider with a gap at the bottom allowing movement between sides. A running wheel was affixed to the floor by the far light-side wall. 1 day after the final habituation session, rats received a single CS presentation and recovery from fear (RFF) behavior was recorded for 1 h. *C1* schematic of the arena and heatmaps illustrating average group location in the chamber during minutes 2–4 of the last habituation session and test session (immediately post-CS). Rats showed clear preferences for zones containing the running wheel and a dark-side corner. Rats mainly froze in the dark side, often in the back corner. Two rats running on the wheel when the CS occurred fled to the dark side and then began freezing. *Panels C2–6*: behavior for the same seven rats during the last habituation session and the test session. Non-defensive behaviors showed a general progression in both sessions from rearing to wheel running to grooming and finally sleeping. The CS presentation strongly suppressed rearing (phase × time: $F_{(14,84)} = 2.7, p < 0.01$) and wheel running ($F_{(14,84)} = 2.8, p < 0.01$), but had no effect on grooming (F values<0.75). Sleeping was mildly suppressed/fragmented, however several rats showed no sleeping so these data are difficult to interpret. “Immobility” is shown instead of freezing in C2 because freezing may not be a reliable indicator of fear level in long sessions. Effects in this figure require replication and verification in females and should be considered preliminary

4.3 Non-associative Learning

The success of the FC paradigm is primarily due to its track record for advancing neurobiological understanding of associative memory processes (Bouton et al. 2001), not necessarily integrated emotional responses that persist beyond the threat. These studies usually take great pains to isolate associative conditioning effects from non-associative changes like habituation, sensitization, and pseudoconditioning. Non-associative factors can have profound effects on behavioral and physiological DRs and perhaps human emotion, but these are infrequently the focus of conditioning studies (Rosen and Schulkin 1998). Three research programs in different stages of development highlight the need for more work on non-associative factors contributing to pathological anxiety.

Learned helplessness is the most thoroughly studied phenomenon (Maier and Seligman 2016). In this paradigm, subjects receive strong, unpredictable shocks that are either escapable (Masters) or inescapable (Yokes). A third group receives no shocks. The major effect of this treatment is that Yoked subjects show a severe deficit in subsequent shock-escape learning in a different context (e.g., shuttleboxes) compared to Masters and No-shock controls (who do not differ from each other) (Seligman and Maier 1967). Yoked animals were originally hypothesized to show an impairment because they learned that their actions failed to control shock exposure. However, recent work is more consistent with the notion that "helplessness" is a non-associative default reaction to extreme, prolonged stress—perhaps due to energy depletion (Conoscenti et al. 2017; Minor et al. 1991). Yoked animals show a wide array of impairments including exaggerated startle, sleep disruptions, behaviors indicative of anxiety and depression in classical assays like EPM and the forced-swim test, and cognitive impairments (Maier and Seligman 2016). Neurobiological studies suggest that the helplessness phenotype is primarily due to sensitization of serotonergic signaling in the DRN that contributes to anxiety (Maier and Watkins 2005). Protection from helplessness is conferred by prelimbic cortex learning processes that register control over threats and inhibit DRN (Amat et al. 2005, 2008).

Another line of research, referred to as stress-enhanced fear learning (SEFL), demonstrates that non-associative effects of strong shock stressors can have long-lasting (>90 day) effects on physiology, anxiety-like behaviors, reactivity to salient stimuli and especially new fear conditioning (Perusini et al. 2016; Conoscenti and Fanselow 2019). Thus, 15 strong shocks dramatically increase the strength of conditioning to a single shock in a separate context even when memory for the original "trauma" is blocked with drugs or suppressed by extinction. Interestingly, the stress phase of SEFL leads to persistent pre-encounter DRs in novel situations where no explicit threat is presented (Hoffman et al. 2022). With stronger conditioning in a paradigm where reward, fear, and safety conditioning occur concurrently to different stimuli, SEFL-stress reduces reward-seeking and impairs fear extinction (Woon et al. 2020). The SEFL phenotype depends on corticosterone in BLA that produces hyperexcitability via changes in AMPA receptor signaling (Perusini et al. 2016; Poulos et al. 2014).

Very recent work using an ecologically-relevant one-trial fear conditioning task suggests that non-associative factors make a strong contribution to defense in naturalistic settings (Zambetti et al. 2022). Here, rats in a large arena foraging for food received FC using different CS–US combinations, with some groups experiencing a realistic model owl that plunged toward the rats simulating a predatory attack (groups: tone-shock, tone-owl, owl-shock, tone/owl-shock). During subsequent tone tests in the foraging arena, no freezing was observed in any of the groups. This is surprising, as even a single tone-shock FC trial reliably produces freezing when the tone is tested in small chambers. There were, however, other robust behavioral changes in some groups. Rats in the owl-shock and tone/owl-shock groups took longer to leave the nest area, vigorously fled the tone, and took many trials before procuring a pellet. In contrast, the tone-shock and tone-owl groups were unaffected by conditioning, despite clear flight reactions during conditioning trials. These and other data show that owl-shock combinations were perceived as significantly more aversive than either stimulus alone. Further, owl-shock and tone/owl-shock rats displayed similar defensive reactions to the tone, even though the tone was novel for the owl-shock group. This suggests that pseudoconditioning makes a stronger contribution to defense in naturalistic settings than associative FC.

Together, these findings illustrate how non-associative processes can affect panic-, fear-, and anxiety-like responding long-term. Similar non-associative processes likely contribute to disorders such as PTSD, especially symptoms reflecting hyperarousal, hypervigilance, and possibly intrusions/nightmares. Progress in understanding the neurobiological mechanisms of non-associative defense is critical, as symptoms caused by such processes are untethered to other trauma-predictive cues. Thus, non-associative symptoms cannot be reduced by therapies relying on extinction or active coping that change anticipatory responses (Markowitz and Fanselow 2020). Instead, drugs or other treatments that reverse or mitigate the biological alterations in threat processing circuits may be required.

5 Beyond Freezing

Conscious distress is the primary problem in human anxiety (Taschereau-Dumouchel et al. 2022). Therefore, preclinical efforts to develop better treatments should have a logical path to predicting distress in humans. This section explores how aligning animal and human aversive conditioning research may enhance translation by identifying stable DR patterns. Note that preclinical researchers should focus on behavioral and physiological DR patterns that produce similar patterns in humans under similar conditions. Studies in humans can then determine which aversive conditioning protocols produce clinically relevant anxiety feelings.

5.1 General Factors Affecting Response Concordance

Desynchrony refers to differences in response patterns over time and *discordance* usually refers to uncorrelated responding, including desynchronous responding. Discordant behavioral, physiological, and experiential responding is a longstanding issue in emotion research (Hollenstein and Lanteigne 2014). Research over many decades has identified factors that affect response concordance in important ways. For instance, self-reports correlate better with other measures if collected in real time vs. after a delay (Robinson and Clore 2002; Barrett et al. 1998). Concordance is generally higher within-subjects than between subjects (Butler et al. 2014). Strongly activated emotions typically produce more concordant responses (Davidson 1992). Rapid response components may also correlate better as these are less affected by slower emotion regulation processes (Evers et al. 2014). This suggests that some degree of discordance is the norm, and very strong concordance may even indicate impaired capacity for emotion regulation (Hollenstein and Lanteigne 2014). Failure to account for different response thresholds and temporal dynamics may also help explain weak concordance in past studies (Davidson 1992). Importantly, different response demands require different patterns of autonomic nervous system (ANS) support (Siegel et al. 2018; Davidson 1992). And since emotions likely evolved to promote adaptive/flexible responding based on environmental options, concordance should not be expected in all situations even if the underlying emotion state is invariant (Sznycer and Cohen 2021). This is a key reason that tasks, threats, and measures in animal and human studies should be as similar as possible in translational projects, at least initially to gain some traction on the problem.

5.2 Discordance in Avoidance Conditioning

Discordance has been cited as a major impediment for translation work in human anxiety (Taschereau-Dumouchel et al. 2022; LeDoux and Pine 2016; Siegel et al. 2018). This is certainly an area that needs more attention from researchers, though the problem may not be as bad as it seems. First, many of the cited papers refer to discordance between subjective fear and avoidance behavior (Lang 1968; Rachman and Hodgson 1974; Ost et al. 1982; Hodgson and Rachman II 1974). However, avoidance is better conceived as a pre-encounter anxiety-related behavior (Moscarello and Penzo 2022; Cain 2019). Weak or inverse relationships between instrumental avoidance and fear/post-encounter DRs are now well-documented depending on stage of training (LeDoux et al. 2017; Cain 2019). Interestingly, many early observations of discordant fear and avoidance were made by therapists treating patients. In these settings, moderate concordance between conscious fear, avoidance, and ANS measures collapsed when therapists placed high demand on patients to control their emotions and maintain exposure (Hodgson and Rachman 1974; Miller and Bernstein 1972). Similar patterns of discordance have been observed in human studies of courage, where subjects elect to approach threats despite strong reports of fear (Nili et al. 2010). Also, at least some of these studies involved patients who were selected for high anxiety symptoms, which restricts the response range and produces weaker correlations than would be observed in an unselected sample (Zinbarg 1998). Further, many studies used behavioral avoidance tests (BATs) that can be influenced by passive avoidance, active avoidance, emotion regulation, and non-avoidance to various degrees (Hodgson and Rachman 1974; Miller and Bernstein 1972). In BATs, patients are asked to approach and endure intense threats (e.g., snake, dark closet) for a significant period of time (e.g., 10 min). Some cannot, some can for a time, and some complete the test. Predictions about fear would be different in each scenario depending on when it was assessed, though this was not always carefully considered. Further, *fear* was often measured with *anxiety* scales that were validated in situations where imminent harm was not at all likely (Husek and Alexander 1963). There is also an unresolved question about the level of concordance that may be considered acceptable for translational work. In Lang's initial papers highlighting the problem of discordance in human anxiety, statistically significant correlations of 0.4 and higher were observed for behavioral, self-report, and observer ratings of fear (Lang 1968). Similar work routinely finds correlations in the 0.5 range or greater (Zinbarg 1998). More recent emotion studies consider correlations in the 0.5 range to be good evidence of response concordance given all the sources of measurement error and moderation working against it (Evers et al. 2014; Friedman et al. 2014). For these and other reasons, response concordance in avoidance remains an open question and should be re-evaluated.

5.3 Discordance vs. Misalignment in Translational Studies of Aversive Conditioning

Concordant responding may be more common in studies of Pavlovian conditioning (Taschereau-Dumouchel et al. 2022; Bouton and Bolles 1980; Leaton and Borszcz 1985), though there are many exceptions (Taschereau-Dumouchel et al. 2022; Frijda 1986). Issues related to discordance may be mitigated as described above. However, translation problems likely arise from even more fundamental misalignment problems. In addition to the misaligned threat levels discussed earlier, human and especially rodent studies often rely on single outcome measures that are categorically different. Rodent FC studies typically measure changes in behavior, especially freezing. Human studies often rely on physiological measures of ANS arousal, especially skin-conductance responses (SCRs). Although these are components of a coordinated defensive response, they are mediated by different effector regions that serve other functions and have unique thresholds of activation and connectivity. Freezing is largely restricted to post-encounter defense mode. SCRs occur in a variety of situations and can be reasonably interpreted as defensive reactions based on other factors, however, they occur at all levels of threat imminence and cannot differentiate post-encounter from other modes of defense (Low et al. 2015; Wendt et al. 2017). Human studies often include subjective measures of distress or contingency awareness to disambiguate the meaning of SCRs, but there is reason to doubt that these index true fear (vs. anxiety or mild aversion) in safe laboratory settings. Early studies on rodent FC routinely employed multiple behavioral and physiological measures and were better positioned to address questions related to emotion (e.g., LeDoux et al. 1988). In the past few decades, however, FC research has focused more on associative learning phenomena that do not require multiple outcome measures (Bouton et al. 2001). Further, even modern studies reporting multiple measures in different response systems rarely report correlations between measures. For these and other reasons it is exceedingly difficult to determine how big the discordance problem really is for translational anxiety research, or to suggest specific solutions. The field must prioritize new research using similar paradigms, threats, and measures in animals and humans to amass a database that can inform the development of better translational strategies. Variance common to multiple measures from different response classes should produce the most accurate estimates of latent fear or anxiety (Zinbarg 1998).

5.4 *Aligning the Measurement of Multiple Responses in Animals and Humans*

An important first step toward improving translational research for anxiety will be duplicating key animal findings assessed with behavior and physiology in humans using similar measures plus reports of subjective distress. This will require selecting responses that differentiate between defense modes and can be measured in both species. Freezing is difficult to measure in humans reacting to weak threats, though recently developed methods assessing body sway appear promising (Roelofs and Dayan 2022). Freezing is also more indicative of post-encounter defense. Conditioned SCRs occur in rats and correlate highly with behavioral measures of fear, but are currently not possible in freely behaving animals (Roberts and Young 1971). Some anxiety-relevant responses that can be measured in rodents and humans include: active avoidance (Collins et al. 2014; Lazaro-Munoz et al. 2010), heart rate (Klumpers et al. 2017; Schipper et al. 2019), respiratory rate (Bagur et al. 2021; Castegnetti et al. 2017), startle (Walker et al. 2009; Norrholm et al. 2015), and blood chemistry (hormones, growth factors, etc., (Graham et al. 2017)). A thorough review of these measures is beyond the scope of this essay, but the examples below illustrate how measuring multiple responses can help identify distinct response modes and better align translational anxiety research in animals and humans.

Exciting work evaluating dynamic threat responses in human conditioning paradigms demonstrates shifts in DR patterns related to threat imminence (Mobbs et al. 2007, 2009; Wendt et al. 2017; Hashemi et al. 2019). Bradycardia was observed when shock was imminent, though this is clearest in studies using continuous

(Gladwin et al. 2016; Rosler and Gamer 2019) rather than looming (Low et al. 2015; Wendt et al. 2017) threats. This "attentive freezing" stage of defense is also associated with increased attention to visual threats, reduced body sway, and enhanced subsequent reaction times, consistent with the idea that freezing is a state of covert action preparation (Roelofs and Dayan 2022; Wendt et al. 2017; Hashemi et al. 2019; Rosler and Gamer 2019). Tachycardic responses were observed in putative pre-encounter and circa-strike phases, both of which support the execution of defensive actions. Interestingly, startle amplitudes increased linearly with imminence when shock was inevitable but were strongly suppressed late in trial sequences when shock was avoidable (Low et al. 2015; Wendt et al. 2017). In contrast, SCRs increased linearly with threat imminence whether or not avoidance was possible. These startle and SCR patterns are very similar to those observed in BATs with phobic patients that endure vs. escape fear-inducing situations (Richter et al. 2012). Respiratory rate has received less attention in human studies (Ojala and Bach 2020) but may also differentiate between stages of defense. In mice, imminent threats reduce respiratory rate, whereas weak, diffuse threats increase respiratory rate (Bagur et al. 2021; Kim et al. 2013). Although respiratory and heart rate are intimately related and follow a similar pattern during threat, variability in these measures appear to follow divergent patterns. At high levels of threat characterized by bradycardia and freezing, heart rate variability increases and respiratory rate variability decreases (Bagur et al. 2021; Hashemi et al. 2019; Signoret-Genest et al. 2023). In human studies that included verbal measures, subjective distress increased with threat imminence and panic was associated with the highest levels of threat when avoidance was not, or no longer, possible (Mobbs et al. 2007, 2009, 2010; Richter et al. 2012). Together these data fit well within a Predatory Imminence Theory framework where sympathetic activation to distal threat supports vigilance, risk assessment, and defensive movements (pre-encounter). When threat imminence increases (post-encounter), parasympathetic activation interrupts movement, focuses attention on threats, and primes circuits for vigorous action. At the final stage, the parasympathetic brake is removed and a surge in sympathetic activity supports fight/flight (circa-strike) or active avoidance (suppression of post-encounter and circa-strike, enabling a return to pre-encounter) (Roelofs and Dayan 2022). A better understanding of these dynamic response patterns may be essential for effective translational studies of anxiety. Designing conditioning paradigms to match perceived threat levels in animals and humans may prove difficult. However, if dynamic responses using multiple measures are included, it may be possible to align response patterns and compare results even if perceived threat is unequal.

5.5 *Interdependent Defensive Responses*

Measuring multiple responses of different classes is also necessary for understanding how response integration relates to central defensive states and perhaps emotions. The relationship between behavior, physiology, and feelings is a critical and unresolved debate dating back to James and Cannon (Damasio and Carvalho 2013). At present, most aversive conditioning studies in animals rely on behavior alone and are unable to contribute to this debate. However, recent findings from rodent studies measuring behavioral and physiological DRs suggest the extra effort is worthwhile. An elegant mouse FC study found that conditioned 4-Hz respiration entrains PL neural activity via the olfactory bulb to control the duration of freezing bouts (Bagur et al. 2021). Another study found that negative feedback from conditioned heart rate responses (bradycardia) in insular cortex is necessary for the calibration of freezing responses to threat-appropriate set points (Klein et al. 2021). And a very recent study found that integrated cardio-behavioral responses distinguish between rapid defensive "microstates" and slower defensive "macrostates" even when behavioral responses like freezing appear uniform (Signoret-Genest et al. 2023). Macrostates control heart rate ceilings and range. This is evident in the slow increase in heart rate ceiling and range over the course of long conditioning sessions. On this background, heart rate is adjusted to a set point for particular cue-induced microstates associated with specific behaviors. This explains why aversive CSs can sometimes result in tachycardia (e.g., in the homecage) when the typical response is bradycardia, and why bradycardia is more pronounced during late conditioning trials. This analysis also led to the identification of a specific population of ventrolateral PAG cells responsible for generating a defensive microstate characterized by freezing and bradycardia. Together these studies demonstrate the value of measuring multiple behavioral and physiological responses and considering the interdependence of DRs. Matching integrated response patterns in animal and human studies will help determine if non-conscious DRs can predict clinically relevant anxiety states in humans. Measuring multicomponent DR patterns may also help reveal distinct anxiety phenotypes and better treatments tailored to individuals (Norrholm and Jovanovic 2018).

6 Conclusion

Animal studies of FC, extinction, and post-encounter freezing played a vital role in establishing the neuroscience of emotion as a legitimate scientific discipline. Although it is hard to blame conditioning paradigms for failures of translation that relied heavily on benzodiazepine-sensitive spontaneous anxiety assays, it is also hard to ignore the fact that intricate knowledge of FC brain circuits has not led to any significant breakthroughs in the treatment of anxiety. One possibility is that threat processing in subcortical survival circuits has minimal or unpredictable influence on cortical circuits mediating conscious emotions like fear and anxiety. Another possibility is that fear and anxiety are distinct emotions serving related but different functions. If so, it is perhaps unsurprising that FC studies of post-encounter defensive states that last seconds to minutes have not identified cures for aberrant pre-encounter anxiety states that operate on much longer time scales. To address these possibilities, studies of conditioned anxiety that match protocols, threats, and measures in animals and humans are necessary. Paradigms that evaluate state transitions across varying levels of perceived threat imminence may be particularly informative, especially if volitional action is possible and multiple DRs from different response classes are measured simultaneously. Work in this vein may help resolve debates between emotion theorists and determine whether translational neuroscience for mental disorders is a viable enterprise.

Acknowledgments Thanks to Rob Sears for helpful comments on drafts of this manuscript and Joseph LeDoux for permission to publish data collected in his lab during my post-doctoral fellowship (Fig. 5b). The project described was supported by Award Number R01MH114931 to C.K.C. from the U.S. National Institutes of Health. The content is solely the responsibility of the author and does not necessarily represent the official views of the National Institute of Health and National Institute of Mental Health.

References

Abiri D et al (2014) Fear extinction learning can be impaired or enhanced by modulation of the CRF system in the basolateral nucleus of the amygdala. Behav Brain Res 271:234–239

Adolphs R, Anderson DJ (2018) The neuroscience of emotion. Princeton University Press, Princeton

Amat J et al (2005) Medial prefrontal cortex determines how stressor controllability affects behavior and dorsal raphe nucleus. Nat Neurosci 8:365–371

Amat J et al (2008) Activation of the ventral medial prefrontal cortex during an uncontrollable stressor reproduces both the immediate and long-term protective effects of behavioral control. Neuroscience 154:1178–1186

Anderson KC, Insel TR (2006) The promise of extinction research for the prevention and treatment of anxiety disorders. Biol Psychiatry 60:319–321

Bagur S et al (2021) Breathing-driven prefrontal oscillations regulate maintenance of conditioned-fear evoked freezing independently of initiation. Nat Commun 12:2605

Bailey JE et al (2011) Preliminary evidence of anxiolytic effects of the CRF1 receptor antagonist R317573 in the 7.5% CO2 proof-of-concept experimental model of human anxiety. J Psychopharmacol 25:1199–1206

Bale TL, Vale WW (2004) CRF and CRF receptors. Annu Rev Pharmacol Toxicol 44:525–557

Barlow DH (2002) Anxiety and its disorders, 2nd edn. Guilford Press, New York, p 704

Barrett LF, Satpute AB (2019) Historical pitfalls and new directions in the neuroscience of emotion. Neurosci Lett 693:9–18

Barrett G et al (1998) Are women who have abortions different from those who do not? Public Health 112:157–163

Beckers T et al (2013) What's wrong with fear conditioning? Biol Psychol 92:90–96

Belke TW, Wagner JP (2005) The reinforcing property and the rewarding aftereffect of wheel running in rats. Behav Process 68:165–172

Blanchard RJ, Blanchard DC (1989) Antipredator defensive behaviors in a visible burrow system. J Comp Psychol 103:70–82

Blanchard DC, Blanchard RJ (2008) Chapter 2.4: Defensive behaviors, fear, and anxiety. In: Blanchard DC et al (eds) Handbook of behavioral neuroscience, vol 17. Elsevier, Netherlands, pp 63–79

Blanchard DC et al (2003) The mouse defense test battery. Eur J Pharmacol 463:97–116

Boddez Y et al (2020) Tackling fear. Neurosci Biobehav Rev 112:410–419

Bolles RC (1960) Species-specific defense reactions. In: Brush FR (ed) Aversive conditioning and learning. Academic Press, New York, pp 183–233
Bouton ME, Bolles RC (1980) Conditioned fear assessed by freezing and by the suppression of three different baselines. Anim Learn Behav 8:429–434
Bouton ME et al (2001) A modern learning theory perspective on the etiology of panic disorder. Psychol Rev 108:4–32
Bouton ME et al (2021) Behavioral and neurobiological mechanisms of Pavlovian and instrumental extinction learning. Physiol Rev 101:611–681
Bowers ME, Ressler KJ (2015) An overview of translationally informed treatments for PTSD. Biol Psychiatry 78:E15–E27
Brown R, LeDoux J (2020) Higher-order memory schema and consciousness experience. Cogn Neuropsychol 37:213–215
Bush DE et al (2007) Individual differences in fear. J Trauma Stress 20:413–422
Butler EA et al (2014) Testing the effects of suppression and reappraisal on emotional concordance using a multivariate multilevel model. Biol Psychol 98:6–18
Cain CK (2019) Avoidance problems reconsidered. Curr Opin Behav Sci 26:9–17
Cain CK, LeDoux JE (2007) Escape from fear. J Exp Psychol Anim Behav Process 33:451–463
Cain CK et al (2004) Adrenergic transmission facilitates extinction of conditional fear in mice. Learn Mem 11:179–187
Cain CK et al (2012) Targeting memory processes with drugs to prevent or cure PTSD. Expert Opin Investig Drugs 21:1323–1350
Cain CK et al (2013) Chapter 41: The neurobiology of fear and anxiety. In: Charney DS et al (eds) Neurobiology of mental illness. Oxford University Press, New York, pp 549–566
Calhoon GG, Tye KM (2015) Resolving the neural circuits of anxiety. Nat Neurosci 18:1394–1404
Castegnetti G et al (2017) Assessing fear learning via conditioned respiratory amplitude responses. Psychophysiology 54:215–223
Choi JS, Kim JJ (2010) Amygdala regulates risk of predation in rats foraging in a dynamic fear environment. PNAS 107:21773–21777
Coker-Appiah DS et al (2013) Looming animate and inanimate threats. Soc Neurosci 8:621–630
Collins KA et al (2014) Taking action in the face of threat. J Neurosci 34:14733–14738
Conoscenti MA, Fanselow MS (2019) Dissociation in effective treatment and behavioral phenotype between stress-enhanced fear learning and learned helplessness. Front Behav Neurosci 13:104
Conoscenti MA et al (2017) Temporal parameters of post-stress prophylactic glucose treatment in rats. Stress 20:265–276
Constantinou E et al (2021) Measuring fear. J Behav Ther Exp Psych 70:101618
Corchs F, Schiller D (2019) Threat-related disorders as persistent motivational states of defense. Curr Opin Behav Sci 26:62–68
Coric V et al (2010) Multicenter, randomized, double-blind, active comparator and placebo-controlled trial of a corticotropin-releasing factor receptor-1 antagonist in GAD. Depress Anxiety 27:417–425
Cox WR et al (2022) Interfering with contextual fear memories by post-reactivation administration of propranolol in mice: a series of null findings. Front Behav Neurosci 16:893572
Craske MG et al (2009) What is an anxiety disorder? Depress Anxiety 26:1066–1085
Craske M et al (2013) Severity measure for specific phobia adult
Craske MG et al (2018) State-of-the-art and future directions for extinction as a translational model for fear and anxiety. Philos Trans R Soc Lond Ser B Biol Sci 373
Damasio A, Carvalho GB (2013) The nature of feelings. Nat Rev Neurosci 14:143–152
Davidson RJ (1992) Prolegomenon to the structure of emotion. Cognit Emot 6:245–268
Davidson RJ (1998) Affective style and affective disorders. Cognit Emot 12:307–330
Davis M et al (2010) Phasic vs sustained fear in rats and humans. Neuropsychopharmacology 35: 105–135
De Franceschi G et al (2016) Vision guides selection of freeze or flight defense strategies in mice. Curr Biol 26:2150–2154

Debiec J, LeDoux JE (2004) Disruption of reconsolidation but not consolidation of auditory fear conditioning by noradrenergic blockade in the amygdala. Neuroscience 129:267–272

Delgado MR et al (2006) Extending animal models of fear conditioning to humans. Biol Psychol 73:39–48

Delgado MR et al (2008) Neural circuitry underlying the regulation of conditioned fear and its relation to extinction. Neuron 59:829–838

Dunlop BW et al (2017) CRF1 antagonism is ineffective for women with PTSD. Biol Psychiatry 82: 866–874

Dunsmoor JE et al (2022) Laboratory models of post-traumatic stress disorder. Neuron 110:1754–1776

Dymond S et al (2015) Fear generalization in humans. Behav Ther 46:561–582

Evers C et al (2014) Emotion response coherence. Biol Psychol 98:43–49

Fadok JP et al (2018) New perspectives on central amygdala function. Curr Opin Neurobiol 49: 141–147

Fanselow MS (1997) Chapter 16: Species-specific defense reactions: retrospect and prospect. In: Bolles RC, Bouton ME, Fanselow MS (eds) Learning, motivation, and cognition: the functional behaviorism. American Psychological Association, Washington, pp 321–341

Fanselow MS (2018a) Emotion, motivation and function. Curr Opin Behav Sci 19:105–109

Fanselow MS (2018b) The role of learning in threat imminence and defensive behaviors. Curr Opin Behav Sci 24:44–49

Fanselow MS (2022) Negative valence systems. Emerg Top Life Sci

Fanselow MS, Lester LS (1988) A functional behavioristic approach to aversively motivated behavior: predatory imminence as a determinant of the topography of defensive behavior. In: Bolles RC, Beecher MD (eds) Evolution and learning. Erlbaum, Hillsdale, pp 185–211

Fanselow MS, Pennington ZT (2018) A return to the psychiatric dark ages with a two-system framework for fear. Behav Res Ther 100:24–29

Foa EB, McLean CP (2016) The efficacy of exposure therapy for anxiety-related disorders and its underlying mechanisms: the case of OCD and PTSD. Annu Rev Clin Psychol 12:1–28

Friedman BH et al (2014) Redundancy analysis of autonomic and self-reported, responses to induced emotions. Biol Psychol 98:19–28

Frijda N (1986) The emotions. Cambridge University Press, Cambridge

Frysztak RJ, Neafsey EJ (1991) The effect of medial frontal cortex lesions on respiration, "freezing," and ultrasonic vocalizations during conditioned emotional responses in rats. Cereb Cortex 1:418–425

Fullana MA et al (2020) Human fear conditioning. Behav Res Ther 124:103528

Furlong T, Carrive P (2007) Neurotoxic lesions centered on the perifornical hypothalamus abolish the cardiovascular and behavioral responses of conditioned fear to context but not of restraint. Brain Res 1128:107–119

Gerlicher AMV et al (2022) Better, worse, or different than expected. Sci Rep 12:5862

Geyer MA, Markou A (2002) Chapter 33: The role of preclinical models in the development of psychotropic drugs. In: Davis KL et al (eds) Neuropsychopharmacology: the fifth generation of progress. Lippincott, Williams, & Wilkins, Philadelphia, pp 445–455

Giustino TF et al (2016) Revisiting propranolol and PTSD. Neurobiol Learn Mem 130:26–33

Gladwin TE et al (2016) Ready and waiting. Neurosci Lett 619:182–188

Goswami S et al (2010) Impact of predatory threat on fear extinction in Lewis rats. Learn Mem 17: 494–501

Gozzi A et al (2010) A neural switch for active and passive fear. Neuron 67:656–666

Graham BM, Milad MR (2011) The study of fear extinction. Am J Psychiatry 168:1255–1265

Graham BM et al (2017) Low endogenous fibroblast growth factor 2 levels are associated with heightened conditioned fear expression in rats and humans. Biol Psychiatry 82:601–607

Gray JA, McNaughton N (2000) The neuropsychology of anxiety. Oxford University Press, New York

Griebel G, Holmes A (2013) 50 years of hurdles and hope in anxiolytic drug discovery. Nature Rev Drug Disc 12:667–687
Grillon C (2009) D-cycloserine facilitation of fear extinction and exposure-based therapy might rely on lower-level, automatic mechanisms. Biol Psychiatry 66:636–641
Grillon C, Ernst M (2020) A way forward for anxiolytic drug development. Neurosci Biobehav Rev 119:348–354
Grillon C et al (1998) Effects of experimental context and explicit threat cues on acoustic startle in Vietnam veterans with posttraumatic stress disorder. Biol Psychiatry 44:1027–1036
Grillon C et al (2004) Effects of the β-blocker propranolol on cued and contextual fear conditioning in humans. Psychopharmacologia 175:342–352
Grillon C et al (2015) The CRH1 antagonist GSK561679 increases human fear but not anxiety as assessed by startle. Neuropsychopharmacology 40:1064–1071
Haaker J et al (2019) Making translation work. Neurosci Biobehav Rev 107:329–345
Haller J et al (2013) Classical and novel approaches to the preclinical testing of anxiolytics. Neurosci Biobehav Rev 37:2318–2330
Hashemi MM et al (2019) Neural dynamics of shooting decisions and the switch from freeze to fight. Sci Rep 9:4240
Headley DB et al (2019) Embracing complexity in defensive networks. Neuron 103:189–201
Hefner K et al (2008) Impaired fear extinction learning and cortico-amygdala circuit abnormalities in a common genetic mouse strain. J Neurosci 28:8074–8085
Helmstetter FJ, Fanselow MS (1993) Aversively motivated changes in meal patterns of rats in a closed economy: the effects of shock density. Anim Learn Behav 21:168–175
Hertenstein E et al (2021) Augmentation of psychotherapy with neurobiological methods. Neuropsychobiology 80:437–453
Hodgson R, Rachman S (1974) II. Desynchrony in measures of fear. Behav Res Ther 12:319–326
Hoffman AN et al (2022) Anxiety, fear, panic. Learn Behav
Hofmann SG, Hayes SC (2019) The future of intervention science: process-based therapy. Clin Psychol Sci 7:37–50
Hoge EA et al (2012) Effect of acute posttrauma propranolol on PTSD outcome and physiological responses during script-driven imagery. CNS Neurosci Ther 18:21–27
Hollenstein T, Lanteigne D (2014) Models and methods of emotional concordance. Biol Psychol 98:1–5
Holmes A, Quirk GJ (2010) Pharmacological facilitation of fear extinction and the search for adjunct treatments for anxiety disorders – the case of yohimbine. Trends Pharmacol Sci 31:2–7
Huang L et al (2017) A retinoraphe projection regulates serotonergic activity and looming-evoked defensive behaviour. Nat Commun 8:14908
Hugdahl K (1981) The 3-systems-model of fear and emotion-a critical examination. Beh Res Ther 19:75–85
Husek TR, Alexander S (1963) The effectiveness of the anxiety differential in examination stress situations. Educ Psychol Meas 23:309–318
Hyman SE (2013) Psychiatric drug development: diagnosing a crisis. Cerebrum 2013:5
Ising M et al (2007) High-affinity CRF1 receptor antagonist NBI-34041. Neuropsychopharmacology 32:1941–1949
Janak PH, Tye KM (2015) From circuits to behaviour in the amygdala. Nature 517:284–292
Jean-Richard-Dit-Bressel P et al (2018) Behavioral and neurobiological mechanisms of punishment. Neuropsychopharmacology 43:1639–1650
Jelen P et al (2003) 22-kHz ultrasonic vocalization in rats as an index of anxiety but not fear. Behav Brain Res 141:63–72
Johansen JP et al (2011) Molecular mechanisms of fear learning and memory. Cell 147:509–524
Josselyn SA, Tonegawa S (2020) Memory engrams. Science 367
Jovanovic T et al (2009) PTSD may be associated with impaired fear inhibition. Psychiatry Res 167:151–160

Jovanovic T et al (2010) Impaired fear inhibition is a biomarker of PTSD but not depression. Depress Anxiety 27:244–251
Jovanovic T et al (2011) Cortisol suppression by dexamethasone reduces exaggerated fear responses in posttraumatic stress disorder. Psychoneuroendocrinology 36:1540–1552
Jovanovic T et al (2020) Psychophysiological treatment outcomes: CRF1 antagonist increases inhibition of fear-potentiated startle in PTSD patients. Psychophysiology 57:e13356
Kaffman A et al (2019) Enhancing the utility of preclinical research in neuropsychiatry drug development. Methods Mol Biol 2011:3–22
Kehne JH, Cain CK (2010) Therapeutic utility of non-peptidic CRF1 receptor antagonists in anxiety, depression, and stress-related disorders. Pharmacol Ther 128:460–487
Kida S (2020) Function and mechanisms of memory destabilization and reconsolidation after retrieval. Proc Jpn Acad Ser B Phys Biol Sci 96:95–106
Kim SY et al (2013) Diverging neural pathways assemble a behavioural state from separable features in anxiety. Nature 496:219–223
Klein AS et al (2021) Fear balance is maintained by bodily feedback to the insular cortex in mice. Science 374:1010–1015
Klumpers F et al (2017) How human amygdala and bed nucleus of the stria terminalis may drive distinct defensive responses. J Neurosci 37:9645–9656
Lang PJ (1968) Fear reduction and fear behavior: problems in treating a construct. In: Shlien JM (ed) Research in psychotherapy. American Psychological Association, pp 90–102
Lazaro-Munoz G et al (2010) Sidman instrumental avoidance initially depends on lateral and basal amygdala and is constrained by central amygdala-mediated Pavlovian processes. Biol Psychiatry 67:1120–1127
Leaton RN, Borszcz GS (1985) Potentiated startle. J Exp Psychol 11:421–428
LeDoux JE (2012) Rethinking the emotional brain. Neuron 73:653–676
LeDoux JE (2015) Anxious. Viking, New York
LeDoux JE (2019) The deep history of ourselves. Viking, New York
LeDoux JE, Brown R (2017) A higher-order theory of emotional consciousness. PNAS 114: E2016–E2E25
LeDoux J, Daw ND (2018) Surviving threats. Nat Rev Neurosci 19:269–282
LeDoux JE, Pine DS (2016) Using neuroscience to help understand fear and anxiety: a two-system framework. Am J Psychiatry 173:1083–1093
LeDoux JE et al (1988) Different projections of the central amygdaloid nucleus mediate autonomic and behavioral correlates of conditioned fear. J Neurosci 8:2517–2529
LeDoux JE et al (2017) The birth, death and resurrection of avoidance. Mol Psychiatry 22:24–36
Lee SC et al (2017) Differential recruitment of competing valence-related amygdala networks during anxiety. Neuron 96:81–88 e5
Liang KC et al (1986) Modulating effects of posttraining epinephrine on memory. Brain Res 368: 125–133
Lissek S et al (2005) Classical fear conditioning in the anxiety disorders. Behav Res Ther 43:1391–1424
Lissek S et al (2006) The strong situation. Biol Psychol 72:265–270
Liu D et al (2022) Different coding characteristics between flight and freezing in dorsal periaqueductal gray of mice during exposure to innate threats. Animal Model Exp Med
Low A et al (2015) When threat is near, get out of here. Psychol Sci 26:1706–1716
Luyten L et al (2021) Lack of drug-induced post-retrieval amnesia for auditory fear memories in rats. BMC Biol 19:17
Maier SF, Seligman ME (2016) Learned helplessness at fifty. Psychol Rev 123:349–367
Maier SF, Watkins LR (2005) Stressor controllability and learned helplessness. Neurosci Biobehav Rev 29:829–841
Maren S (2005) Synaptic mechanisms of associative memory in the amygdala. Neuron 47:783–786
Maren S (2022) Unrelenting fear under stress. Front Syst Neurosci 16:888461

Markou A et al (2009) Removing obstacles in neuroscience drug discovery. Neuropsychopharmacology 34:74–89
Markowitz S, Fanselow M (2020) Exposure therapy for PTSD. Brain Sci 10
McCue MG et al (2014) Medial amygdala lesions selectively block aversive pavlovian-instrumental transfer in rats. Front Behav Neurosci 8:329
McGhee LL et al (2009) The effect of propranolol on posttraumatic stress disorder in burned service members. J Burn Care Res 30:92–97
Meyerbroeker K et al (2012) Does yohimbine hydrochloride facilitate fear extinction in virtual reality treatment of fear of flying? Psychother Psychosom 81:29–37
Milad MR, Quirk GJ (2012) Fear extinction as a model for translational neuroscience. Annu Rev Psychol 63:129–151
Milad MR et al (2009) Neurobiological basis of failure to recall extinction memory in PTSD. Biol Psychiatry 66:1075–1082
Miller BV, Bernstein DA (1972) Instructional demand in a behavioral avoidance test for claustrophobic fears. J Abnorm Psychol 80:206–210
Minor TR et al (1991) Inverting the traditional view of “learned helplessness”. In: Denny MR (ed) Fear, avoidance and phobias. Erlbaum, Hillsdale, pp 87–133
Mobbs D (2018) The ethological deconstruction of fear(s). Curr Opin Behav Sci 24:32–37
Mobbs D et al (2007) When fear is near. Science 317:1079–1083
Mobbs D et al (2009) From threat to fear. J Neurosci 29:12236–12243
Mobbs D et al (2010) Neural activity associated with monitoring the oscillating threat value of a tarantula. PNAS 107:20582–20586
Mobbs D et al (2015) The ecology of human fear. Front Neurosci 9:55
Mobbs D et al (2019) Viewpoints: approaches to defining and investigating fear. Nat Neurosci 22: 1205–1216
Mobbs D et al (2020) Space, time, and fear. Trends Cogn Sci 24:228–241
Morris RW, Bouton ME (2007) The effect of yohimbine on the extinction of conditioned fear: a role for context. Behav Neurosci 121:501–514
Moscarello JM, Hartley CA (2017) Agency and the calibration of motivated behavior. Trends Cogn Sci 21:725–735
Moscarello JM, Maren S (2018) Flexibility in the face of fear. Curr Opin Behav Sci 19:44–49
Moscarello JM, Penzo MA (2022) The central nucleus of the amygdala and the construction of defensive modes across the threat-imminence continuum. Nat Neurosci 25:999–1008
Murrough JW, Charney DS (2017) Corticotropin-releasing factor type 1 receptor antagonists for stress-related disorders: time to call it quits? Biol Psychiatry 82:858–860
Neuhoff JG (2018) Chapter 12: Adaptive biases in visual and auditory looming perception. In: Hubbard TL (ed) Spatial biases in perception and cognition. Cambridge University Press, pp 180–192
Nili U et al (2010) Fear thou not. Neuron 66:949–962
Norrholm SD, Jovanovic T (2018) Fear processing, psychophysiology, and PTSD. Harv Rev Psychiatry 26:129–141
Norrholm SD et al (2015) Fear load: the psychophysiological over-expression of fear as an intermediate phenotype associated with trauma reactions. Int J Psychophysiol 98:270–275
Nugent NR et al (2010) The efficacy of early propranolol administration at reducing PTSD symptoms in pediatric injury patients: a pilot study. J Trauma Stress 23:282–287
Ohman A, Mineka S (2001) Fears, phobias, and preparedness: toward an evolved module of fear and fear learning. Psychol Rev 108:483–522
Ojala KE, Bach DR (2020) Measuring learning in human classical threat conditioning. Neurosci Biobehav Rev 114:96–112
Ost LG et al (1982) Individual response patterns and the effects of different behavioral methods in the treatment of claustrophobia. Behav Res Ther 20:445–460
Peng Y et al (2022) Threat neurocircuitry predicts the development of anxiety and depression symptoms in a longitudinal study. Biol Psychiatry Cogn Neurosci Neuroimaging

Perusini JN, Fanselow MS (2015) Neurobehavioral perspectives on the distinction between fear and anxiety. Learn Mem 22:417–425

Perusini JN et al (2016) Induction and expression of fear sensitization caused by acute traumatic stress. Neuropsychopharmacology 41:45–57

Pigeon S et al (2022) Impairing memory reconsolidation with propranolol in healthy and clinical samples: a meta-analysis. J Psychiatry Neurosci 47:E109–EE22

Pine DS, LeDoux JE (2017) Elevating the role of subjective experience in the clinic: response to Fanselow and Pennington. Am J Psychiatry 174:1121–1122

Pitman RK et al (2002) Pilot study of secondary prevention of posttraumatic stress disorder with propranolol. Biol Psychiatry 51:189–192

Poulos AM et al (2014) Amnesia for early life stress does not preclude the adult development of posttraumatic stress disorder symptoms in rats. Biol Psychiatry 76:306–314

Powers MB et al (2009) Facilitation of fear extinction in phobic participants with a novel cognitive enhancer. J Anxiety Disord 23:350–356

Priebe K et al (2013) Frequency of intrusions and flashbacks in patients with PTSD related to childhood sexual abuse. Psychol Assess 25:1370–1376

Quinn JJ et al (2002) Post-training excitotoxic lesions of the dorsal hippocampus attenuate forward trace, backward trace, and delay fear conditioning in a temporally specific manner. Hip 12:495–504

Rachman S, Hodgson R (1974) I. Synchrony and desynchrony in fear and avoidance. Behav Res Ther 12:311–318

Rau V, Fanselow MS (2009) Exposure to a stressor produces a long lasting enhancement of fear learning in rats. Stress 12:125–133

Raut SB et al (2022) Effects of propranolol on the modification of trauma memory reconsolidation in PTSD patients. J Psychiatr Res 150:246–256

Reist C et al (2001) β-Adrenergic blockade and emotional memory in PTSD. Neuropsychopharmacology 4:377–383

Ressler KJ (2020) Translating across circuits and genetics toward progress in fear- and anxiety-related disorders. Am J Psychiatry 177:214–222

Richter J et al (2012) Dynamics of defensive reactivity in patients with panic disorder and agoraphobia. Biol Psychiatry 72:512–520

Risbrough V (2010) Behavioral correlates of anxiety. Curr Top Behav Neurosci 2:205–228

Riskind JH et al (2014) Influence of anxiety, depression and looming cognitive style on auditory looming perception. J Anxiety Disord 28:45–50

Roberts LE, Young R (1971) Electrodermal responses are independent of movement during aversive conditioning in rats, but heart rate is not. J Comp Physiol Psychol 77:495–512

Robinson MD, Clore GL (2002) Episodic and semantic knowledge in emotional self-report. J Pers Soc Psychol 83:198–215

Roelofs K, Dayan P (2022) Freezing revisited. Nat Rev Neurosci 23:568–580

Rolls ET (2008) Emotion, higher-order syntactic thoughts, and consciousness. In: Weiskrantz L, Davies M (eds) Frontiers of consciousness: Chichele lectures. Oxford University Press, Oxford, pp 131–167

Rosen JB, Schulkin J (1998) From normal fear to pathological anxiety. Psychol Rev 105:325–350

Rosenberg L et al (2018) Does acute propranolol treatment prevent posttraumatic stress disorder, anxiety, and depression in children with burns? J Child Adolesc Psychopharmacol 28:117–123

Rosler L, Gamer M (2019) Freezing of gaze during action preparation under threat imminence. Sci Rep 9:17215

Sakamoto IM (2020) Individual differences in ultrasonic vocalizations and freezing during fear learning and extinction in female rats. In: Pharmacology, physiology and neuroscience. USC, Columbia, p 50

Sanderson WC, Barlow DH (1990) A description of patients diagnosed with DSM-III-R generalized anxiety disorder. J Nerv Ment Dis 178:588–591

Schaffner KF (2020) Chapter 32: Approaches to multilevel models of fear. In: Kendler KS, Parnas J, Zachlar P (eds) Levels of analysis in psychopathology. Cambridge University Press, New York, pp 384–409

Schipper P et al (2019) The association between serotonin transporter availability and the neural correlates of fear bradycardia. PNAS 116:25941–25947

Seligman ME, Maier SF (1967) Failure to escape traumatic shock. J Exp Psychol 74:1–9

Sewart A et al (2021) Pre-treatment hippocampal functioning impacts context renewal for cholinergic modulated exposure therapy. Biol Psychol 165:108167

Shackman AJ, Fox AS (2016) Contributions of the central extended amygdala to fear and anxiety. J Neurosci 36:8050–8063

Shang C et al (2015) A parvalbumin-positive excitatory visual pathway to trigger fear responses in mice. Science 348:1472–1477

Shear MK et al (1997) Multicenter collaborative panic disorder severity scale. Am J Psychiatry 154: 1571–1575

Siegel EH et al (2018) Emotion fingerprints or emotion populations? Psychol Bull 144:343–393

Signoret-Genest J et al (2023) Integrated cardio-behavioral responses to threat define defensive states. Nat Neurosci 26:447–457

Smits JA et al (2014) Yohimbine enhancement of exposure therapy for social anxiety disorder. Biol Psychiatry 75:840–846

Sotres-Bayon F, Quirk GJ (2010) Prefrontal control of fear. Curr Opin Neurobiol 20:231–235

Southwick SM et al (1993) Abnormal noradrenergic function in posttraumatic stress disorder. Arch Gen Psychiatry 50:266–274

Stein MB et al (2007) Pharmacotherapy to prevent PTSD. J Trauma Stress 20:923–932

Stein MB et al (2021) Randomized, placebo-controlled trial of the angiotensin receptor antagonist losartan for posttraumatic stress disorder. Biol Psychiatry 90:473–481

Stemerding LE et al (2022) Demarcating the boundary conditions of memory reconsolidation: an unsuccessful replication. Sci Rep 12:2285

Stewart AM et al (2015) The failure of anxiolytic therapies in early clinical trials. Expert Opin Investig Drugs 24:543–556

Sznycer D, Cohen AS (2021) Are emotions natural kinds after all? Evol Psychol 19: 14747049211016009

Taschereau-Dumouchel V et al (2020) Multivoxel pattern analysis reveals dissociations between subjective fear and its physiological correlates. Mol Psychiatry 25:2342–2354

Taschereau-Dumouchel V et al (2022) Putting the "mental" back in "mental disorders". Mol Psychiatry

Tovote P et al (2015) Neuronal circuits for fear and anxiety. Nat Rev Neurosci 16:317–331

Tuerk PW et al (2018) Augmenting treatment efficiency in exposure therapy for PTSD. Cogn Beh Ther 47:351–371

Vaiva G et al (2003) Immediate treatment with propranolol decreases posttraumatic stress disorder two months after trauma. Biol Psychiatry 54:947–949

Waddell J et al (2006) Effects of bed nucleus of the stria terminalis lesions on conditioned anxiety. Behav Neurosci 120:324–336

Walker DL et al (2002) Facilitation of conditioned fear extinction by systemic administration or intra-amygdala infusions of D-cycloserine as assessed with fear-potentiated startle in rats. J Neurosci 22:2343–2351

Walker D et al (2009) Differential effects of the CRF-R1 antagonist GSK876008 on fear-potentiated, light- and CRF-enhanced startle suggest preferential involvement in sustained vs phasic threat responses. Neuropsychopharmacology 34:1533–1542

Wei P et al (2015) Processing of visually evoked innate fear by a non-canonical thalamic pathway. Nat Commun 6:6756

Wen Z et al (2021) Fear extinction learning modulates large-scale brain connectivity. NeuroImage 238:118261

Wendt J et al (2017) Active avoidance and attentive freezing in the face of approaching threat. NeuroImage 158:196–204
White KS et al (2006) Avoidance behavior in panic disorder: the moderating influence of perceived control. Behav Res Ther 44:147–157
White LK et al (2017) Complementary features of attention bias modification therapy and cognitive-behavioral therapy in pediatric anxiety disorders. Am J Psychiatry 174:775–784
Woods AM, Bouton ME (2006) D-cycloserine facilitates extinction but does not eliminate renewal of the conditioned emotional response. Behav Neurosci 120:1159–1162
Woon EP et al (2020) Differential effects of prior stress on conditioned inhibition of fear and fear extinction. Behav Brain Res 381:112414
Yilmaz M, Meister M (2013) Rapid innate defensive responses of mice to looming visual stimuli. Curr Biol 23:2011–2015
Young KS et al (2021) Dysregulation of threat neurocircuitry during fear extinction: the role of anhedonia. Neuropsychopharmacology 46:1650–1657
Zambetti PR et al (2022) Ecological analysis of Pavlovian fear conditioning in rats. Commun Biol 5:830
Zhou F et al (2021) A distributed fMRI-based signature for the subjective experience of fear. Nat Commun 12:6643
Zinbarg RE (1998) Concordance and synchrony in measures of anxiety and panic reconsidered. Behav Ther 29:301–323

Understanding Human Fear Extinction: Insights from Psychophysiology

Jessica Woodford, Manessa Riser, and Seth Davin Norrholm

Contents

Abstract The study of fear extinction has been driven largely by Pavlovian fear conditioning methods across the translational spectrum. The primary methods used to study these processes in humans have been recordings of skin conductance (historically termed galvanic skin response) and fear-potentiation of the acoustic startle reflex. As outlined in the following chapter, the combined corpus of this work has demonstrated the value of psychophysiology in better understanding the underlying neurobiology of extinction learning in healthy humans as well as those with psychopathologies. In addition, psychophysiological approaches, which allow for

J. Woodford and M. Riser
Department of Psychiatry and Behavioral Neurosciences, Neuroscience Center for Anxiety, Stress, and Trauma, Wayne State University School of Medicine, Detroit, MI, USA

S. D. Norrholm (✉)
Department of Psychiatry and Behavioral Neurosciences, Neuroscience Center for Anxiety, Stress, and Trauma, Wayne State University School of Medicine, Detroit, MI, USA

Department of Behavioral Sciences and Leadership, United States Air Force Academy, Colorado Springs, CO, USA
e-mail: Seth.Norrholm@afacademy.af.edu; SNorrholm@wayne.edu

Curr Topics Behav Neurosci (2023) 64: 59–78
https://doi.org/10.1007/7854_2023_435
Published Online: 2 August 2023

the preservation of methods between species, have shown their applicability to the assessment of wide-ranging treatment effects. The chapter concludes with potential trajectories for future study in this area.

Keywords Acoustic startle · Fear conditioning · Posttraumatic stress disorder · Psychophysiology · Skin conductance

1 Introduction

Impairments in the process of fear extinction, or the ability to reduce one's level of fear to a cue or context that is no longer threatening, are considered to be a primary mechanism underlying pathological fear and anxiety. For this reason, the onset, progression, treatment, and potential symptomatic relapse of pathological fear and anxiety can be translationally modeled through Pavlovian fear-conditioning paradigms that test acquisition, extinction training, extinction recall, and return of fear (Briscione et al. 2014). A focus of this book and some of its content is on the psychobiological processes underlying the development and expression of fear extinction learning. Neuroscientists and clinicians with an interest in fear extinction have a host of methodological and analytical tools available to study extinction processes in humans. In fact, many of these tools are discussed at length throughout this volume. A significant body of knowledge regarding human fear extinction learning has been accumulated through the use of psychophysiological techniques that index the activity of several neural and peripheral systems (i.e., fear-potentiated startle, skin conductance, heart-rate variability, or pupillometry, to name a few). In this chapter, we will focus on two experimental methods that have formed a significant portion of the empirical corpus linked to human fear extinction: electrodermal recordings from the sweat glands of the skin (skin conductance; SC) and electromyographic (EMG) measurements of the muscles that control eyeblink and facial expression (via fear-potentiated startle techniques).

2 Psychophysiological Tools

2.1 Electrodermal Indices: Skin Conductance (SC)

Responsivity of electrodermal systems is often assessed through the measurement of sweat gland activity from the skin which, in turn, affects conductivity at the skin's surface (Cacciopo et al. 2004). Measurement typically includes capturing a baseline level of skin conductance level followed by determination of evoked skin conductance responses to presented stimuli. Skin conductance is an excellent measure of arousal as it reflects underlying sympathetic nervous system activation (Lang et al.

1998). As a measure of arousal, skin conductance is well suited to explore learning phenomena such as habituation, exposure to novelty, and fear or threat conditioning (Orr et al. 2000). Responses are reliably collected from electrodes placed on the fingers or surface of the palm of the hand. Evoked responses typically peak within 3–6 s after stimulus onset following a latency of 1–4 s, and, as such, the skin conductance response (SCR) is defined as the difference in skin conductance during a baseline, pre-stimulus window, and a peak post-stimulus response. Following successful fear acquisition, the SCR is stronger in the presence of the reinforced CS+ as compared to a non-reinforced CS−. Based on the reliability of these acquisition effects, numerous translational research groups have used SCR as a primary outcome measure to investigate the neural processes underlying fear extinction learning as well as its clinical manifestations (for a representative example, see Milad et al. 2009).

2.2 Electromyographic Indices: Acoustic Startle Response

A highly translational tool that has been used to study the extinction of learned fear in humans is the assessment of the acoustic startle response within cued or contextual conditioning paradigms. The acoustic startle response is a reflexive contraction of the facial muscles (notably the *orbicularis oculi* muscle that encircles each eye) following exposure to a sudden auditory stimulus (e.g., 108 dB burst of white noise; (Blumenthal et al. 2005; Landis and Hunt 1939; Lang et al. 1990)). Noise-generated startle responses occur quickly within 20–200 ms after the stimulus is presented and can be modulated by emotional states such as fear (or threat) given the connectivity between the amygdala and the pons, which initiates the blink response (Davis 1997). As will be discussed in this chapter, *fear-potentiated startle*, or the relative increase in the magnitude of the acoustic startle response in the presence of a stimulus (i.e., conditioned stimulus, CS) or context that has been repeatedly paired with an aversive outcome (e.g., unconditioned stimulus, US; (Davis and Astrachan 1978)), has been used to better understand human fear conditioning and extinction; processes that can be altered in psychiatric conditions such as posttraumatic stress disorder, PTSD (Briscione et al. 2014; VanElzakker et al. 2014).

Unlike SCR, fear-potentiated startle (FPS) allows for the comparison of elicited responses to the CS+ versus CS− as well as enhanced startle to the CS+ as compared to a baseline startle response (termed noise-alone or inter-trial intervals, ITIs). SCR typically uses responses to a CS− or an unextinguished CS− as a comparator in fear acquisition and extinction learning, respectively. It is this latter feature that affords one the capacity to observe within-session extinction effects (i.e., reductions in the conditioned response) that can be obscured with SCR due to a number of factors including rapid habituation.

A growing body of work has demonstrated a strong association between physiological SCR and FPS discrimination between a reinforced CS+ and a non-reinforced CS− and this same discrimination in US-expectancy measures during fear conditioning, or acquisition (Norrholm et al. 2006; Constantinou et al. 2021). However, this tight relationship between these two frequently used physiological indices is not as clear during extinction training and extinction recall. Fear extinction studies that have used SCR as a primary psychophysiological outcome have shown strong reliability in illustrating between-session extinction effects (Milad et al. 2006, 2007, 2009) while FPS extinction studies have shown a capacity to observe both within-session extinction and between-session extinction effects (Norrholm et al. 2011a, b; Warren et al. 2014).

Psychophysiological techniques can be advantageous for the study of fear extinction (and its prerequisite acquisition phases) due to a high degree of translation (i.e., relatively small translational gap that exists) between rodent and human applications. Conditioned fear can be developed in both animal models and human studies with similar methods (e.g., repeatedly pairing a previously neutral stimulus with an aversive outcome) and, as is the case with fear-potentiated startle, very similar behavioral indices. As a direct result of the preliminary findings obtained with pre-clinical animal studies, fear-potentiated startle methods have been successfully developed to reliably assess enhanced startle to discrete cues (Norrholm and Jovanovic 2018), to general contexts (Schmitz and Grillon 2012), and to fear-relevant virtual reality-based cues (Maples-Keller et al. 2019).

3 Within-Session Extinction

In most fear extinction experimental paradigms, fear acquisition (i.e., the repeated pairing of a previously neutral stimulus with an aversive, unconditioned stimulus to elicit a conditioned response) is typically followed by an extinction training session (also referred to as within-session extinction) during which non-reinforced trials of the CS+ are presented repeatedly. Extinction training may occur within minutes (e.g., Norrholm et al. 2006, 2008) or several hours up to a day later (e.g., Milad et al. 2009; Lonsdorf and Merz 2017) and is characterized by a decrement in an acquired conditioned response as the previously reinforced CS loses its predictive value. The initial set of trials during extinction training (sometimes termed early extinction) can serve as a test of acquisition recall and may reflect an extended state of excitation to the previously reinforced CS+ that should diminish as more non-reinforced CS+ trials are presented. It is the within-session extinction training phase that has been most closely matched with the in-session process of exposure therapy (Foa et al. 2007). As such, there has been a call for greater use of at least 24 h passage of time between acquisition and extinction training as this allows for optimized fear memory consolidation and may be more ecologically valid (Adolph et al. 2022; Lonsdorf and Merz 2017). However, for logistical reasons, many human extinction studies utilize immediate extinction (within minutes of the end of acquisition) while most pre-clinical rodent studies use delayed extinction (about a day) and this fact represents both a translational gap between species and a gap between human laboratory and clinic (Lonsdorf and Merz 2017).

4 Between-Session Extinction

Between-session extinction, or extinction recall, is said to have successfully occurred if the previously extinguished conditioned response remains at a minimal level or is undetectable when the test organism is presented with the extinguished CS again. If the conditioned response re-appears at the time of an extinction recall test, the conditioned response is said to have returned through the process of spontaneous recovery (Bouton and Swartzentruber 1991) or renewal if there was a shift in context from the extinction training context (Bouton and Ricker 1994). Regardless of the psychophysiological modality utilized, the duration of an extinction recall test is relatively short, with a small number of CS trials, as extending this test becomes an assessment of re-extinction rather than return of fear.

5 Return of Fear After Extinction Learning

The prevailing view on fear extinction is that the original fear acquisition memory is not erased but rather a new extinction learning memory trace is formed which competes with the original fear memory (Bouton 2004; Myers and Davis 2007; Pavlov 1927). As such, the original fear memory can be recalled through experimental manipulations that elicit a return of fear through the aforementioned spontaneous recovery (passage of time since extinction training), renewal (change in context), or reinstatement (brief unsignaled reintroduction of the US). Numerous applications of psychophysiologically-based examinations of fear extinction have included an assessment of the return of fear through the latter processes. Accordingly, when the expression of the conditioned response (e.g., fear-potentiated startle, SCR) is low or absent, we view this as a dominance of the more recently encoded extinction memory and when the conditioned response is strong we view this as a dominance, or return, of the original fear memory. An assessment of return of fear following successful extinction is important as it models clinical relapse of fear-related symptoms (Bouton 2004; Craske et al. 2008a; Vervliet et al. 2013) and allows for the investigation of inhibitory memory recall (Haaker et al. 2014; Norrholm and Jovanovic 2018). As will be discussed below, psychophysiological studies have shown that fear extinction, as described by Vervliet et al. (2013), is easy to "learn" but difficult to "remember" (Vervliet et al. 2013). In other words, the return of fear in many human psychophysiological fear conditioning and extinction applications is quite commonly found.

Psychophysiological paradigms employing both fear-potentiated startle and SCR have included return of fear tests as soon as 24 h after extinction training. As noted by Lonsdorf and colleagues (2017), this test of retention simultaneously measures extinction recall and spontaneous recovery depending on the directionality of the result (increase/maintenance or decrease/absence of the conditioned response signifying return of fear or extinction retention, respectively). It is generally accepted that the degree of spontaneous recovery is proportional to the time that has elapsed between the end of extinction training and the return of fear test (Quirk 2002), although this is based almost exclusively on rodent data. Spontaneous recovery of previously extinguished conditioned fear has been observed by several groups using SCR (Guastella et al. 2007; Huff et al. 2009; Schiller et al. 2008) and fear-potentiated startle (Norrholm et al. 2008). In a manner that differs from their sensitivity with respect to within-session extinction learning (i.e., startle is sensitive to within-session extinction decrements whereas SCR is not as much), both fear-potentiated startle and SCR are sensitive to retention test effects be they spontaneous recovery or extinction recall/retention.

The return of conditioned fear due to a change in context, or renewal, is commonly examined in rodent fear conditioning and extinction studies due to the relative ease of switching contextual cues in this type of study. Contextual shifts are less common in human psychophysiological studies of fear extinction and have most reliably occurred through the use of virtual reality techniques (Zabik et al. 2023). For example, Milad et al. (2005) demonstrated renewal effects after extinction learning such that there were minimal SCRs upon reintroduction of the previously reinforced CS+ in the extinction training context (context B) but renewed fear responses when the CS+ was presented in the original conditioning context (context A) (Milad et al. 2005). Effting and Kindt (2007), using skin conductance with expectancy ratings, furthered this by showing that renewal effects using an ABA design (context A for acquisition, context B for extinction training followed by a return to context A for extinction retention test) were stronger than renewal with the introduction of a novel context during extinction recall (e.g., an ABC design; (Effting and Kindt 2007)). The reliability, stability, and potential clinical applications of ABA renewal procedures (using skin conductance as a primary outcome) have been most consistently demonstrated by the Milad research group (see Zeidan et al. 2012).

Reinstatement refers to the return of a conditioned fear response following brief, repeated, unsignaled presentations of the US after extinction training is completed (see Bouton and Bolles 1979; Rescorla and Heth 1975). Several groups have demonstrated the reinstatement of previously extinguished conditioned fear using psychophysiological means (see Vervliet et al. 2013; Cisler et al. 2020). For example, LaBar and Phelps (2005), using skin conductance measures, showed reinstatement in humans to be context (and by extension) hippocampally dependent (LaBar and Phelps 2005). Additionally, human fear-potentiated startle methods have successfully included assessments of reinstatement in both proof-of-concept studies (e.g., Norrholm et al. 2006) and as part of the reconsolidation update techniques (Warren et al. 2014) first introduced by the Monfils and Schiller groups (e.g., (Schiller et al. 2010); see Chapter in this volume). The latter reconsolidation update studies aided in the demonstration that return of fear through reinstatement could be attained through the use of skin conductance (Schiller et al. 2010) or fear-potentiated startle measures (Soeter and Kindt 2011; Warren et al. 2014) individually, or with the concurrent assessment of both psychophysiological indices (Kindt and Soeter 2013).

Psychophysiological studies have also shown that the return of fear is not always specific to the previously reinforced CS+. Many human fear-conditioning paradigms include a non-reinforced CS− typically within the same modality of the CS+ (e.g., visual geometric shape) and many groups have reported equivalent conditioned fear responses to the CS− during a return of fear test (for examples, see Cisler et al. 2020; Dirikx et al. 2009; Sokol and Lovibond 2012). This lack of differential responding to a CS+ and CS− during the assessment of return of fear may be related to factors such as anxious disposition or pathological anxiety or fear generalization (Grillon 2002; Lissek et al. 2010).

6 Experimental Design and Extinction Learning

It is clear based on the voluminous data available in the existing literature regarding extinction learning that the majority of translational studies use exposure therapy as the clinical analog for laboratory extinction paradigms. In fact, in many studies, significant efforts are made to tailor pre-clinical human studies toward the goal of informing exposure therapy applications. For example, a recent study by Adolph et al. (2022) included a number of modifications directly aimed at closing the gap between laboratory paradigms and clinical exposure treatment. These modifications included: (1) the use of delayed extinction such that there was a time lapse of 24 h between acquisition and extinction training sessions, (2) informing participants of the CS/US contingencies *before* the initiation of fear conditioning to "standardize" fear levels in participants following acquisition, (3) a reinforcement rate of 60% during acquisition aimed at assuring variance in extinction levels, and (4) the inclusion of human facial pictures as conditioned stimuli to increase ecological validity.

Many of the experimental decisions made by the Adolph team were informed by a recent methodological position paper by Lonsdorf and colleagues (2017) focused on the assessment of human fear acquisition, extinction, and return of fear. The latter Lonsdorf paper was the culmination of discussions among 14 major human fear conditioning laboratories from across Europe. This chapter will not recapitulate the observations from that group and we refer interested readers to that reference. We can, however, summarize some of the relevant material discussed by the Lonsdorf compendium. The authors identified several factors that significantly contribute to the expression of human fear extinction, including how it is assessed psychophysiologically, and to the interpretation of these collective findings. As examples, the relevant factors include: the number and modality of conditioned and unconditioned stimuli, fear acquisition reinforcement rate, immediate versus delayed extinction, the presence or absence of verbal instructions, type of return of fear assessment, outcome measures employed (e.g., physiological, cognitive, verbal), the number of sessions and days that comprise a selected fear extinction paradigm, and data processing techniques.

Related to the experimental features alluded to in the previous section, the fear-relevance of conditioned stimuli plays a role in extinction learning. Fear extinction learning is delayed when the CSs used in an experimental procedure are fear-relevant such as spiders or snakes as opposed to neutral shapes (for review see Ohman and Mineka 2001). In addition, despite the protraction of extinction learning with fear-relevant, salient stimuli, the post-extinction return of fear to these emotionally charged stimuli is robust and has been seen in tests of renewal, as an example (Kindt and Soeter 2013; Kindt et al. 2009; Soeter and Kindt 2010).

7 Facilitation of Extinction Learning

The potential pharmacological enhancement of extinction learning (and by extension of prolonged exposure therapy) has been significantly informed with the use of psychophysiologically-based fear acquisition and extinction paradigms (please see Rothbaum and Davis 2003; Briscione et al. 2014). To date, the most promising agent aimed at enhancing extinction learning has been D-cycloserine (DCS), a N-methyl-D-aspartate glutamate receptor partial agonist. Translationally speaking, DCS enhanced extinction learning in rodent applications (Ledgerwood et al. 2003; Walker et al. 2002) and showed efficacy in enhancing exposure therapy in anxious patients when administered as a single dose just prior to a treatment session (Difede et al. 2022; Guastella et al. 2007; Kushner et al. 2007; Otto et al. 2010; Ressler et al. 2004; Rothbaum et al. 2014; Wilhelm et al. 2008). Notably, DCS has also shown ineffectiveness in some treatment regimens including as an adjunct to exposure therapy for social anxiety disorder (social phobia) (Hofmann et al. 2006; Rodrigues et al. 2014). A more detailed discussion of D-cycloserine can be found in the chapter by the Ressler author group included in this volume and here (Mataix-Cols et al. 2017; Rosenfield et al. 2019; Vervliet 2008).

A significant body of literature emerging from both clinical (Feduccia and Mithoefer 2018) and pre-clinical rodent (Young et al. 2015, 2017) studies pointed to 3,4-methylenedioxymethamphetamine (MDMA; Ecstasy) as a candidate agent with which to enhance extinction learning. More specifically, MDMA has shown promise as a treatment for enhancing PTSD exposure therapies while also facilitating extinction retention in Pavlovian fear-conditioning applications. For these reasons, there have been a few notable investigations in human fear acquisition and extinction paradigms with psychophysiological outcomes. Maples-Keller and colleagues (2022b) employed a well-established fear-potentiated startle paradigm to further explore the capability of MDMA to enhance extinction (Maples-Keller et al. 2022b). In their randomized, placebo-controlled trial, Maples-Keller and co-authors observed no between-group differences in either within-session extinction during training or between-session extinction at a recall test. The authors did, however, find that significantly more participants retained extinction learning in the MDMA group as compared to the placebo group. Vizeli a et al. (2022) employed similar methods, with the addition of skin conductance measures to fear-potentiated startle, to also investigate the potential for MDMA to enhance extinction learning. The Vizeli study yielded mixed results as SCR responses showed an enhancing effect of MDMA during extinction training and at recall (i.e., lower SCR) while fear-potentiated startle measures revealed no significant group differences. Clearly at this point it is too soon to determine the degree to which MDMA facilitates human extinction learning but its clinical effects on extinction's clinical analog, exposure therapy, have been quite robust (e.g., Mithoefer et al. 2011).

Translational investigations of potential pharmacological enhancers of extinction learning are ongoing with a focus on agents such as endocannabinoids, monoamine modulators, glucocorticoids, neuropeptides, histone deacetylase (HDAC) inhibitors, estrogens, and methylene blue. There are also efforts to enhance extinction learning and its consolidation via other approaches such as device-based neuromodulations, exercise, as well as meditation – some of these tools and findings will be discussed in other chapters in this book.

8 Fear Extinction: Clinical Applications and Implications

As evidenced by several of the contributing chapters, fear extinction remains a critical area of psychophysiological human research given: (1) the high relevance of the shared limbic neurobiological underpinnings that mediate fear- and anxiety-related psychopathologies and fear learning as well as (2) its analogous laboratory representation of the clinical process of prolonged exposure therapy. A meta-analysis by Lissek et al. (2005) that specifically targeted studies that employed psychophysiological techniques indicated that anxiety disorder patients consistently show a pattern of fearful reactions to stimuli that no longer predict danger (Lissek et al. 2005). Consistent with this meta-analysis, there have been several reports in the literature of impaired fear extinction, as evidenced through psychophysiology, in sample populations with panic disorder (Michael et al. 2007), PTSD (Blechert et al. 2007; Linnman et al. 2011; Milad et al. 2009; Norrholm et al. 2011a, b, 2014), and pediatric anxiety (Craske et al. 2008b). Further, in PTSD specifically, diagnostic status and symptomatology, in both clinical and civilian populations, has been associated with at least three extinction learning trajectories that are not necessarily mutually exclusive: (1) elevated or "over-expressed" conditioned fear at the time of extinction learning assessment, (2) impaired, or incomplete, learning during extinction training, and (3) impaired extinction retention as evidenced by maintained conditioned fear to an extinguished CS+ as compared to post-extinction learning or post-acquisition levels of fear (see summary depicted in Fig. 1).

The enhanced expression, or over-expression, of learned fear to a previously reinforced CS+ has been termed *fear load* and has been observed in primarily civilian traumatized populations (Norrholm et al. 2011a, b, 2014; Orcutt et al. 2016). Accumulating empirical evidence has shown that fear load: (1) is associated with increased negative bias toward threat (Fani et al. 2012), (2) is responsive to fluctuations in activity within the hypothalamic–pituitary–adrenal (HPA axis;

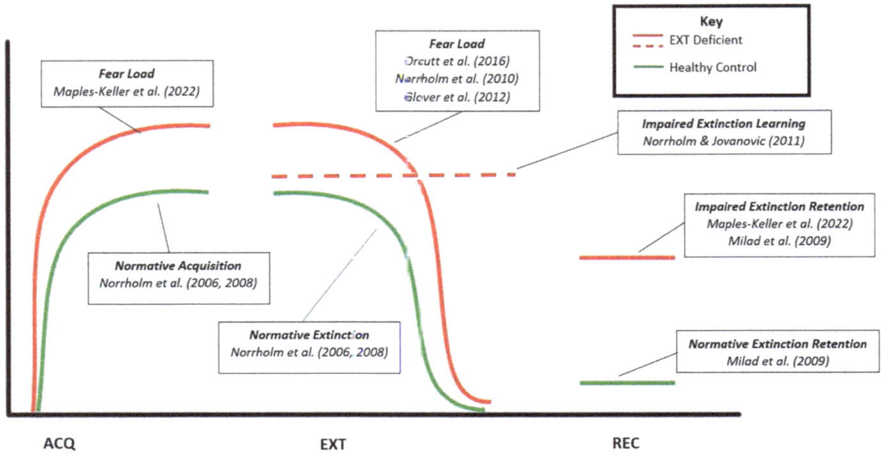

Fig. 1 Schematic representation of the fear extinction trajectories that have been observed in psychiatrically healthy and previously traumatized human populations with PTSD. Key empirical references are included in text boxes accompanying each trajectory type. *ACQ* fear acquisition, *EXT* within-session extinction training, *REC* between-session extinction recall (also termed retention)

(Michopoulos et al. 2017)), (3) is linked to lower circulating estrogen levels at the time of extinction learning (Glover et al. 2012), (4) is linked to amygdala reactivity to fearful faces in fMRI (France et al. 2022), (5) may be attenuated through self-medication with illicit substances (Davis et al. 2013), and (6) may be related to impaired fear inhibition as it is predicted by terminal levels of fear to the CS− at the end of acquisition (Norrholm et al. 2011a, b).

Impaired extinction learning, as evidenced by an incomplete reduction in acquired fear as the previously reinforced CS+ is presented alone (i.e., extinction training or learning), has been observed in military populations of veterans and service members with PTSD. For example, Norrholm and Jovanovic (2011) reported that a cohort of veterans from Operations Enduring (OEF) and Iraqi Freedom (OIF) displayed a "shallow" extinction curve using fear-potentiated startle measures (Norrholm and Jovanovic 2011). In addition, Acheson et al. (2015) showed that Marine and Navy service members who were part of the Marine Resiliency Study II and had PTSD symptoms pre-deployment had incomplete extinction of self-reported anxiety toward the CS+ during extinction training (Acheson et al. 2015). Impaired extinction learning, as evidenced by altered recall, has been repeatedly shown by Milad et al. (2008, 2009) in PTSD patient samples. More specifically, individuals with PTSD exhibit a reduced ability to recall extinction learning when assessed 1 day after extinction training using virtual context and cues (Milad et al. 2008, 2009).

A very recent study provided further support for the theoretical link between the emotional learning phenomena underlying extinction learning and the foundational bases supporting exposure therapy as a first-line psychotherapy for fear-, anxiety-, and trauma-related disorders. Using many of the fear-potentiated startle methods described throughout this chapter, Maples-Keller and colleagues (2022a) assessed fear acquisition, extinction training, and extinction recall before and after treatment in a cohort of U.S. Veterans and servicemembers enrolled in an intensive outpatient treatment program for PTSD and related disorders (Maples-Keller et al. 2022a). This study replicated an earlier finding from this group that conditioned fear was higher in PTSD patients (e.g., fear load) at pre-treatment as compared to trauma-exposed controls. At the conclusion of treatment, fear-potentiated startle measures revealed that extinction learning was maintained (i.e., successful extinction recall) in those patients who were high responders to the extinction-based exposure treatments as compared to low responders who showed a return of conditioned fear at the recall test. These results represent the very limited amount of data regarding extinction learning as it relates to ongoing treatment.

The revelation afforded by commonly used psychophysiological techniques that conditioned fear responses persist over sequential phases of extinction learning is in parallel with clinically available data suggesting that impaired within-session (i.e., extinction learning) and between-session (i.e., extinction recall) extinction learning underlie many of the fear-related symptoms of PTSD (Briscione et al. 2014; Rothbaum and Davis 2003). Psychophysiologically derived extinction data have also illustrated extinction learning trajectories that are consistent with those observed in pre-clinical animal models (Galatzer-Levy et al. 2013) as well as posttraumatic stress disorder symptom outcomes (Galatzer-Levy et al. 2017).

9 Extinction Learning: State or Trait?

There is an open question with regard to human extinction learning that has been highlighted through psychophysiological methods and paradigms discussed in this chapter. That question refers to the degree to which extinction learning ability (and fear acquisition learning as well) reflects trait or state effects. As revealed in studies of psychiatrically healthy humans (e.g., Norrholm et al. 2006, 2008) and in research samples administered a potential extinction-enhancing manipulation (e.g., (Smits et al. 2013a, b; Telch et al. 2014)), there are pre-existing, pre-manipulation individual differences in one's ability to extinguish fear. These individual, "trait" differences often lead investigators to include a cut-off criterion to identify "extinguishers" versus "non-extinguishers" which, in turn, raises important questions regarding replicability, generalizability, and standardization. On the other hand, there are significant findings in the literature, some very recent, that show extinction learning to be a malleable, dynamic process that is responsive to external manipulation (e.g., (Maples-Keller et al. 2022a; Reist et al. 2021)) that appears to be more experiential than genetic. For example, in a monozygotic twin study, Milad et al. (2008) used skin conductance measures to provide evidence that failure of extinction learning is acquired. This extinction impairment was seen in the traumatized twin but not in his non-traumatized co-twin.

There have been several noteworthy findings from the field to suggest that impaired fear extinction, as evidenced by psychophysiological techniques, represents an etiological risk factor for the development of anxiety and trauma- and stressor-related disorders. For example, Craske and co-authors (2008b) reported that the children of anxious parents showed slower extinction learning than psychiatrically healthy control children (Craske et al. 2008b). In addition, pre-existing, pre-trauma impairments in extinction learning in both firefighters and soldiers deployed to Afghanistan predicted a significant portion of the variance (~30%) in those who went on to develop clinical PTSD (Guthrie and Bryant 2006; Lommen et al. 2013). Interestingly, van Rooij et al. (2021) found that hippocampal activation during extinction was associated with higher resilience and lower future PTSD occurrence in a prospective emergency room study (van Rooij et al. 2021). Lastly, neural correlates of impaired fear extinction (e.g., sustained amygdala activation and reduced anterior cingulate cortex activity), assessed during the extinction phase of a cued fear-conditioning task, have been observed in individuals with subclinical anxiety at risk for progression to full-blown anxiety disorder (Sehlmeyer et al. 2011).

10 Conclusions

The impetus behind the creation of this text and its constituent chapters was to generate a comprehensive discussion of current trends in fear extinction from a widely translational view from molecules to behavior and from bench to bedside. With respect to psychophysiological approaches to the study of human fear extinction, we can make the following statements regarding the current state of the field: (1) the primarily employed psychophysiological methods have provided paradigmatic platforms that allow for the study of fear extinction processes among psychiatrically healthy adults as well as psychiatric samples of interest including those with pathological fear and anxiety; (2) these platforms have afforded in-depth investigation into the acquisition, within-session and between-session extinction learning, and the return of conditioned fear post-extinction; (3) the available platforms have shown to be sensitive to treatment effects, both psychological and pharmacological; (4) psychophysiologically-based paradigms have shown some discriminative success in identifying differences in extinction learning moderated and mediated by intrinsic factors such as genomics and hormonal states as well as extrinsic factors such as exposure to severe stress and psychological trauma, and (5) future endeavors will include a combination of clinical and pre-clinical investigations that can inform one another reciprocally while maintaining direct applicability and relatively small translational gaps in methodology.

References

Acheson DT, Geyer MA, Baker DG, Nievergelt CM, Yurgil K, Risbrough VB, Team M-I (2015) Conditioned fear and extinction learning performance and its association with psychiatric symptoms in active duty marines. Psychoneuroendocrinology 51:495–505. https://doi.org/10.1016/j.psyneuen.2014.09.030

Adolph D, Flasinski T, Lippert MW, Pflug V, Hamm AO, Richter J, Margraf J, Schneider S (2022) Measuring extinction learning across the lifespan – adaptation of an optimized paradigm to closely match exposure treatment procedures. Biol Psychol 170:108311. https://doi.org/10.1016/j.biopsycho.2022.108311

Blechert J, Michael T, Vriends N, Margraf J, Wilhelm FH (2007) Fear conditioning in posttraumatic stress disorder: evidence for delayed extinction of autonomic, experiential, and behavioural responses. Behav Res Ther 45(9):2019–2033. http://www.sciencedirect.com/science/article/B6V5W-4N7RWB1-3/2/d9dae6ee8f5ed0a9175de49a72cbea5f

Blumenthal TD, Cuthbert BN, Filion DL, Hackley S, Lipp OV, Van Boxtel A (2005) Committee report: guidelines for human startle eyeblink electromyographic studies. Psychophysiology 42:1–15

Bouton ME (2004) Context and behavioral process in extinction. Learn Mem 11:485–494

Bouton ME, Bolles RC (1979) Contextual control of the extinction of conditioned fear. Learn Motiv 10:455–466

Bouton ME, Ricker ST (1994) Renewal of extinguished responding in a second context. Anim Learn Behav 22(3):317–324

Bouton ME, Swartzentruber D (1991) Sources of relapse in Pavlovian and instrumental learning. Clin Psychol Rev 11:123–140

Briscione MA, Jovanovic T, Norrholm SD (2014) Conditioned fear associated phenotypes as robust, translational indices of trauma-, stressor-, and anxiety-related behaviors. Front Psych

Cacciopo J, Tassinary LG, Bernston GG (2004) Handbook of psychophysiology, 2nd edn. Cambridge University Press

Cisler JM, Privratsky AA, Sartin-Tarm A, Sellnow K, Ross M, Weaver S, Hahn E, Herringa RJ, James GA, Kilts CD (2020) L-DOPA and consolidation of fear extinction learning among women with posttraumatic stress disorder. Transl Psychiatry 10(1):287. https://doi.org/10.1038/s41398-020-00975-3

Constantinou E, Purves KL, McGregor T, Lester KJ, Barry TJ, Treanor M, Craske MG, Eley TC (2021) Measuring fear: association among different measures of fear learning. J Behav Ther Exp Psychiatry 70:101618. https://doi.org/10.1016/j.jbtep.2020.101618

Craske MG, Kircanski K, Zelikowsky M, Mystkowski J, Chowdhury N, Baker A (2008a) Optimizing inhibitory learning during exposure therapy. Behav Res Ther 46(1):5–27. https://doi.org/10.1016/j.brat.2007.10.003

Craske MG, Waters AM, Lindsey Bergman R, Naliboff B, Lipp OV, Negoro H, Ornitz EM (2008b) Is aversive learning a marker of risk for anxiety disorders in children? Behav Res Ther 46(8):954–967. https://doi.org/10.1016/j.brat.2008.04.011

Davis M (1997) Neurobiology of fear responses: the role of the amygdala. J Neuropsychiatry Clin Neurosci 9(3):382–402

Davis M, Astrachan DI (1978) Conditioned fear and startle magnitude: effects of different footshock or backshock intensities used in training. J Exp Psychol Anim Behav Process 4:95–103

Davis TA, Jovanovic T, Norrholm SD, Glover EM, Swanson M, Spann S, Bradley B (2013) Substance use attenuates physiological responses associated with PTSD among individuals with co-morbid PTSD and SUDs. J Psychol Psychother Suppl 7. https://doi.org/10.4172/2161-0487.S7-006

Difede J, Rothbaum BO, Rizzo AA, Wyka K, Spielman L, Reist C, Roy MJ, Jovanovic T, Norrholm SD, Cukor J, Olden M, Glatt CE, Lee FS (2022) Enhancing exposure therapy for posttraumatic stress disorder (PTSD): a randomized clinical trial of virtual reality and imaginal exposure with a cognitive enhancer. Transl Psychiatry 12(1):299. https://doi.org/10.1038/s41398-022-02066-x

Dirikx T, Vansteenwegen D, Eelen P, Hermans D (2009) Non-differential return of fear in humans after a reinstatement procedure. Acta Psychol 130(3):175–182. https://doi.org/10.1016/j.actpsy.2008.12.002

Effting M, Kindt M (2007) Contextual control of human fear associations in a renewal paradigm. Behav Res Ther 45(9):2002–2018. https://doi.org/10.1016/j.brat.2007.02.011

Fani N, Tone EB, Phifer J, Norrholm SD, Bradley B, Ressler KJ, Kamkwalala A, Jovanovic T (2012) Attention bias toward threat is associated with exaggerated fear expression and impaired extinction in PTSD. Psychol Med 42(3):533–543. https://doi.org/10.1017/S0033291711001565

Feduccia AA, Mithoefer MC (2018) MDMA-assisted psychotherapy for PTSD: are memory reconsolidation and fear extinction underlying mechanisms? Prog Neuro-Psychopharmacol Biol Psychiatry 84(Pt A):221–228. https://doi.org/10.1016/j.pnpbp.2018.03.003

Foa EB, Hembree E, Rothbaum BO (2007) Prolonged exposure therapy for PTSD: emotional processing of traumatic experiences, therapist guide. Oxford University Press

France JM, Reda M, Marusak HA, Riser M, Wiltshire C, Davie W, Grasser L, Wanna C, Stenson A, Ely T, Norrholm S, Stevens J, Jovanovic T (2022) Anxiety, fear extinction, and threat-related amygdala reactivity in children exposed to urban trauma. J Exp Psychopathol 13(4). https://doi.org/10.1177/20438087221132501

Galatzer-Levy IR, Ankri Y, Freedman S, Israeli-Shalev Y, Roitman P, Gilad M, Shalev A (2013) Early PTSD symptom trajectories: persistence, recovery, and response to treatment: results from the Jerusalem Trauma Outreach and Prevention Study (J-TOPS). PLoS One 8(8)

Galatzer-Levy IR, Andero R, Sawamura T, Jovanovic T, Papini S, Ressler KJ, Norrholm SD (2017) A cross species study of heterogeneity in fear extinction learning in relation to FKBP5 variation and expression: implications for the acute treatment of posttraumatic stress disorder. Neuropharmacology 116:188–195. https://doi.org/10.1016/j.neuropharm.2016.12.023

Glover EM, Jovanovic T, Mercer KB, Kerley K, Bradley B, Ressler KJ, Norrholm SD (2012) Estrogen levels are associated with extinction deficits in women with posttraumatic stress disorder [research support, N.I.H.]. Biol Psychiatry 72(1):19–24. https://doi.org/10.1016/j.biopsych.2012.02.031

Grillon C (2002) Startle reactivity and anxiety disorders: aversive conditioning, context, and neurobiology. Biol Psychiatry 52(10):958–975. http://www.ncbi.nlm.nih.gov/entrez/query.fcgi?cmd=Retrieve&db=PubMed&dopt=Citation&list_uids=12437937

Guastella AJ, Lovibond PF, Dadds MR, Mitchell P, Richardson R (2007) A randomized controlled trial of the effect of d-cycloserine on extinction and fear conditioning in humans. Behav Res Ther

Guthrie RM, Bryant RA (2006) Extinction learning before trauma and subsequent posttraumatic stress. Psychosom Med 68:307–311

Haaker J, Golkar A, Hermans D, Lonsdorf TB (2014) A review on human reinstatement studies: an overview and methodological challenges. Learn Mem 21(9):424–440. https://doi.org/10.1101/lm.036053.114

Hofmann SG, Meuret AE, Smits JA, Simon NM, Pollack MH, Eisenmenger K, Shiekh M, Otto MW (2006) Augmentation of exposure therapy with D-cycloserine for social anxiety disorder [comparative study randomized controlled trial]. Arch Gen Psychiatry 63(3):298–304. https://doi.org/10.1001/archpsyc.63.3.298

Huff NC, Hernandez JA, Blanding NQ, LaBar KS (2009) Delayed extinction attenuates conditioned fear renewal and spontaneous recovery in humans. Behav Neurosci 123(4):834–843. https://doi.org/10.1037/a0016511

Kindt M, Soeter M (2013) Reconsolidation in a human fear conditioning study: a test of extinction as updating mechanism. Biol Psychol 92:43–50

Kindt M, Soeter M, Vervliet B (2009) Beyond extinction: erasing human fear responses and preventing the return of fear. Nat Neurosci 12(3):256–258. https://doi.org/10.1038/nn.2271

Kushner MG, Kim SW, Donahue C, Thuras P, Adson D, Kotlyar M, McCabe J, Peterson J, Foa EB (2007) D-cycloserine augmented exposure therapy for obsessive-compulsive disorder [randomized controlled trial research support, non-U.S. gov't]. Biol Psychiatry 62(8):835–838. https://doi.org/10.1016/j.biopsych.2006.12.020

LaBar KS, Phelps EA (2005) Reinstatement of conditioned fear in humans is context dependent and impaired in amnesia. Behav Neurosci 119(3):677–686

Landis C, Hunt W (1939) The startle pattern. Farrar and Rinehart

Lang PJ, Bradley MM, Cuthbert BN (1990) Emotion, attention, and the startle reflex. Psychol Rev 9:377–395

Lang PJ, Bradley MM, Cuthbert BN (1998) Emotion, motivation, and anxiety: brain mechanisms and psychophysiology. Biol Psychiatry 44:1248–1263

Ledgerwood L, Richardson R, Cranney J (2003) Effects of D-cycloserine on extinction of conditioned freezing. Behav Neurosci 117(2):341–349

Linnman C, Rougemont-Bucking A, Beucke JC, Zeffiro TA, Milad MR (2011) Unconditioned responses and functional fear networks in human classical conditioning. Behav Brain Res. https://doi.org/10.1016/j.bbr.2011.02.045

Lissek S, Powers AS, McClure EB, Phelps EA, Woldehawariat G, Grillon C, Pine DS (2005) Classical fear conditioning in the anxiety disorders: a meta-analysis. Behav Res Ther 43(11): 1391–1424

Lissek S, Rabin S, Heller RE, Lukenbaugh D, Geraci M, Pine DS, Grillon C (2010) Overgeneralization of conditioned fear as a pathogenic marker of panic disorder. Am J Psychiatry 167(1):47–55. https://doi.org/10.1176/appi.ajp.2009.09030410

Lommen MJ, Engelhard IM, Sijbrandij M, van den Hout MA, Hermans D (2013) Pre-trauma individual differences in extinction learning predict posttraumatic stress. Behav Res Ther 51(2): 63–67. https://doi.org/10.1016/j.brat.2012.11.004

Lonsdorf TB, Merz CJ (2017) More than just noise: inter-individual differences in fear acquisition, extinction and return of fear in humans – biological, experiential, temperamental factors, and methodological pitfalls. Neurosci Biobehav Rev 80:703–728. https://doi.org/10.1016/j.neubiorev.2017.07.007

Maples-Keller JL, Rauch SAM, Jovanovic T, Yasinski CW, Goodnight JM, Sherrill A, Black K, Michopoulos V, Dunlop BW, Rothbaum BO, Norrholm SD (2019) Changes in trauma-potentiated startle, skin conductance, and heart rate within prolonged exposure therapy for PTSD in high and low treatment responders. J Anxiety Disord 68:102147. https://doi.org/10.1016/j.janxdis.2019.102147

Maples-Keller J, Watkins LE, Nylocks KM, Yasinski C, Coghlan C, Black K, Jovanovic T, Rauch SAM, Rothbaum BO, Norrholm SD (2022a) Acquisition, extinction, and return of fear in veterans in intensive outpatient prolonged exposure therapy: a fear-potentiated startle study. Behav Res Ther 154:104124. https://doi.org/10.1016/j.brat.2022.104124

Maples-Keller JL, Norrholm SD, Burton M, Reiff C, Coghlan C, Jovanovic T, Yasinski C, Jarboe K, Rakofsky J, Rauch S, Dunlop BW, Rothbaum BO (2022b) A randomized controlled trial of 3,4-methylenedioxymethamphetamine (MDMA) and fear extinction retention in healthy adults. J Psychopharmacol 36(3):368–377. https://doi.org/10.1177/02698811211069124

Mataix-Cols D, Fernandez de la Cruz L, Monzani B, Rosenfield D, Andersson E, Perez-Vigil A, Frumento P, de Kleine RA, Difede J, Dunlop BW, Farrell LJ, Geller D, Gerardi M, Guastella AJ, Hofmann SG, Hendriks GJ, Kushner MG, Lee FS, Lenze EJ, Levinson CA, McConnell H, Otto MW, Plag J, Pollack MH, Ressler KJ, Rodebaugh TL, Rothbaum BO, Scheeringa MS, Siewert-Siegmund A, Smits JAJ, Storch EA, Strohle A, Tart CD, Tolin DF, van Minnen A, Waters AM, Weems CF, Wilhelm S, Wyka K, Davis M, Ruck C et al (2017) D-cycloserine augmentation of exposure-based cognitive behavior therapy for anxiety, obsessive-compulsive, and posttraumatic stress disorders: a systematic review and meta-analysis of individual participant data. JAMA Psychiatry 74(5):501–510. https://doi.org/10.1001/jamapsychiatry.2016.3955

Michael T, Blechert J, Vriends N, Margraf J, Wilhelm FH (2007) Fear conditioning in panic disorder: enhanced resistance to extinction [research support, non-U.S. gov't]. J Abnorm Psychol 116(3):612–617. 10.1037/0021-843X.116.3.612

Michopoulos V, Norrholm SD, Stevens JS, Glover EM, Rothbaum BO, Gillespie CF, Schwartz AC, Ressler KJ, Jovanovic T (2017) Dexamethasone facilitates fear extinction and safety discrimination in PTSD: a placebo-controlled, double-blind study. Psychoneuroendocrinology 83:65–71. https://doi.org/10.1016/j.psyneuen.2017.05.023

Milad MR, Orr SP, Pitman RK, Rauch SL (2005) Context modulation of memory for fear extinction in humans. Psychophysiology 42:456–464

Milad MR, Goldstein JM, Orr SP, Wedig MM, Klibanski A, Pitman RK, Rauch SL (2006) Fear conditioning and extinction: influence of sex and menstrual cycle in healthy humans. Behav Neurosci 120(6):1196–1203

Milad MR, Wright CI, Orr SP, Pitman RK, Quirk GJ, Rauch SL (2007) Recall of fear extinction in humans activates the ventromedial prefrontal cortex and hippocampus in concert. Biol Psychiatry 62(5):446–454. http://www.sciencedirect.com/science/article/B6T4S-4MSHTFT-9/2/d9117cc8e1f3d7e26de88618b57d04c3

Milad MR, Orr SP, Lasko NB, Chang Y, Rauch SL, Pitman RK (2008) Presence and acquired origin of reduced recall for fear extinction in PTSD: results of a twin study. J Psychiatr Res 42(7):515–520. http://www.sciencedirect.com/science/article/B6T8T-4RY91Y4-2/2/b26b6246e120dd5ef8b8005b9fcda7e4

Milad MR, Pitman RK, Ellis CB, Gold AL, Shin LM, Lasko N, Zeidan MA, Handwerger K, Orr SP, Rauch SL (2009) Neurobiological basis of failure to recall extinction memory in posttraumatic stress disorder. Biol Psychiatry 66(12):1075–1082. http://www.sciencedirect.com/science/article/B6T4S-4X6VB0M-3/2/86951097ab9f64a51c6c87f86372bcbc

Mithoefer MC, Wagner MT, Mithoefer AT, Jerome L, Doblin R (2011) The safety and efficacy of {+/-}3,4-methylenedioxymethamphetamine-assisted psychotherapy in subjects with chronic, treatment-resistant posttraumatic stress disorder: the first randomized controlled pilot study [clinical trial, phase II randomized controlled trial]. J Psychopharmacologia 25(4):439–452. https://doi.org/10.1177/0269881110378371

Myers KM, Davis M (2007) Mechanisms of fear extinction. Mol Psychiatry 12(2):120–150. https://doi.org/10.1038/sj.mp.4001939

Norrholm SD, Jovanovic T (2011) Translational fear inhibition models as indicies of trauma-related psychopathology. Curr Psychiatr Rev 7

Norrholm SD, Jovanovic T (2018) Fear processing, psychophysiology, and PTSD. Harv Rev Psychiatry 26(3):129–141. https://doi.org/10.1097/HRP.0000000000000189

Norrholm SD, Jovanovic T, Vervliet B, Myers KM, Davis M, Rothbaum BO, Duncan EJ (2006) Conditioned fear extinction and reinstatement in a human fear-potentiated startle paradigm. Learn Mem 13(6):681–685

Norrholm SD, Vervliet B, Jovanovic T, Boshoven W, Myers KM, Davis M, Rothbaum BO, Duncan EJ (2008) Timing of extinction relative to acquisition: a parametric analysis of fear extinction in humans. Behav Neurosci 122(5):1016–1030

Norrholm SD, Anderson KM, Olin IW, Jovanovic T, Kwon C, Warren VT, McCarthy A, Bosshardt L, Sabree J, Duncan EJ, Rothbaum BO, Bradley B (2011a) Versatility of fear-potentiated startle paradigms for assessing human conditioned fear extinction and return of fear. Front Behav Neurosci 5:77. https://doi.org/10.3389/fnbeh.2011.00077

Norrholm SD, Jovanovic T, Olin IW, Sands LA, Karapanou I, Bradley B, Ressler KJ (2011b) Fear extinction in traumatized civilians with posttraumatic stress disorder: relation to symptom severity. Biol Psychiatry 69(6):556–563. https://doi.org/10.1016/j.biopsych.2010.09.013

Norrholm SD, Glover EM, Stevens JS, Fani N, Galatzer-Levy IR, Bradley B, Ressler KJ, Jovanovic T (2014) Fear load: the psychophysiological over-expression of fear as an intermediate phenotype associated with trauma reactions. Int J Psychophysiol. https://doi.org/10.1016/j.ijpsycho.2014.11.005

Ohman A, Mineka S (2001) Fears, phobias, and preparedness: toward an evolved module of fear and fear learning. Psychol Rev 108:483–522

Orcutt HK, Hannan SM, Seligowski AV, Jovanovic T, Norrholm SD, Ressler KJ, McCanne T (2016) Fear-potentiated startle and fear extinction in a sample of undergraduate women exposed to a campus mass shooting. Front Psychol 7:2031. https://doi.org/10.3389/fpsyg.2016.02031

Orr SP, Metzger LJ, Lasko NB, Macklin ML, Peri T, Pitman RK (2000) De novo conditioning in trauma-exposed individuals with and without posttraumatic stress disorder. J Abnorm Psychol 109(2):290–298. http://www.ncbi.nlm.nih.gov/entrez/query.fcgi?cmd=Retrieve&db=PubMed&dopt=Citation&list_uids=10895567

Otto MW, Tolin DF, Simon NM, Pearlson GD, Basden S, Meunier SA, Hofmann SG, Eisenmenger K, Krystal JH, Pollack MH (2010) Efficacy of d-cycloserine for enhancing response to cognitive-behavior therapy for panic disorder. Biol Psychiatry 67(4):365–370. https://doi.org/10.1016/j.biopsych.2009.07.036

Pavlov IP (1927) Conditioned reflexes. Oxford University Press

Quirk GJ (2002) Memory for extinction of conditioned fear is long-lasting and persists following spontaneous recovery. Learn Mem 9(6):402–407

Reist C, Jovanovic T, Kantarovich D, Weingast L, Hollifield M, Novin M, Khalaghizadeh S, Jafari B, George R, Riser M, Woodford J, Norrholm SD (2021) An analysis of fear inhibition and fear extinction in a sample of veterans with obstructive sleep apnea (OSA): implications for co-morbidity with post-traumatic stress disorder (PTSD). Behav Brain Res 404:113172. https://doi.org/10.1016/j.bbr.2021.113172

Rescorla RA, Heth CD (1975) Reinstatement of fear to an extinguished conditioned stimulus. J Exp Psychol Anim Behav Process 1(1):88–96. http://www.ncbi.nlm.nih.gov/entrez/query.fcgi?cmd=Retrieve&db=PubMed&dopt=Citation&list_uids=1151290

Ressler KJ, Rothbaum BO, Tannenbaum L, Anderson P, Graap K, Zimand E, Hodges L, Davis M (2004) Cognitive enhancers as adjuncts to psychotherapy: use of D-cycloserine in phobic individuals to facilitate extinction of fear. Arch Gen Psychiatry 61(11):1136–1144. http://www.ncbi.nlm.nih.gov/entrez/query.fcgi?cmd=Retrieve&db=PubMed&dopt=Citation&list_uids=15520361

Rodrigues H, Figueira I, Lopes A, Goncalves R, Mendlowicz MV, Coutinho ES, Ventura P (2014) Does d-cycloserine enhance exposure therapy for anxiety disorders in humans? A meta-analysis. PLoS One 9(7):e93519. https://doi.org/10.1371/journal.pone.0093519

Rosenfield D, Smits JAJ, Hofmann SG, Mataix-Cols D, de la Cruz LF, Andersson E, Ruck C, Monzani B, Perez-Vigil A, Frumento P, Davis M, de Kleine RA, Difede J, Dunlop BW, Farrell LJ, Geller D, Gerardi M, Guastella AJ, Hendriks GJ, Kushner MG, Lee FS, Lenze EJ, Levinson CA, McConnell H, Plag J, Pollack MH, Ressler KJ, Rodebaugh TL, Rothbaum BO, Storch EA, Strohle A, Tart CD, Tolin DF, van Minnen A, Waters AM, Weems CF, Wilhelm S, Wyka K, Altemus M, Anderson P, Cukor J, Finck C, Geffken GR, Golfels F, Goodman WK, Gutner CA, Heyman I, Jovanovic T, Lewin AB, McNamara JP, Murphy TK, Norrholm S, Thuras P, Turner C, Otto MW (2019) Changes in dosing and dose timing of D-cycloserine explain its apparent declining efficacy for augmenting exposure therapy for anxiety-related disorders: an individual participant-data meta-analysis. J Anxiety Disord 68:102149. https://doi.org/10.1016/j.janxdis.2019.102149

Rothbaum BO, Davis M (2003) Applying learning principles to the treatment of post-trauma reactions. Ann N Y Acad Sci 1008:112–121

Rothbaum BO, Price M, Jovanovic T, Norrholm SD, Gerardi M, Dunlop B, Davis M, Bradley B, Duncan EJ, Rizzo A, Ressler KJ (2014) A randomized, double-blind evaluation of d-cycloserine or alprazolam combined with virtual reality exposure therapy for posttraumatic stress disorder in Iraq and Afghanistan War veterans. Am J Psychiatry. https://doi.org/10.1176/appi.ajp.2014.13121625

Schiller D, Cain CK, Curley NG, Schwartz JS, Stern SA, Ledoux JE, Phelps EA (2008) Evidence for recovery of fear following immediate extinction in rats and humans. Learn Mem 15(6):394–402. https://doi.org/10.1101/lm.909208

Schiller D, Monfils M-H, Raio CM, Johnson DC, LeDoux JE, Phelps EA (2010) Preventing the return of fear in humans using reconsolidation update mechanisms. Nature 463(7277):49–53. https://doi.org/10.1038/nature08637. http://www.nature.com/nature/journal/v463/n7277/suppinfo/nature08637_S1.html

Schmitz A, Grillon C (2012) Assessing fear and anxiety in humans using the threat of predictable and unpredictable aversive events (the NPU-threat test). Nat Protoc 7(3):527–532. https://doi.org/10.1038/nprot.2012.001

Sehlmeyer C, Dannlowski U, Schoning S, Kugel H, Pyka M, Pfleiderer B, Zwitserlood P, Schiffbauer H, Heindel W, Arolt V, Konrad C (2011) Neural correlates of trait anxiety in fear extinction. Psychol Med 41(4):789–798. https://doi.org/10.1017/S0033291710001248

Smits JA, Rosenfield D, Otto MW, Marques L, Davis ML, Meuret AE, Simon NM, Pollack MH, Hofmann SG (2013a) D-cycloserine enhancement of exposure therapy for social anxiety disorder depends on the success of exposure sessions [randomized controlled trial research support, N.I.H., extramural]. J Psychiatr Res 47(10):1455–1461. https://doi.org/10.1016/j.jpsychires.2013.06.020

Smits JA, Rosenfield D, Otto MW, Powers MB, Hofmann SG, Telch MJ, Pollack MH, Tart CD (2013b) D-cycloserine enhancement of fear extinction is specific to successful exposure sessions: evidence from the treatment of height phobia. Biol Psychiatry 73(11):1054–1058. https://doi.org/10.1016/j.biopsych.2012.12.009

Soeter M, Kindt M (2010) Dissociating response systems: erasing fear from memory. Neurobiol Learn Mem 94(1):30–41. https://doi.org/10.1016/j.nlm.2010.03.004

Soeter M, Kindt M (2011) Disrupting reconsolidation: pharmacological and behavioral manipulation. Learn Mem 18:357–366

Sokol N, Lovibond PF (2012) Cross-US reinstatement of human conditioned fear: return of old fears or emergence of new ones? Behav Res Ther 50(5):313–322. https://doi.org/10.1016/j.brat.2012.02.005

Telch MJ, Bruchey AK, Rosenfield D, Cobb AR, Smits J, Pahl S, Gonzalez-Lima F (2014) Effects of post-session administration of methylene blue on fear extinction and contextual memory in adults with claustrophobia [randomized controlled trial research support, N.I.H., extramural]. Am J Psychiatry 171(10):1091–1098. https://doi.org/10.1176/appi.ajp.2014.13101407

van Rooij SJH, Ravi M, Ely TD, Michopoulos V, Winters SJ, Shin J, Marin MF, Milad MR, Rothbaum BO, Ressler KJ, Jovanovic T, Stevens JS (2021) Hippocampal activation during contextual fear inhibition related to resilience in the early aftermath of trauma. Behav Brain Res 408:113282. https://doi.org/10.1016/j.bbr.2021.113282

VanElzakker MB, Dahlgren MK, Davis FC, Dubois S, Shin LM (2014) From Pavlov to PTSD: the extinction of conditioned fear in rodents, humans, and anxiety disorders. Neurobiol Learn Mem 113:3–18. https://doi.org/10.1016/j.nlm.2013.11.014

Vervliet B (2008) Learning and memory in conditioned fear extinction: effects of D-cycloserine. Acta Psychol 127(3):601–613

Vervliet B, Craske MG, Hermans D (2013) Fear extinction and relapse: state of the art. Annu Rev Clin Psychol 9:215–248. https://doi.org/10.1146/annurev-clinpsy-050212-185542

Vizeli P, Straumann I, Duthaler U, Varghese N, Eckert A, Paulus MP, Risbrough V, Liechti ME (2022) Effects of 3,4-methylenedioxymethamphetamine on conditioned fear extinction and retention in a crossover study in healthy subjects. Front Pharmacol 13:906639. https://doi.org/10.3389/fphar.2022.906639. eCollection 2022. PMID: 35910354

Walker DL, Ressler KJ, Lu KT, Davis M (2002) Facilitation of conditioned fear extinction by systemic administration or intra-amygdala infusions of D-cycloserine as assessed with fear-potentiated startle in rats. J Neurosci 22(6):2343–2351. http://www.ncbi.nlm.nih.gov/entrez/query.fcgi?cmd=Retrieve&db=PubMed&dopt=Citation&list_uids=11896173

Warren VT, Anderson KM, Kwon C, Bosshardt L, Jovanovic T, Bradley B, Norrholm SD (2014) Human fear extinction and return of fear using reconsolidation update mechanisms: the contribution of on-line expectancy ratings. Neurobiol Learn Mem 113:165–173. https://doi.org/10.1016/j.nlm.2013.10.014

Wilhelm S, Buhlmann U, Tolin DF, Meunier SA, Pearlson GD, Reese HE, Cannistraro P, Jenike MA, Rauch SL (2008) Augmentation of behavior therapy with D-cycloserine for obsessive-compulsive disorder. Am J Psychiatry 165(3):335–341.; quiz 409. https://doi.org/10.1176/appi.ajp.2007.07050776

Young MB, Andero R, Ressler KJ, Howell LL (2015) 3,4-Methylenedioxymethamphetamine facilitates fear extinction learning. Transl Psychiatry 5:e634. https://doi.org/10.1038/tp.2015.138

Young MB, Norrholm SD, Khoury LM, Jovanovic T, Rauch SAM, Reiff CM, Dunlop BW, Rothbaum BO, Howell LL (2017) Inhibition of serotonin transporters disrupts the enhancement of fear memory extinction by 3,4-methylenedioxymethamphetamine (MDMA). Psychopharmacology 234(19):2883–2895. https://doi.org/10.1007/s00213-017-4684-8

Zabik NL, Peters C, Iadipaolo A, Marusak HA, Rabinak CA (2023) Comparison of behavioral and brain indices of fear renewal during a standard vs. novel immersive reality Pavlovian fear extinction paradigm in healthy adults. Behav Brain Res 437:114154. https://doi.org/10.1016/j.bbr.2022.114154

Zeidan MA, Milad KL, Thompson-Hollands J, Im JJY, Dougherty DD, Holt DJ, Orr SP, Milad MR (2012) Test–retest reliability during fear acquisition and fear extinction in humans. CNS Neurosci Ther 18(4):313–317

Neuroimaging of Fear Extinction

Kevin S. LaBar

Contents

Abstract Extinguishing fear and defensive responses to environmental threats when they are no longer warranted is a critical learning ability that can promote healthy self-regulation and, ultimately, reduce susceptibility to or maintenance of affective-, trauma-, stressor-,and anxiety-related disorders. Neuroimaging tools provide an important means to uncover the neural mechanisms of effective extinction learning that, in turn, can abate the return of fear. Here I review the promises and pitfalls of functional neuroimaging as a method to investigate fear extinction circuitry in the healthy human brain. I discuss the extent to which neuroimaging has validated the core circuits implicated in rodent models and has expanded the scope of the brain regions implicated in extinction processes. Finally, I present new advances made possible by multivariate data analysis tools that yield more refined insights into the brain–behavior relationships involved.

Keywords Amygdala · Extinction · Fear conditioning · Fear reinstatement · Functional magnetic resonance imaging

---

K. S. LaBar (✉)
Center for Cognitive Neuroscience, Duke University, Durham, NC, USA
e-mail: klabar@duke.edu

Curr Topics Behav Neurosci (2023) 64: 79–102
https://doi.org/10.1007/7854_2023_429
Published Online: 17 July 2023

1 Introduction

Extinction is a learning process by which conditioned responses to cues that predict appetitive or aversive reinforcers diminish following a change in stimulus or environmental contingencies that reduce the predictive cue–reinforcer relationship. Although extinction can occur across both aversive and appetitive contexts, here I will only consider situations in which conditioned responses have been acquired in response to threats. In a typical laboratory threat paradigm, extinction is assessed when a conditioned stimulus (CS), which previously had a predictive relationship with delivery of an aversive unconditioned reinforcer (US), is later encountered repeatedly without reinforcement. Behavioral or physiological indices are taken as evidence that defensive responding to the predictive CS has been extinguished. Although there is debate regarding how changes in defensive neurophysiological responding relate to the subjective experience of fear or anxiety to the cues or contexts involved (LeDoux 2014), I will use the traditional terms "fear conditioning" or "fear extinction" for the experimental procedures and underlying learning processes, and I will present some evidence that addresses this relationship directly. Extinction learning is the key mechanism implicated in exposure therapies, which have demonstrated the most success in the treatment of phobias and posttraumatic stress disorder (PTSD) (Bohnlein et al. 2020; McLean et al. 2022). Because of its therapeutic potential, there is much interest in identifying the brain circuits that support extinction learning in humans. This effort has implications for developing targeted interventions that restore functionality in the relevant neural pathways.

With the advent of functional neuroimaging – especially functional magnetic resonance imaging (fMRI) in the 1990s – it became possible to visualize blood flow and oxygenation changes related to neuronal activity with millimeter-level spatial precision across the entire human brain. Contemporaneous lesion and electrophysiological studies in rodents identified the core neural circuitry involved in extinction learning, which characterized the important contributions of the amygdala, ventromedial prefrontal cortex (vmPFC), and hippocampus (Milad and Quirk 2012). Accordingly, initial work in humans sought to apply the methodological advance of fMRI to investigate whether this circuitry is conserved across species. Here I review this literature and show how its rigor has evolved over the past 25 years to ask more sophisticated questions regarding the mechanisms of fear extinction in humans. I begin with an assessment of methodological issues commonly encountered in neuroimaging research using fear extinction paradigms, including experimental design and analytical, technical, and interpretational challenges. From there, neuroimaging studies of fear extinction in healthy subjects will be reviewed along with recent advances and suggestions for future work that can improve upon our current understanding of the underlying mechanisms involved in extinction learning.

2 Methodological Issues with Neuroimaging as a Tool for Studying Fear Extinction

It is important to acknowledge up front that every research methodology has idiosyncratic strengths and weaknesses and that it is critical to combine information across multiple neuroscientific tools to characterize the brain–behavior relationships involved. FMRI offers several advantages to studying fear extinction in humans compared to other available neuroscientific tools. These include (1) the ability to sample activity across the whole brain simultaneously, which is important for fear extinction, given that the structures involved span cortical, limbic, and subcortical brain areas; (2) spatial resolution at the single-millimeter scale; (3) its suitability for recording concurrent psychophysiological and behavioral measures on single trials with minimal interference; (4) its compatibility with multisensory and virtual reality (VR) platforms to present conditioned cues and contexts in a more ecologically valid manner; and (5) the availability of advanced analytic tools that assess not only magnitude estimates in individual brain regions but also distributed activity patterns over space and time, functional connectivity across regions, and the configuration of network architectures using graph-theoretic approaches.

Despite these appealing facets of fMRI as a methodology, there are also several limitations that impact the inferences that can be drawn. One overlooked issue is that the MRI scanning environment is not affectively neutral and may create conditions that can interfere with extinction learning for some participants. Claustrophobia is a risk factor for MRI scanning; the scanner bore is isolating, dark, and noisy; and fear/anxiety are enhanced when participants enter the scanner as measured by self-report and by spontaneous brain activity patterns that index specific emotions (Kragel et al. 2016). Given these considerations, the scanner itself can serve as a contextual reminder of a prior fear acquisition experimental phase or threats in general, and the ambient noise level limits the effectiveness of auditory cues as CSs or startle probes, which are common applications in animal research (it also limits the use of loud noises as USs). Due to practical issues including cost and human subjects scheduling logistics, fMRI studies of extinction learning are usually conducted within a single testing session. This design makes comparisons to animal research challenging, as animal studies typically spread out extinction trials across several days, and the mechanisms of massed (within-session) vs. spaced (between-session) extinction learning are not equivalent (Plendl and Wotjak 2010).

Additional significant limitations relate to the hemodynamic signal itself. The blood oxygen level-dependent (BOLD) signal extracted and quantified in fMRI research is an indirect measure of neuronal activity (Logothetis and Wandell 2004). The BOLD signal elicited in subcortical structures like the amygdala is weak (typically less than 1% signal change is observed), which biases fMRI toward cortical activation sites. Increasing the number of data samples from a participant can improve the signal-to-noise ratio. However, fear acquisition, extinction, and recovery all tend to happen quickly in healthy human subjects, which makes capturing these processes challenging with neuroimaging tools. The BOLD signal is also prone to magnetic susceptibility artifacts in structures that lie adjacent to ventricles, sinus cavities, or bone (Ojemann et al. 1997). Unfortunately, many brain areas implicated in animal studies of extinction are subcortical and/or reside in such susceptibility zones, including the amygdala, hippocampus, vmPFC, periaqueductal gray, and the hypothalamus. These issues are partially overcome by using special imaging protocols that minimize the signal distortions (Morawetz et al. 2008; Song 2001), by using more lenient small-volume corrections for multiple comparisons in subcortical regions rather than correcting uniformly across all voxels in the brain, and by relating brain and behavioral signals on a trial-by-trial basis (Lim et al. 2009; Visser et al. 2016).

Furthermore, the BOLD signal cannot readily distinguish between excitatory and inhibitory processes. The BOLD signal may be inherently biased toward capturing excitatory processes because it is most closely tied to local field potentials, and the spatial distribution, geometric alignment, and neuronal proportions of post-synaptic inputs contributing to local field potentials tend to favor excitation (Logothetis 2001; Logothetis and Wandell 2004). Nonetheless, both inhibitory and excitatory neuronal processes require oxygen. Thus, an increase in BOLD signal could be seen in either case. The BOLD signal is further increased when excitatory and inhibitory inputs are compared locally to determine axonal firing compared to excitation/inhibition alone (Logothetis and Wandell 2004). Depending on how it is implemented neuronally, inhibition may even generate a pattern of both increases and decreases in BOLD signal over time (Logothetis 2001; Moon et al. 2021). Electrophysiological studies in animals have revealed a complex series of inhibitory and excitatory changes in intra-amygdala pathways and between select subregions of the amygdala and other structures (such as the prelimbic vs. infralimbic PFC) as being critical for understanding fear memory processes, including extinction (Herry and Johansen 2014; Milad and Quirk 2012). The lack of clear differentiation between inhibitory and excitatory signals hinders a thorough analysis of extinction learning in humans and complicates predictions (e.g., should the BOLD signal in the amygdala increase for both fear acquisition and extinction?). It is also important to keep in mind that many electrophysiological studies in rodents measure action potentials (suprathreshold axonal outputs) rather than local field potentials (primarily reflecting graded and summed dendritic inputs), further complicating direct cross-species comparisons.

Issues also arise with respect to the temporal and spatial resolution of the BOLD signal. The peak of the hemodynamic response typically occurs around 6 s after stimulus onset. This response profile is adequate for many research questions, particularly as it conveniently mimics typical CS durations and physiological response profiles such as skin conductance responses (SCRs). But the temporal sluggishness of fMRI hinders the ability of researchers to definitively address some interesting questions that require a more acute temporal resolution. For instance, animal studies have demonstrated that CSs engage the amygdala through both a direct subcortical route from the thalamus and indirect cortically mediated pathways (LeDoux 1993). However, there are debates regarding the importance and functionality of the direct subcortical pathway to the amygdala in humans (Pessoa and Adolphs 2010). To try to address this issue, fMRI researchers resort to examining functional connectivity patterns across brain regions and relating these trial-wise pattern associations to behavior, as more direct evidence requires millisecond-level temporal resolution that can only be obtained using invasive intracranial recordings in human patients with epilepsy (Mendez-Bertolo et al. 2016).

The millimeter-level spatial resolution of fMRI is better than that obtained using other non-invasive human neuroscientific methods, yet it is not possible to directly resolve neuronal circuits to the level of subnuclei implicated in animal models of extinction learning (e.g., detecting signal in the intercalated amygdalar nuclei, or clearly delineating central vs. medial amygdala, basal vs. lateral amygdala, or subfields of the hippocampus). Although probabilistic atlases have been created to subdivide structures like the amygdala and hippocampus into their constituent parts – with guidance from thinly sliced post-mortem brain tissue (Amunts et al. 2005) – there are controversies surrounding the reliability of some of these parcellations given the lower resolution present in the MRI acquisition parameters, even in the case of extracting hippocampal subfield volumes from 1-mm^3 voxels (Wisse et al. 2020). The evidence from probabilistic atlases can be bolstered by using high-resolution fMRI, computational modeling, or functional connectivity approaches. For instance, the basolateral vs. centromedial amygdalar complexes can sometimes be differentiated on the basis of their differing computational roles in associative learning (Boll et al. 2013) or their differential connections to cortical areas (Brown et al. 2014). Recent advances in multivoxel pattern analysis (MVPA) and other multivariate decoding approaches enable a more detailed look at distributed patterns of activity within structures like the amygdala and hippocampus that do not require resolving subnuclear boundaries (Norman et al. 2006). In this case, researchers do not spatially smooth the original fMRI data and the analysis relaxes the statistical criteria to include voxels that are not the most highly activated by the task. In this way, MVPA can reveal distributed patterns with finer sensitivity than the standard univariate analytic approach. This increased precision can help to delineate neural circuits for fear vs. safety relevant for extinction learning because neurons that signal negative vs. positive value during affective learning tend to be intermixed locally in the primate amygdala and orbitofrontal cortex (Belova et al. 2008; Morrison and Salzman 2009) and are not necessarily segregated by subnuclear boundaries.

Like most neuroscientific measurements, evoked BOLD responses are always quantified relative to a control stimulus and/or a preceding pre-stimulus baseline. For this reason, many human fear conditioning studies use a within-subjects, differential-conditioning paradigm where statistical contrasts compare brain activity to a reinforced stimulus (CS+) relative to an unreinforced control stimulus (CS-). This paradigm differs from many between-subjects animal studies that compare CS-evoked responses from a group that received paired CS–US presentations against a group where the CS and US were delivered in a pseudorandom order as a non-associative learning control. Because the CS- may become a safety signal in the context of within-subjects designs (given that it is never followed by the US), it is sometimes argued that the CS+ vs. CS- comparison is really a threat vs. safety cue comparison rather than a threat vs. neutral cue comparison. This consideration makes interpreting extinction challenging because it becomes unclear in these contrasts whether extinction-related changes are signaling a reduction in the threat/fear value of the CS+, an increase in the safety value of the CS+, or a change in the relative safety value of the CS- (if both the CS+ and CS- become "safe" as a consequence of extinction learning, there may even be no detectable signal that differentiates them). It is helpful to visualize the individual parameter estimates from each stimulus in these contrasts to help determine whether changes in processing the CS+ or CS- are driving them, although changes in fear vs. safety value remain challenging to dissociate. To circumvent this issue, some research groups present subjects with two CSs during conditioning, and then they only extinguish fear to one of these cues. In this case, BOLD responses are compared between the extinguished CS+ and the unextinguished CS+. This comparison avoids the safety confound in the CS+ vs. CS- extinction contrasts but introduces another confound because the unextinguished CS+ has had additional time for the acquired fear memory to undergo consolidation prior to the extinction test. Recent advances in MVPA may help to partly resolve some of these issues (see "Advances using multivariate approaches" below). These differing design features are important to keep in mind when reading the literature and attempting to compare results to animal models.

3 Testing the Core Fear Extinction Circuit with Univariate Analytic Approaches

With these methodological considerations in mind, I will now turn to examining how human fMRI studies have contributed to validating animal models of fear extinction. Although fear extinction circuits have been elaborated extensively over the years, three core components consist of the vmPFC (infralimbic cortex in rodents), hippocampus, and amygdala (Maren et al. 2013; Milad and Quirk 2012). These brain regions interact through a series of complex pathways to mediate extinction learning, the recall of extinction learning after a delay, and the recovery of fear after extinction has taken place. The balance of activity within these circuits, along with those that mediate fear expression – including the anterior cingulate gyrus (ACC)/dorsomedial PFC (prelimbic cortex in rodents) and their connections with the amygdala and hippocampus – determine whether fear memories are expressed or not following extinction training. The distinction between infralimbic vs. prelimbic cortex in fear extinction vs. fear expression may not be as clear cut in rodents as the standard model suggests, which is partly due to the overlapping bidirectional signaling between these regions and the amygdala (Giustino and Maren 2015). In addition, the homology between the prelimbic/infralimbic cortex in rodents and the mPFC in humans is not straightforward.

These caveats notwithstanding, initial fMRI studies in humans began to identify the neural correlates of within-session extinction learning. The first such study showed that activity in the amygdala to a CS+ transiently appears during early extinction trials, and then dissipates as extinction learning proceeds (LaBar et al. 1998). This study was limited in spatial extent to a few brain slices centered on the amygdala and was less informative than subsequent whole-brain studies in determining the contributions of other brain regions. A study using aversive smells as USs also found transient amygdala activation during extinction learning, along with the vmPFC (Gottfried and Dolan 2004). Phelps and colleagues reported activity in the amygdala, vmPFC, and hippocampus during extinction learning, but each of these regions more strongly signaled the CS- than the CS+ (Phelps et al. 2004). In the Phelps et al. (2004) study, the extent of amygdala activity during extinction learning correlated with reduced physiological arousal to the CS+, as measured by SCR. Yet another seminal study (Milad et al. 2007) reported activity in the amygdala and vmPFC in a CS+ > CS- contrast during within-session extinction learning.

Given that imaging studies on this topic involved small sample sizes of subjects with mixed findings, Fullana and colleagues conducted a meta-analysis of whole-brain activation maps from 31 fear extinction fMRI studies (Fullana et al. 2018). When the data from these studies were averaged across all extinction learning trials, the CS+ > CS- contrast revealed activation in the vmPFC but not in the amygdala or hippocampus. Unfortunately, it was not possible to conduct a separate meta-analysis to confirm the reports of transient amygdala responses during extinction learning, due to the paucity of studies that contained a contrast comparing early vs. late extinction trials. Furthermore, this meta-analytic approach did not apply small-volume correction and thus may have under-reported activation in subcortical regions like the amygdala.

Other analyses examined fMRI responses during the recall of extinction learning after a delay, rather than initial within-session extinction learning. The motivation for these studies is to establish any differences in mediating shorter-term vs. longer-term extinction learning, and to test the hypothesis that the core fear extinction network activity during initial extinction training establishes plasticity needed for long-term consolidation of extinction memories (Santini et al. 2004). In general, these studies simply repeated another session of extinction training 24 h later, although the time interval has varied somewhat across studies – up to 1 week later (Lonsdorf et al. 2014).

The Phelps et al. (2004) study mentioned above also investigated 24-h delayed extinction recall. Activity was found in both the vmPFC and parahippocampal cortex, although again it was greater for the CS- than the CS+. The authors further showed that a reduction in vmPFC activity from early to late extinction learning during the initial extinction training session predicted smaller SCRs to the CS+ during extinction recall. This latter finding suggests that, similar to rodent models, the vmPFC may play a key role in initiating the consolidation of extinction learning into long-term storage. Other studies compared responses to a CS+ that had undergone extinction the day before to another CS+ that was unextinguished. The first study of this kind showed greater vmPFC and hippocampal activation to the extinguished CS+ (relative to the unextinguished CS+) during extinction recall, as well as greater vmPFC–hippocampal and vmPFC–amygdala functional connectivity

(Milad et al. 2007). Activity in both the hippocampus and vmPFC at extinction recall correlated with a physiological index of extinction retention, as measured by SCR.

Fullana et al. (2018) conducted a meta-analysis of fMRI studies of extinction recall (N = 16). The meta-analysis initially pooled data across studies that used a CS + vs. CS- comparison along with those that used an extinguished CS + vs. unextinguished CS+ comparison. This pooled analysis did not reveal reliable activity across studies in the vmPFC, amygdala, or hippocampus, although it may be inappropriate to pool data across studies with such different statistical contrasts. A separate sub-analysis on the 9 studies using the latter contrast type did reveal consistent activity in the vmPFC and hippocampus but not the amygdala. Similar limitations of this meta-analysis apply as mentioned above for the meta-analysis on initial extinction learning with respect to the null findings in the amygdala.

In sum, neuroimaging studies have provided some support for the core model of fear extinction. The studies conducted to date have more consistently implicated the vmPFC than the hippocampus or amygdala during both initial extinction learning and extinction recall, although methodological limitations have hindered a thorough analysis of the amygdala's involvement, which may be transient and/or lower in signal amplitude compared to cortical regions. Functional connectivity patterns suggest that these regions act in concert during initial extinction learning and extinction recall, a point which will be elaborated later in the discussion of fear renewal and reinstatement.

4 Context Sensitivity of Extinguished Fears: Methodological Considerations for Renewal and Reinstatement Tests

The analyses summarized in the prior section focused on cue-driven activity signifying extinguished representations of prior threats. However, the core fear extinction circuit is also hypothesized to play a key role in the contextual disambiguation of threat from safety signals. A major tenet of associative learning theory is that extinction is more fragile and context-dependent than fear acquisition (Bouton 2000; for alternate theoretical views, see Dunsmoor et al. 2015). According to this perspective, extinction learning creates a new, safe memory trace for conditioned threats that competes for expression of the original fear acquisition memory and that contextual cues can bias the expression of either the threat or safe memories associated with a cue (Bouton 1993). Context effects on conditioning and their neurobiological mechanisms have been well documented in animal models, which emphasize the role of the hippocampal system and its interactions with fronto-amygdala circuits that mediate either fear expression, including the dorsal anterior cingulate (dACC) and dorsomedial prefrontal cortex (dmPFC), or safety signaling in the vmPFC (Maren et al. 2013).

In animal studies, contextual manipulations are typically done in a holistic, multimodal fashion by moving the research subject from one cage to another that varies in visual, tactile, olfactory, and/or auditory background cues. Behavioral and neural responses to conditioned threat cues can then be compared across experimental phases conducted in different cages – the initial fear acquisition context, the extinction context, and novel contexts. When extinguished threat cues are subsequently re-encountered in the original threat context or a novel context, fear responses tend to recover (termed "fear renewal"). Similarly, when extinguished threat cues are presented in a context where a threat reminder was recently encountered, fear responses tend to recover (termed "fear reinstatement"). Both of these cases illustrate the fragility and context dependency of extinction learning. The return of fear in these paradigms is typically short-lived, as the conditioned responses re-extinguish quickly when the cue continues to be presented without aversive reinforcement.

The translation of these paradigms to humans is complicated by the ability to effectively manipulate environmental contexts. In a direct analogy to rodent studies, LaBar and Phelps (2005) physically moved human participants from one testing room to another that differed in multimodal background cues to demonstrate that fear reinstatement was context-specific and impaired in amnesic patients. This kind of manipulation, however, is not feasible in an MRI scanner. Consequently, some researchers investigate fear reinstatement without demonstrating its context specificity. Others use relatively simple context manipulations like changing the visual features of the background screen in which the CSs are presented. Such 2-D unimodal context manipulations, though, are weaker and less likely to engage hippocampal-dependent mechanisms than multimodal, holistic environment shifts. The hippocampal system particularly engages with more complex forms of relational processing across both spatial and temporal cues, along with navigation-based coding of environmental layouts (Bird and Burgess 2008). Cell assemblies in the medial temporal lobes are sensitive to self-generated movement parameters that link environmental features to the location of the organism in space via head-direction cells, grid cells, place cells, and path integration mechanisms (Aery Jones and Giocomo 2023).

To engage these relational and navigational mechanisms, neuroimaging investigators are increasingly integrating 3-D virtual reality (VR) technology into their extinction paradigms to enhance the ecological validity and robustness of context manipulations. Using VR, participants navigate through 3-D virtual worlds in which they experience a feeling of immersion ("presence") in the environments while encountering dynamic conditioned threats, like slithering snakes, or safety cues along their path. Context manipulations then involve a holistic environmental shift where the same CSs are re-encountered in a completely different virtual world (e.g., shifting from a house interior to a wooded outdoor scene (Huff et al. 2010)), sometimes accompanied by changes in background auditory cues. Behavioral studies using 3-D VR have shown that conducting extinction training in multiple virtual contexts reduces subsequent fear reinstatement relative to single-context extinction training (Dunsmoor et al. 2014) and that extinction learning takes longer when remaining in the acquisition context, rather than shifting to a novel context, on a delayed extinction test (Huff et al. 2011). Although the confines of an MRI scanner prevent movement and limit the field of view relative to behavioral work, researchers can simulate passive navigation along a forward-moving path through a 3-D virtual world in a first-person ("egocentric") perspective. In the following sections, I selectively review the evolution of fMRI studies from 2-D to 3-D context manipulations that evaluate the role of the core fear extinction circuit on fear renewal and reinstatement.

5 Fear Renewal

The first fMRI study on contextual fear renewal following extinction superimposed CSs on different 2-D colored screens that were paired with (or without) background sounds (Kalisch et al. 2006). The vmPFC and anterior hippocampus were preferentially active and functionally coupled during extinction recall when the CS+ was re-experienced in the extinction context but not the acquisition context. At a more relaxed statistical threshold, the amygdala was more engaged in the CS+ > CS- contrast during extinction recall in the original acquisition context than the extinction context.

Although all studies reviewed so far have used aversive electric stimulation as the unconditioned stimulus (US), other paradigms have used an interoceptive conditioning manipulation that involves painful rectal distensions as a US. It is unclear whether the conditioning mechanisms for internal visceral pain are similar to those for external threats. Nonetheless, one study of this type examined contextual effects on fear renewal by changing the background color of the CS and the corresponding illumination of the scanner bore (Icenhour et al. 2015). There were no significant changes in the hippocampus or amygdala, but vmPFC signaling of the CS- was boosted upon returning back to the original acquisition context after extinction. This study, however, failed to find statistically significant renewal effects behaviorally or physiologically, suggesting that the contextual manipulation was relatively ineffective.

Hermann and colleagues used 2-D pictures of rooms with embedded color CSs to investigate fear renewal in both the original acquisition context and a novel context, compared to extinction recall in the extinction context (Hermann et al. 2016). To account for individual differences in the effectiveness of the experimental manipulation, the imaging data were further broken down into subgroups of participants who demonstrated higher vs. lower SCRs in each renewal context. Participants who exhibited higher SCR evidence for renewal in the acquisition context had greater amygdalar, dACC, insular, and hippocampal activity in the CS+ > CS- contrast compared to extinction recall. Across all participants, fear renewal in the novel context was associated with greater left hippocampal activity in the CS+ > CS- contrast compared to extinction recall, whereas right hippocampal activity was greater during extinction recall than in the novel context. The differential hippocampal effects were also associated with differences in functional connectivity patterns in the same contrasts: relative to low-SCR participants, high-SCR participants exhibited greater functional connectivity between the left hippocampus and the dACC, insula, and vmPFC, whereas the right hippocampus was more strongly coupled with the amygdala and other portions of the hippocampus.

The first fMRI study applying 3-D VR to investigate context-dependent fear renewal displayed dynamic snakes and spiders as CSs during passive navigation through one virtual world for acquisition, another virtual world for extinction, and then both of these virtual worlds on a 24-h delayed extinction recall/fear renewal test (Åhs et al. 2015). Fear renewal was established physiologically through SCR analyses and behaviorally through subjective anxiety and shock expectancy ratings. Region-of-interest analyses focused on the amygdala, hippocampus, vmPFC, and dmPFC, and results were broken down by early vs. late portions of the test sessions on Day 2, given that fear renewal tends to be transient. Activity in the vmPFC was greater to the CS- than the CS+ on Day 2 overall but was especially greater in this contrast during the early portion of extinction recall relative to fear renewal. The hippocampus also showed greater responses to the CS- than the CS+ during the later portion of extinction recall. By contrast, activity in the dmPFC was greater to the CS + than CS- on Day 2 overall but was especially greater in this contrast during the later portion of fear renewal relative to extinction recall. Individual differences in SCRs generated to the CS+ during fear renewal were correlated with the dmPFC activity (and amygdala at a lower threshold). Structural equation modeling further elucidated how activation in the hippocampus and amygdala in the CS+ > CS- contrast was statistically mediated through the vmPFC/dmPFC as a function of the context manipulation on Day 2. Results from this analysis indicated that during extinction recall, the vmPFC partially mediated the relationship between right hippocampal and amygdala signaling, whereas during fear renewal, the dmPFC fully mediated this relationship. These functional relationships provided additional insights into the reorganization of the network from extinction recall to fear renewal than what was observed from the mean activity levels in the regions.

Finally, Zabik and colleagues directly compared 3-D vs. 2-D VR effects on the neural correlates of fear renewal and extinction recall (Zabik et al. 2023). For the 2-D task, the CSs were colored lights embedded in outdoor pictures, and the US was a loud noise. For the 3-D task, the CSs were colored bags embedded in outdoor virtual worlds and the US was a hissing snake that popped out of the bag. Note that the US types were not the same across the two studies, and they differed from all other fMRI studies discussed so far. There were no physiological indices of conditioned fear taken during this study; participants only reported whether they felt the US delivery was likely or not. During both extinction recall and fear renewal, hippocampal activity was greater overall on the 3-D task than on the 2-D task (main effect of task). The amygdala showed a main effect of time during fear renewal, becoming larger over time in both tasks. The dACC showed different effects during fear renewal, with larger responses to the CS- on the 2-D task but larger responses to the CS+ on the 3-D task.

Overall, these initial studies on fear renewal indicate context-dependent changes in the core components of the fear conditioning and extinction circuits identified in animal models. However, results to date are inconsistent across studies. Differences in the experimental designs make comparisons across studies challenging. From the single study that directly compared 2-D vs. 3-D context manipulations (Zabik et al. 2023), it does appear that hippocampal engagement is stronger overall when conditioned cues are presented in 3-D environments, suggesting enhanced cue-context binding in VR simulations. These studies also indicate that it is important to consider not just the overall magnitude of activity in the brain regions but, perhaps more importantly, their functional connectivity, mediated patterns of co-activation across structures, and relationships with individual differences in physiological expression.

6 Fear Reinstatement

Lonsdorf and colleagues examined reinstatement to both cues and background contexts across two cohorts of participants (Lonsdorf et al. 2014). They reported more robust fear reinstatement for contexts than cues, which was associated with amygdala activity (that decreased over time in one of the cohorts) and the anterior hippocampus, which increased in strength from the extinction recall rest to reinstatement. Reinstatement of cued fear was associated with vmPFC activity, but at a reduced statistical threshold in the replication cohort. In a follow-up study on cued fear, Scharfenort and Lonsdorf (2016) found that reinstatement was associated with increased hippocampal activation to the CS+ and decreased activation to the CS- in the vmPFC, relative to the end of extinction training. Reinstatement-induced activity in the vmPFC negatively correlated with SCRs to the CS-. This study also examined generalized reinstatement (reinstatement effects to both the CS+ and CS-) as a potential model of overgeneralization relevant for anxiety disorders, but those results are not reported here.

A large, multi-site study reported increased activation in the hippocampus and ACC to the CS+ following reinstatement but no significant effects in the vmPFC or amygdala (Ridderbusch et al. 2021). It should be noted that this latter study differed from other reinstatement studies in multiple ways: it implemented an instructed fear paradigm (participants were told ahead of time which CS predicted the US), it used 24-h delayed extinction training, and it did not obtain physiological evidence for conditioned learning. Finally, Hermann and colleagues (2020) compared both renewal and reinstatement of fear following extinction learning in either a single extinction context or multiple extinction contexts. Compared to single-context extinction, multi-context extinction reduced amygdala activation to the CS+ on both renewal and reinstatement tests. These results are consistent with theoretical accounts suggesting that conducting extinction in multiple contexts should reduce its context dependency and lead to more robust extinction (for a discussion of different mechanisms that might account for the beneficial effects of multi-context extinction, see Dunsmoor et al. (2014)).

As with the discussion on fear renewal, fear reinstatement paradigms broadly implicate the core extinction circuitry, but results vary across studies. Due to paradigmatic differences, it is difficult to compare these studies directly. More studies using basic neuroscience approaches are clearly warranted on these topics. This line of work can advance knowledge of fear recovery mechanisms that have translational relevance for understanding the effectiveness of exposure therapy and persistence of acquired threat associations in disorders of fear and anxiety. As introduced above, the studies to date have largely relied on traditional univariate approaches to neuroimaging and, in some cases, have been focused on targeting the core structures implicated in animal models. Thus, there is a need to consider other contributing structures to human extinction learning and return-of-fear phenomena beyond the amygdala, hippocampus, and vmPFC. In addition, sophisticated multivariate analysis tools may provide unique insights into understanding how distributed representations of cues and contexts change during extinction learning, extinction recall, and fear recovery. Finally, considering individual differences may uncover important brain–behavior relationships and moderating influences that are obscured when conducting group-averaged analyses. Recent advances on these fronts are reviewed next. Return-of-fear mechanisms are also discussed in greater detail in the Monfils and Woodford chapters in this volume.

7 Expanding the Core Model

Many of the studies reviewed above reported activation in regions outside the core extinction circuit. The cerebellum is one brain region that has been implicated in fear learning and extinction processes across species. In their meta-analysis, Fullana and colleagues demonstrated reliable anterior cerebellar activation during extinction learning (Fullana et al. 2018). A high-field imaging study further indicated a role for the cerebellum in fear renewal (Batsikadze et al. 2022). Using an MRI-compatible VR stimulation that manipulated the distance between the conditioned stimuli and the research participant, Faul and colleagues reported that the cerebellum plays a particular role in the extinction and reinstatement of nearby threats (Faul et al. 2020). This finding was interpreted as part of a reactive circuit specialized for threat imminence. Cerebellar pathways in fear learning, fear memory retention, and extinction learning have been established in rodent research via connections that modulate processing in the amygdala, periaqueductal gray, and prefrontal cortex (Frontera et al. 2020; Jung et al. 2022; Lawrenson et al. 2022; Sacchetti et al. 2007).

In their meta-analysis, Fullana and colleagues (2018) noted that the dACC and insula were also reliably activated during extinction learning and extinction recall. These regions are engaged in some fear renewal and reinstatement paradigms (see the prior section), albeit less consistently. Rodent studies do implicate the prelimbic cortex in conditioned fear acquisition and expression, but there is more evidence for its involvement in appetitive extinction (Moorman and Aston-Jones 2023) than in fear extinction. Santini and colleagues further showed that inhibiting protein synthesis in the infralimbic cortex – but not in the insula – impaired consolidation of extinction learning in rodents (Santini et al. 2004). Although earlier rodent work had identified the insula as a structure important for conveying US information to the amygdala during fear conditioning, more recent studies suggest that the periaqueductal gray region may instead play this role (Herry and Johansen 2014). Thus, the human fear extinction circuit may be broader in scope than in rodents (for an even broader implication of large-scale networks using functional connectivity metrics, see Wen et al. 2021). Recent studies reveal contributions of other brain regions in rodents, but these tend to be adjacent regions in the medial temporal lobe, such as the entorhinal cortex (Baldi and Bucherelli 2015), or subcortical areas, such as the periaqueductal gray (Herry and Johansen 2014) or thalamus (Silva et al. 2021). In humans, the amygdala tends to be co-activated along with portions of the ACC and insula as part of the salience network, as revealed in resting-state fMRI studies (Menon 2015). Large-scale cognitive and affective functions are generally more widely distributed in humans than in rodents, which parallels the expansion of neocortex (and its subcortical connections) in primates and more recent adaptations in the hippocampus, insula, ACC, and other association cortices in hominids (Allman et al. 2001; Craig 2011; Kaas 2012; Pine et al. 2021). That being said, neuroimaging studies can only reveal brain activity that correlates with the task design and cannot identify which neural components are critical for task performance, which may explain why imaging studies indicate participation of a broader set of brain regions. Neurologic patients with damage to the amygdala (and surrounding cortex), vmPFC, and cerebellum exhibit deficits in conditioned fear learning (Battaglia et al. 2020; Bechara et al. 1995; LaBar et al. 1995; Maschke et al. 2000; Peper et al. 2001; Phelps et al. 1998; Weike et al. 2005) but see (Åhs et al. 2010)), and amnesic patients with hippocampal damage are impaired in fear reinstatement (LaBar and Phelps 2005). Establishing the essential role of these and other brain regions in extinction learning is challenging, however, given that fear acquisition is either partially or fully impaired prior to extinction taking place. Based on these considerations, an expanded model of the fear extinction circuit is presented in Fig. 1.

8 Advances Using Multivariate Approaches

Multivariate analyses have advanced knowledge about how conditioned fear associations are maintained or recovered over time, how the patterns of brain activity relate to behavior in a more refined manner, and how representations of conditioned stimuli dynamically change as a function of extinction learning (for a review, see Hennings et al. 2022a). For instance, Visser and colleagues captured multivoxel patterns of trial-to-trial correlations in responding to face and house CSs during fear conditioning and a ~4-week delayed memory test (Visser et al. 2011, 2013), followed by extinction learning (Visser et al. 2013). They compared these

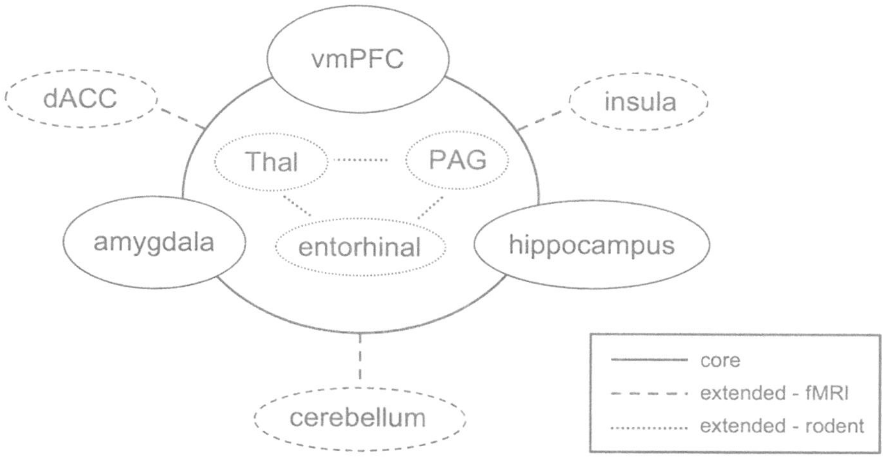

Fig. 1 An expanded model of fear extinction circuitry. Core regions are based on the traditional model initially developed from rodent studies; this core circuit has been partially validated in human imaging studies. Expanded regions include additional brain areas implicated in meta-analyses of human neuroimaging studies, as well as areas identified in rodent studies but not as well validated in humans. *dACC* dorsal anterior cingulate gyrus, *PAG* periaqueductal gray, *thal* thalamus, *vmPFC* ventromedial prefrontal cortex

trial-wise correlated brain activation patterns as a function of both stimulus identity (correlations between successive presentations of the same stimulus class, such as face CS-s) as well as correlations between different classes of stimuli that shared a reinforcement history (e.g., correlations between presentations of face and house CS +s). As fear learning progressed, the both the within-class (face–face) and between-class (face–house) correlations increased to the CS + s across many brain regions whereas those to the CS-s did not. These trial-wise correlations subsequently decreased during extinction learning. The correlation patterns emerged even in brain regions that showed no univariate difference in activation, indicating that the MVPA approach is more sensitive as a neural index of fear learning and extinction. In addition, the MVPA-derived fear learning index predicted individual differences in subsequent fear retention, such that individuals with greater between-stimulus brain correlations to the CS+ in frontolimbic brain areas during conditioning had higher pupillary responses to the CS+ during the delayed retention test. Again, these results could not be explained by differences in univariate activation during conditioning in the same brain regions.

As introduced earlier, one central tenet of associative learning theory is that experimental contexts across conditioning phases establish different memory traces that compete for expression, such as threat associations to the CS+ established during acquisition and safety associations to the same stimulus during extinction (Bouton 1993). It has been challenging to determine how the representation of a cued threat changes as a function of extinction learning, or whether it is possible to distinguish the acquisition and extinction representations. MVPA methods provide a solution by quantifying how the correlated patterns of brain activity change to conditioned stimuli from the late acquisition phase to the early and late portions of extinction training (Faul et al. 2020; Graner et al. 2020). To facilitate the contextual distinction between experimental phases, immersive VR was used in the MRI scanner to create different 3-D background contexts for acquisition and extinction training. In one study, ROI analyses showed that activity patterns in the amygdala, hippocampus, and vmPFC become more dissimilar to the CS+ as extinction progresses, with analogous whole-brain results in the cingulate gyrus, insula, cerebellum, visual cortex, and somatosensory cortex (Graner et al. 2020). By contrast, correlated activity patterns to the CS- were flat during extinction, which was interpreted to indicate a selective time-dependent carryover of threat representations into the early extinction trials for the CS+. Similar to the Visser and colleagues' results described above (Visser et al. 2011, 2013), the findings could not be accounted for by mean univariate activity in the ROIs. Using a similar analysis and a 3-D VR paradigm in which the virtual distance to conditioned threats was manipulated, Faul et al. (2020) showed that the correlated patterns in the cerebellum during extinction were less differentiated compared to late acquisition when the CS+ was encountered in close proximity to the participant. Moreover, this lack of differential cerebellar patterning between experimental phases predicted subsequent fear reinstatement to the CS+ the following day for both near and far threats.

The separability of acquisition and extinction memories has also been explored using a trial-unique conditioning approach in which different exemplars of animals and tools were used as conditioned stimuli during acquisition and extinction training (Hennings et al. 2022b). The multivariate fMRI activity patterns between these trial-unique events were then compared to an explicit memory retrieval task for the same items the following day. The representational similarity was greater for CS+ than for CS- items from the conditioning phase in the dACC, insula, cerebellum, and several frontoparietal regions, whereas the same analysis from the extinction phase revealed greater similarity in the vmPFC. Other cortical regions exhibited pattern similarity across both acquisition and extinction phases in the same CS+ > CS- comparison. The posterior hippocampus exhibited greater representational similarity for stimuli from the fear conditioning phase (irrespective of CS type), whereas the anterior hippocampus exhibited greater representational similarity for stimuli from the extinction phase (irrespective of CS type). These results extend the MVPA analyses from prior studies by showing how MVPA-based representations at the individual-trial level from acquisition and extinction phases of learning carry over to subsequent re-encounters to the same stimuli after extinction has taken place. For additional analyses of this paradigm that demonstrate the value of MVPA in revealing mental representations of the extinction context during a subsequent fear renewal test, see Hennings et al. (2020).

9 Relating Fear Extinction Correlates to Individual Differences in Affect

Most fMRI research averages the data across all participants in the study to characterize the neural correlates of extinction learning and recall. However, individual differences in the subjective experience of fear and anxiety, physiological expression, emotion regulation abilities, and related genetic and neurobiological constructs can be leveraged to better understand brain–behavior relationships in extinction. A central question in the field is the extent to which conditioning and extinction processes pertain to the subjective experience of fear and anxiety. Although there are arguments that the experience of fear and conditioned learning may be dissociable in terms of brain systems (LeDoux 2014; LeDoux and Pine 2016), these constructs are inter-related. When subjective experience is queried intermittently during conditioning procedures, fear and anxiety ratings increase during fear acquisition/renewal/reinstatement and decrease during extinction learning/recall, along with reports of negative valence and arousal (Åhs et al. 2015; Batsikadze et al. 2022; Braem et al. 2017; Dirikx et al. 2004; Feng et al. 2016; Hermans et al. 2005; Lonsdorf et al. 2014; Raes et al. 2014; Schiele et al. 2016).

Individual differences in these subjective experiences and trait affect constructs impact the neural correlates of extinction. One study found that subjective fear ratings correlated with the multivariate amygdala response during extinction to a CS+ paired with an actual shock but not to a CS+ that was paired with a visual image of a shock (Braem et al. 2017). In addition, state anxiety ratings taken at the end of acquisition training predict a persistence in the amygdala's multivariate response to the CS+, relative to a CS-, throughout extinction (Graner et al. 2020). Individual differences in trait anxiety are associated with increased amygdala activation and reduced dACC activation during extinction (Sehlmeyer et al. 2011). Conversely, individuals who have better emotion regulation abilities as indexed by cognitive reappraisal exhibit enhanced vmPFC engagement during extinction recall (Hermann et al. 2014). Although beyond the scope of this review, these findings have implications for more in-depth investigations of extinction processing as a model system for disorders of fear and anxiety (Cooper and Dunsmoor 2021; Craske et al. 2018; Milad and Quirk 2012) and as a predictor of treatment response in clinical settings (Scheveneels et al. 2021).

10 Conclusions and Future Directions

Since its discovery in the early 1990s, fMRI has provided an unprecedented look into whole-brain activation patterns, connectivity, and dynamics involved in extinction learning. While the method has some limitations in terms of temporal resolution and sensitivity to smaller subcortical brain areas, it has provided important insights into fear extinction circuits and their role in extinction recall and return-of-fear phenomena. Some of the brain regions implicated in rodent models have been partially validated in humans, while others such as the ACC, insula, and cerebellum have gained new prominence in meta-analytic summaries. Recent advances in multivariate analytic tools have demonstrated greater sensitivity than traditional univariate methods to reveal contributions of the core extinction circuit in the amygdala, vmPFC, and hippocampus.

Future work can build upon these advances by developing more ecologically valid paradigms (i.e., through the use of VR/augmented reality) to characterize how contextual factors bias threat vs. safety signaling in the brain. Computational modeling, multivariate analyses, and machine learning tools can further an understanding of the component processes involved in extinction, relate the brain data to key theoretical constructs, and disentangle the mnemonic representations established by different phases of conditioned learning. Few studies to date have combined fMRI with neuromodulatory approaches, such as transcranial magnetic stimulation, pharmacologic manipulations, or neurofeedback, to unpack which components of the broader circuitry are critical for effective learning. Larger sample sizes are needed to reliably assess how individual difference variables such as subjective fear/anxiety, intolerance of uncertainty, or other genetic/trait predispositions of affect provide susceptibility or resilience to the neural mechanisms of extinction and fear recovery. Finally, brain imaging can be combined with renewed efforts to behaviorally optimize and personally tailor extinction learning to prevent fear overgeneralization as a precursor to developing novel exposure-based interventions for clinical affective disorders.

References

Aery Jones EA, Giocomo LM (2023) Neural ensembles in navigation: from single cells to population codes. Curr Opin Neurobiol 78:102665. https://doi.org/10.1016/j.conb.2022.102665

Åhs F, Frans O, Tibblin B, Kumlien E, Fredrikson M (2010) The effects of medial temporal lobe resections on verbal threat and fear conditioning. Biol Psychol 83(1):41–46. https://doi.org/10.1016/j.biopsycho.2009.10.002

Åhs F, Kragel PA, Zielinski DJ, Brady R, LaBar KS (2015) Medial prefrontal pathways for the contextual regulation of extinguished fear in humans. Neuroimage 122:262–271. https://doi.org/10.1016/j.neuroimage.2015.07.051

Allman JM, Hakeem A, Erwin JM, Nimchinsky E, Hof P (2001) The anterior cingulate cortex. The evolution of an interface between emotion and cognition. Ann N Y Acad Sci 935:107–117. http://www.ncbi.nlm.nih.gov/entrez/query.fcgi?cmd=Retrieveanddb=PubMedanddopt=Citationandlist_uids=11411161

Amunts K, Kedo O, Kindler M, Pieperhoff P, Mohlberg H, Shah NJ, Habel U, Schneider F, Zilles K (2005) Cytoarchitectonic mapping of the human amygdala, hippocampal region and entorhinal cortex: intersubject variability and probability maps. Anat Embryol 210(5–6):343–352. https://doi.org/10.1007/s00429-005-0025-5

Baldi E, Bucherelli C (2015) Brain sites involved in fear memory reconsolidation and extinction of rodents. Neurosci Biobehav Rev 53:160–190. https://doi.org/10.1016/j.neubiorev.2015.04.003

Batsikadze G, Diekmann N, Ernst TM, Klein M, Maderwald S, Deuschl C, Merz CJ, Cheng S, Quick HH, Timmann D (2022) The cerebellum contributes to context-effects during fear extinction learning: a 7T fMRI study. Neuroimage 253:119080. https://doi.org/10.1016/j.neuroimage.2022.119080

Battaglia S, Garofalo S, di Pellegrino G, Starita F (2020) Revaluing the role of vmPFC in the acquisition of Pavlovian threat conditioning in humans. J Neurosci 40(44):8491–8500. https://doi.org/10.1523/JNEUROSCI.0304-20.2020

Bechara A, Tranel D, Damasio H, Adolphs R, Rockland C, Damasio AR (1995) Double dissociation of conditioning and declarative knowledge relative to the amygdala and hippocampus in humans. Science 269(5227):1115–1118. http://www.ncbi.nlm.nih.gov/entrez/query.fcgi?cmd=Retrieveanddb=PubMedanddopt=Citationandlist_uids=7652558

Belova MA, Paton JJ, Salzman CD (2008) Moment-to-moment tracking of state value in the amygdala. J Neurosci 28(40):10023–10030. http://www.ncbi.nlm.nih.gov/entrez/query.fcgi?cmd=Retrieveanddb=PubMedanddopt=Citationandlist_uids=18829960

Bird CM, Burgess N (2008) The hippocampus and memory: insights from spatial processing. Nat Rev Neurosci 9(3):182–194. https://doi.org/10.1038/nrn2335

Bohnlein J, Altegoer L, Muck NK, Roesmann K, Redlich R, Dannlowski U, Leehr EJ (2020) Factors influencing the success of exposure therapy for specific phobia: a systematic review. Neurosci Biobehav Rev 108:796–820. https://doi.org/10.1016/j.neubiorev.2019.12.009

Boll S, Gamer M, Gluth S, Finsterbusch J, Buchel C (2013) Separate amygdala subregions signal surprise and predictiveness during associative fear learning in humans. Eur J Neurosci 37(5): 758–767. https://doi.org/10.1111/ejn.12094

Bouton ME (1993) Context, time, and memory retrieval in the interference paradigms of Pavlovian learning. Psychol Bull 114(1):80–99. http://www.ncbi.nlm.nih.gov/entrez/query.fcgi?cmd=Retrieveanddb=PubMedanddopt=Citationandlist_uids=8346330

Bouton ME (2000) A learning theory perspective on lapse, relapse, and the maintenance of behavior change. Health Psychol 19(1 Suppl):57–63. http://www.ncbi.nlm.nih.gov/entrez/query.fcgi?cmd=Retrieveanddb=PubMedanddopt=Citationandlist_uids=10709948

Braem S, De Houwer J, Demanet J, Yuen KSL, Kalisch R, Brass M (2017) Pattern analyses reveal separate experience-based fear memories in the human right amygdala. J Neurosci 37(34): 8116–8130. https://doi.org/10.1523/JNEUROSCI.0908-17.2017

Brown, V. M., LaBar, K. S., Haswell, C. C., Gold, A. L., Mid-Atlantic, M. W., Beall, S. K., Van Voorhees, E., Marx, C. E., Calhoun, P. S., Fairbank, J. A., Green, K. T., Tupler, L. A., Weiner, R. D., Beckham, J. C., Brancu, M., Hoerle, J. M., Pender, M., Kudler, H., Swinkels, C. M., . . . Morey, R. A. (2014). Altered resting-state functional connectivity of basolateral and centromedial amygdala complexes in posttraumatic stress disorder. Neuropsychopharmacology, 39(2), 361–369. https://doi.org/10.1038/npp.2013.197

Cooper SE, Dunsmoor JE (2021) Fear conditioning and extinction in obsessive-compulsive disorder: a systematic review. Neurosci Biobehav Rev 129:75–94. https://doi.org/10.1016/j.neubiorev.2021.07.026

Craig AD (2011) Significance of the insula for the evolution of human awareness of feelings from the body. Ann N Y Acad Sci 1225:72–82. https://doi.org/10.1111/j.1749-6632.2011.05990.x

Craske MG, Hermans D, Vervliet B (2018) State-of-the-art and future directions for extinction as a translational model for fear and anxiety. Philos Trans R Soc Lond B Biol Sci 373(1742). https://doi.org/10.1098/rstb.2017.0025

Dirikx T, Hermans D, Vansteenwegen D, Baeyens F, Eelen P (2004) Reinstatement of extinguished conditioned responses and negative stimulus valence as a pathway to return of fear in humans. Learn Mem 11(5):549–554. http://www.ncbi.nlm.nih.gov/entrez/query.fcgi?cmd=Retrieveanddb=PubMedanddopt=Citationandlist_uids=15466307

Dunsmoor JE, Ahs F, Zielinski DJ, LaBar KS (2014) Extinction in multiple virtual reality contexts diminishes fear reinstatement in humans. Neurobiol Learn Mem 113:157–164. https://doi.org/10.1016/j.nlm.2014.02.010

Dunsmoor JE, Niv Y, Daw N, Phelps EA (2015) Rethinking extinction. Neuron 88(1):47–63. https://doi.org/10.1016/j.neuron.2015.09.028

Faul L, Stjepanovic D, Stivers JM, Stewart GW, Graner JL, Morey RA, LaBar KS (2020) Proximal threats promote enhanced acquisition and persistence of reactive fear-learning circuits. Proc Natl Acad Sci U S A 117(28):16678–16689. https://doi.org/10.1073/pnas.2004258117
Feng P, Zheng Y, Feng T (2016) Resting-state functional connectivity between amygdala and the ventromedial prefrontal cortex following fear reminder predicts fear extinction. Soc Cogn Affect Neurosci:991–1001. https://doi.org/10.1093/scan/nsw031
Frontera JL, Baba Aissa H, Sala RW, Mailhes-Hamon C, Georgescu IA, Lena C, Popa D (2020) Bidirectional control of fear memories by cerebellar neurons projecting to the ventrolateral periaqueductal grey. Nat Commun 11(1):5207. https://doi.org/10.1038/s41467-020-18953-0
Fullana MA, Albajes-Eizagirre A, Soriano-Mas C, Vervliet B, Cardoner N, Benet O, Radua J, Harrison BJ (2018) Fear extinction in the human brain: a meta-analysis of fMRI studies in healthy participants. Neurosci Biobehav Rev 88:16–25. https://doi.org/10.1016/j.neubiorev.2018.03.002
Giustino TF, Maren S (2015) The role of the medial prefrontal cortex in the conditioning and extinction of fear. Front Behav Neurosci 9:298. https://doi.org/10.3389/fnbeh.2015.00298
Gottfried JA, Dolan RJ (2004) Human orbitofrontal cortex mediates extinction learning while accessing conditioned representations of value. Nat Neurosci 7(10):1144–1152. https://doi.org/10.1038/nn1314
Graner JL, Stjepanovic D, LaBar KS (2020) Extinction learning alters the neural representation of conditioned fear. Cogn Affect Behav Neurosci 20(5):983–997. https://doi.org/10.3758/s13415-020-00814-4
Hennings AC, McClay M, Lewis-Peacock JA, Dunsmoor JE (2020) Contextual reinstatement promotes extinction generalization in healthy adults but not PTSD. Neuropsychologia 147:107573. https://doi.org/10.1016/j.neuropsychologia.2020.107573
Hennings AC, Cooper SE, Lewis-Peacock JA, Dunsmoor JE (2022a) Pattern analysis of neuroimaging data reveals novel insights on threat learning and extinction in humans. Neurosci Biobehav Rev 142:104918. https://doi.org/10.1016/j.neubiorev.2022.104918
Hennings AC, McClay M, Drew MR, Lewis-Peacock JA, Dunsmoor JE (2022b) Neural reinstatement reveals divided organization of fear and extinction memories in the human brain. Curr Biol 32(2):304–314 e305. https://doi.org/10.1016/j.cub.2021.11.004
Hermann A, Keck T, Stark R (2014) Dispositional cognitive reappraisal modulates the neural correlates of fear acquisition and extinction. Neurobiol Learn Mem 113:115–124. https://doi.org/10.1016/j.nlm.2014.03.008
Hermann A, Stark R, Milad MR, Merz CJ (2016) Renewal of conditioned fear in a novel context is associated with hippocampal activation and connectivity. Soc Cogn Affect Neurosci 11(9):1411–1421. https://doi.org/10.1093/scan/nsw047
Hermann A, Stark R, Muller EA, Kruse O, Wolf OT, Merz CJ (2020) Multiple extinction contexts modulate the neural correlates of context-dependent extinction learning and retrieval. Neurobiol Learn Mem 168:107150. https://doi.org/10.1016/j.nlm.2019.107150
Hermans D, Dirikx T, Vansteenwegen D, Baeyens F, Van den Bergh O, Eelen P (2005) Reinstatement of fear responses in human aversive conditioning. Behav Res Ther 43(4):533–551. http://www.ncbi.nlm.nih.gov/entrez/query.fcgi?cmd=Retrieveanddb=PubMedanddopt=Citationandlist_uids=15701362
Herry C, Johansen JP (2014) Encoding of fear learning and memory in distributed neuronal circuits. Nat Neurosci 17(12):1644–1654. https://doi.org/10.1038/nn.3869
Huff NC, Zielinski DJ, Fecteau ME, Brady R, LaBar KS (2010) Human fear conditioning conducted in full immersion 3-dimensional virtual reality. J Vis Exp 42:e1993. https://doi.org/10.3791/1993
Huff N, Alba Hernandez J, Fecteau M, Zielinski D, Brady R, LaBar KS (2011) Revealing context-specific conditioned fear memories with full immersion virtual reality [original research]. Front Behav Neurosci 5. https://doi.org/10.3389/fnbeh.2011.00075

Icenhour A, Kattoor J, Benson S, Boekstegers A, Schlamann M, Merz CJ, Forsting M, Elsenbruch S (2015) Neural circuitry underlying effects of context on human pain-related fear extinction in a renewal paradigm. Hum Brain Mapp 36(8):3179–3193. https://doi.org/10.1002/hbm.22837

Jung SJ, Vlasov K, D'Ambra AF, Parigi A, Baya M, Frez EP, Villalobos J, Fernandez-Frentzel M, Anguiano M, Ideguchi Y, Antzoulatos EG, Fioravante D (2022) Novel Cerebello-Amygdala connections provide missing link between cerebellum and limbic system. Front Syst Neurosci 16:879634. https://doi.org/10.3389/fnsys.2022.879634

Kaas JH (2012) The evolution of neocortex in primates. Prog Brain Res 195:91–102. https://doi.org/10.1016/B978-0-444-53860-4.00005-2

Kalisch R, Korenfeld E, Stephan KE, Weiskopf N, Seymour B, Dolan RJ (2006) Context-dependent human extinction memory is mediated by a ventromedial prefrontal and hippocampal network. J Neurosci 26(37):9503–9511. https://doi.org/10.1523/Jneurosci.2021-06.2006

Kragel PA, Knodt AR, Hariri AR, LaBar KS (2016) Decoding spontaneous emotional states in the human brain. PLoS Biol 14(9):e2000106. https://doi.org/10.1371/journal.pbio.2000106

LaBar KS, Phelps EA (2005) Reinstatement of conditioned fear in humans is context dependent and impaired in amnesia. Behav Neurosci 119(3):677. http://sfx.lib.duke.edu:9003/sfx_local?sid=Entrez%3APubMed;id=pmid%3A15998188

LaBar KKS, LeDoux JJE, Spencer DDD, Phelps EEA (1995) Impaired fear conditioning following unilateral temporal lobectomy in humans. J Neurosci 15(10):6846. http://sfx.lib.duke.edu:9003/sfx_local?sid=Entrez%3APubMed;id=pmid%3A7472442

LaBar KS, Gatenby JC, Gore JC, LeDoux JE, Phelps EA (1998) Human amygdala activation during conditioned fear acquisition and extinction: a mixed-trial fMRI study. Neuron 20(5):937–945. http://www.ncbi.nlm.nih.gov/entrez/query.fcgi?cmd=Retrieveanddb=PubMedanddopt=Citationandlist_uids=9620698

Lawrenson C, Paci E, Pickford J, Drake RAR, Lumb BM, Apps R (2022) Cerebellar modulation of memory encoding in the periaqueductal grey and fear behaviour. eLife 11. https://doi.org/10.7554/eLife.76278

LeDoux JE (1993) Emotional memory: in search of systems and synapses. Ann N Y Acad Sci 702:149–157. http://www.ncbi.nlm.nih.gov/entrez/query.fcgi?cmd=Retrieveanddb=PubMedanddopt=Citationandlist_uids=8109874

LeDoux JE (2014) Coming to terms with fear. Proc Natl Acad Sci U S A 111(8):2871–2878. https://doi.org/10.1073/pnas.1400335111

LeDoux JE, Pine DS (2016) Using neuroscience to help understand fear and anxiety: a two-system framework. Am J Psychiatry 173(11):1083–1093. https://doi.org/10.1176/appi.ajp.2016.16030353

Lim SL, Padmala S, Pessoa L (2009) Segregating the significant from the mundane on a moment-to-moment basis via direct and indirect amygdala contributions. Proc Natl Acad Sci U S A 106(39):16841–16846. https://doi.org/10.1073/pnas.0904551106

Logothetis NK (2001) What we can do and what we cannot do with fMRI. Nature 453. https://doi.org/10.1038/nature06976

Logothetis NK, Wandell BA (2004) Interpreting the BOLD signal. Annu Rev Physiol 66:735–769. https://doi.org/10.1146/annurev.physiol.66.082602.092845

Lonsdorf TB, Haaker J, Kalisch R (2014) Long-term expression of human contextual fear and extinction memories involves amygdala, hippocampus and ventromedial prefrontal cortex: a reinstatement study in two independent samples. Soc Cogn Affect Neurosci 9(12):1973–1983. https://doi.org/10.1093/scan/nsu018

Maren S, Phan KL, Liberzon I (2013) The contextual brain: implications for fear conditioning, extinction and psychopathology. Nat Rev Neurosci 14(6):417–428. https://doi.org/10.1038/nrn3492

Maschke M, Drepper J, Kindsvater K, Kolb FP, Diener HC, Timmann D (2000) Fear conditioned potentiation of the acoustic blink reflex in patients with cerebellar lesions. J Neurol Neurosurg Psychiatry 68(3):358–364. https://doi.org/10.1136/jnnp.68.3.358

McLean CP, Levy HC, Miller ML, Tolin DF (2022) Exposure therapy for PTSD: a meta-analysis. Clin Psychol Rev 91:102115. https://doi.org/10.1016/j.cpr.2021.102115

Mendez-Bertolo C, Moratti S, Toledano R, Lopez-Sosa F, Martinez-Alvarez R, Mah YH, Vuilleumier P, Gil-Nagel A, Strange BA (2016) A fast pathway for fear in human amygdala. Nat Neurosci 19(8):1041-+. https://doi.org/10.1038/nn.4324

Menon V (2015) Salience network. Brain mapping: an encyclopedic reference, vol 2, pp 597–611

Milad MR, Quirk GJ (2012) Fear extinction as a model for translational neuroscience: ten years of Progress. Annu Rev Psychol 63(63):129–151. https://doi.org/10.1146/annurev.psych.121208.131631

Milad MR, Wright CI, Orr SP, Pitman RK, Quirk GJ, Rauch SL (2007) Recall of fear extinction in humans activates the ventromedial prefrontal cortex and hippocampus in concert. Biol Psychiatry 62(5):446–454. https://doi.org/10.1016/j.biopsych.2006.10.011

Moon HS, Jiang H, Vo TT, Jung WB, Vasquez AL, Kim S-G (2021) Contribution of excitatory and inhibitory neuronal activity to BOLD fMRI. Cereb Cortex 31:4053–4067. https://doi.org/10.1093/cercor/bhab068

Moorman DE, Aston-Jones G (2023) Prelimbic and infralimbic medial prefrontal cortex neuron activity signals cocaine seeking variables across multiple timescales. Psychopharmacology (Berl) 240(3):575–594. https://doi.org/10.1007/s00213-022-06287-2

Morawetz C, Holz P, Lange C, Baudewig J, Weniger G, Irle E, Dechent P (2008) Improved functional mapping of the human amygdala using a standard functional magnetic resonance imaging sequence with simple modifications. Magn Reson Imaging 26(1):45–53. https://doi.org/10.1016/j.mri.2007.04.014

Morrison SE, Salzman CD (2009) The convergence of information about rewarding and aversive stimuli in single neurons. J Neurosci 29(37):11471–11483. https://doi.org/10.1523/JNEUROSCI.1815-09.2009

Norman KA, Polyn SM, Detre GJ, Haxby JV (2006) Beyond mind-reading: multi-voxel pattern analysis of fMRI data. Trends Cogn Sci 10(9):424–430. S1364-6613(06)00184-7 [pii]. https://doi.org/10.1016/j.tics.2006.07.005

Ojemann JG, Akbudak E, Snyder AZ, McKinstry RC, Raichle ME, Conturo TE (1997) Anatomic localization and quantitative analysis of gradient refocused echo-planar fMRI susceptibility artifacts. Neuroimage 6(3):156–167. http://www.ncbi.nlm.nih.gov/entrez/query.fcgi?cmd=Retrieveanddb=PubMedanddopt=Citationandlist_uids=9344820

Peper M, Karcher S, Wohlfarth R, Reinshagen G, LeDoux JE (2001) Aversive learning in patients with unilateral lesions of the amygdala and hippocampus. Biol Psychol 58(1):1–23. http://www.ncbi.nlm.nih.gov/entrez/query.fcgi?cmd=Retrieveanddb=PubMedanddopt=Citationandlist_uids=11473792

Pessoa L, Adolphs R (2010) Emotion processing and the amygdala: from a 'low road' to 'many roads' of evaluating biological significance. Nat Rev Neurosci 11(11):773–783. nrn2920 [pii]. https://doi.org/10.1038/nrn2920

Phelps EA, LaBar KS, Anderson AK, O'Connor KJ, Fulbright RK, Spencer DD (1998) Specifying the contributions of the human amygdala to emotional memory: a case study. Neurocase 4:527–540

Phelps EAEA, Delgado MRMR, Nearing KIKI, LeDoux JEJE (2004) Extinction learning in humans: role of the amygdala and vmPFC. Neuron 43(6):897. http://sfx.lib.duke.edu:9003/sfx_local?sid=Entrez%3APubMed;id=pmid%3A15363399

Pine DS, Wise SP, Murray EA (2021) Evolution, emotion, and episodic engagement. Am J Psychiatry 178(8):701–714. https://doi.org/10.1176/appi.ajp.2020.20081187

Plendl W, Wotjak CT (2010) Dissociation of within- and between-session extinction of conditioned fear. J Neurosci 30(14):4990–4998. https://doi.org/10.1523/JNEUROSCI.6038-09.2010

Raes AK, De Houwer J, De Schryver M, Brass M, Kalisch R (2014) Do CS-US pairings actually matter? A within-subject comparison of instructed fear conditioning with and without actual CS-US pairings. PloS One 9(1):e84888. https://doi.org/10.1371/journal.pone.0084888

Ridderbusch IC, Wroblewski A, Yang Y, Richter J, Hollandt M, Hamm AO, Wittchen HU, Strohle A, Arolt V, Margraf J, Lueken U, Herrmann MJ, Kircher T, Straube B (2021) Neural adaptation of cingulate and insular activity during delayed fear extinction: a replicable pattern across assessment sites and repeated measurements. Neuroimage 237:118157. https://doi.org/10.1016/j.neuroimage.2021.118157

Sacchetti B, Sacco T, Strata P (2007) Reversible inactivation of amygdala and cerebellum but not perirhinal cortex impairs reactivated fear memories. Eur J Neurosci 25(9):2875–2884. https://doi.org/10.1111/j.1460-9568.2007.05508.x

Santini E, Ge H, Ren K, Pena de Ortiz S, Quirk GJ (2004) Consolidation of fear extinction requires protein synthesis in the medial prefrontal cortex. J Neurosci 24(25):5704–5710. http://www.ncbi.nlm.nih.gov/entrez/query.fcgi?cmd=Retrieveanddb=PubMedanddopt=Citationandlist_uids=15215292

Scharfenort R, Lonsdorf TB (2016) Neural correlates of and processes underlying generalized and differential return of fear. Soc Cogn Affect Neurosci 11(4):612–620. https://doi.org/10.1093/scan/nsv142

Scheveneels S, Boddez Y, Hermans D (2021) Predicting clinical outcomes via human fear conditioning: a narrative review. Behav Res Ther 142:103870. https://doi.org/10.1016/j.brat.2021.103870

Schiele MA, Reinhard J, Reif A, Domschke K, Romanos M, Deckert J, Pauli P (2016) Developmental aspects of fear: comparing the acquisition and generalization of conditioned fear in children and adults. Dev Psychobiol 58(4):471–481. https://doi.org/10.1002/dev.21393

Sehlmeyer C, Dannlowski U, Schoning S, Kugel H, Pyka M, Pfleiderer B, Zwitserlood P, Schiffbauer H, Heindel W, Arolt V, Konrad C (2011) Neural correlates of trait anxiety in fear extinction. Psychol Med 41(4):789–798. https://doi.org/10.1017/S0033291710001248

Silva BA, Astori S, Burns AM, Heiser H, van den Heuvel L, Santoni G, Martinez-Reza MF, Sandi C, Graff J (2021) A Thalamo-Amygdalar circuit underlying the extinction of remote fear memories. Nat Neurosci 24(7):964–974. https://doi.org/10.1038/s41593-021-00856-y

Song AW (2001) Single-shot EPI with signal recovery from the susceptibility-induced losses. Magn Reson Med 46(2):407–411. http://www.ncbi.nlm.nih.gov/entrez/query.fcgi?cmd=Retrieveanddb=PubMedanddopt=Citationandlist_uids=11477647

Visser RM, Scholte HS, Kindt M (2011) Associative learning increases trial-by-trial similarity of BOLD-MRI patterns. J Neurosci 31(33):12021–12028. https://doi.org/10.1523/JNEUROSCI.2178-11.2011

Visser RM, Scholte HS, Beemsterboer T, Kindt M (2013) Neural pattern similarity predicts long-term fear memory. Nat Neurosci 16(4):388–390. https://doi.org/10.1038/nn.3345

Visser RM, de Haan MIC, Beemsterboer T, Haver P, Kindt M, Scholte HS (2016) Quantifying learning-dependent changes in the brain: single-trial multivoxel pattern analysis requires slow event-related fMRI. Psychophysiology 53(8):1117–1127. https://doi.org/10.1111/psyp.12665

Weike AI, Hamm AO, Schupp HT, Runge U, Schroeder HW, Kessler C (2005) Fear conditioning following unilateral temporal lobectomy: dissociation of conditioned startle potentiation and autonomic learning. J Neurosci 25(48):11117–11124. http://www.ncbi.nlm.nih.gov/entrez/query.fcgi?cmd=Retrieveanddb=PubMedanddopt=Citationandlist_uids=16319311

Wen Z, Chen ZS, Milad MR (2021) Fear extinction learning modulates large-scale brain connectivity. Neuroimage 238:118261. https://doi.org/10.1016/j.neuroimage.2021.118261

Wisse LEM, Chetelat G, Daugherty AM, de Flores R, la Joie R, Mueller SG, Stark CEL, Bakker A, Olsen RK, Carr VA (2020) Hippocampal subfield volumetry from structural isotropic 1 mm3 MRI scans: a note of caution. Hum Brain Mapp 42:539–550. https://doi.org/10.1002/hbm.25234

Zabik NL, Peters C, Iadipaolo A, Marusak HA, Rabinak CA (2023) Comparison of behavioral and brain indices of fear renewal during a standard vs. novel immersive reality Pavlovian fear extinction paradigm in healthy adults. Behav Brain Res 437:114154. https://doi.org/10.1016/j.bbr.2022.114154

Part II
Individual Differences in Fear Extinction

The Impacts of Sex Differences and Sex Hormones on Fear Extinction

Eric Raul Velasco, Antonio Florido, Laura Perez-Caballero, Ignacio Marin, and Raul Andero

Contents

E. R. Velasco and I. Marin
Institut de Neurociències, Universitat Autònoma de Barcelona, Barcelona, Spain

A. Florido and L. Perez-Caballero
Institut de Neurociències, Universitat Autònoma de Barcelona, Barcelona, Spain

Departament de Psicobiologia i de Metodologia de les Ciències de la Salut, Universitat Autònoma de Barcelona, Barcelona, Spain

R. Andero (✉)
Institut de Neurociències, Universitat Autònoma de Barcelona, Barcelona, Spain

Departament de Psicobiologia i de Metodologia de les Ciències de la Salut, Universitat Autònoma de Barcelona, Barcelona, Spain

Centro de Investigación Biomédica en Red de Salud Mental (CIBERSAM), Instituto de Salud Carlos III, Madrid, Spain

Unitat de Neurociència Traslacional, Parc Taulí Hospital Universitari, Institut d'Investigació i Innovació Parc Taulí (I3PT), Institut de Neurociències, Universitat Autònoma de Barcelona, Bellaterra, Spain

ICREA, Pg. Lluís Companys 23, Barcelona, Spain
e-mail: raul.andero@uab.cat

Curr Topics Behav Neurosci (2023) 64: 105–132
https://doi.org/10.1007/7854_2023_426
Published Online: 2 August 2023

Abstract Fear extinction memories are strongly modulated by sex and hormonal status, but the exact mechanisms are still being discovered. In humans, there are some basal and task-related features in which male and female individuals differ in fear conditioning paradigms. However, analyses considering the effects of sex hormones demonstrate a role for estradiol in fear extinction memory consolidation. Translational studies are taking advantage of the convergent findings between species to understand the brain structures implicated. Nevertheless, the human brain is complex and the transfer of these findings into the clinics remains a challenge. The promising advances in the field together with the standardization of fear extinction methodologies in humans will benefit the design of new personalized therapies.

Keywords Estradiol · Fear extinction · Progesterone · Sex differences · Sex hormones

1 Introduction

According to classical Fear Conditioning (FC) paradigms, the repeated association of an initially neutral stimulus with an unconditioned stimulus (US) turns the former into a conditioned stimulus (CS) capable of eliciting a conditioned response (CR). Fear extinction (FE) refers to the decrement of a conditioned fear response that occurs with the repeated presentation of a conditioned fear stimulus that is unreinforced (Milad and Quirk 2012). The initial presentation of a series of unreinforced stimuli during a session is termed Fear Extinction Learning (FEL) or Fear Extinction Training. While recognizing that both terms are used in the literature, for this chapter, we will use the former term, learning. The retention, or strength, of FE memories is evaluated with further presentations of unreinforced stimuli in a session termed Fear Extinction Recall (FER). Salient temporal features of FE can be studied by subjecting individuals to repeated exposures to FE sessions, computing FE indexes, and assessing the spontaneous recovery of fear with time. Additionally, contextual features and memory reconsolidation processes are evaluated by fear renewal or fear reinstatement tasks (Myers and Davis 2007). The tasks where FE is separated from FC by minutes are known as immediate-FE paradigms, while the ones that include a memory consolidation window are termed delayed-FE paradigms. During FE, two time-constrained processes are thought to interact: the erasure of a previously acquired fear memory and the formation of a new inhibitory memory (Myers et al. 2006).

Classical conditioning models, and especially FE models, are valuable due to their high translational potential. Alterations in the acquisition of the conditioned response, the learning of FE, or the inhibition of fear are potential behavioral biomarkers for diseases like Alzheimer's, schizophrenia, or posttraumatic stress disorder (PTSD; Hoefer et al. 2008; Holt et al. 2009; Jovanovic et al. 2010; Milad et al. 2009b). Notably, impairments in FE processes have been mapped to neuronal signatures that are shared among species (Milad and Quirk 2012). Pathological FE can be studied in animal models by exposure to different stressors like immobilization or single prolonged stress (Deslauriers et al. 2018; Stockhorst and Antov 2015) or by the use of rodent models based on genetic differences or polymorphisms (Lonsdorf and Kalisch 2011; McCaughran et al. 2000; Stiedl et al. 1999).

2 Sex Differences in FE

The study of sex differences in FE is relevant given the differential vulnerability of women to suffer from stressor- and fear-based disorders (Bangasser and Valentino 2014). Women are not only more affected but also experience greater morbidity, suggesting an important role in the persistence of pathological fear for psychiatric symptomatology (Maeng and Milad 2015). Further, studies have documented organic and functional brain differences among the sexes that can influence stress and emotional processing (Bangasser et al. 2010; de Vries 2008). Indeed, some of these functional differences are a product of shifting hormonal levels that facilitate aversive internal states (Kanwal et al. 2021). Thus, the study of the hormonal and neuropeptide systems will shed light on the mechanisms implicated in the regulation of fear and stress processing (Florido et al. 2021; Ressler et al. 2011). In the following sections, we will review the specific effects of the menstrual/estrous cycle and gonadal hormones over FE in adulthood.

2.1 Menstrual/Estrous Cycle and FE

In humans, the menstrual cycle lasts 28 days (±5 days) and is divided into four phases. The Early Follicular (EF) phase (~0–7 days) has low levels of sex hormones that are followed by rising estradiol levels that peak during the Late Follicular (LF) phase (~10–15 days). After ovulation, mid-estradiol and high progesterone levels are detected in the Mid-Luteal (ML) phase (~17–25 days) until a sharp drop in hormonal levels occurs in the Late Luteal phase (~26–28 days). In rodents, the estrous cycle lasts 4–5 days and is composed of four phases. The proestrus is characterized by high peaks of estradiol and progesterone followed by ovulation. In estrus, hormonal levels are low and extend into the initial part of the metestrus. In metestrus, progesterone levels increase and remain high until mid-diestrus returning to low basal levels before the next proestrus phase (Lebron-Milad and Milad 2012). Despite some similarities, the estrous and menstrual cycles are not equivalent. For research purposes, the proestrus is considered a "high hormone state" and the metestrus/diestrus a "low hormone state" (Lebron-Milad and Milad 2012).

In rodent cued-FE paradigms, no clear sex differences are found if only the sex factor is considered. However, analyses including the estrous cycle show that animals subjected to FEL during proestrus have lower freezing rates at FER compared to metestrus, estrus, or diestrus (Gruene et al. 2015; Milad et al. 2009a; Rey et al. 2014). Notably, the effects of the estrous cycle in contextual-FC are opposite to cued-FC paradigms, with diestrus females having lower freezing rates during FEL compared to proestrus (Blume et al. 2017). However, other studies have found lower freezing levels for proestrus females during both FC and FEL (Blume et al. 2017; Chang et al. 2009; Gruene et al. 2015).

Human studies describe functional changes in the brain across the distinct phases of the menstrual cycle as evidenced by tasks evaluating socio-emotional behaviors and cognition (Sundström-Poromaa and Gingnell 2014; Van Wingen et al. 2011). Arousal and brain reactivity to threat is greater during EF compared to LF, but the greatest amygdala reactivity is seen during the ML phase (Goldstein et al. 2005, 2010; Sundström-Poromaa and Gingnell 2014). Analyses using sex as a between-group factor find greater skin conductance response (SCR) to the reinforced conditioned stimulus (CS+) in men during FC and immediate-FE paradigms, but no additional differences are detected if the menstrual cycle phase is not considered (Antov and Stockhorst 2014; Merz et al. 2012a; Milad et al. 2006). Unlike in rodent models, the effects of the menstrual cycle phases over FE are less clear. Human analyses are largely focused on the effects of absolute sex hormone levels and use them as continuous variables or as thresholds to determine experimental groups. The comparison of findings between both approaches – menstrual cycle segmentation and absolute hormonal levels – is problematic and often results in non-conclusive findings. For example, low estrogen levels were found to be associated with greater fear responses during FEL without any effect over FC, but analyses considering the menstrual cycle phase found between-group differences during FC (Wegerer et al. 2014). Another study using an immediate-FE paradigm found that EF women have smaller SCR and better FE retention index (100−(100*[first 2 CS+ trials − first 2 CS− trials in FER/max differential SCR in FC])) compared to LF women or men (Milad et al. 2006). This finding seems contrary to the robust literature showing that low estrogen levels are detrimental to FE. However, another study failed to replicate these differences and reported no effects for EF or LF in neither an immediate-FE or a delayed-FE task (Antov and Stockhorst 2014).

The mechanisms underlying FE improvements during the proestrus phase in rodents are likely related to estradiol dynamics and its effects on the threat detection system. High estradiol levels are associated with attenuations in the threat detection system, including greater recruitment of prefrontal structures, and greater inhibition of the amygdala (Blume et al. 2017; Goldstein et al. 2005, 2010; Hwang et al. 2015; Zeidan et al. 2011). Moreover, increasing estradiol levels during the estrous cycle positively regulate hippocampal dendritic spines (Woolley and McEwen 1993). There are additional effects of estradiol over cholinergic, monoaminergic, and neuropeptide functions that will be reviewed in detail in the following sections. These heterogeneous mechanisms converge on a greater prefrontal function during FEL and the recruitment of protein synthesis and memory consolidation processes

(Cover et al. 2014). The influence of the menstrual cycle on human FE is still unclear and the mechanisms behind these effects remain to be systematically explored. There are specific features of the menstrual cycle that are worth covering in future studies, such as the effects of the late luteal phase in FE or the influence of inter-individual differences in hormonal levels.

In sum, proestrus seems to be associated with lower freezing rates during cued-FER in rodents, but the effects of the menstrual cycle in humans are not conclusive. This divergence arises from the inherent limitations in these studies including the use of different methodologies, a lack of systematic segmentation into menstrual cycle phases, the use of different control groups to detect experimental effects (EF vs ML; EF vs LF), and the use of heterogeneous measures and indexes to assess FE memory. In addition, menstrual cycle effects are not reported or tested systematically possibly because the analyses are secondary to the main hypothesis and rely on low sample sizes. Notably, the effects of the menstrual cycle phase over FE seem subtle and influenced by actual or prior stress exposure. To date, it remains unknown whether the impairments in FER during low estrogen states arise during a specific timeframe in the menstrual cycle of all women or from individual differences in sex hormone dynamics (Fig. 1). These research gaps will be addressed by future studies using sufficiently powered experimental designs and correct monitoring of the menstrual cycle (Schmalenberger et al. 2021).

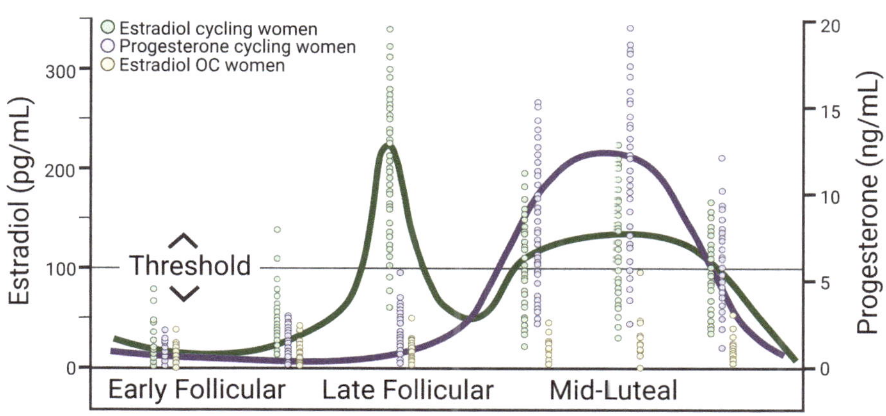

Fig. 1 Inter-individual variability in absolute sex hormone levels. Estradiol (green) and progesterone (violet) levels in cycling women 18–42 years. Assumed estradiol levels (yellow) in women taking monophasic oral contraceptives (OC). An approximate threshold commonly used for median sample-split approaches in estradiol and FE studies is depicted. Figure adapted with permission from the authors Sundström-Poromaa, I., & Gingnell, M. (2014). Front Neurosci. 2014; 8: 380

2.2 Estradiol and FE

Estrogens are one of the main sex hormones in female organisms. Estradiol, the most potent estrogen, is secreted mainly by the ovary along with other weaker estrogens but their synthesis can also take place to a lesser extent in the adrenal glands, peripheral tissues, and brain (Melmed et al. 2019). In males, testosterone is aromatized to estradiol by 5α-reductase. Estrogen receptors (ERs) are located in intracellular compartments or the membrane, along with the G protein-coupled estrogen receptor (GPER). The binding of estradiol to membrane ERs increases intracellular calcium and activates cyclic AMP/Protein kinase A (cAMP/PKA), mitogen-activated protein kinases, extracellular signal-regulated kinases (MAPK/ERK), and the phosphoinositide 3-kinase/protein kinase B (PI3K/AKT) pathways to converge at nuclear targets (Cover et al. 2014). There are differences in the distribution of ERs among sexes with additional fluctuations in ERs expression for females throughout the estrous cycle (Mendoza-Garcés et al 2011). The effects of estrogens on the brain are widespread involving positive effects on neuronal survival, pain, motor control, mood, cognition, and anxiety (Cersosimo and Benarroch 2015). In addition, estradiol promotes hippocampal-mediated memory and some forms of associative memory (Leuner et al. 2004; McEwen 2002).

Estradiol is a strong modulator of FE memories in animals and humans. Rodents undergoing FEL under a high estrogenic state (proestrus) have better FER compared to subjects with a low estrogenic state (metestrus-diestrus; Graham and Daher 2016; Graham and Milad 2013; Milad et al. 2009a; Zeidan et al. 2011). Human studies generally support the notion that women with high estradiol levels have better FER than women with low estradiol levels, as documented by lower differential SCR during FER, higher FE retention indexes (100−[100*(differential SCR in FER 2 initial trials/largest differential SCR in FC)]) or lower recovery of fear (100* (average SCR to first 4 CS+ trials in FER/largest SCR to CS+ FC); Graham and Milad 2013; Milad et al. 2010; Zeidan et al. 2011). However, the use of unstandardized sample-split approaches and the heterogeneity of methods can result in null effects for estradiol levels (Lebron-Milad et al. 2012; White and Graham 2016).

Similar to endogenous modulation, FE memory can also be affected by exogenous estrogen administration. Dosing estradiol or progesterone to metestrus females improves FER, while the blockade of ERs or progesterone receptors (PRs) during proestrus impairs FER (Milad et al. 2009a). Further studies showed that an ERβ agonist but not an ERα agonist could improve deficient FER in metestrus females. Notably, the positive effects of estradiol seem to be allocated to an early phase of FE memory consolidation, since immediate post-FEL doses of estradiol improve FER but not doses that are given 4 h post-FEL (Zeidan et al. 2011). Still, these beneficial effects of estradiol seem to be constrained to cued-FE paradigms because high doses of estradiol increase contextual fear memory and impair contextual FE (McDermott et al. 2015). In humans, oral estradiol dosed to EF women shortly before FEL results in lower recovery of fear compared to placebo. Interestingly, in this study, it was observed that women displayed a differential response to the drug, with those having the highest increases in estradiol (>80% basal) benefiting the most in terms of fear recovery. Thus, relative increases in estradiol levels may be a relevant factor driving the improvements in FE (Graham and Milad 2013).

The mechanisms by which estradiol influences FE include genomic and non-genomic effects. The activation of membrane ERs increases intracellular calcium leading to the activation of kinase pathways (McEwen 1999). Specifically, the MAPK/ERK pathway and PI3K/AKT pathway modulate synaptic plasticity and synaptic potentiation and are implicated in the retrieval of fear memories, acquisition of FE, and consolidation of FE (Chen et al. 2005; Herry et al. 2006; Hugues et al. 2006; Lu et al. 2001). These pathways converge at nuclear targets where they can promote CREB phosphorylation or trigger the transcription of genes relevant to memory consolidation (Cover et al. 2014). Therefore, estradiol can induce stronger memory consolidation by increasing long-term potentiation (LTP; Smith and McMahon 2005).

Estradiol can also promote structural and functional states with enhanced neurotransmission. The cyclical fluctuations of estradiol in females are followed by changes in ERs expression, a different density of dendritic spines, and enhanced glutamatergic transmission at the hippocampus and prefrontal cortex (PFC; Chen et al. 2009; Mitterling et al. 2010; Woolley and McEwen 1993). Notably, the effects of estradiol over hippocampal spines seem to be differentially driven by ERs isoforms. ERα agonists promote spine formation in the hippocampus while ERβ agonists halt spinogenesis but promote fast excitatory postsynaptic potentials, synaptic potentiation, and spine maturation (McEwen et al. 2012). Thus, systemic ERβ agonists may improve FER in metestrus females by increasing NMDA-mediated signaling and enhancing LTP of prefrontal and hippocampal synapses (Galvin and Ninan 2014; Hasegawa et al. 2015; Zeidan et al. 2011). There are additional effects of estradiol during neurodevelopment related to structural and functional sex differences in the hippocampus, hypothalamus, and bed nucleus of the stria terminalis (BNST) for which their specific relation with FE remains unknown (McCarthy 2008).

The PFC and the amygdala are implicated in FE and are under monoaminergic and cholinergic neuromodulation. Estrogen effects on prefrontal function and FE seem to follow an inverted U-shape (Barha et al. 2010; Graham and Scott 2018; Sheppard et al. 2019). Moreover, estrogens facilitate noradrenergic activity in the locus coeruleus and modulate serotoninergic and cholinergic systems at the midbrain raphe and basal forebrain (Bangasser and Valentino 2014; McEwen 2002). Variations in these neurotransmitter systems result in changes in arousal and attention with additional effects on learning and memory (McEwen 2002). For example, researchers have described important sex differences in the regulation of locus coeruleus by CRF. In females, the CRF1 receptor preferentially couples to the GTP-binding protein Gs and promotes cAMP/PKA pathway signaling. In turn, the association of CRF1 with β-arrestin, a molecule that promotes receptor internalization, occurs preferentially in males (Bangasser and Cuarenta 2021). Estradiol can also modulate neuropeptide systems and impact FE. The upregulation of MAPK/

ERK and PI3K pathways converges on CREB phosphorylation and promotes BDNF transcription, which in turn enhances spinogenesis and lowers the threshold for memory consolidation (Andero and Ressler 2012; Cover et al. 2014).

Given that FE impairments are characteristic of fear- and stressor-based disorders, there is a strong interest in researching their mechanisms. Estradiol exerts a stimulatory effect over the hypothalamic-pituitary-adrenal (HPA) axis, promotes lower activations in the threat response system, and modulates glucocorticoid receptors (GRs) and corticotropin-releasing factor (CRF) transcription (Goldstein et al. 2010; Kudielka and Kirschbaum 2005). Human studies consistently report an increased risk for PTSD in women even when controlling for confounding variables (Breslau et al. 1998). However, the notion of low estrogen is associated with a greater risk to develop PTSD has not been empirically validated in humans. Furthermore, animal models of traumatic stress do not find a clear vulnerability for females to develop a more severe posttraumatic behavioral phenotype compared to males (Keller et al. 2015; Wingo et al. 2018; Velasco et al. 2022). One study in women and mice observed that traumatic stress experienced during menstrual/estrous cycle phases with low estrogen was not associated with a specific vulnerability to develop greater posttraumatic symptoms in humans or more severe deficits in FE in rodents (Velasco et al. 2022).

Estrogen levels may not be related to a greater likelihood to develop PTSD, but they modulate FE memories in healthy, traumatized, and clinical populations. A paradigm using violent films as US found that lower estradiol levels during the EF phase in healthy women were associated with greater differential SCR during FEL and stronger intrusive memories (Wegerer et al. 2014). Similarly, healthy EF women and traumatized low-estradiol women have impaired fear potentiated startle (FPS) inhibition (Glover et al. 2013). Another study found that the detrimental effects of low estradiol could extend into clinical populations since PTSD low-estrogen women showed greater FPS at FEL and greater symptom severity compared to PTSD high-estrogen women. However, in this study, there were no statistically significant differences in FPS among non-PTSD groups (Glover et al. 2012). In contrast to these findings, another study using SCR as an outcome measure found that low estradiol levels (EF) in PTSD women were related to better FE retention (differential SCR in FER – differential SCR in early FEL) compared to ML women (high progesterone, mid estrogen; Pineles et al. 2016).

Gross comparisons between men and women with PTSD show that women have greater conditioned responses during FC whereas men have greater impairments in FER (Inslicht et al. 2013; Shvil et al. 2014). Notably, the detrimental effects of low estradiol extend to women with other fear disorders as women with spider phobia and low estradiol have slower rates of symptom improvement and greater fear in a FER test (100* (average SCR 2 last FEL/largest SCR to CS in FC); Graham et al. 2018; Li and Graham 2016). Despite these promising effects of estradiol over FE, no studies have explored the efficacy of its exogenous administration in clinical populations. A study using a putative animal model of PTSD showed that estradiol given to ovariectomized females could rescue some of the detrimental effects of the conditioned responses elicited by single prolonged stress (Mirshekar et al. 2013).

The PACAP-PAC1R (*ADCYAP1-ADCYAP1R*) system is a key point for the modulation of stress and FE memories by estradiol. A single nucleotide polymorphism (SNP; rs2267735) in the PAC1R gene located in an estrogen response element (ERE) is associated with impairments in FE, greater FPS responses, greater amygdala reactivity to threat, and greater PTSD symptoms in traumatized women but not men (Almli et al. 2013; Ressler et al. 2011; Stevens et al. 2014). Under physiological conditions, PACAP-PAC1R modulates central and neuroendocrine responses to stress and *ADCYAP1R* transcription is enhanced by estradiol (Hammack and May 2015). The SNP in the PAC1R gene compromises the binding of estradiol to the ERE sequence resulting in lower *ADCYAP1R* mRNA transcripts and greater PTSD symptoms. Hence, high estrogen levels are thought to compensate for the disbalances in *ADCYAP1R* transcription in subjects with the risk genotype, but during low estrogen states, they render vulnerable to the effects of stress (Mercer et al. 2016). A recent study showed that traumatic stress in female mice elicits impairments in FE related to an upregulation of PACAP-PAC1R gene transcripts and increased PACAP neuronal activity in the hypothalamus and amygdala. The inhibition of the medial amygdala to ventromedial hypothalamus (VMH) circuit, which is part of a medial hypothalamic defensive system, rescued the impairments in FE and the upregulation of PACAP after stress (Velasco et al. 2022). Along with other studies implicating PACAP signaling in the hippocampus and BNST, this neuropeptide system is shown to be crucial for the regulation of behavior and FE memories in face of stress (Hammack and May 2015). The sex differences in stressor- and fear-related disorders may relate to differential regulation of the PACAP-PAC1R system, especially at targets under sex hormone influence like the BNST, VMH, or hippocampus.

In sum, estradiol is a strong modulator of FE memories with important effects on their consolidation, but also promoting their encoding and retrieval. Individuals with high estrogen states show greater inhibitory control that is related to enhancements in prefrontal and hippocampal function along with facilitated LTP. In turn, low estrogen states are detrimental to FE consolidation affecting healthy and traumatized populations. The interaction of estradiol with other signaling systems may account for the greater likelihood of women experiencing adverse and long-lasting consequences after stress. Much research remains to be carried out to reconcile the divergent findings on FE between absolute estrogen levels and menstrual cycle phases. Some promising results suggest that estradiol may be beneficial for subjects in vulnerable conditions but randomized controlled trials with sufficiently powered samples are needed to answer these questions.

2.3 Progesterone and FE

Progesterone is a steroid hormone produced cyclically by the corpus luteum in females, and independently from the cycle, in the adrenal gland and brain of both sexes. It is tightly regulated by gonadotrophins and serves as a substrate for cortisol and allopregnanolone, a positive allosteric modulator of the GABA-A receptors ($GABA_ARs$; Melmed et al. 2019). Progesterone dynamics are different between humans and rodents mainly due to a distinct regulation of the corpus luteum between species (Accialini et al. 2017). The specific dynamics of progesterone and its secondary effects through its metabolites give rise to conflicting reports about the role of progesterone in memory and fear. For example, progesterone enhances the positive effects of estradiol on hippocampal spinogenesis in the short term, but after 24 h spinogenesis returns to pre-estradiol levels (Woolley and McEwen 1993). Furthermore, progesterone can promote fear responses, increase amygdala reactivity, and increase the functional coupling with the dorsal anterior cingulate cortex when facing a threat (Ferree et al. 2011; van Wingen et al. 2008). However, anxiolysis is generally reported if the analyses are focused on its metabolite allopregnanolone (Nillni et al. 2011).

In rodent FE studies, progesterone or estradiol administration during metestrus improves FER with synergistic effects observed if both hormones are given together (Milad et al. 2009a). Further, the administration of mifepristone (PRs, GRs, and androgen receptor (AR) antagonist) or fulvestrant (non-selective ERs antagonist) blocks the enhancement of FER observed in females taking FEL during proestrus (Milad et al. 2009a). Similarly, the deficits in FE observed during metestrus are rescued if mifepristone is dosed during proestrus (Graham and Daher 2016). In this same study, it was shown that estradiol and progesterone effects over FE are related to a temporal sequence of administration. Ovariectomized rats receiving estradiol had improved freezing during FER, but progesterone doses exerted biphasic effects. Progesterone given shortly before FEL (6 h) resulted in further improvements in FER, but progesterone given 24 h before FEL abolished the initial improvements obtained by estradiol (Graham and Daher 2016). Another study found that allopregnanolone was able to induce state-dependent improvements in contextual FE through actions in the BNST, and state-independent effects through actions in the basolateral amygdala (Acca et al. 2017). However, systemic Ganaxolone (synthetic allopregnanolone) did not improve between-session FE (Pinna and Rasmusson 2014). Despite these promising findings in animals, no study in humans has found any effect for progesterone over FE (Milad et al. 2010).

The interactions of progesterone with prior or actual stress exposure are relevant for fear and traumatic memories. Progesterone is secreted from the adrenals in face of stress and exerts a negative modulatory effect over the HPA axis (Crowley and Girdler 2014). Traumatic stress exposure during a high progesterone state does not facilitate the development of a more severe posttraumatic phenotype in animals nor greater overall posttraumatic symptoms in humans (Velasco et al. 2022). However, high progesterone states during trauma are associated with a greater encoding of negative memories as evidenced by enhanced retention and recollection of emotional images, greater emotional reactivity, and increased flashbacks and intrusive memories (Bryant et al. 2011; Ertman et al. 2011). Nevertheless, high flashbacks and intrusive memories during high progesterone states may be secondary to a state-dependent effect (Bryant et al. 2011).

Allopregnanolone interacts with stress and FE, being beneficial for FE consolidation only in stressed but not in control animals (Pinna and Rasmusson 2014). Notably, reduced central and peripheral allopregnanolone conversion is observed in men and women with PTSD, male war veterans, and an animal model of social isolation (Kilts et al. 2010; Pibiri et al. 2008; Pineles et al. 2018; Rasmusson et al. 2006, 2019). Interestingly, despite the deficits in allopregnanolone metabolism observed in both sexes, they may arise from different enzymatic pathways (Pineles et al. 2018; Rasmusson et al. 2019). One study in humans found that PTSD women had worse FE retention (differential SCR FER – differential SCR early FEL) during a high progesterone state (ML) as compared to PTSD women in a low progesterone/ low estradiol state (EF). However, this interaction was reversed in control women who had worse FE retention scores during EF (Pineles et al. 2016). There is still not a clear explanation for these discrepancies between PTSD and non-PTSD subjects, but a small follow-up study showed that FE deficits in PTSD women in the ML phase were associated with low resting levels of allopregnanolone (Pineles et al. 2020).

The are several mechanisms by which progesterone can influence FE. The effects of allopregnanolone over the HPA axis and memory are attributed to its capacity to modulate $GABA_A$Rs. Extrasynaptic $GABA_A$Rs are thought to maintain an inhibitory tone and are crucial in individuals exposed to stress who experience synaptic $GABA_A$Rs downregulations (Pibiri et al. 2008; Pinna et al. 2008). The state-dependent effects of progesterone in FE may be related to an enhanced encoding or recall of fear memories by the modulation of the amygdala's inhibitory tone (Bryant et al. 2011; Ertman et al. 2011). Also, progesterone can reverse some of the events triggered by estradiol by altering the phosphorylation of MAP kinases and CREB, as in the case of the biphasic effects of progesterone on hippocampal spinogenesis (Harburger et al. 2009; Murphy and Segal 2000). The interactions of progesterone with stress and FE are considerably more complex and possibly combine state-dependent effects and changes in steroid metabolism. For example, progesterone can reduce cortisol responses in acute stress tasks and promote a divergent modulation of threat responses depending on the timing of the doses (Childs et al. 2010; Frye and Walf 2008; Graham and Daher 2016). In addition, progesterone by itself does not seem to facilitate the appearance of a posttraumatic phenotype, but an increased risk may be a product of pre-existing congenital or acquired alterations in steroid dynamics. It is still unknown whether the female-specific risk of PTSD arises from a reduced regulation by estradiol of on-demand neurosteroidogenesis, as could occur in women with the risk genotype in the *ADCYAP1R1* SNP (Pineles et al. 2020; Ressler et al. 2011).

In sum, progesterone and FE are fruitful but complex fields of research. Progesterone can enhance the positive effects of estradiol on FE under constrained circumstances. Moreover, fear memories can undergo direct or indirect modulation by progesterone and its metabolites. The inter-species differences in progesterone dynamics call for caution when designing translational studies, which may benefit from studying estradiol and progesterone together. Importantly, researchers should not foresee the temporal sequence of events triggered by hormonal exposure since rapid hormonal shifts could be carrying vulnerability or protection windows, rather than absolute high or low hormonal levels. Still, much research is needed in females since many of the studies covered here are focused on the dynamics of progesterone in male rodents and the findings do not necessarily extend to females or humans.

2.4 Testosterone and FE

Testosterone is the dominating sex hormone in male organisms with a circadian pattern of secretion and anabolic and androgenic effects in both sexes. It binds to AR and triggers genomic and non-genomic effects although many of its functions are mediated through its conversion to dihydrotestosterone or estradiol in target tissues (Melmed et al. 2019). Testosterone can have anxiolytic effects in animals and humans (Aikey et al. 2002; Hermans et al. 2007), but the results vary among species and paradigms (Suarez-Jimenez et al. 2013). In FC-FE studies there are conflicting results reported for the role of testosterone in contextual (Anagnostaras et al. 1998; McDermott et al. 2012) and cued fear conditioning (Chen et al. 2014; McDermott et al. 2012). However, some studies show that testosterone is a positive modulator of FE consolidation. Males that receive a GnRH agonist before FEL, which transiently increases testosterone levels, have better FE memory consolidation (Maeng et al. 2017). Likewise, the blockade of aromatase before FEL impairs FER 24 h later and these effects are rescued with the co-administration of 17-β estradiol, suggesting that estradiol may be necessary for FE consolidation in males and females (Graham and Milad 2013).

The mechanisms by which these effects arise are still research opportunities but include effects mediated by its binding to AR or aromatization to estradiol, increases in LTP in the amygdala, changes in centro-lateral amygdala reactivity, and modulation of hippocampal function through aromatase-dependent mechanisms (Chen et al. 2014; Fester et al. 2012; Graham and Milad 2013; Maeng et al. 2017). Testosterone may play a role in female FE as well. Testosterone administration results in lower FPS during FC and lower SCR responses to negative pictures in women (Hermans et al. 2006, 2007). In addition, women with social anxiety disorder that received testosterone before an exposure therapy session had initial increases in the subjective reactivity to fear but steeper declines. Notably, this pattern was maintained in a second non-pharmacologically enhanced exposure therapy session, although these effects were only documented in women with high basal testosterone levels (Hutschemaekers et al. 2021).

Sex hormones modulate central stress responses and HPA reactivity. In turn, stressors can exert a positive or negative modulation of testosterone secretion related to the type of stressor and possibly an anticipatory challenge effect (Hermans et al. 2007; Sapolsky 2005; Van Honk et al. 2005). By studying testosterone and cortisol dynamics together, researchers found that high testosterone/cortisol ratios were associated with better FEL in men, although this improvement was only observed in morning assessments. To note, neither absolute hormonal levels nor low testosterone/cortisol ratios were associated with differences in FE (Pace-Schott et al.

2013). In a similar line, another study showed that pre-deployment hormonal profiles characterized by blunted cortisol and testosterone responses to stress were related to increased PTSD symptom severity (Josephs et al. 2017). The role of testosterone in PTSD is not clear, high testosterone levels in the blood or cerebrospinal fluid of PTSD men are reported (Karlović et al. 2012; Rahe et al. 1990), but also lower levels have been observed (Mulchahey et al. 2001). This discrepancy may result from differences in steroid regulation and divergent response trajectories within men (Gillespie et al. 2013; Mulchahey et al. 2001).

In sum, testosterone seems to positively influence FE. It can decrease conditioned responses in males and females but also improve FE memory consolidation, especially in the early phases. The mechanisms of its effects include direct actions through AR and its aromatization to estradiol. However, some studies suggest that these effects could arise from its interaction with glucocorticoids and the modulation of the stress response system. Studies using subgroup trajectories or subgroup response profiles will provide a clearer picture of the effects of testosterone on FE. There are relevant within-male differences in testosterone response profiles that can be additionally modulated by the type of stimulus or the type of threat that is being faced. Finally, there are promising positive effects of testosterone on exposure therapy over populations with anxiety or stress-related disorders that warrant further research.

2.5 *Exogenous Hormones and FE*

Exogenous hormonal therapy is used worldwide to target endocrine, metabolic, and reproductive processes with up to 80% of women who have ever used hormonal contraceptives (Daniels and Mosher 2013). Oral contraceptives differ in their composition, strength, route of administration, and effector mechanisms but the combined oral contraceptives, estrogen (ethinylestradiol) and progestin, are the most used (Melmed et al. 2019). Also, oral contraceptives act by blocking ovulation through hypothalamic mechanisms, although other formulations based on progestin-only or other progestins have different target effects (Pletzer and Kerschbaum 2014). Despite well-known health risks and benefits for metabolic, neoplastic, and gynecological processes, the effects of oral contraceptives on the brain are not well-defined despite they may carry important risks (Gingnell et al. 2013; Skovlund et al. 2016, 2018). Oral contraceptives intake induces structural and functional changes in the brain as evidenced by resting-state functional connectivity studies of the amygdala, hippocampus, PFC, and hypothalamus (Chen et al. 2021; Lisofsky et al. 2016). Notably, oral contraceptives users have lower amygdala reactivity and greater dorsal anterior cingulate cortex and insular reactivity to traumatic stimuli when compared to normally cycling women (Petersen et al. 2015; Petersen and Cahill 2015). Further effects on cognition, emotional reactivity, social reward, and socio-emotional behaviors are reported for oral contraceptives users (Lewis et al. 2019; Montoya and Bos 2017). However, some contradictory findings point out that the effects of oral contraceptives on brain function depend on their subtype and divergent androgenic profile (Gingnell et al. 2013). In general, oral contraceptives intake is related to deficits in emotional processing with enhanced emotional reactivity and impaired emotion recognition. Of note, these negative effects of oral contraceptives may pertain mostly to women with a history of previous side effects or psychiatric comorbidity (Gingnell et al. 2013).

In FE literature, oral contraceptives and low-estrogen states are detrimental to FE. Heterogeneous etiologies for low-estrogen states are often combined to approach FE questions. Therefore, low estrogen groups can be constituted by women with estrogen levels below an arbirtrary sample-dependent threshold, by women completing the study in a specific phase of the menstrual cycle or by women taking oral contraceptives. It is relevant to note that despite these etiologies often being considered as a whole, they are not equivalent and have important functional differences that will be further addressed. Translational evidence shows that oral contraceptives intake can impair FER. Rats chronically treated with levonorgestrel, or women taking oral contraceptives, have impairments in FER as evidenced by greater freezing scores or greater fear recovery (100*(average SCR in FER/largest SCR to CS+ in FC)) and (100*(average SCR to four first FER/largest SCR to CS+ in FC); Graham and Milad 2013; White and Graham 2016). These impairments in FER are similar to those observed in metestrus rats and women with low estrogens (median sample split), although women taking oral contraceptives seem to have greater impairments (Graham and Milad 2013; Milad et al. 2009a, 2010).

A series of experiments have demonstrated that the FE deficits secondary to low-estrogen states can be rescued if estradiol signaling is reestablished during FEL. In detail, levonorgestrel-induced impairments in FER in rats were rescued by the termination of hormonal therapy before FEL, or by an ERα or ERβ agonist (Graham and Milad 2013). In metestrus rats, an ERβ agonist, but not an ERα agonist, before FEL was able to decrease freezing levels during FER (Zeidan et al. 2011). Similarly, estradiol given to oral women taking contraceptives and women in the EF phase can decrease FER as evidenced by lower fear recovery (100*(average SCR in FER/largest SCR to CS+ in FC)) and recovery of fear (first four trials of CS + U – CS + E); Graham and Milad 2013; Wen et al. 2021). Therefore, the general picture shows that low-estrogen states are detrimental to FE processes, but they can be rescued by increasing estradiol signaling during FEL.

According to functional studies, low-estrogen states converge on the improvement of FE by estradiol but use different neuronal substrates. A study that found no differences in SCR during FEL showed that oral contraceptives users had greater activations in the amygdala, thalamus, dorsal anterior cingulate cortex, and ventromedial prefrontal cortex when compared to ML women (Merz et al. 2012a). However, these differences are attenuated as control groups and methodologies differ between studies (Hwang et al. 2015). Another study showed that the positive effect of exogenous estradiol over FE was related to strong activations in the ventromedial prefrontal cortex of EF women during FEL and FER. In contrast, the ventromedial prefrontal cortex of women taking oral contraceptives was only activated during late FEL and they showed additional activations in the cuneus and dorsal anterior cingulate cortex during early FEL and FER, respectively (Wen et al. 2021). These differential activation patterns were possibly related to the chronic effects of oral contraceptives administration on the brain but could also be related to the greater increases in blood estradiol observed in women taking oral contraceptives after exogenous estradiol doses. In addition, researchers observed a specific resting-state functional connectivity network component that was strongly correlated with estradiol levels during a post-FEL period, when FE memory consolidation is thought to occur (Wen et al. 2021). In the clinical extension of FE, oral contraceptives intake can be detrimental to exposure therapy in spider-phobic women by increasing behavioral avoidance and subjective fear levels compared to non-oral contraceptives users (Graham et al. 2018; Raeder et al. 2019).

The interactions of stress and oral contraceptives intake over FE are not well studied. Oral contraceptives intake alters cortisol responses to stress, with estradiol-containing formulations blunting cortisol responses but levonorgestrel-only formulations potentiating neuroendocrine responses (Aleknaviciute et al. 2017; Kirschbaum et al. 1999; Kuhlmann and Wolf 2005; Merz 2017). Moreover, the increased glucocorticoid function in oral contraceptives users is associated with lower hippocampal volumes, analogous to the effects of chronic psychogenic stressors (Hertel et al. 2017; McEwen 1998). Cortisol doses before FC result in null effects for SCR but exert effects on the hippocampus, increasing its activation in oral contraceptives users and decreasing it in men and ML women (Merz et al. 2012a, b). With all the detrimental effects of oral contraceptives over fear and stress processing, it would be reasonable to suspect that oral contraceptives intake would be related to an increased vulnerability to psychopathology after exposure to stressors. Nevertheless, no studies to date have described a risk for the appearance or increased severity of PTSD in chronic oral contraceptives users. In turn, a study exploring the use of emergency contraception shortly after sexual abuse showed that women taking a combination of estrogen and progesterone were protected compared to naturally cycling women that declined any hormonal treatment. Notably, researchers in this study found no statistically significant differences in symptom severity scores between women taking oral contraceptives before sexual abuse and women taking emergency contraception after sexual assault. This suggests that exogenous hormones may be associated with better outcomes after trauma (Ferree et al. 2012).

The mechanisms by which oral contraceptives impacts FE are possibly related to secondary decreases in estrogen levels. For example, studies have found that sex hormone suppression is associated with lower allopregnanolone and decreased GABAergic inhibitory tone (Andréen et al. 2009). FE memory is strongly related to hippocampal function and chronic oral contraceptives intake may alter the dynamics of hippocampal spine density. In ovariectomized animals, blunted sex hormones are paralleled with declines in cholinergic function, impairments in memory tasks, and a decrease in spine density in the hippocampus (Albert and Newhouse 2019; Woolley and McEwen 1993). Estrogen can rescue spine density declines in the hippocampus, but the cholinergic function is only restored in animals with previously intact cholinergic systems (Woolley and McEwen 1993). However, no studies have explored the effects of hormonal contraception in the cholinergic system or the hippocampus at such a microscopic level.

Low estrogen states converge on the impairments in FER but with associated functional differences. For example, the improvements of FER by an ERα or ERβ agonist in animals with chronic hormonal treatment are not mirrored in low estrogen cycling rats that only benefit from an ERβ agonist (Graham and Milad 2013; Zeidan et al. 2011). Furthermore, oral contraceptives users show differences in basal and task-related neural and neurophysiological responses that are not described in cycling women with low estrogens. Oral contraceptives women have greater basal and unconditioned SCR, lower insular activations to US, blunted SCR and FPS during FC, and altered neuronal activation during FC (Armbruster et al. 2017; Hwang et al. 2015; Merz et al. 2012a, b, 2013). Together, these date evidence the widespread impact of oral contraceptives intake over emotional processing. A theoretical model of emotion appraisal and regulation recapitulates how estradiol may be able to influence brain function by shifting the balance between the dorsal and ventral systems. The dorsal system includes lateral parts of the PFC and promotes cognitive processes, selective attention, and the voluntary regulation of emotional states. In turn, the ventral system includes the amygdala, insula, ventral striatum, orbitofrontal cortex, and ventromedial prefrontal cortex and is related to the identification of emotionally salient stimuli and the generation of automatic responses and emotional states (Phillips et al. 2003, 2008). High estradiol states would therefore promote dorsal system activity, whereas low estradiol states would shift the balance toward ventral system function (Albert and Newhouse 2019).

In sum, the reviewed data suggest that oral contraceptives intake causes structural and functional changes that can lead to impairments in FER. The impairments in FER observed in low estrogen states can be rescued by restoring estrogen signaling before FEL. Furthermore, the effects of oral contraceptives over FE are not exclusive to fear processing but are included in the context of an overall increased emotional reactivity and impaired emotional processing. Theoretical models suggest that estradiol promotes the activity of top-down regulating structures, whereas low estrogen states facilitate the function of automatic emotion recognition systems. The study of the detrimental effects of oral contraceptives over FE requires the design of studies that dissect the specific conditions and underlying mechanisms. Different types of oral contraceptives are often combined in study groups despite their diverging androgenic profiles and hormonal concentrations. Moreover, the individual responses to oral contraceptives intake are still understudied and may hold a clue about how some individuals are at greater risk for the detrimental effects of oral contraceptives intake. Notably, oral contraceptives use is growing among young populations, but it is still unknown whether early use is associated with arrests in brain maturation that are further evidenced in adulthood.

3 Areas of Future Research

Future studies still have much to add to our understanding of the sex differences in FE. Large hormonal shifts in sensitive periods of women's life may represent windows of vulnerability or protection (Maeng and Milad 2015; Rehbein et al. 2021). The female-biased prevalence of stressor- and fear-based disorders emerges around adolescence, which is an important developmental period with hormonal shifts and a generally impaired FE (Baker et al. 2014; Patton et al. 1996). Notably, more studies are needed to determine the time and mechanisms by which sex differences and the gonadal modulation of FE memory appear, given the contradictory findings of estradiol effects over FE on adolescents and adults (Perry et al. 2020). Moreover, the influence of estradiol over FE can change during a lifetime, as the positive relationship between estradiol and FE is only observed in nulliparous females but not in women or rodents with reproductive experience (Milligan-Saville and Graham 2016; Tang and Graham 2020). Thus, pregnancy can induce permanent brain changes that mitigate the effects of estrogen over FE, but the mechanisms of this effect remain unknown.

To date, cross-sectional studies haven't been able to distinguish if the specific vulnerability for FER impairments arises in all women during low-estrogen phases of the menstrual cycle or if the vulnerability pertains mostly to a subset of women experiencing chronic low-estrogen states or large hormonal shifts. Evidence from rodent studies supports the role of the specific phases of the estrous cycle but these findings have not been replicated in humans. Notably, there are large inter-individual differences in the levels of sex hormones during equivalent phases of the menstrual cycle (Sundström-Poromaa and Gingnell 2014) (Fig. 1). Longitudinal studies are needed to answer whether women experiencing large hormonal shifts during the menstrual cycle are at risk for impaired FE. The inclusion of subgroup trajectories into FE analyses may help to answer these questions, improve the translation of findings, and enrich our understanding of individual features not evidenced in group analyses (Duits et al. 2021; Galatzer-Levy et al. 2017; Pöhlchen et al. 2020). There are reports of specific trajectory behavioral phenotypes emerging during FC and FE that are more likely associated with specific sex (Gruene et al. 2015; Leen et al. 2021). These trajectory phenotypes found during FE could be compared to those observed in traumatized populations, which have shown that sex along with other individual or environmental factors are crucial factors leveraging individuals toward resilience or a chronic disease course (Galatzer-Levy et al. 2013, 2018; Orcutt et al. 2004).

There are studies reporting sex differences in FC but no studies link them to the findings in FE (Day and Stevenson 2020; Merz et al. 2018). Despite most of the studies covered reported null effects for estradiol or menstrual cycle on the neurophysiological measures during FC, there are large underlying differences in the patterns of brain activation. These patterns of brain activity during FC may be able to predict FE success, as the structures involved in memory encoding are more likely to be recruited during subsequent trials of memory expression (Phillips et al. 2003, 2008). An example of this effect is given by the large differences in brain activation patterns observed during FC and FEL in women taking oral contraceptives that only manifest at the behavioral level until FER assessments. Finally, there are plenty of opportunities to improve the quality of the evidence from FE studies on sex differences. The inclusion of absolute and relative hormonal measures, or the computation of progesterone/estradiol ratios, can minimize the bias introduced by sample-split approaches (Seligowski et al. 2020). Future studies with larger sample sizes and enough statistical power will allow us to delineate the specific effects of each phase of the menstrual cycle over FE. In consequence, studies approaching these questions with inadequate sample sizes and an uneven distribution of individuals at different menstrual cycle phases among experimental groups will complicate our capacity to compare findings between studies.

4 Conclusions

Gonadal hormones exert important effects on FE memory (Fig. 2). However, these effects do not seem to be exclusive to fear learning processes, but rather a product of widespread influences of gonadal hormones over emotional processing. Periods with low estrogen levels in females are detrimental to FE memory consolidation. Still, the discrepancy in the findings of the menstrual/estrous cycle effects in animal and human studies has not allowed for the elucidation of the specific temporal window of vulnerability for FE impairments in women. Longitudinal studies that consider inter-individual differences in their design will be able to fill these gaps in the literature. In light of personalized medicine, there is a promising use for drug targeting processes related to sex hormones to enhance exposure therapy outcomes but the specific details about dose and target populations remain to be determined. Lessons can be learned from other sex hormone-targeting drugs that have paved their way into the clinics, as in the case of allopregnanolone indicated for postpartum depression ("Brexanolone (Zulresso) for Postpartum Depression" 2019). Many FE studies are taking advantage of a translational approach to elucidate the mechanisms by which stressor- and fear-based disorders appear and sustain. Nevertheless, the field urges a standardization of the methods used for quantification and analyses given that the translation of animal and preclinical findings into the clinics remains a challenge.

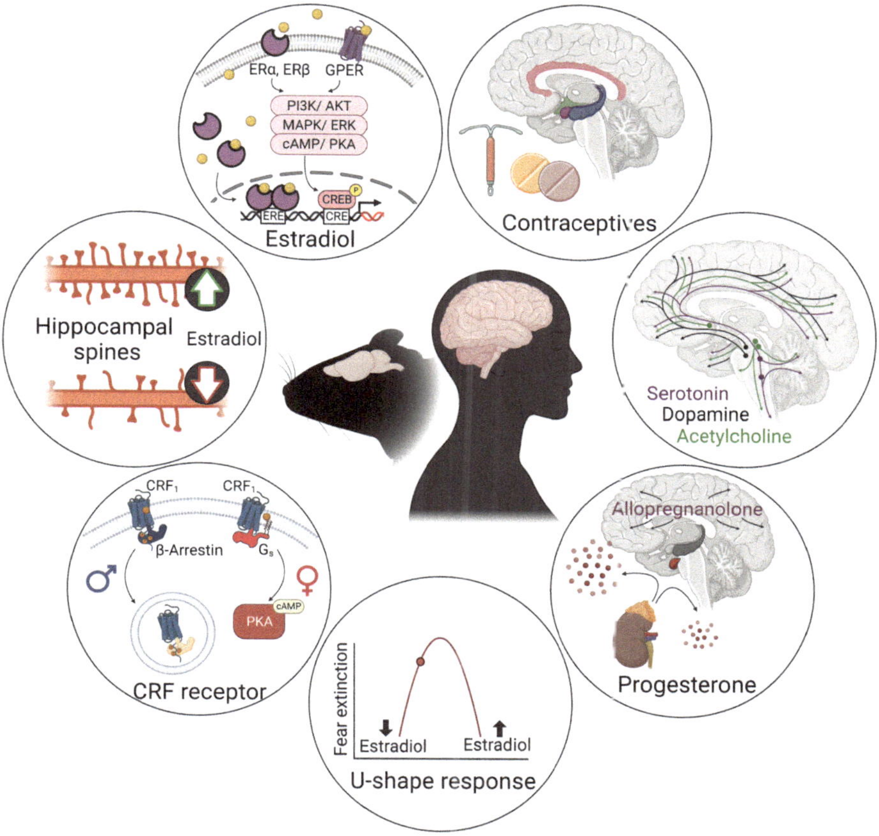

Fig. 2 Summary depicting some mechanisms implicated in the sex differences and sex hormone regulation of FE. *cAMP* cyclic AMP, *cAMP/PKA* cyclic AMP/protein kinase A pathway, *CREB* cyclic AMP response element-binding protein, CRF_1 corticotropin-releasing hormone receptor 1, *ERE* estrogen response element, *ERα* estrogen receptor α, *ERβ* estrogen receptor β, *GPER* G protein-coupled estrogen receptor 1, G_s G_s alpha subunit of heterotrimeric G protein receptor, *MAPK/ERK* mitogen-activated protein kinases, extracellular signal-regulated kinases pathway, *PI3K/AKT* phosphoinositide 3-kinase/protein kinase B pathway, *PKA* protein kinase A

References

Acca GM, Mathew AS, Jin J, Maren S, Nagaya N (2017) Allopregnanolone induces state-dependent fear via the bed nucleus of the stria terminalis. Horm Behav 89:137–144

Accialini P, Hernandez SF, Abramovich D, Tesone M (2017) The rodent corpus luteum. In: Life cycle corpus luteum, pp 117–131

Aikey JL, Nyby JG, Anmuth DM, James PJ (2002) Testosterone rapidly reduces anxiety in male house mice (Mus musculus). Horm Behav 42:448–460

Albert KM, Newhouse PA (2019) Estrogen, stress, and depression: cognitive and biological. Interactions 15:399–423. https://doi.org/10.1146/annurev-clinpsy-050718-095557

Aleknaviciute J, Tulen JHM, De Rijke YB, Bouwkamp CG, van der Kroeg M, Timmermans M, Wester VL, Bergink V, Hoogendijk WJG, Tiemeier H, van Rossum EFC, Kooiman CG, Kushner SA (2017) The levonorgestrel-releasing intrauterine device potentiates stress reactivity. Psychoneuroendocrinology 80:39–45

Almli LM, Mercer KB, Kerley K, Feng H, Bradley B, Conneely KN, Ressler KJ (2013) ADCYAP1R1 genotype associates with post-traumatic stress symptoms in highly traumatized African-American females. Am J Med Genet B Neuropsychiatr Genet 162B:262–272

Anagnostaras SG, Maren S, Decola JP, Lane NI, Gale GD, Schlinger BA, Fanselow MS (1998) Testicular hormones do not regulate sexually dimorphic Pavlovian fear conditioning or perforant-path long-term potentiation in adult male rats. Behav Brain Res 92:1–9

Andero R, Ressler KJ (2012) Fear extinction and BDNF: translating animal models of PTSD to the clinic. Genes Brain Behav 11:503–512

Andréen L, Nyberg S, Turkmen S, van Wingen G, Fernández G, Bäckström T (2009) Sex steroid induced negative mood may be explained by the paradoxical effect mediated by GABAA modulators. Psychoneuroendocrinology 34:1121–1132

Antov MI, Stockhorst U (2014) Stress exposure prior to fear acquisition interacts with estradiol status to alter recall of fear extinction in humans. Psychoneuroendocrinology 49:106–118

Armbruster D, Kirschbaum C, Strobel A (2017) The not-so-bitter pill: effects of combined oral contraceptives on peripheral physiological indicators of emotional reactivity. Horm Behav 94: 97–105

Baker KD, Den ML, Graham BM, Richardson R (2014) A window of vulnerability: impaired fear extinction in adolescence. Neurobiol Learn Mem 113:90–100

Bangasser DA, Cuarenta A (2021) Sex differences in anxiety and depression: circuits and mechanisms. Nat Rev Neurosci 22:674–684

Bangasser DA, Valentino RJ (2014) Sex differences in stress-related psychiatric disorders: neurobiological perspectives. Front Neuroendocrinol 35:303–319

Bangasser DA, Curtis A, Reyes BAS, Bethea TT, Parastatidis I, Ischiropoulos H, Van Bockstaele EJ, Valentino RJ (2010) Sex differences in corticotropin-releasing factor receptor signaling and trafficking: potential role in female vulnerability to stress-related psychopathology. Mol Psychiatry 15(877):896–904

Barha CK, Dalton GL, Galea LAM (2010) Low doses of 17alpha-estradiol and 17beta-estradiol facilitate, whereas higher doses of estrone and 17alpha- and 17beta-estradiol impair, contextual fear conditioning in adult female rats. Neuropsychopharmacology 35:547–559

Blume SR, Freedberg M, Vantrease JE, Chan R, Padival M, Record MJ, De Joseph MR, Urban JH, Rosenkranz JA (2017) Sex- and estrus-dependent differences in rat basolateral amygdala. J Neurosci 37:10567–10586

Breslau N, Kessler RC, Chilcoat HD, Schultz LR, Davis GC, Andreski P (1998) Trauma and posttraumatic stress disorder in the community: the 1996 Detroit Area Survey of Trauma. Arch Gen Psychiatry 55:626–632

Brexanolone (Zulresso) for Postpartum Depression (2019) J Am Med Assoc

Bryant RA, Felmingham KL, Silove D, Creamer M, O'Donnell M, McFarlane AC (2011) The association between menstrual cycle and traumatic memories. J Affect Disord 131:398–401

Cersosimo MG, Benarroch EE (2015) Estrogen actions in the nervous system. Neurology 85:263–273

Chang Y-J, Yang C-H, Liang Y-C, Yeh C-M, Huang C-C, Hsu K-S (2009) Estrogen modulates sexually dimorphic contextual fear extinction in rats through estrogen receptor β. Hippocampus 19:1142–1150

Chen X, Garelick MG, Wang H, Li V, Athos J, Storm DR (2005) PI3 kinase signaling is required for retrieval and extinction of contextual memory. Nat Neurosci 8:925–931

Chen J-R, Yan Y-T, Wang T-J, Chen L-J, Wang Y-J, Tseng G-F (2009) Gonadal hormones modulate the dendritic spine densities of primary cortical pyramidal neurons in adult female rat. Cereb Cortex 19:2719–2727

Chen LS, Tzeng WY, Chuang JY, Cherng CG, Gean PW, Yu L (2014) Roles of testosterone and amygdaloid LTP induction in determining sex differences in fear memory magnitude. Horm Behav 66:498–508

Chen KX, Worley S, Foster H, Edasery D, Roknsharifi S, Ifrah C, Lipton ML (2021) Oral contraceptive use is associated with smaller hypothalamic and pituitary gland volumes in healthy women: A structural MRI study. PLoS One 16

Childs E, Van Dam NT, de Wit H (2010) Effects of acute progesterone administration upon responses to acute psychosocial stress in men. Exp Clin Psychopharmacol 18:78–86

Cover KK, Maeng LY, Lebrón-Milad K, Milad MR (2014) Mechanisms of estradiol in fear circuitry: implications for sex differences in psychopathology. Transl Psychiatry

Crowley SK, Girdler SS (2014) Neurosteroid, GABAergic and hypothalamic pituitary adrenal (HPA) axis regulation: what is the current state of knowledge in humans? Psychopharmacology 231:3619–3634

Daniels K, Mosher WD (2013) Contraceptive methods women have ever used: United States, 1982–2010. Natl Health Stat Rep:1–15

Day HLL, Stevenson CW (2020) The neurobiological basis of sex differences in learned fear and its inhibition. Eur J Neurosci 52:2466–2486

de Vries GJ (2008) Sex differences in vasopressin and oxytocin innervation of the brain. Prog Brain Res 170:17–27

Deslauriers J, Toth M, Der-Avakian A, Risbrough VB (2018) Current status of animal models of posttraumatic stress disorder: behavioral and biological phenotypes, and future challenges in improving translation. Biol Psychiatry 83:895–907

Duits P, Baas JMP, Engelhard IM, Richter J, Huisman-van Dijk HM, Limberg-Thiesen A, Heitland I, Hamm AO, Cath DC (2021) Latent class growth analyses reveal overrepresentation of dysfunctional fear conditioning trajectories in patients with anxiety-related disorders compared to controls. J Anxiety Disord 78:102361

Ertman N, Andreano JM, Cahill L (2011) Progesterone at encoding predicts subsequent emotional memory. Learn Mem 18:759

Ferree NK, Kamat R, Cahill L (2011) Influences of menstrual cycle position and sex hormone levels on spontaneous intrusive recollections following emotional stimuli. Conscious Cogn 20:1154–1162

Ferree NK, Wheeler M, Cahill L (2012) The influence of emergency contraception on post-traumatic stress symptoms following sexual assault. J Forensic Nurs 8:122–130

Fester L, Prange-Kiel J, Zhou L, Blittersdorf BV, Böhm J, Jarry H, Schumacher M, Rune GM (2012) Estrogen-regulated synaptogenesis in the hippocampus: sexual dimorphism in vivo but not in vitro. J Steroid Biochem Mol Biol 131:24–29

Florido A, Velasco ER, Soto-Faguás CM, Gomez-Gomez A, Perez-Caballero L, Molina P, Nadal R, Pozo OJ, Saura CA, Andero R (2021) Sex differences in fear memory consolidation via Tac2 signaling in mice. Nat Commun 12:1–19

Frye CA, Walf AA (2008) Progesterone to ovariectomized mice enhances cognitive performance in the spontaneous alternation, object recognition, but not placement, water maze, and contextual and cued conditioned fear tasks. Neurobiol Learn Mem 90:171

Galatzer-Levy IR, Ankri Y, Freedman S, Israeli-Shalev Y, Roitman P, Gilad M, Shalev AY (2013) Early PTSD symptom trajectories: persistence, recovery, and response to treatment: results from the Jerusalem Trauma Outreach and Prevention Study (J-TOPS). PLoS One 8

Galatzer-Levy IR, Andero R, Sawamura T, Jovanovic T, Papini S, Ressler KJ, Norrholm SD (2017) A cross species study of heterogeneity in fear extinction learning in relation to FKBP5 variation and expression: Implications for the acute treatment of posttraumatic stress disorder. Neuropharmacology 116:188–195

Galatzer-Levy IR, Huang SH, Bonanno GA (2018) Trajectories of resilience and dysfunction following potential trauma: a review and statistical evaluation. Clin Psychol Rev

Galvin C, Ninan I (2014) Regulation of the mouse medial prefrontal cortical synapses by endogenous estradiol. Neuropsychopharmacology 39:2086–2094

Gillespie CF, Almli LM, Smith AK, Bradley B, Kerley K, Crain DF, Mercer KB, Weiss T, Phifer J, Tang Y, Cubells JF, Binder EB, Conneely KN, Ressler KJ (2013) Sex dependent influence of a functional polymorphism in steroid 5-α-reductase type 2 (SRD5A2) on post-traumatic stress symptoms. Am J Med Genet B Neuropsychiatr Genet 0:283

Gingnell M, Engman J, Frick A, Moby L, Wikström J, Fredrikson M, Sundström-Poromaa I (2013) Oral contraceptive use changes brain activity and mood in women with previous negative affect on the pill – a double-blinded, placebo-controlled randomized trial of a levonorgestrel-containing combined oral contraceptive. Psychoneuroendocrinology 38:1133–1144

Glover EM, Jovanovic T, Mercer KB, Kerley K, Bradley B, Ressler KJ, Norrholm SD (2012) Estrogen levels are associated with extinction deficits in women with posttraumatic stress disorder. Biol Psychiatry 72:19–24

Glover EM, Mercer KB, Norrholm SD, Davis M, Duncan E, Bradley B, Ressler KJ, Jovanovic T (2013) Inhibition of fear is differentially associated with cycling estrogen levels in women. J Psychiatry Neurosci 38:341

Goldstein JM, Jerram M, Poldrack R, Ahern T, Kennedy DN, Seidman LJ, Makris N (2005) Hormonal cycle modulates arousal circuitry in women using functional magnetic resonance imaging. J Neurosci 25:9309–9316

Goldstein JM, Jerram M, Abbs B, Whitfield-Gabrieli S, Makris N (2010) Sex differences in stress response circuitry activation dependent on female hormonal cycle. J Neurosci 30:431–438

Graham BM, Daher M (2016) Estradiol and progesterone have opposing roles in the regulation of fear extinction in female rats. Neuropsychopharmacology 41:774

Graham BM, Milad MR (2013) Blockade of estrogen by hormonal contraceptives impairs fear extinction in female rats and women. Biol Psychiatry 73:371–378

Graham BM, Scott E (2018) Effects of systemic estradiol on fear extinction in female rats are dependent on interactions between dose, estrous phase, and endogenous estradiol levels. Horm Behav 97:67–74

Graham BM, Li SH, Black MJ, Öst LG (2018) The association between estradiol levels, hormonal contraceptive use, and responsiveness to one-session-treatment for spider phobia in women. Psychoneuroendocrinology 90:134–140

Gruene TM, Roberts E, Thomas V, Ronzio A, Shansky RM (2015) Sex-specific neuroanatomical correlates of fear expression in prefrontal-amygdala circuits. Biol Psychiatry 78:186–193

Hammack SE, May V (2015) Pituitary adenylate cyclase activating polypeptide in stress-related disorders: data convergence from animal and human studies. Biol Psychiatry 78:167

Harburger LL, Saadi A, Frick KM (2009) Dose-dependent effects of post-training estradiol plus progesterone treatment on object memory consolidation and hippocampal extracellular signal-regulated kinase activation in young ovariectomized mice. Neuroscience 160:6–12

Hasegawa Y, Hojo Y, Kojima H, Ikeda M, Hotta K, Sato R, Ooishi Y, Yoshiya M, Chung BC, Yamazaki T, Kawato S (2015) Estradiol rapidly modulates synaptic plasticity of hippocampal neurons: involvement of kinase networks. Brain Res 1621:147–161

Hermans EJ, Putman P, Baas JM, Koppeschaar HP, van Honk J (2006) A single administration of testosterone reduces fear-potentiated startle in humans. Biol Psychiatry 59:872–874

Hermans EJ, Putman P, Baas JM, Gecks NM, Kenemans JL, van Honk J (2007) Exogenous testosterone attenuates the integrated central stress response in healthy young women. Psychoneuroendocrinology 32:1052–1061

Herry C, Trifilieff P, Micheau J, Lüthi A, Mons N (2006) Extinction of auditory fear conditioning requires MAPK/ERK activation in the basolateral amygdala. Eur J Neurosci 24:261–269

Hertel J, König J, Homuth G, Van Der Auwera S, Wittfeld K, Pietzner M, Kacprowski T, Pfeiffer L, Kretschmer A, Waldenberger M, Kastenmüller G, Artati A, Suhre K, Adamski J, Langner S, Völker U, Völzke H, Nauck M, Friedrich N, Grabe HJ (2017) Evidence for stress-like alterations in the HPA-axis in women taking oral contraceptives. Sci Rep 7

Hoefer M, Allison SC, Schauer GF, Neuhaus JM, Hall J, Dang JN, Weiner MW, Miller BL, Rosen HJ (2008) Fear conditioning in frontotemporal lobar degeneration and Alzheimer's disease. Brain 131:1646–1657

Holt DJ, Lebron-Milad K, Milad MR, Rauch SL, Pitman RK, Orr SP, Cassidy BS, Walsh JP, Goff DC (2009) Extinction memory is impaired in schizophrenia. Biol Psychiatry 65:455

Hugues S, Chessel A, Lena I, Marsault R, Garcia R (2006) Prefrontal infusion of PD098059 immediately after fear extinction training blocks extinction-associated prefrontal synaptic plasticity and decreases prefrontal ERK2 phosphorylation. Synapse 60 280–287

Hutschemaekers MHM, de Kleine RA, Hendriks GJ, Kampman M, Roelofs K (2021) The enhancing effects of testosterone in exposure treatment for social anxiety disorder: a randomized proof-of-concept trial. Transl Psychiatry 11:1–7

Hwang MJ, Zsido RG, Song H, Pace-Schott EF, Miller KK, Lebron-Milad K, Marin M-F, Milad MR (2015) Contribution of estradiol levels and hormonal contraceptives to sex differences within the fear network during fear conditioning and extinction. BMC Psychiatry 15:295

Inslicht SS, Metzler TJ, Garcia NM, Pineles SL, Milad MR, Orr SP, Marmar CR, Neylan TC (2013) Sex differences in fear conditioning in posttraumatic stress disorder. J Psychiatr Res 47:64–71

Josephs RA, Cobb AR, Lancaster CL, Lee HJ, Telch MJ (2017) Dual-hormone stress reactivity predicts downstream war-zone stress-evoked PTSD. Psychoneuroendocrinology 78:76–84

Jovanovic T, Norrholm SD, Blanding NQ, Davis M, Duncan E, Bradley B, Ressler KJ (2010) Impaired fear inhibition is a biomarker of PTSD but not depression. Depress Anxiety 27:244–251

Kanwal JK, Coddington E, Frazer R, Limbania D, Turner G, Davila KJ, Givens MA, Williams V, Datta SR, Wasserman S (2021) Internal state: dynamic, interconnected communication loops distributedacross body, brain, and time. Integr Comp Biol 61:867

Karlović D, Serretti A, Marčinko D, Martinac M, Silić A, Katinić K (2012) Serum testosterone concentration in combat-related chronic posttraumatic stress disorder. Neuropsychobiology 65: 90–95

Keller SM, Schreiber WB, Staib JM, Knox D (2015) Sex differences in the single prolonged stress model. Behav Brain Res 286:29–32

Kilts JD, Tupler LA, Keefe FJ, Payne VM, Hamer RM, Naylor JC, Calnaido RP, Morey RA, Strauss JL, Parke G, Massing MW, Youssef NA, Shampine LJ, Marx CE (2010) Neurosteroids and self-reported pain in veterans who served in the U.S. Military after September 11, 2001. Pain Med 11:1469

Kirschbaum C, Kudielka BM, Gaab J, Schommer NC, Hellhammer DH (1999) Impact of gender, menstrual cycle phase, and oral contraceptives on the activity of the hypothalamus-pituitary-adrenal axis. Psychosom Med 61:154–162

Kudielka BM, Kirschbaum C (2005) Sex differences in HPA axis responses to stress: a review. Biol Psychol 69:113–132

Kuhlmann S, Wolf OT (2005) Cortisol and memory retrieval in women: influence of menstrual cycle and oral contraceptives. Psychopharmacology 183:65–71

Lebron-Milad K, Milad MR (2012) Sex differences, gonadal hormones and the fear extinction network: implications for anxiety disorders. Biol Mood Anxiety Disord 2:3

Lebron-Milad K, Abbs B, Milad MR, Linnman C, Rougemount-Bücking A, Zeidan MA, Holt DJ, Goldstein JM (2012) Sex differences in the neurobiology of fear conditioning and extinction: a preliminary fMRI study of shared sex differences with stress-arousal circuitry. Biol Mood Anxiety Disord 2

Leen NA, Duits P, Baas JMP (2021) Trajectories of fear learning in healthy participants are able to distinguish groups that differ in individual characteristics, chronicity of fear and intrusions. J Behav Ther Exp Psychiatry 72:101653

Leuner B, Mendolia-Loffredo S, Shors TJ (2004) High levels of estrogen enhance associative memory formation in ovariectomized females. Psychoneuroendocrinology 29:883–890

Lewis SJ, Arseneault L, Caspi A, Fisher HL, Matthews T, Moffitt TE, Odgers CL, Stahl D, Teng JY, Danese A (2019) The epidemiology of trauma and post-traumatic stress disorder in a representative cohort of young people in England and Wales. Lancet Psychiatry 6:247–256

Li S, Graham BM (2016) Estradiol is associated with altered cognitive and physiological responses during fear conditioning and extinction in healthy and spider phobic women. Behav Neurosci 130:614–623

Lisofsky N, Riediger M, Gallinat J, Lindenberger U, Kühn S (2016) Hormonal contraceptive use is associated with neural and affective changes in healthy young women. NeuroImage 134:597–606

Lonsdorf TB, Kalisch R (2011) A review on experimental and clinical genetic associations studies on fear conditioning, extinction and cognitive-behavioral treatment. Transl Psychiatry 1

Lu KT, Walker DL, Davis M (2001) Mitogen-activated protein kinase cascade in the basolateral nucleus of amygdala is involved in extinction of fear-potentiated startle. J Neurosci 21

Maeng LY, Milad MR (2015) Sex differences in anxiety disorders: interactions between fear, stress, and gonadal hormones. Horm Behav 76:106

Maeng LY, Taha MB, Cover KK, Glynn SS, Murillo M, Lebron-Milad K, Milad MR (2017) Acute gonadotropin-releasing hormone agonist treatment enhances extinction memory in male rats. Psychoneuroendocrinology 82:164–172

McCarthy MM (2008) Estradiol and the developing brain. Physiol Rev 88:91

McCaughran JA, Bell J, Hitzemann RJ (2000) Fear-potentiated startle response in mice: genetic analysis of the C57BL/6J and DBA/2J intercross. Pharmacol Biochem Behav 65:301–312

McDermott CM, Liu D, Schrader LA (2012) Role of gonadal hormones in anxiety and fear memory formation and inhibition in male mice. Physiol Behav 105:1168–1174

McDermott CM, Liu D, Ade C, Schrader LA (2015) Estradiol replacement enhances fear memory formation, impairs extinction and reduces COMT expression levels in the hippocampus of ovariectomized female mice. Neurobiol Learn Mem 118:167–177

McEwen BS (1998) Stress, adaptation, and disease allostasis and allostatic load. Ann N Y Acad Sci 840:33–44

McEwen BS (1999) Clinical review 108: the molecular and neuroanatomical basis for estrogen effects in the central nervous system. J Clin Endocrinol Metab

McEwen B (2002) Estrogen actions throughout the brain. Recent Prog Horm Res 57:357–384

McEwen BS, Akama KT, Spencer-Segal JL, Milner TA, Waters EM (2012) Estrogen effects on the brain: actions beyond the hypothalamus via novel mechanisms. Behav Neurosci 126:4

Melmed S, Auchus RJ, Goldfine AB, Koenig R, Rosen CJ, Williams RH (2019) Williams textbook of endocrinology, 14th edn. Elsevier

Mendoza-Garcés L, Mendoza-Rodríguez CA, Jiménez-Trejo F, Picazo O, Rodríguez MC, Cerbón M (2011) Differential expression of estrogen receptors in two hippocampal regions during the estrous cycle of the rat. Anat Rec (Hoboken) 294:1913–1919

Mercer KB, Dias B, Shafer D, Maddox SA, Mulle JG, Hu P, Walton J, Ressler KJ (2016) Functional evaluation of a PTSD-associated genetic variant: estradiol regulation and ADCYAP1R1. Transl Psychiatry 6:e978–e978

Merz CJ (2017) Contribution of stress and sex hormones to memory encoding. Psychoneuroendocrinology 82:51–58

Merz CJ, Tabbert K, Schweckendiek J, Klucken T, Vaitl D, Stark R, Wolf OT (2012a) Neuronal correlates of extinction learning are modulated by sex hormones. Soc Cogn Affect Neurosci 7: 819–830

Merz CJ, Tabbert K, Schweckendiek J, Klucken T, Vaitl D, Stark R, Wolf OT (2012b) Oral contraceptive usage alters the effects of cortisol on implicit fear learning. Horm Behav 62: 531–538

Merz CJ, Wolf OT, Schweckendiek J, Klucken T, Vaitl D, Stark R (2013) Stress differentially affects fear conditioning in men and women. Psychoneuroendocrinology 38:2529–2541

Merz CJ, Kinner VL, Wolf OT (2018) Let's talk about sex … differences in human fear conditioning. Curr Opin Behav Sci 23:7–12

Milad MR, Quirk GJ (2012) Fear extinction as a model for translational neuroscience: ten years of progress. Annu Rev Psychol

Milad MR, Orr SP, Klibanski A, Goldstein JM, Wedig MM, Pitman RK, Rauch SL (2006) Fear conditioning and extinction: Influence of sex and menstrual cycle in healthy humans. Behav Neurosci 120:1196–1203

Milad MR, Igoe SA, Lebron-Milad K, Novales JE (2009a) Estrous cycle phase and gonadal hormones influence conditioned fear extinction. Neuroscience 164:887–895

Milad MR, Pitman RK, Ellis CB, Gold AL, Shin LM, Lasko NB, Zeidan MA, Handwerger K, Orr SP, Rauch SL (2009b) Neurobiological basis of failure to recall extinction memory in posttraumatic stress disorder. Biol Psychiatry 66:1075–1082

Milad MR, Zeidan MA, Contero A, Pitman RK, Klibanski A, Rauch SL, Goldstein JM (2010) The influence of gonadal hormones on conditioned fear extinction in healthy humans. Neuroscience 168:652–658

Milligan-Saville JS, Graham BM (2016) Mothers do it differently: reproductive experience alters fear extinction in female rats and women. Transl Psychiatry 6:e928–e928

Mirshekar M, Abrari K, Goudarzi I, Rashidy-Pour A (2013) Systemic administrations of β-estradiol alleviate both conditioned and sensitized fear responses in an ovariectomized rat model of post-traumatic stress disorder. Neurobiol Learn Mem 102:12–19

Mitterling KL, Spencer JL, Dziedzic N, Shenoy S, McCarthy K, Waters EM, McEwen BS, Milner TA (2010) Cellular and subcellular localization of estrogen and progestin receptor immunoreactivities in the mouse hippocampus. J Comp Neurol 518:2729–2743

Montoya ER, Bos PA (2017) How oral contraceptives impact social-emotional behavior and brain function. Trends Cogn Sci 21:125–136

Mulchahey JJ, Ekhator NN, Zhang H, Kasckow JW, Baker DG, Geracioti TD (2001) Cerebrospinal fluid and plasma testosterone levels in post-traumatic stress disorder and tobacco dependence. Psychoneuroendocrinology 26:273–285

Murphy DD, Segal M (2000) Progesterone prevents estradiol-induced dendritic spine formation in cultured hippocampal neurons. Neuroendocrinology 72:133–143

Myers KM, Davis M (2007) Mechanisms of fear extinction. Mol Psychiatry 12:120–150

Myers KM, Ressler KJ, Davis M (2006) Different mechanisms of fear extinction dependent on length of time since fear acquisition. Learn Mem 13:216–223

Nillni YI, Toufexis DJ, Rohan KJ (2011) Anxiety sensitivity, the menstrual cycle, and panic disorder: a putative neuroendocrine and psychological interaction. Clin Psychol Rev 31:1183

Orcutt HK, Erickson DJ, Wolfe J (2004) The course of PTSD symptoms among Gulf War veterans: a growth mixture modeling approach. J Trauma Stress 17:195–202

Pace-Schott EF, Spencer RMC, Vijayakumar S, Ahmed NAK, Verga PW, Orr SP, Pitman RK, Milad MR (2013) Extinction of conditioned fear is better learned and recalled in the morning than in the evening. J Psychiatr Res 47:1776

Patton GC, Hibbert ME, Carlin J, Shao Q, Rosier M, Caust J, Bowes G (1996) Menarche and the onset of depression and anxiety in Victoria, Australia. J Epidemiol Community Health 50:661–666

Perry CJ, Ganella DE, Nguyen LD, Du X, Drummond KD, Whittle S, Pang TY, Kim JH (2020) Assessment of conditioned fear extinction in male and female adolescent rats. Psychoneuroendocrinology 116

Petersen N, Cahill L (2015) Amygdala reactivity to negative stimuli is influenced by oral contraceptive use. Soc Cogn Affect Neurosci 10:1266–1272

Petersen N, Touroutoglou A, Andreano JM, Cahill L (2015) Oral contraceptive pill use is associated with localized decreases in cortical thickness. Hum Brain Mapp 36:2644

Phillips ML, Drevets WC, Rauch SL, Lane R (2003) Neurobiology of emotion perception I: the neural basis of normal emotion perception. Biol Psychiatry 54:504–514

Phillips ML, Ladouceur CD, Drevets WC (2008) A neural model of voluntary and automatic emotion regulation: implications for understanding the pathophysiology and neurodevelopment of bipolar disorder. Mol Psychiatry 13:829

Pibiri F, Nelson M, Guidotti A, Costa E, Pinna G (2008) Decreased corticolimbic allopregnanolone expression during social isolation enhances contextual fear: a model relevant for posttraumatic stress disorder. Proc Natl Acad Sci U S A 105:5567–5572

Pineles SL, Nillni YI, King MW, Patton SC, Bauer MR, Mostoufi SM, Gerber MR, Hauger R, Resick PA, Rasmusson AM, Orr SP (2016) Extinction retention and the menstrual cycle: different associations for women with posttraumatic stress disorder. J Abnorm Psychol 125: 349–355

Pineles SL, Nillni YI, Pinna G, Irvine J, Webb A, Arditte Hall KA, Hauger R, Miller MW, Resick PA, Orr SP, Rasmusson AM (2018) PTSD in women is associated with a block in conversion of progesterone to the GABAergic neurosteroids allopregnanolone and pregnanolone measured in plasma. Psychoneuroendocrinology 93:133–141

Pineles SL, Nillni YI, Pinna G, Webb A, Arditte Hall KA, Fonda JR, Irvine J, King MW, Hauger RL, Resick PA, Orr SP, Rasmusson AM (2020) Associations between PTSD-Related extinction retention deficits in women and plasma steroids that modulate brain GABA A and NMDA receptor activity. Neurobiol Stress 13

Pinna G, Rasmusson AM (2014) Ganaxolone improves behavioral deficits in a mouse model of post-traumatic stress disorder. Front Cell Neurosci 8:256

Pinna G, Agis-Balboa RC, Pibiri F, Nelson M, Guidotti A, Costa E (2008) Neurosteroid biosynthesis regulates sexually dimorphic fear and aggressive behavior in mice. Neurochem Res 33: 1990–2007

Pletzer BA, Kerschbaum HH (2014) 50 years of hormonal contraception-time to find out, what it does to our brain. Front Neurosci 8

Pöhlchen D, Leuchs L, Binder FP, Blaskovich B, Nantawisarakul T, Topalidis P, Brückl TM, Norrholm SD, Jovanovic T, Spoormaker VI, Binder EB, Czisch M, Erhardt A, Grandi NC, Ilic-Cocic S, Lucae S, Sämann P, Tontsch A (2020) No robust differences in fear conditioning between patients with fear-related disorders and healthy controls. Behav Res Ther 129

Raeder F, Heidemann F, Schedlowski M, Margraf J, Zlomuzica A (2019) No pills, more skills: the adverse effect of hormonal contraceptive use on exposure therapy benefit. J Psychiatr Res 119: 95–101

Rahe RH, Karson S, Howard NS, Rubin RT, Poland RE (1990) Psychological and physiological assessments on American hostages freed from captivity in Iran. Psychosom Med 52:1–16

Rasmusson AM, Pinna G, Paliwal P, Weisman D, Gottschalk C, Charney D, Krystal J, Guidotti A (2006) Decreased cerebrospinal fluid allopregnanolone levels in women with posttraumatic stress disorder. Biol Psychiatry 60:704–713

Rasmusson AM, King MW, Valovski I, Gregor K, Scioli-Salter E, Pineles SL, Hamouda M, Nillni YI, Anderson GM, Pinna G (2019) Relationships between cerebrospinal fluid GABAergic neurosteroid levels and symptom severity in men with PTSD. Psychoneuroendocrinology 102:95–104

Rehbein E, Hornung J, Sundström Poromaa I, Derntl B (2021) Shaping of the female human brain by sex hormones: a review. Neuroendocrinology 111:183–206

Ressler KJ, Mercer KB, Bradley B, Jovanovic T, Mahan A, Kerley K, Norrholm SD, Kilaru V, Smith AK, Myers AJ, Ramirez M, Engel A, Hammack SE, Toufexis D, Braas KM, Binder EB, May V (2011) Post-traumatic stress disorder is associated with PACAP and the PAC1 receptor. Nature 470:492–497

Rey CD, Lipps J, Shansky RM (2014) Dopamine D1 receptor activation rescues extinction impairments in low-estrogen female rats and induces cortical layer-specific activation changes in prefrontal–amygdala circuits. Neuropsychopharmacology 39:1282–1289

Sapolsky RM (2005) The influence of social hierarchy on primate health. Science 308:648–652

Schmalenberger KM, Tauseef HA, Barone JC, Owens SA, Lieberman L, Jarczok MN, Girdler SS, Kiesner J, Ditzen B, Eisenlohr-Moul TA (2021) How to study the menstrual cycle: Practical tools and recommendations. Psychoneuroendocrinology 123:104895

Seligowski AV, Hurly J, Mellen E, Ressler KJ, Ramikie TS (2020) Translational studies of estradiol and progesterone in fear and PTSD. https://doi.org/10.1080/20008198.2020.172385711

Sheppard PAS, Choleris E, Galea LAM (2019) Structural plasticity of the hippocampus in response to estrogens in female rodents. Mol Brain 12

Shvil E, Sullivan GM, Schafer S, Markowitz JC, Campeas M, Wager TD, Milad MR, Neria Y (2014) Sex differences in extinction recall in posttraumatic stress disorder: a pilot fMRI study. Neurobiol Learn Mem 113:101

Skovlund CW, Mørch LS, Kessing LV, Lidegaard O (2016) Association of hormonal contraception with depression. JAMA Psychiatry 73:1154–1162

Skovlund CW, Mørch LS, Kessing LV, Lange T, Lidegaard J (2018) Association of hormonal contraception with suicide attempts and suicides. Am J Psychiatry 175:336–342

Smith CC, McMahon LL (2005) Estrogen-induced increase in the magnitude of long-term potentiation occurs only when the ratio of NMDA transmission to AMPA transmission is increased. J Neurosci 25:7780–7791

Stevens JS, Almli LM, Fani N, Gutman DA, Bradley B, Norrholm SD, Reiser E, Ely TD, Dhanani R, Glover EM, Jovanovic T, Ressler KJ (2014) PACAP receptor gene polymorphism impacts fear responses in the amygdala and hippocampus. Proc Natl Acad Sci U S A 111:3158–3163

Stiedl O, Radulovic J, Lohmann R, Birkenfeld K, Palve M, Kammermeier J, Sananbenesi F, Spiess J (1999) Strain and substrain differences in context- and tone-dependent fear conditioning of inbred mice. Behav Brain Res 104:1–12

Stockhorst U, Antov MI (2015) Modulation of fear extinction by stress, stress hormones and estradiol: a review. Front Behav Neurosci 9:359

Suarez-Jimenez B, Gore HE, Hachey J, King HM, Lacreuse A (2013) Testosterone modulation of anxiety in gonadally-suppressed male rhesus monkeys: a role for gonadotropins? Pharmacol Biochem Behav 104:97–104

Sundström-Poromaa I, Gingnell M (2014) Menstrual cycle influence on cognitive function and emotion processing from a reproductive perspective. Front Neurosci

Tang S, Graham BM (2020) Hormonal, reproductive, and behavioural predictors of fear extinction recall in female rats. Horm Behav 121

Van Honk J, Peper JS, Schutter DJLG (2005) Testosterone reduces unconscious fear but not consciously experienced anxiety: implications for the disorders of fear and anxiety. Biol Psychiatry 58:218–225

van Wingen GA, van Broekhoven F, Verkes RJ, Petersson KM, Bäckström T, Buitelaar JK, Fernández G (2008) Progesterone selectively increases amygdala reactivity in women. Mol Psychiatry 13:325–333

Van Wingen GA, Ossewaarde L, Bäckström T, Hermans EJ, Fernández G (2011) Gonadal hormone regulation of the emotion circuitry in humans. Neuroscience 191:38–45

Velasco ER, Florido A, Flores A, Senabre E, Gomez-Gomez A, Torres A, Roca A, Norrholm S, Newman EL, Das P, Ross RA, Lori A, Pozo OJ, Ressler KJ, Garcia-Esteve LL, Jovanovic T, Andero R (2022) PACAP-PAC1R modulates fear extinction via the ventromedial hypothalamus. Nat Commun 13

Wegerer M, Kerschbaum H, Blechert J, Wilhelm FH (2014) Low levels of estradiol are associated with elevated conditioned responding during fear extinction and with intrusive memories in daily life. Neurobiol Learn Mem 116:145–154

Wen Z, Hammoud MZ, Scott JC, Jimmy J, Brown L, Marin MF, Asnaani A, Gur RC, Foa EB, Milad MR (2021) Impact of exogenous estradiol on task-based and resting-state neural signature during and after fear extinction in healthy women. Neuropsychopharmacology 46:2278–2287

White EC, Graham BM (2016) Estradiol levels in women predict skin conductance response but not valence and expectancy ratings in conditioned fear extinction. Neurobiol Learn Mem 134:339–348

Wingo AP, Velasco ER, Florido A, Lori A, Choi DC, Jovanovic T, Ressler KJ, Andero R (2018) Expression of the PPM1F gene is regulated by stress and associated with anxiety and depression. Biol Psychiatry 83:284–295

Woolley CS, McEwen BS (1993) Roles of estradiol and progesterone in regulation of hippocampal dendritic spine density during the estrous cycle in the rat. J Comp Neurol 336:293–306

Zeidan MA, Igoe SA, Linnman C, Vitalo A, Levine JB, Klibanski A, Goldstein JM, Milad MR (2011) Estradiol modulates medial prefrontal cortex and amygdala activity during fear extinction in women and female rats. Biol Psychiatry 70:920–927

The Impact of Sleep on Fear Extinction

Ryan Bottary, Laura D. Straus, and Edward F. Pace-Schott

Contents

R. Bottary
Department of Psychology and Neuroscience, Boston College, Chestnut Hill, MA, USA

Division of Sleep Medicine, Harvard Medical School, Boston, MA, USA
e-mail: ryan.bottary@bc.edu

L. D. Straus
Department of Research, San Francisco VA Health Care System, San Francisco, CA, USA

Department of Psychiatry and Behavioral Sciences, University of California San Francisco, San Francisco, CA, USA
e-mail: laura.straus@ucsf.edu

E. F. Pace-Schott (✉)
Department of Psychiatry, Harvard Medical School, Boston, MA, USA

Department of Psychiatry, Massachusetts General Hospital, Charlestown, MA, USA

Athinoula A. Martinos Center for Biomedical Imaging, Charlestown, MA, USA
e-mail: epace-schott@mgh.harvard.edu

Curr Topics Behav Neurosci (2023) 64: 133–156
https://doi.org/10.1007/7854_2023_431
Published Online: 27 July 2023

Abstract Sleep plays a crucial role in the consolidation of memories, including those for fear acquisition and extinction training. This chapter reviews findings from studies testing this relationship in laboratory, naturalistic, and clinical settings. While evidence is mixed, several studies in humans have linked fear and extinction recall/retention to both rapid eye-movement and slow wave sleep. Sleep appears to further aid in the processing of both simulated and actual trauma and improves psychotherapeutic treatment outcomes in those with anxiety and trauma- and stressor-related disorders. This chapter concludes with a discussion of the current challenges facing sleep and emotional memory research in addition to suggestions for improving future research.

Keywords Extinction · Fear conditioning · Memory · REM · Sleep

1 Introduction

How we remember both acquired fear and its extinction is considered a critical process for the development and maintenance of anxiety and trauma- and stressor-related disorders (Pace-Schott et al. 2015a). Sleep has been demonstrated to influence our ability to regulate our emotions and determine whether and how intensely we will later remember emotional experiences (Goldstein and Walker 2014; van der Helm and Walker 2009). Growing evidence supports sleep's role in the consolidation and generalization of fear and extinction memories (Colvonen et al. 2019; Davidson and Pace-Schott 2020; Pace-Schott et al. 2015b). Sleep has been further shown to play a critical role in the processing of both simulated and actual traumatic experiences (Davidson and Marcusson-Clavertz 2022; Davidson and Pace-Schott 2020). While findings remain mixed, studies to date have highlighted the potential importance of both rapid eye-movement (REM) sleep and slow wave sleep (SWS), as well as specific sleep oscillations present during these sleep stages, for processing emotional experiences (Kim and Payne 2020; Pace-Schott et al. 2015a; Schäfer et al. 2020). Findings to date have direct translational relevance as (1) recall/retention of fear and extinction memories influence the development and treatment of anxiety and trauma- and stressor-related disorders and (2) sleep disturbances are common in these patients (Pace-Schott et al. 2015a). In this chapter, we will review the current literature linking sleep to fear acquisition, extinction training, and fear and extinction recall/retention. We will discuss basic findings that suggest sleep amount and quality determines how well individuals retain experimentally induced fear and extinction memories. We will further discuss how sleep manipulations have been employed to uncover both sleep-based memory mechanisms and to enhance extinction under experimental and therapeutic conditions. This chapter will conclude with a discussion of the present challenges when conducting this type of research and offer some potential solutions to be implemented in future work.

2 Sleep

Sleep is a transient, reversible state of relative behavioral inactivity, marked by progressively reduced responsiveness to, and perception of, the external world (Carskadon and Dement 2017). While sleep and wake states can be differentiated broadly with self-report measures and actigraphy methods, the gold standard for measuring human sleep physiology is polysomnography (PSG), a montage of electrodes applied to the scalp to monitor brain activity (i.e., electroencephalography, EEG) and to the face to monitor eye movements (i.e., electrooculography, EOG) and skeletal muscle tone (i.e., electromyography: EMG). While standardized criteria have changed over time, sleep scoring largely relies on the qualitative assessment of the frequency and amplitude of dominant EEG oscillations co-occurring with characteristic EOG and EMG activity (Iber et al. 2007).

Using the above-mentioned signals, human sleep is divided into rapid eye-movement (REM) and non-REM (NREM) sleep (Iber et al. 2007; see Fig. 1). REM sleep is characterized by low amplitude, high frequency EEG activity paired with rapid, saccadic eye movements and reduced skeletal muscle tone. NREM sleep is subdivided into stages N1, N2, and N3 (commonly referred to as slow wave sleep, SWS), corresponding to progressive deepening of sleep. N1 is characterized by

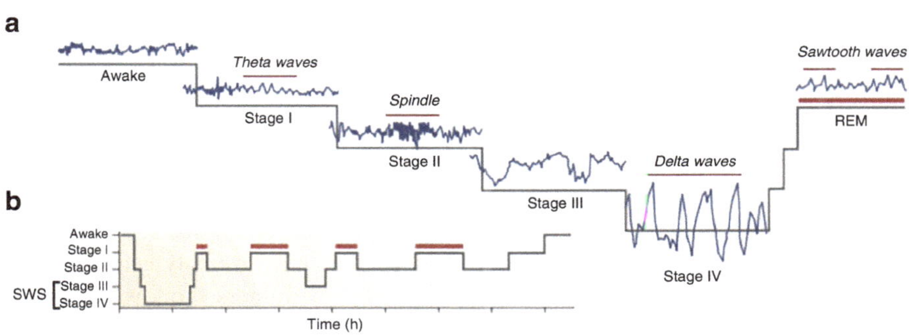

Fig. 1 Human sleep composition (figure adapted from Hobson and Pace-Schott 2002). (**a**) Electroencephalographic (EEG) traces for wake, non-rapid eye-movement (NREM) sleep stages I–IV and rapid eye-movement (REM) sleep. NREM stage I (N1) coincides with sleep onset with theta waves being the dominant EEG waveform. NREM stage II (N2) sleep is characterized by two EEG signatures; K-Complexes, a high amplitude sharp wave (not shown here), and Sleep Spindles, a ~12–15 Hz burst of activity that lasts 0.5–2 s. Slow wave sleep (SWS or N3), the combination of NREM stages III and IV based on updated criteria, is dominated by high amplitude, low frequency (0.5–4 Hz) delta waves. REM sleep is dominated by low amplitude, high frequency waveforms and sawtooth waves. (**b**) Sleep hypnogram representing a typical night of sleep. Hypnograms are generated by scoring sleep stages based on visual inspection of the above-described EEG waveforms. The amount of time spent in SWS tends to be greater in the first half of the night and lesser in the second half. REM sleep tends to occur every 90–120 min with REM periods becoming progressively longer across the night. N2 sleep accounts for about half of all sleep obtained during the night and tends to be equally distributed across the first and second halves of the night. N1 and wake normally account for a small proportion of the sleep period

replacement of wake-related alpha frequency (8–13 Hz) EEG activity, primarily expressed in occipital electrodes, with theta-frequency (4–7 Hz) and slow rolling eye movements. N2 is characterized by two distinct EEG oscillations: (1) the sleep spindle, a ~12–15 Hz burst of activity that persists for 0.5–2 s, and (2) the k-complex, a high amplitude sharp wave. Both oscillations occur maximally in central electrodes. SWS is characterized by slow, high amplitude 0.5–4 Hz (delta) oscillations occurring maximally in frontal electrodes. In addition to visual inspection of EEG traces, sleep scientists are increasingly applying signal processing techniques to understand sleep and its role in waking behavior. For example, spectral analysis can be performed to decompose complex EEG waveforms into their component frequencies (Prerau et al. 2017; Walczak and Chokroverty 2009). This technique helps sleep scientists identify dominant frequencies in the EEG signal and determine whether greater frequency-specific power benefits or impairs learning and memory.

Healthy young adults generally spend 75–80% of total sleep time (TST) in NREM sleep (N1, 2–5%; N2, 45–50%; SWS, 13–23%) and the remaining 20–25% in REM sleep (Carskadon and Dement 2017). Wakefulness during the sleep period, referred to as wake after sleep onset (WASO), tends to comprise less than 5% of TST (Carskadon and Dement 2017). Sleep stage distribution changes across the night, with the first half rich in SWS and the second half rich in REM sleep with REM periods cycling roughly every 90–120 min and REM duration becoming progressively longer across the night (Brown et al. 2012; Pace-Schott 2009). N2 sleep amount is generally consistent across the night.

3 Sleep and Memory Consolidation

Sleep has been shown to be critical for the consolidation of recently acquired memories. Specifically, we tend to remember what we have recently learned better across time when sleep, rather than continued wakefulness, follows learning (Girardeau and Lopes-dos-Santos 2021; Klinzing et al. 2019; Rasch and Born 2013). Further, sleep is theorized to actively consolidate and transform, rather than passively protect, memories (Klinzing et al. 2019). For non-emotional declarative memories, evidence supports sleep's role in systems-level memory consolidation, a process by which newly acquired memories in the hippocampus are transferred to long-term storage in the cortex (Born and Wilhelm 2012; Diekelmann and Born 2010; Klinzing et al. 2019). This process is theorized to be supported by memory replay during NREM sleep, which, in turn, is orchestrated by the coordination of hippocampal sharp wave ripples (~80 Hz) and thalamic sleep spindles (~12–15 Hz) occurring in the "up states" (i.e., periods of synchronized neuronal depolarization) of cortical slow oscillations (i.e., <1 Hz slow waves) (Klinzing et al. 2019; see Fig. 2).

Evidence for sleep's role in the consolidation of emotional episodic memories remains mixed. Early findings implicated sleep, and REM sleep specifically, in the preferential consolidation and reduction of emotional reactivity to emotional,

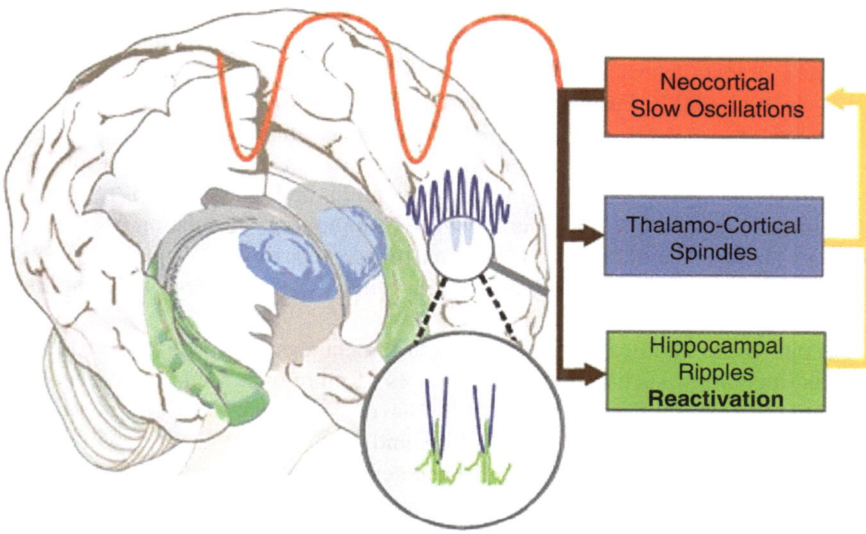

Fig. 2 Sleep facilitates the replay and active consolidation of recently acquired memories through coordination of cortical slow oscillations (SO; <1 Hz), thalamocortical sleep spindles (~12–15 Hz) occurring during the "up states" of SOs, and hippocampal ripples (~80 Hz) occurring in the troughs of spindles (figure adapted from Born and Wilhelm 2012)

episodic memories (Goldstein and Walker 2014; van der Helm and Walker 2009). REM sleep's theorized role for emotion processing was sparked by early observations that vivid, emotional dreams occurred most often during REM sleep, perhaps representing a replay and consolidation of emotional memory traces (Perogamvros et al. 2013; Stickgold 2001). Subsequent neuroimaging studies of sleep demonstrated that nodes in the emotional memory network are either more active (e.g., amygdala and hippocampus) or less active (lateral prefrontal cortex, LPFC) during REM sleep (Dang-Vu et al. 2010). Further, acetylcholine activity, the neurochemical associated with learning-dependent neural plasticity, is elevated during REM sleep (Hobson and Pace-Schott 2002), suggesting this sleep stage's possible memory function. Behavioral studies supported this REM-sleep-dependent emotional memory processing theory by demonstrating that periods of sleep enriched with REM sleep were associated with better retention of emotional content (Cartwright et al. 1975; Wagner 2001), while REM sleep deprivation impaired retention of emotional memories (Greenberg et al. 1983). Correlational studies have further shown that REM duration (Nishida et al. 2009; Payne et al. 2012) and spectral features, including theta activity (Kim et al. 2019; Nishida et al. 2009; Sopp et al. 2017), predict the retention of emotional memories.

However, theories about sleep's role in emotional episodic memory processing have been greatly tempered in light of recent evidence (Davidson et al. 2021). For example, two recent meta-analyses found no benefit of sleep compared to wake for the selective consolidation of emotional over neutral episodic memories (Lipinska et al. 2019; Schäfer et al. 2020). Also, a recent meta-analysis reported that subjective, but not objective (i.e., psychophysiological), measures of arousal to negative stimuli *increased* across delays including sleep but not across delays including wake (Lipinska et al. 2022). Further, whereas the meta-analysis by Schäfer et al. (2020) mentioned above determined that REM-rich sleep may be critical for sleep-dependent effects to be expressed, a large-scale study of 929 participants found no association between emotional memory retention and REM sleep physiology (Ackermann et al. 2015). Similarly, studies employing stimulation methods to suppress (Johnson and Durrant 2018) or enhance (Harrington et al. 2021) REM sleep theta spectral activity did not produce the expected effects on emotional memory retention.

Despite these conflicting findings, an argument can be made that emotional episodic memories, such as those for pictures of emotional scenes or objects, are qualitatively different from the types of low-level associative memories involved in fear acquisition and extinction. With this in mind, the following sections will explore the current evidence linking sleep to these latter types of emotional memories.

4 Sleep Impacts the Consolidation of Memories for Fear Acquisition, Extinction Training, and Safety Signals

In contrast to studies of emotional episodic memories, sleep's beneficial role for consolidating both fear and extinction-relevant memories has been observed more consistently. However, if sleep preserves both fear and extinction memories, is it better to sleep or remain awake after emotional experiences, such as traumas, to promote adaptive emotion processing? That is the question for the current section. Developing conditions of generalized fear, such as anxiety and trauma- and stressor-related disorders, and specific fears, such as phobias, rely on a combination of experiencing and retaining specific aversive memories (i.e., fear memories) and the inability to discriminate those memories from a larger set of non-threatening stimuli that may share a similar context (i.e., safety signals) or qualitative features (stimulus generalization). Treating these conditions requires acquiring and retaining extinction memories which are dissociable from (e.g., Hennings et al. 2021), and compete with (Milad and Quirk 2012), fear memories for expression. While findings remain mixed, what emerges from the current literature is that sleep strengthens and generalizes both fear acquisition and extinction memories (Davidson and Pace-Schott 2020; Pace-Schott et al. 2015a). Further, preliminary evidence suggests that sleep may also be critical for retaining memories for learned safety signals (Straus et al. 2018a).

4.1 Sleep, Fear Acquisition, and Recall/Retention

Several studies support sleep's role in acquiring, recalling/retaining, and generalizing fear memories. For example, partial sleep deprivation on the night prior to fear acquisition reduced skin conductance response (SCR) habituation to both subsequently aversively reinforced and unreinforced stimuli (Peters et al. 2014). Participants deprived of sleep for a full night prior to fear acquisition similarly displayed exaggerated subjective fear ratings and increased SCRs during fear acquisition compared to participants who slept (Feng et al. 2018a, b). The heightened fear acquisition resulting from sleep deprivation was paralleled by increased functional brain activity in fear-associated regions (i.e., amygdala and insula) and reductions in fear-regulating regions (i.e., ventromedial prefrontal cortex, vmPFC; Feng et al. 2018a, b). REM sleep on the night prior to fear acquisition may be particularly important for determining fear acquisition. For example, a study by Lerner et al. (2017) reported that participants who obtained more REM sleep on the night prior to fear acquisition exhibited attenuated functional brain activity in nodes of both fear and extinction networks, including the vmPFC, hippocampus, and amygdala, during fear acquisition.

Sleep, and REM sleep specifically, following fear acquisition may be equally important for the consolidation of fear memories. In a study by Menz et al. (2013), participants acquired fear to two stimuli (CS + E and CS + N) using a mild electric shock but not to two control stimuli (CS-E and CS-N, respectively). Following fear acquisition, one of the CS+ (CS + E), but not the other (CS + N), was immediately extinguished through repeated presentation without shock reinforcement (i.e., extinction training). Participants were then either deprived of one night of sleep or slept normally. Compared to sleep-deprived participants, participants who slept exhibited greater fear recall/retention as reflected in SCRs, self-reported shock expectancy, and amygdala reactivity to the CS + N compared to CS-N. Further, in those who slept, SCR-measured fear recall/retention was positively associated with REM sleep duration during the night. In a separate study, Menz et al. (2016) showed that fear recall/retention was greater in participants who slept during a period rich in REM sleep compared to those who remained awake during an equivalent delay. No such differences were observed when comparing participants who slept during a period rich in SWS to those who remained awake. Marshall et al. (2014) also showed that increased REM sleep on the night following fear acquisition, compared to the night before fear acquisition, predicted greater differentiation between threat versus safety signals.

With respect to fear generalization, Davidson et al. (2018) showed that fear may generalize to otherwise non-threatening stimuli following a daytime nap. In this study, participants acquired fear to one of two differently sized circles (i.e., one large circle and one small circle), using shock reinforcement, prior to a delay that included a daytime nap or an equivalent time spent awake. Following the delay, participants were presented with the reinforced stimulus (CS+; e.g., large circle) and non-reinforced stimulus (CS-; e.g., small circle), in addition to new circle stimuli that varied in size between the size extremes of the CS+ and CS-. Participants who stayed awake during the delay maintained higher SCRs to the CS+ compared to CS-, while those who slept displayed similar SCRs to both the CS+ and CS-. Further, sleep participants exhibited heightened SCR to the CS- at fear recall/retention compared to fear acquisition, which the authors interpreted as indicating fear generalization. However, these same participants also showed reduced SCR to the CS+ from fear acquisition to recall/retention and group differences in SCR between the sleep and wake participants were not observed. It is therefore unclear in this study whether sleep's impact on generalization was adaptive (i.e., overall fear reduction) or maladaptive (i.e., overall fear increase).

In contrast to the findings above, other work has found either no benefit of continued wake compared to sleep for fear recall/retention and generalization (Davidson et al. 2016) or even increased fear generalization following continued wake (Zenses et al. 2020). At least one study reported that sleep may, in fact, be beneficial for the inhibition of fear recall/retention. In this study, Spoormaker et al. (2014) conditioned participants, using mild electric shocks, to geometric shapes (CS +) interspersed with shapes never paired with a shock (CS-). When participants were shown these objects 1 day later, those who exhibited greater REM sleep duration (i.e., minutes spent in REM sleep) during the intervening sleep period exhibited the lowest amount of fear recall/retention, as measured by SCR reactivity to the CS+. Further, REM sleep duration explained the positive association between vmPFC activity to the CS+ during fear acquisition and reduced SCR to the CS+ at fear recall/retention. Given the role of the vmPFC in inhibiting emotional brain structures, such as the amygdala (Akirav and Maroun 2007), this latter finding might suggest that REM sleep plays a critical role in maintaining or strengthening limbic connections responsible for fear inhibition.

4.2 Sleep, Extinction Training, and Recall/Retention

Like fear memories, memories for learned extinction (i.e., through extinction training) may benefit from both pre-training and post-training sleep. For example, actigraphy-measured baseline sleep quality (Pace-Schott et al. 2015b) and sleep timing regularity (Bottary et al. 2020a) have both been shown to predict better extinction recall/retention. Straus et al. (2017) demonstrated that participants deprived of a night of sleep between fear acquisition and extinction training exhibited impaired extinction recall/retention after a subsequent night of recovery sleep, despite showing normal fear reduction during extinction training. In a separate study, Spoormaker et al. (2010) found that participants who obtained REM sleep during a daytime nap that followed extinction training displayed greater extinction recall/retention, measured by SCRs and functional brain activity in the vmPFC, compared to participants who did not obtain REM sleep. In the previously mentioned study by Menz et al. (2016), participants who slept during a period rich in REM sleep, compared to those who remained awake, demonstrated better extinction recall/retention and reduced functional brain activity in several fear network regions, including the amygdala and anterior cingulate cortex (although, surprisingly, vmPFC activity was also lower). Correlational studies have similarly found that the percentage of time spent in REM sleep (Bottary et al. 2020b; Pace-Schott et al. 2014), REM continuity (Bottary et al. 2020b; Straus et al. 2017), and REM theta spectral activity (Bottary et al. 2020b) predict successful extinction recall/retention. To probe the causal role of REM sleep for successful extinction recall/retention, Spoormaker et al. (2012) selectively deprived participants of REM sleep following extinction training. Compared to control participants who were intermittently awoken during NREM sleep, yet had intact REM sleep, REM-deprived participants displayed deficits in extinction recall/retention. Seo et al. (2021b) further found that both partial sleep restriction and total sleep deprivation had deleterious effects on engagement of fear-regulating and extinction promoting functional brain regions during extinction training. Lastly, sleep may play a role in the generalization of learned extinction. For example, Pace-Schott et al. (2009) conditioned healthy young adults to two color stimuli (CS + E, CS + U), but not a third (CS-), using finger-shock reinforcement, and then immediately extinguished one CS+ (CS + E), but not the other (CS + U). Extinction recall/retention and generalization were tested 12-h later, following a delay that included overnight sleep or daytime wake. In the group that slept, SCRs to the CS + U were significantly reduced such that they did not differ from those to the CS + E, whereas, in the wake group, responses to the CS + U remained higher than to the CS + E. This led the authors to conclude that learned extinction to the CS + E had generalized to the CS + U, which was not extinguished.

4.3 Sleep, Safety Signal Learning, and Recall/Retention

As mentioned above, the ability to discriminate between cues that predict aversive outcomes (i.e., threat signals) and those that share a similar spatial or temporal context but do not predict an aversive outcome (i.e., safety signals) is a critical component of fear acquisition and generalization. Sleep prior to and following safety learning has been shown to predict safety signal recall/retention. In a recent meta-analysis by Schenker et al. (2021), reduced SWS on the night prior to extinction training was associated with greater SCRs to previously learned safety signals (i.e., CS-), suggesting that SWS may prime the brain to successfully encode safety signals. Further, REM sleep following safety signal learning has been shown to be critical for consolidating safety signal memories (Marshall et al. 2014). As we will discuss in a later section, REM sleep's role in the consolidation of safety signals is particularly relevant for psychiatric conditions characterized by overgeneralized fear, such as posttraumatic stress disorder (PTSD).

5 Altering Fear and Extinction Memories during Sleep

A growing body of work suggests that reactivating and further strengthening new memories is possible during sleep (Hu et al. 2020; Oudiette and Paller 2013). Using targeted memory reactivation, researchers have successfully used non-intrusive memory cues, such as odors or quiet tones, to reactivate memories for learned content during sleep. These cued memories, in turn, are better remembered than memories that are not cued, especially when cueing occurs during SWS (Hu et al. 2020). While this approach has largely been studied using non-emotional declarative memory tasks, a few reports have provided evidence for its potential in modulating fear and extinction memories. For example, Hauner et al. (2013) conditioned participants to two different CS+ stimuli, each paired with a distinct odor, and then gave them an opportunity to nap. During the nap, one of the two odors was presented during periods of SWS. When participants were later re-exposed to both CS+, they exhibited reduced fear responses specifically to the CS+ that was cued during the nap. In a follow-up study, He et al. (2015) replicated this finding, using non-sleep disturbing auditory tone cues in place of odor stimuli. Importantly, this study demonstrated that extinction of conditioned fear during SWS was as effective as extinction training during wake. Nonetheless, several studies that attempted to cue emotion regulatory memories during REM sleep did not observe enhanced recall (Borghese et al. 2022; Halonen et al. 2021; Hu et al. 2020).

Interestingly, the positive effects of sleep cueing on fear acquisition and extinction recall/retention may critically depend on how established one of these memories is prior to cueing. For example, Ai et al. (2015) conditioned participants to one colored shape (CS+), but not another (CS–), and then, 1 day later, extinguished the CS+ in the presence of an extinction tone cue. Following extinction, participants either napped or remained awake for an equivalent amount of time. In both conditions, participants were presented with the extinction tone cue, a control tone, or no tone. Tones in the nap condition were timed to occur during SWS. The results showed that participants presented with the extinction tone cues during SWS exhibited significantly higher fear recall/retention than participants presented with the control tone or no tone. The opposite was true for wake participants; fear recall/retention was significantly lower in participants presented with the extinction tone compared to those presented with the control tone or no tone. While seemingly counterintuitive, as presentation of the extinction tone may be predicted to enhance extinction memory reactivation and consolidation, the researchers speculated that the night of sleep following fear acquisition played a critical role. In an attempt to reactivate and strengthen the labile extinction memory, the auditory tone presented during the nap may have facilitated the reactivation and strengthening of the fear memory, which had an entire night to be consolidated prior to extinction training.

6 Sleep, Extinction Training, and Recall/Retention in Clinical Populations

Given evidence for sleep's role in consolidating fear acquisition and extinction memories in healthy, good sleeping adults, a small but growing number of studies have attempted to track this relationship in clinical populations for which retention of fear and extinction memories are of pathophysiological relevance. For example, Richards et al. (2022) conditioned trauma-exposed participants with and without PTSD to one colored circle (CS+), but not another (CS-), using an aversive air-puff to the throat. Fear acquisition was measured using fear-potentiated startle (FPS), an alternative to SCR that measures the blink-startle response to a loud tone presented soon after CS presentation onset. Participants were then given an opportunity to nap and then, during extinction training, were re-presented both the CS+, absent the aversive air-puff, and CS-. Those who obtained more REM sleep during the nap exhibited more rapid extinction, expressed as steeper declines in blink-startle responses to the conditioned stimulus during extinction training. However, this was not influenced by the presence or absence of PTSD. In a separate study, Straus et al. (2018b) fear conditioned veterans with PTSD using an FPS procedure similar to the above-mentioned study. The CS+ was then extinguished 1 day later. Extinction recall/retention was tested on a third day by presenting both stimuli and measuring a participant's blink-startle response to a paired loud tone. Individual differences in sleep physiology did not correlate with measures of extinction recall/retention. However, REM sleep percentage was positively associated with safety signal recall/retention, as measured by the slope of the decline in reactivity to the CS- during extinction recall/retention. Lastly, Bottary et al. (2020b) observed that REM sleep may favor the consolidation of either fear or extinction memories based on the presence or absence of disordered sleep. In this study, healthy, good sleeping participants and participants with insomnia disorder (ID) were conditioned to two colored lights (CS + E and CS + U), but not a third (CS-), appearing on-screen in a virtual fear acquisition context (an office). One of the CS+ (CS + E), but not the other (CS + U), was then immediately extinguished within a different virtual extinction training context (a conference room). One day later, participants viewed both CS + s and the CS- in the extinction context. In healthy, good sleepers, extinction recall/retention was positively associated with the quality of REM sleep (i.e., its continuity and theta spectral power) during the night between extinction training and extinction recall/retention. Interestingly, the opposite was true for the participants with ID; those who entered REM sleep more quickly spent a greater percentage of time in REM and who had less fragmented REM exhibited impaired extinction recall/retention. These findings may suggest that, when fear and extinction memories are learned in close temporal proximity, REM prioritizes consolidation of the more salient memory. In healthy individuals, this may be the adaptive extinction memory, while in individuals with ID it may be the fear memory. Interestingly, these findings align with another study that demonstrated that individuals with insomnia symptoms showed impaired overnight emotion regulation (Wassing et al. 2019). Further, they may help to explain the bidirectional relationship of insomnia with anxiety and trauma- and stressor-related disorders reported in epidemiological studies (e.g., Hertenstein et al. 2019) and in experimental work linking anxiety to threat-related memory biases (Herrera et al. 2017). REM sleep dysregulation, which has been observed in ID (Riemann et al. 2012) may impair the consolidation of naturalistically or therapeutically acquired extinction memories, thereby contributing to the development or perpetuation of anxiety disorders or PTSD (for a review, see Pace-Schott et al. 2022). In fact, Seo et al. (2021a) showed that individuals with PTSD exhibited both significantly reduced subjective sleep quality and less engagement of emotion regulatory brain regions during extinction training compared with trauma-exposed participants without PTSD. Moreover, a general pattern of delayed engagement of brain regions associated with extinction training has been described in 3 different comparisons of sleep compromised groups relative to their controls (Pace-Schott et al. 2022).

7 The Role of Sleep and Circadian Rhythms in Treatment for Anxiety and Trauma- and Stressor-Related Disorders

Sleep has been shown to play an important role during the course of successful treatment for anxiety and trauma- and stressor-related disorders. In these observational studies, patient sleep quality was tracked during the course of psychotherapy to determine its relationship to treatment outcomes. An early study by Zalta et al. (2013) observed that poorer pretreatment sleep quality resulted in both slower treatment response and poorer treatment outcomes in individuals undergoing cognitive-behavioral therapy for social anxiety disorder (SAD). Further, subjective feeling of "restedness" on the day of the therapy session predicted reduced symptom severity during the session. These effects were replicated and extended in a follow-up study by Dutcher et al. (2021) that found that poorer pre-treatment sleep quality predicted slower therapeutic improvement in SAD symptoms and worse outcomes both at the termination of therapy and at a 3-month follow-up. Interestingly, greater total sleep time prior to therapy sessions predicted less severe symptoms during sessions. While evidence remains preliminary, at least one study found that improvement in sleep quality, operationalized as reduced insomnia severity, predicted better PTSD treatment outcomes (Zalta et al. 2020). Moreover, a meta-analysis has shown that sleep-focused treatments such as cognitive behavioral therapy for insomnia (CBT-I) or imagery rehearsal therapy (IRT) for posttraumatic nightmares can improve daytime PTSD symptoms with a moderate effect size (Ho et al. 2016). In addition to sleep quality, circadian factors, such as the morning cortisol acrophase, may account for a morning advantage observed in experimental extinction learning and generalization (Pace-Schott et al. 2013, 2015b) as well as the therapeutic extinction underlying exposure therapy (Lass-Hennemann and Michael 2014; Meuret et al. 2016).

8 Sleep's Role in Processing Simulated and Real-Life Trauma

Following a traumatic event, naturalistic extinction may occur in the absence of explicit attempts to diminish fear via exposure. Under these circumstances, adaptive emotion processing may be expressed in individuals as the absence or relative attenuation of posttraumatic symptoms, including the involuntary experiencing of trauma-related intrusive thoughts. Several attempts have been made to study how sleep might facilitate the natural processing of emotional experiences both in the lab and in cases of actual trauma. Laboratory experiments have typically used a paradigm in which exposure to a highly upsetting film, the "analog trauma," is followed by monitoring of intrusive memories of the film across several subsequent days (Holmes and Bourne 2008). For example, Kleim et al. (2016) presented participants with a neutral film and a film containing traumatic content, including physical and sexual violence. Half of the participants were then given an opportunity to sleep while the other half remained awake. Participants who slept after watching the films, compared to those who did not, reported significantly fewer and less intense trauma-film-related intrusive thoughts during the subsequent week. Using a series of short trauma films that contained fatal transportation accidents, Zeng et al. (2021) similarly found that sleep, compared to sleep deprivation, reduced the number of trauma-film-related intrusive thoughts and reduced emotional arousal to the trauma memories across the subsequent week. Similar sleep-related reductions in intrusive thought occurrence have been observed when comparing napping to an equivalent period of daytime wake (Woud et al. 2018) and overnight sleep to partial sleep deprivation (Sopp et al. 2019). It is worth noting, however, that sleep deprivation following simulated trauma has also been shown to reduce next-day trauma-film-related intrusive thoughts (Porcheret et al. 2015) and trauma memory generalization (Kuriyama et al. 2010). In the former, such reduction was hypothesized to be due to reduced sleep-dependent consolidation of the memory of the film (Porcheret et al. 2015). Despite such conflicting results, a recent meta-analysis of 7 such studies has suggested that sleep compared with overnight sleep deprivation (5 studies) or a period of wakefulness (2 studies) has the effect of reducing the number of intrusive memories but not the intensity of such memories (Davidson and Marcusson-Clavertz 2022). With respect to the impact of REM sleep on trauma film-related intrusions, Kleim et al. (2016) showed that a higher frequency of rapid eye movements during REM sleep (i.e., REM density), but not REM sleep duration, following trauma film viewing *increased* the number of intrusions experienced. Sopp et al. (2019) further found that greater post-encoding REM sleep duration was associated with *reduced* analog PTSD symptoms, such as re-experiencing, hyperarousal, and avoidance. Reduced re-experiencing was specifically correlated with *greater* REM sleep theta spectral power in this study.

A small number of studies to date have explored the relationship between sleep in the early aftermath of actual traumas and its impact on subsequent trauma-related intrusive thoughts and posttraumatic symptoms. For example, Luik et al. (2019)

recruited participants from the hospital emergency department after they either experienced or witnessed a traumatic motor vehicle accident and had them self-report intrusive thoughts and sleep quality over the subsequent week. They observed that individuals experiencing poorer sleep quality also tended to experience a greater number of intrusive trauma-related thoughts. In a separate study by Porcheret et al. (2019), sleep on the night immediately following a traumatic event predicted the number of intrusive trauma-related thoughts over the subsequent week in a u-shaped pattern; those who slept less than ~2 h or greater than ~11 h experienced the greatest number of subsequent intrusive thoughts. REM sleep in the early aftermath of a trauma has specifically been linked to the development of posttraumatic symptomatology. Mellman et al. (2002) showed that individuals with fragmented REM sleep following a trauma were more likely to go on to develop PTSD. However, in a follow-up study, Mellman et al. (2007) showed that beta spectral activity during REM sleep predicted reduced PTSD symptomatology, including nightmare severity, at a two-month follow-up. This latter finding is of interest given a growing number of studies that have aimed to identify sleep biomarkers of PTSD development and resilience. For example, trauma-exposed individuals who were resilient to PTSD exhibited higher REM theta spectral power compared to those who developed PTSD (Cowdin et al. 2014). Similar to Mellman et al. (2007), a recent report linked increased beta spectral power during sleep to reduced PTSD symptomatology and nightmare frequency in trauma-exposed individuals with and without PTSD (Denis et al. 2021).

9 Uses for Sleep in Enhancing Therapeutic Extinction

Given evidence for sleep's role in enhancing extinction memory consolidation and the positive impact of good sleep on psychotherapy outcomes for anxiety and traumatic stress disorders, a few studies have attempted to augment exposure-based psychotherapy with sleep interventions. In an early study by Pace-Schott et al. (2012), sleep was shown to promote the consolidation and generalization of extinction learning following simulated exposure therapy for participants with spider phobia (see Fig. 3). In this study, spider-fearing young women repeatedly viewed a 1-min video of a spider followed by a 12-h delay including continued daytime wake or overnight sleep. Following the delay, participants viewed videos of the original spider as well as a novel spider. To control for time-of-day effects, two additional groups viewed the videos in the morning or evening and were tested after a two-hour delay including continued wake. Despite successful extinction learning (reduced physiological and subjective reactivity) in all groups, only participants who slept between extinction training and extinction recall/retention phases retained reduced psychophysiological reactivity to the original spider and did not show sensitization to the novel spider. This indicated that sleep may enhance both the consolidation and generalization of extinction memories. In a separate study, Kleim et al. (2014) tested whether sleep following virtual reality spider exposure in individuals with spider

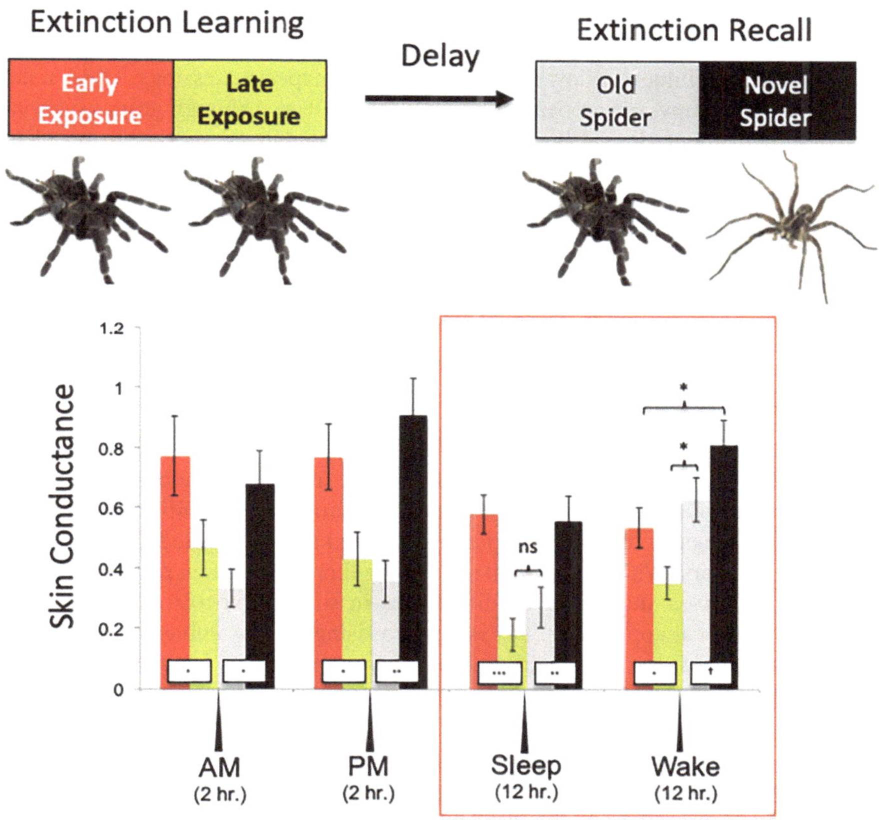

Fig. 3 Sleep promotes the retention of learned extinction and prevents sensitization to similar fear-promoting stimuli (figure adapted from Pace-Schott et al. 2012). Pace-Schott et al. (2012) exposed spider-fearing participants to a video of a spider and then had participants stay awake for 2 h in the morning ("AM") or evening ("PM"), remain awake during a 12-h delay across the day ("Wake"), or sleep overnight ("Sleep"). While both the Sleep and Wake condition showed significant reductions in SCR to the spider stimulus during extinction learning, only the Sleep condition retained this extinction across a 12-h delay. Further, the Wake condition exhibited a significantly larger SCR to a novel spider video following the delay (i.e., sensitization), a response not observed in the Sleep condition

phobia would reduce fear during later exposure to a real spider. Participants meeting clinical criteria for spider phobia were exposed to spiders using virtual reality and then either took a short nap or remained awake. Those who slept, compared to those who remained awake, showed significantly diminished self-reported spider-related fearful thoughts when they were exposed to a real spider, 1 week later. Finally, Pace-Schott et al. (2018) demonstrated that brief naps following exposure-based psychotherapy may enhance the physiological expression of extinction learning resulting from exposure-based therapy for social anxiety disorder (SAD). In this study, patients underwent a modified Trier social stress test prior to a 5-week course of exposure-based psychotherapy for SAD. Participants were randomized to either nap or continued wakefulness following two of the social-exposure sessions. Then, at the conclusion of therapy, they repeated the social stress test. Although groups demonstrated an equivalent clinical improvement, those who napped showed lower electrodermal and cortisol reactivity to the post-therapy social stress test, suggesting that the naps may have reduced physiological reactivity to social fear. More recently, Azza et al. (2022) showed that sleep compared to wake following an adaptive updating of an aversive memory may aid in reducing reactivity to the subsequent recall of that memory.

10 Contemporary Challenges

Several challenges currently exist when aiming to study the relationship between sleep and consolidation of fear and extinction memories. Replicability and generalizability of findings are the foremost issues that need to be evaluated and addressed in the coming years. While the current state of the literature is often perplexing in its inconsistency across studies, given differences in study protocols, outcome measures, and statistical approaches, this is not surprising. For example, studies examining sleep effects have differed in their basic technique for measuring fear acquisition and extinction training, viz. electrodermal response or fear-potentiated startle (Lonsdorf et al. 2017). Moreover, they have varied in their choice of conditioned and aversive stimuli, intervals between the fear acquisition, extinction training, and extinction recall/retention phases and methods by which learning and memory are measured (e.g., by self-report, psychophysiology, functional brain activity). For example, the exclusion of physiologically non-conditioning subjects or those showing excessive movement in the MRI scanner may dramatically impact final sample sizes. Even when similar protocols are used, it is the rule, rather than the exception, that outcome variable calculations and how they are subsequently analyzed differ widely between studies (Lonsdorf et al. 2019; Ney et al. 2018, 2020). In addition to recent meta-analyses (e.g., Davidson and Marcusson-Clavertz 2022; Lipinska et al. 2019, 2022; Schäfer et al. 2020; Schenker et al. 2021), a novel within-study approach, "multiverse analysis," tests for any specific group comparison (e.g., PTSD vs. trauma-exposed controls) all permutations of potentially confounding variables (e.g., methods of calculating of extinction retention, exclusion of non-conditioners; Lewis et al. 2022; Silberzahn et al. 2018; Steegen et al. 2016). Multiverse analyses then examine the distribution of outcomes across these different permutations to assess the degree of consistency of results from these different "universes" within a dataset. A recent such analysis by Lonsdorf et al. (2022) showed considerable impact of different experimental methods and indices on the outcome of fear conditioning and extinction protocols.

Finally, while by no means unique to the study of sleep and fear and extinction recall/retention, most studies to date have examined small samples resulting in low statistical power and replicability (Button et al. 2013; Ioannidis 2005). This problem has become apparent in recent meta-analyses (Lipinska et al. 2019, 2022; Schäfer et al. 2020; Schenker et al. 2021) and large-scale studies (e.g., Ackermann et al. 2015), results of which have challenged several axioms of the sleep and emotional memory field. While sample size is dictated by a number of factors that are often outside of the control of the experimenter, future efforts to increase sample size without increasing experimenter burden will be a critical step forward. Possible solutions might involve large-scale, multi-site prospective studies such as the "AURORA" (Advancing Understanding of RecOvery afteR traumA) study (Neylan et al. 2021; Stevens et al. 2021). Inter-institutional collaborations using standardized protocols might also be used to increase sample size (Davidson et al. 2021). The latter might be adapted to reduce participant burden using web-based protocols, ambulatory sleep monitoring, and smart-phone applications (e.g., Hinrichs et al. 2017) to reduce the number of in-laboratory visits. While these approaches may temporarily reduce experimental control, formulation of protocols to adapt to our increasingly remote work styles may, in the long run, have substantial benefits to the reliability and generalizability of our findings.

11 Conclusions

Sleep quality has been shown both experimentally and naturalistically to impact the retention of fear and extinction recall/retention. Evidence from human studies has demonstrated that both REM and NREM sleep may play important roles in diminishing fear responses in the absence of explicit extinction training, as well as retaining memories for experimental or therapeutic extinction using exposure therapy. Findings from basic science research have been successfully translated to understand sleep's role in processing real-world traumas. Further, targeting sleep in the context of therapeutic extinction may be a fruitful area for both basic science and clinical research. Despite current inconsistencies in the literature, sleep remains a vital area of study for understanding both emotion processing on a basic level and under conditions of psychopathology.

References

Ackermann S, Hartmann F, Papassotiropoulos A, de Quervain DJF, Rasch B (2015) No associations between interindividual differences in sleep parameters and episodic memory consolidation. Sleep 38(6):951–959. https://doi.org/10.5665/sleep.4748

Ai S-Z, Chen J, Liu J-F, He J, Xue Y-X, Bao Y-P, Han F, Tang X-D, Lu L, Shi J (2015) Exposure to extinction-associated contextual tone during slow-wave sleep and wakefulness differentially modulates fear expression. Neurobiol Learn Mem 123:159–167. https://doi.org/10.1016/j.nlm.2015.06.005

Akirav I, Maroun M (2007) The role of the medial prefrontal Cortex-Amygdala circuit in stress effects on the extinction of fear. Neural Plast 2007:30873. https://doi.org/10.1155/2007/30873

Azza Y, Wilhelm FH, Seifritz E, Junghanns K, Kleim B, Wilhelm I (2022) Sleep's role in updating aversive autobiographical memories. Transl Psychiatry 12(1) Article 1. https://doi.org/10.1038/s41398-022-01878-1

Borghese F, Henckaerts P, Guy F, Perez Mayo C, Delplanque S, Schwartz S, Perogamvros L (2022) Targeted memory reactivation during REM sleep in patients with social anxiety disorder. Front Psych 13:904704

Born J, Wilhelm I (2012) System consolidation of memory during sleep. Psychol Res 76(2): 192–203. https://doi.org/10.1007/s00426-011-0335-6

Bottary R, Cunningham TJ, Spencer RMC, Pace-Schott EF (2020a) Social Jetlag is independently associated with chronotype and poor memory for extinguished fear. Exp Results 1:e22. https://doi.org/10.1017/exp.2020.26

Bottary R, Seo J, Daffre C, Gazecki S, Moore KN, Kopotiyenko K, Dominguez JP, Gannon K, Lasko NB, Roth B, Milad MR, Pace-Schott EF (2020b) Fear extinction memory is negatively associated with REM sleep in insomnia disorder. Sleep 43(7):zsaa007. https://doi.org/10.1093/sleep/zsaa007

Brown RE, Basheer R, McKenna JT, Strecker RE, McCarley RW (2012) Control of sleep and wakefulness. Physiol Rev

Button KS, Ioannidis JP, Mokrysz C, Nosek BA, Flint J, Robinson ES, Munafò MR (2013) Power failure: why small sample size undermines the reliability of neuroscience. Nat Rev Neurosci 14(5):365–376

Carskadon MA, Dement WC (2017) Normal human sleep: an overview. In: Kryger M, Roth T, Dement WC (eds) Principles and practice of sleep medicine, 6th edn. Elsevier, Philadelphia, pp 15–24

Cartwright RD, Lloyd S, Butters E, Weiner L, Mccarthy L, Hancock J (1975) Effects of REM time on what is recalled. Psychophysiology 12(5):561–568. https://doi.org/10.1111/j.1469-8986.1975.tb00047.x

Colvonen PJ, Straus LD, Acheson D, Gehrman P (2019) A review of the relationship between emotional learning and memory, sleep, and PTSD. Curr Psychiatry Rep 21(1):2. https://doi.org/10.1007/s11920-019-0987-2

Cowdin N, Kobayashi I, Mellman TA (2014) Theta frequency activity during rapid eye movement (REM) sleep is greater in people with resilience versus PTSD. Exp Brain Res 232(5): 1479–1485. https://doi.org/10.1007/s00221-014-3857-5

Dang-Vu TT, Schabus M, Desseilles M, Sterpenich V, Bonjean M, Maquet P (2010) Functional neuroimaging insights into the physiology of human sleep. Sleep 33(12):1589–1603. https://doi.org/10.1093/sleep/33.12.1589

Davidson P, Marcusson-Clavertz D (2022) The effect of sleep on intrusive memories in daily life: a systematic review and meta-analysis of trauma film experiments. Sleep:zsac280. https://doi.org/10.1093/sleep/zsac280

Davidson P, Pace-Schott E (2020) The role of sleep in fear learning and memory. Curr Opin Psychol 34:32–36. https://doi.org/10.1016/j.copsyc.2019.08.016

Davidson P, Carlsson I, Jönsson P, Johansson M (2016) Sleep and the generalization of fear learning. J Sleep Res 25(1):88–95. https://doi.org/10.1111/jsr.12339

Davidson P, Carlsson I, Jönsson P, Johansson M (2018) A more generalized fear response after a daytime nap. Neurobiol Learn Mem 151:18–27. https://doi.org/10.1016/j.nlm.2018.03.005

Davidson P, Jönsson P, Carlsson I, Pace-Schott E (2021) Does sleep selectively strengthen certain memories over others based on emotion and perceived future relevance? Nat Sci Sleep 13:1257–1306. https://doi.org/10.2147/NSS.S286701

Denis D, Bottary R, Cunningham TJ, Zeng S, Daffre C, Oliver KL, Moore K, Gazecki S, Kram Mendelsohn A, Martinez U, Gannon K, Lasko NB, Pace-Schott EF (2021) Sleep power spectral density and spindles in PTSD and their relationship to symptom severity. Front Psych 12: 766647. https://doi.org/10.3389/fpsyt.2021.766647

Diekelmann S, Born J (2010) The memory function of sleep. Nat Rev Neurosci 11(2) Article 2. https://doi.org/10.1038/nrn2762

Dutcher CD, Dowd SM, Zalta AK, Taylor DJ, Rosenfield D, Perrone A, Otto MW, Pollack MH, Hofmann SG, Smits JAJ (2021) Sleep quality and outcome of exposure therapy in adults with social anxiety disorder. Depress Anxiety 38(11):1182–1190. https://doi.org/10.1002/da.23167

Feng P, Becker B, Feng T, Zheng Y (2018a) Alter spontaneous activity in amygdala and vmPFC during fear consolidation following 24 h sleep deprivation. Neuroimage 172:461–469. https://doi.org/10.1016/j.neuroimage.2018.01.057

Feng P, Becker B, Zheng Y, Feng T (2018b) Sleep deprivation affects fear memory consolidation: bi-stable amygdala connectivity with insula and ventromedial prefrontal cortex. Soc Cogn Affect Neurosci 13(2):145–155. https://doi.org/10.1093/scan/nsx148

Girardeau G, Lopes-dos-Santos V (2021) Brain neural patterns and the memory function of sleep. Science 374(6567):560–564

Goldstein AN, Walker MP (2014) The role of sleep in emotional brain function. Annu Rev Clin Psychol 10(1):679–708. https://doi.org/10.1146/annurev-clinpsy-032813-153716

Greenberg R, Pearlman C, Schwartz W, Grossman H (1983) Memory, emotion, and REM sleep. J Abnorm Psychol 92:378–381. https://doi.org/10.1037/0021-843X.92.3.378

Halonen R, Kuula L, Makkonen T, Kauramäki J, Pesonen A-K (2021) Self-conscious affect is modulated by rapid eye movement sleep but not by targeted memory reactivation – a pilot study. Front Psychol 12

Harrington MO, Ashton JE, Ngo H-VV, Cairney SA (2021) Phase-locked auditory stimulation of theta oscillations during rapid eye movement sleep. Sleep 44(4). https://doi.org/10.1093/sleep/zsaa227

Hauner KK, Howard JD, Zelano C, Gottfried JA (2013) Stimulus-specific enhancement of fear extinction during slow-wave sleep. Nat Neurosci 16(11):1553–1555. https://doi.org/10.1038/nn.3527

He J, Sun H-Q, Li S-X, Zhang W-H, Shi J, Ai S-Z, Li Y, Li X-J, Tang X-D, Lu L (2015) Effect of conditioned stimulus exposure during slow wave sleep on fear memory extinction in humans. Sleep 38(3):423–431. https://doi.org/10.5665/sleep.4502

Hennings AC, McClay M, Drew MR, Lewis-Peacock JA, Dunsmoor JE (2021) Neural reinstatement reveals divided organization of fear and extinction memories in the human brain. Curr Biol. https://doi.org/10.1016/j.cub.2021.11.004

Herrera S, Montorio I, Cabrera I, Botella J (2017) Memory bias for threatening information related to anxiety: an updated meta-analytic review. J Cogn Psychol 29(7):832–854. https://doi.org/10.1080/20445911.2017.1319374

Hertenstein E, Feige B, Gmeiner T, Kienzler C, Spiegelhalder K, Johann A, Jansson-Fröjmark M, Palagini L, Rücker G, Riemann D, Baglioni C (2019) Insomnia as a predictor of mental disorders: a systematic review and meta-analysis. Sleep Med Rev 43:96–105. https://doi.org/10.1016/j.smrv.2018.10.006

Hinrichs R, Michopoulos V, Winters S, Rothbaum AO, Rothbaum BO, Ressler KJ, Jovanovic T (2017) Mobile assessment of heightened skin conductance in posttraumatic stress disorder. Depress Anxiety 34(6):502–507. https://doi.org/10.1002/da.22610

Ho FY-Y, Chan CS, Tang KN-S (2016) Cognitive-behavioral therapy for sleep disturbances in treating posttraumatic stress disorder symptoms: A meta-analysis of randomized controlled trials. Clin Psychol Rev 43:90–102

Hobson JA, Pace-Schott EF (2002) The cognitive neuroscience of sleep: neuronal systems, consciousness and learning. Nat Rev Neurosci 3(9) Article 9. https://doi.org/10.1038/nrn915

Holmes EA, Bourne C (2008) Inducing and modulating intrusive emotional memories: A review of the trauma film paradigm. Acta Psychol (Amst) 127(3):553–566. https://doi.org/10.1016/j.actpsy.2007.11.002

Hu X, Cheng LY, Chiu MH, Paller KA (2020) Promoting memory consolidation during sleep: a meta-analysis of targeted memory reactivation. Psychol Bull 146(3):218

Iber C, Ancoli-Israel S, Chesson AL, Quan SF (2007) The AASM manual for the scoring of sleep and associated events: rules, terminology and technical specifications, vol 1. American Academy of Sleep Medicine, Westchester

Ioannidis JPA (2005) Why most published research findings are false. PLoS Med 2(8):e124. https://doi.org/10.1371/journal.pmed.0020124

Johnson JM, Durrant SJ (2018) The effect of cathodal transcranial direct current stimulation during rapid eye-movement sleep on neutral and emotional memory. R Soc Open Sci 5(7):172353

Kim SY, Payne JD (2020) Neural correlates of sleep, stress, and selective memory consolidation. Curr Opin Behav Sci 33:57–64. https://doi.org/10.1016/j.cobeha.2019.12.009

Kim SY, Kark SM, Daley RT, Alger SE, Rebouças D, Kensinger EA, Payne JD (2019) Interactive effects of stress reactivity and rapid eye movement sleep theta activity on emotional memory formation. Hippocampus 30(8):829–841

Kleim B, Wilhelm FH, Temp L, Margraf J, Wiederhold BK, Rasch B (2014) Sleep enhances exposure therapy. Psychol Med 44(7):1511–1519. https://doi.org/10.1017/S0033291713001748

Kleim B, Wysokowsky J, Schmid N, Seifritz E, Rasch B (2016) Effects of sleep after experimental trauma on intrusive emotional memories. Sleep 39(12):2125–2132. https://doi.org/10.5665/sleep.6310

Klinzing JG, Niethard N, Born J (2019) Mechanisms of systems memory consolidation during sleep. Nat Neurosci 22(10) Article 10. https://doi.org/10.1038/s41593-019-0467-3

Kuriyama K, Soshi T, Kim Y (2010) Sleep deprivation facilitates extinction of implicit fear generalization and physiological response to fear. Biol Psychiatry 68(11):991–998. https://doi.org/10.1016/j.biopsych.2010.08.015

Lass-Hennemann J, Michael T (2014) Endogenous cortisol levels influence exposure therapy in spider phobia. Behav Res Ther 60:39–45. https://doi.org/10.1016/j.brat.2014.06.009

Lerner I, Lupkin SM, Sinha N, Tsai A, Gluck MA (2017) Baseline levels of rapid eye movement sleep may protect against excessive activity in fear-related neural circuitry. J Neurosci 37(46):11233–11244. https://doi.org/10.1523/JNEUROSCI.0578-17.2017

Lewis MW, Pace-Schott EF, Rosso IM (2022) Multiverse analysis demonstrates an impact of methodology on results in a study of extinction retention and PTSD. Sleep 1

Lipinska G, Stuart B, Thomas KGF, Baldwin DS, Bolinger E (2019) Preferential consolidation of emotional memory during sleep: a meta-analysis. Front Psychol 10:1014. https://doi.org/10.3389/fpsyg.2019.01014

Lipinska G, Austin H, Moonsamy JR, Henry M, Lewis R, Baldwin DS, Thomas KGF, Stuart B (2022) Preferential consolidation of emotional reactivity during sleep: a systematic review and meta-analysis. Front Behav Neurosci 16:976047. https://doi.org/10.3389/fnbeh.2022.976047

Lonsdorf TB, Menz MM, Andreatta M, Fullana MA, Golkar A, Haaker J, Heitland I, Hermann A, Kuhn M, Kruse O, Meir Drexler S, Meulders A, Nees F, Pittig A, Richter J, Römer S, Shiban Y, Schmitz A, Straube B et al (2017) Don't fear 'fear conditioning': methodological considerations for the design and analysis of studies on human fear acquisition, extinction, and return of fear. Neurosci Biobehav Rev 77:247–285. https://doi.org/10.1016/j.neubiorev.2017.02.026

Lonsdorf TB, Merz CJ, Fullana MA (2019) Fear extinction retention: is it what we think it is? Biol Psychiatry 85(12):1074–1082

Lonsdorf TB, Gerlicher A, Klingelhöfer-Jens M, Krypotos A-M (2022) Multiverse analyses in fear conditioning research. Behav Res Ther 153:104072

Luik AI, Iyadurai L, Gebhardt I, Holmes EA (2019) Sleep disturbance and intrusive memories after presenting to the emergency department following a traumatic motor vehicle accident: an exploratory analysis. Eur J Psychotraumatol 10(1):1556550. https://doi.org/10.1080/20008198.2018.1556550

Marshall AJ, Acheson DT, Risbrough VB, Straus LD, Drummond SPA (2014) Fear conditioning, safety learning, and sleep in humans. J Neurosci 34(35):11754–11760. https://doi.org/10.1523/JNEUROSCI.0478-14.2014

Mellman TA, Bustamante V, Fins AI, Pigeon WR, Nolan B (2002) REM sleep and the early development of posttraumatic stress disorder. Am J Psychiatry 159(10):1696–1701. https://doi.org/10.1176/appi.ajp.159.10.1696

Mellman TA, Pigeon WR, Nowell PD, Nolan B (2007) Relationships between REM sleep findings and PTSD symptoms during the early aftermath of trauma. J Trauma Stress 20(5):893–901. https://doi.org/10.1002/jts.20246

Menz MM, Rihm JS, Salari N, Born J, Kalisch R, Pape HC, Marshall L, Büchel C (2013) The role of sleep and sleep deprivation in consolidating fear memories. Neuroimage 75:87–96. https://doi.org/10.1016/j.neuroimage.2013.03.001

Menz MM, Rihm JS, Büchel C (2016) REM sleep is causal to successful consolidation of dangerous and safety stimuli and reduces return of fear after extinction. J Neurosci 36(7): 2148–2160. https://doi.org/10.1523/JNEUROSCI.3083-15.2016

Meuret AE, Rosenfield D, Bhaskara L, Auchus R, Liberzon I, Ritz T, Abelson JL (2016) Timing matters: endogenous cortisol mediates benefits from early-day psychotherapy. Psychoneuroendocrinology 74:197–202. https://doi.org/10.1016/j.psyneuen.2016.09.008

Milad MR, Quirk GJ (2012) Fear extinction as a model for translational neuroscience: ten years of progress. Annu Rev Psychol 63(1):129–151. https://doi.org/10.1146/annurev.psych.121208.131631

Ney LJ, Wade M, Reynolds A, Zuj DV, Dymond S, Matthews A, Felmingham KL (2018) Critical evaluation of current data analysis strategies for psychophysiological measures of fear conditioning and extinction in humans. Int J Psychophysiol 134:95–107

Ney LJ, Laing PAF, Steward T, Zuj DV, Dymond S, Felmingham KL (2020) Inconsistent analytic strategies reduce robustness in fear extinction via skin conductance response. Psychophysiology 57(11):e13650. https://doi.org/10.1111/psyp.13650

Neylan TC, Kessler RC, Ressler KJ, Clifford G, Beaudoin FL, An X, Stevens JS, Zeng D, Linnstaedt SD, Germine LT, Sheikh S, Storrow AB, Punches BE, Mohiuddin K, Gentile NT, McGrath ME, van Rooij SJH, Haran JP, Peak DA et al (2021) Prior sleep problems and adverse post-traumatic neuropsychiatric sequelae of motor vehicle collision in the AURORA study. Sleep 44(3):zsaa200. https://doi.org/10.1093/sleep/zsaa200

Nishida M, Pearsall J, Buckner RL, Walker MP (2009) REM sleep, prefrontal theta, and the consolidation of human emotional memory. Cereb Cortex 19(5):1158–1166. https://doi.org/10.1093/cercor/bhn155

Oudiette D, Paller KA (2013) Upgrading the sleeping brain with targeted memory reactivation. Trends Cogn Sci 17(3):142–149. https://doi.org/10.1016/j.tics.2013.01.006

Pace-Schott EF (2009) Sleep architecture

Pace-Schott EF, Milad MR, Orr SP, Rauch SL, Stickgold R, Pitman RK (2009) Sleep promotes generalization of extinction of conditioned fear. Sleep 32(1):19–26

Pace-Schott EF, Verga PW, Bennett TS, Spencer RMC (2012) Sleep promotes consolidation and generalization of extinction learning in simulated exposure therapy for spider fear. J Psychiatr Res 46(8):1036–1044. https://doi.org/10.1016/j.jpsychires.2012.04.015

Pace-Schott EF, Spencer RMC, Vijayakumar S, Ahmed NAK, Verga PW, Orr SP, Pitman RK, Milad MR (2013) Extinction of conditioned fear is better learned and recalled in the morning than in the evening. J Psychiatr Res 47(11):1776–1784. https://doi.org/10.1016/j.jpsychires.2013.07.027

Pace-Schott EF, Tracy LE, Rubin Z, Mollica AG, Ellenbogen JM, Bianchi MT, Milad MR, Pitman RK, Orr SP (2014) Interactions of time of day and sleep with between-session habituation and extinction memory in young adult males. Exp Brain Res 232(5):1443–1458. https://doi.org/10.1007/s00221-014-3829-9

Pace-Schott EF, Germain A, Milad MR (2015a) Sleep and REM sleep disturbance in the pathophysiology of PTSD: the role of extinction memory. Biol Mood Anxiety Disord 5:3. https://doi.org/10.1186/s13587-015-0018-9

Pace-Schott EF, Rubin ZS, Tracy LE, Spencer RMC, Orr SP, Verga PW (2015b) Emotional trait and memory associates of sleep timing and quality. Psychiatry Res 229(3):999–1010. https://doi.org/10.1016/j.psychres.2015.05.069

Pace-Schott EF, Bottary RM, Kim S-Y, Rosencrans PL, Vijayakumar S, Orr SP, Lasko NB, Goetter EM, Baker AW, Bianchi MT, Gannon K, Hoeppner SS, Hofmann SG, Simon NM (2018) Effects of post-exposure naps on exposure therapy for social anxiety. Psychiatry Res 270:523–530. https://doi.org/10.1016/j.psychres.2018.10.015

Pace-Schott EF, Seo J, Bottary R (2022) The influence of sleep on fear extinction in trauma-related disorders. Neurobiol Stress:100500

Payne JD, Chambers AM, Kensinger EA (2012) Sleep promotes lasting changes in selective memory for emotional scenes. Front Integr Neurosci 6:108. https://doi.org/10.3389/fnint.2012.00108

Perogamvros L, Dang-Vu TT, Desseilles M, Schwartz S (2013) Sleep and dreaming are for important matters. Front Psychol 4:474. https://doi.org/10.3389/fpsyg.2013.00474

Peters AC, Blechert J, Sämann PG, Eidner I, Czisch M, Spoormaker VI (2014) One night of partial sleep deprivation affects habituation of hypothalamus and skin conductance responses. J Neurophysiol 112(6):1267–1276. https://doi.org/10.1152/jn.00657.2013

Porcheret K, Holmes EA, Goodwin GM, Foster RG, Wulff K (2015) Psychological effect of an analogue traumatic event reduced by sleep deprivation. Sleep 38(7):1017–1025. https://doi.org/10.5665/sleep.4802

Porcheret K, van der Kloet DH, Goodwin GM, Foster RG, Wulff K, Holmes EA (2019) Investigation of the impact of total sleep deprivation at home on the number of intrusive memories to an analogue trauma. Transl Psychiatry 9(1) Article 1. https://doi.org/10.1038/s41398-019-0403-z

Prerau MJ, Brown RE, Bianchi MT, Ellenbogen JM, Purdon PL (2017) Sleep neurophysiological dynamics through the lens of multitaper spectral analysis. Physiology 32(1):60–92

Rasch B, Born J (2013) About sleep's role in memory. Physiol Rev 93(2):681–766. https://doi.org/10.1152/physrev.00032.2012

Richards A, Inslicht SS, Yack LM, Metzler TJ, Russell Huie J, Straus LD, Dukes C, Hubachek SQ, Felmingham KL, Mathalon DH, Woodward SH, Neylan TC (2022) The relationship of fear-potentiated startle and polysomnography-measured sleep in trauma-exposed men and women with and without PTSD: testing REM sleep effects and exploring the roles of an integrative measure of sleep, PTSD symptoms, and biological sex. Sleep 45(1):zsab271. https://doi.org/10.1093/sleep/zsab271

Riemann D, Spiegelhalder K, Nissen C, Hirscher V, Baglioni C, Feige B (2012) REM sleep instability – a new pathway for insomnia? Pharmacopsychiatry 45(5):167–176. https://doi.org/10.1055/s-0031-1299721

Schäfer SK, Wirth BE, Staginnus M, Becker N, Michael T, Sopp MR (2020) Sleep's impact on emotional recognition memory: a meta-analysis of whole-night, nap, and REM sleep effects. Sleep Med Rev 51:101280. https://doi.org/10.1016/j.smrv.2020.101280

Schenker MT, Ney LJ, Miller LN, Felmingham KL, Nicholas CL, Jordan AS (2021) Sleep and fear conditioning, extinction learning and extinction recall: a systematic review and meta-analysis of polysomnographic findings. Sleep Med Rev 59:101501. https://doi.org/10.1016/j.smrv.2021.101501

Seo J, Oliver KI, Daffre C, Moore KN, Gazecki S, Lasko NB, Milad MR, Pace-Schott EF (2021a) Associations of sleep measures with neural activations accompanying fear conditioning and extinction learning and memory in trauma-exposed individuals. Sleep:zsab261. https://doi.org/10.1093/sleep/zsab261

Seo J, Pace-Schott EF, Milad MR, Song H, Germain A (2021b) Partial and total sleep deprivation interferes with neural correlates of consolidation of fear extinction memory. Biol Psychiatry Cogn Neurosci Neuroimaging 6(3):299–309. https://doi.org/10.1016/j.bpsc.2020.09.013

Silberzahn R, Uhlmann EL, Martin DP, Anselmi P, Aust F, Awtrey E, Bahník Š, Bai F, Bannard C, Bonnier E (2018) Many analysts, one data set: making transparent how variations in analytic choices affect results. Adv Methods Pract Psychol Sci 1(3):337–356

Sopp MR, Michael T, Weeß H-G, Mecklinger A (2017) Remembering specific features of emotional events across time: the role of REM sleep and prefrontal theta oscillations. Cogn Affect Behav Neurosci 17(6):1186–1209. https://doi.org/10.3758/s13415-017-0542-8

Sopp MR, Brueckner AH, Schäfer SK, Lass-Hennemann J, Michael T (2019) REM theta activity predicts re-experiencing symptoms after exposure to a traumatic film. Sleep Med 54:142–152. https://doi.org/10.1016/j.sleep.2018.10.030

Spoormaker VI, Sturm A, Andrade KC, Schröter MS, Goya-Maldonado R, Holsboer F, Wetter TC, Sämann PG, Czisch M (2010) The neural correlates and temporal sequence of the relationship between shock exposure, disturbed sleep and impaired consolidation of fear extinction. J Psychiatr Res 44(16):1121–1128. https://doi.org/10.1016/j.jpsychires.2010.04.017

Spoormaker VI, Schröter MS, Andrade KC, Dresler M, Kiem SA, Goya-Maldonado R, Wetter TC, Holsboer F, Sämann PG, Czisch M (2012) Effects of rapid eye movement sleep deprivation on fear extinction recall and prediction error signaling. Hum Brain Mapp 33(10):2362–2376. https://doi.org/10.1002/hbm.21369

Spoormaker VI, Gvozdanovic GA, Sämann PG, Czisch M (2014) Ventromedial prefrontal cortex activity and rapid eye movement sleep are associated with subsequent fear expression in human subjects. Exp Brain Res 232(5):1547–1554. https://doi.org/10.1007/s00221-014-3831-2

Steegen S, Tuerlinckx F, Gelman A, Vanpaemel W (2016) Increasing transparency through a multiverse analysis. Perspect Psychol Sci 11(5):702–712

Stevens JS, Harnett NG, Lebois LAM, van Rooij SJH, Ely TD, Roeckner A, Vincent N, Beaudoin FL, An X, Zeng D, Neylan TC, Clifford GD, Linnstaedt SD, Germine LT, Rauch SL, Lewandowski C, Storrow AB, Hendry PL, Sheikh S et al (2021) Brain-based biotypes of psychiatric vulnerability in the acute aftermath of trauma. Am J Psychiatry 178(11): 1037–1049. https://doi.org/10.1176/appi.ajp.2021.20101526

Stickgold R (2001) Sleep, learning, and dreams: off-line memory reprocessing. Science 294(5544): 1052–1057. https://doi.org/10.1126/science.1063530

Straus LD, Acheson DT, Risbrough VB, Drummond SPA (2017) Sleep deprivation disrupts recall of conditioned fear extinction. Biol Psychiatry Cogn Neurosci Neuroimaging 2(2):123–129. https://doi.org/10.1016/j.bpsc.2016.05.004

Straus LD, Drummond SPA, Risbrough VB, Norman SB (2018a) Sleep disruption, safety learning, and fear extinction in humans: implications for posttraumatic stress disorder. In: Vermetten E, Baker DG, Risbrough VB (eds) Behavioral neurobiology of PTSD. Springer, pp 193–205. https://doi.org/10.1007/7854_2017_31

Straus LD, Norman SB, Risbrough VB, Acheson DT, Drummond SPA (2018b) REM sleep and safety signal learning in posttraumatic stress disorder: a preliminary study in military veterans. Neurobiol Stress 9:22–28. https://doi.org/10.1016/j.ynstr.2018.07.001

van der Helm E, Walker MP (2009) Overnight therapy? The role of sleep in emotional brain processing. Psychol Bull 135(5):731–748. https://doi.org/10.1037/a0016570

Wagner U (2001) Emotional memory formation is enhanced across sleep intervals with high amounts of rapid eye movement sleep. Learn Mem 8(2):112–119. https://doi.org/10.1101/lm.36801

Walczak TS, Chokroverty S (2009) Chapter 12 – electroencephalography, electromyography, and electro-oculography: general principles and basic technology. In: Chokroverty S (ed) Sleep disorders medicine, 3rd edn. W.B. Saunders, pp 157–181. https://doi.org/10.1016/B978-0-7506-7584-0.00012-4

Wassing R, Benjamins JS, Talamini LM, Schalkwijk F, Van Someren EJW (2019) Overnight worsening of emotional distress indicates maladaptive sleep in insomnia. Sleep 42(4). https://doi.org/10.1093/sleep/zsy268

Woud ML, Cwik JC, Blackwell SE, Kleim B, Holmes EA, Adolph D, Zhang H, Margraf J (2018) Does napping enhance the effects of cognitive bias modification-appraisal training? An experimental study. PloS One 13(2):e0192837. https://doi.org/10.1371/journal.pone.0192837

Zalta AK, Dowd S, Rosenfield D, Smits JA, Otto MW, Simon NM, Meuret AE, Marques L, Hofmann SG, Pollack MH (2013) Sleep quality predicts treatment outcome in CBT for social anxiety disorder. Depress Anxiety 30(11):1114–1120

Zalta AK, Pinkerton LM, Valdespino-Hayden Z, Smith DL, Burgess HJ, Held P, Boley RA, Karnik NS, Pollack MH (2020) Examining insomnia during intensive treatment for veterans with posttraumatic stress disorder: does it improve and does it predict treatment outcomes? J Trauma Stress 33(4):521–527

Zeng S, Lau EYY, Li SX, Hu X (2021) Sleep differentially impacts involuntary intrusions and voluntary recognitions of lab-analogue traumatic memories. J Sleep Res 30(3):e13208. https://doi.org/10.1111/jsr.13208

Zenses A-K, Lenaert B, Peigneux P, Beckers T, Boddez Y (2020) Sleep deprivation increases threat beliefs in human fear conditioning. J Sleep Res 29(3):e12873. https://doi.org/10.1111/jsr.12873

Impact of Stress and Exercise on Fear Extinction

Jessie Provencher, Rebecca Cernik, and Marie-France Marin

Contents

Abstract This chapter reviews the literature on the impact of stress and exercise on fear extinction. Given that key brain regions of the fear circuitry (e.g., hippocampus, amygdala and frontal cortex) can be modulated by stress hormones, it is important to investigate how stress influences this process. Laboratory-based studies performed in healthy adults have yielded mixed results, which are most likely attributable to various methodological factors. Among these factors, inter-individual differences modulating the stress response and timing of stressor administration with respect to the task may contribute to this heterogeneity. Given that fear is a core manifestation of various psychopathologies and that exposure-based therapy relies on fear extinction principles, several studies have attempted to assess the role of stress hormones

J. Provencher, R. Cernik, and M.-F. Marin (✉)
Department of Psychology, Université du Québec à Montréal, Montreal, QC, Canada

Research Centre of the Institut Universitaire en Santé Mentale de Montréal, Montreal, QC, Canada
e-mail: marin.marie-france@uqam.ca

Curr Topics Behav Neurosci (2023) 64: 157–178
https://doi.org/10.1007/7854_2023_432
Published Online: 28 July 2023

on exposure-based therapy in patients suffering from post-traumatic stress disorder or anxiety disorders. These studies tend to suggest a beneficial impact of stress hormones (through either natural endogenous variations or synthetic administration) on exposure-based therapy as assessed mostly by subjective fear measures. Similar to stress, exercise can have an impact on many physiological and biological systems in humans. Of note, exercise modulates biomarkers such as brain-derived neurotrophic factor (BDNF) and anandamide (EAE) that act on brain regions implicated in the fear circuitry, supporting the importance of studying the impact of exercise on fear extinction. Overall, the results converge and indicate that fear extinction (tested in the laboratory or via exposure-based therapy in clinical populations) can be enhanced with exercise. Further research is needed to understand the mechanisms by which stress and exercise modulate fear learning and extinction processes, as well as to maximize the applicability to clinical contexts.

Keywords Exposure-based therapy · Extinction recall · Fear conditioning · Fear extinction · Laboratory-based experiments · Physical exercise · Sex differences · Stress

In this chapter, we will discuss the literature regarding the impact of stress and exercise on fear extinction. Several experimental and clinical studies have already addressed these issues. However, before diving into the heart of the matter, it is important to understand the mechanisms underlying the stress response and the context in which the various experimental and clinical studies have been conducted.

1 The Stress Response: Mechanisms and Brain Regions

The stress response is essential for human survival. Stress is a psychobiological response that is triggered when the brain perceives a stimulus as a threat (Mason 1975; Selye 1950). Physiologically, this leads to an important cascade involving several brain regions and the synthesis and release of several hormones. The stress response unfolds via two different hormonal cascades (see Fig. 1): (1) activation of the sympathetic-adrenal-medullary (SAM) axis (Sapolsky et al. 2000), which occurs rapidly and leads to adaptive physiological responses to a stressor and (2) activation of the hypothalamic-pituitary-adrenal (HPA) axis (Sapolsky et al. 2000), a much slower process that persists longer across time. When an individual perceives a threat in their environment, the SAM axis commences with a distress signal sent from the amygdala and hypothalamus that activate the sympathetic nervous system, leading to the secretion of adrenaline (A) and noradrenaline (NA) by the adrenal medulla (Cannon 1915). The secretion of these catecholamines is followed by the activation of multiple brain regions such as the hippocampus, amygdala, and cingulate, insular, and frontal cortices (Reznikoff et al. 1986). It is important to note that these regions are also implicated in fear circuitry and are characterized by their abundance of

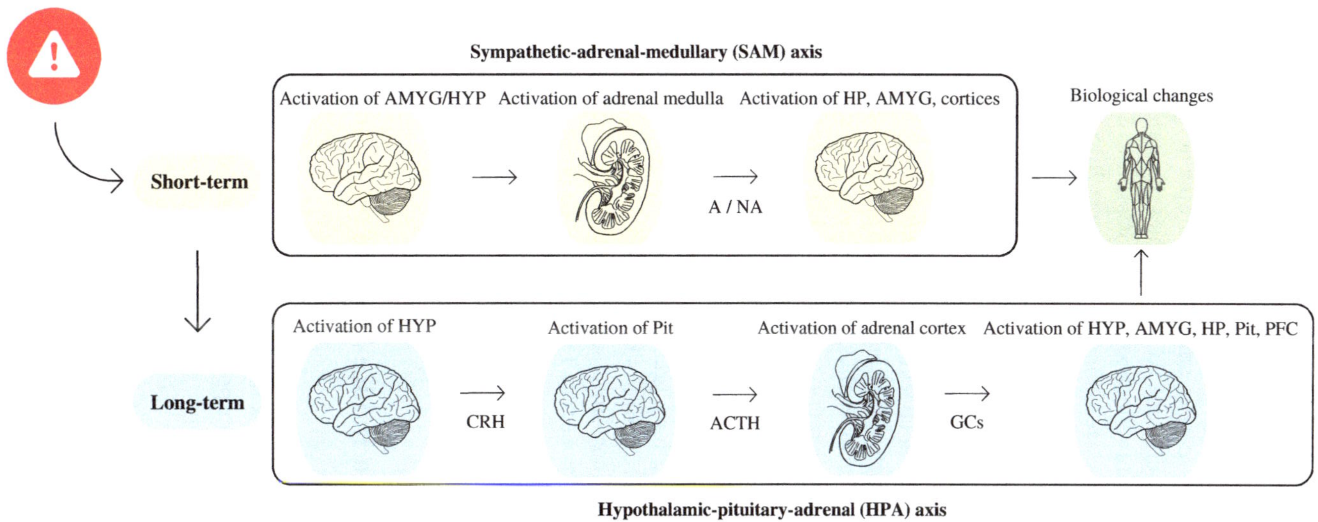

Fig. 1 Schematic representation of the two axes involved during a stress response. *AMYG* amygdala, *HYP* hypothalamus, *HP* hippocampus, *Pit* pituitary, *PFC* prefrontal cortex, *CRH* corticotropin-releasing hormone, *ACTH* adrenocorticotropic hormone, *GCs* glucocorticoids, *A* adrenaline, *NA* noradrenaline

adrenaline receptors (Reznikoff et al. 1986). The activation of the SAM axis in response to a stressor results in the following biological changes: increased blood pressure, heart rate, and respiration (Cannon and Lissák 1939; Selye 1950). The second cascade begins with the activation of the hypothalamus. This brain structure secretes corticotropin-releasing hormone (CRH), which leads to the activation of the pituitary gland. The latter secretes adrenocorticotropic hormone (ACTH), which is then released into the bloodstream. ACTH activates the cortex of the adrenal glands, resulting in the production of glucocorticoids (GCs), where cortisol is the principal GC in humans (for reviews, see Gunnar and Vazquez 2015; Heim et al. 1997; López et al. 1999). GCs can cross the blood–brain barrier and thereafter bind to corticosteroid receptors in the hypothalamus, amygdala, hippocampus, pituitary, and prefrontal cortex (Joëls and Baram 2009; Lupien et al. 2009; Mcewen et al. 1968). These brain regions play an important role in various cognitive processes, notably memory consolidation and retrieval. In general, it has been shown that stress promotes memory consolidation (Beckner et al. 2006; Cahill et al. 2003; for reviews, see de Quervain et al. 2017; Marin et al. 2019; Schwabe et al. 2012), notably for emotional information (Maheu et al. 2004; Payne et al. 2007; for a review, see Marin et al. 2019), but impairs memory retrieval (Buchanan et al. 2006; Buchanan and Tranel 2008; de Quervain et al. 1998; Tollenaar et al. 2009; for reviews, see Het et al. 2005; Marin et al. 2019). Rich in stress hormone receptors, these regions are also implicated in the fear circuitry (for a review, see Shin and Liberzon 2010). The binding of GCs to corticosteroid receptors leads to a decrease in the production of CRH and ACTH, allowing for a gradual cessation of the stress response (de Kloet et al. 2008; McEwen et al. 2016). As noted, several regions are involved in both the stress and fear response. Thus, stress seems to have a considerable influence on brain regions involved in fear processes. Furthermore, as noted and discussed in subsequent chapters in this book, dysregulation of this stress response has a significant impact on the brain circuits involved in emotion regulation and in turn is heavily intertwined with the etiology of certain psychopathologies.

2 Different Ways to Induce Stress in a Laboratory Setting

To investigate the effects of stress and its mechanisms in the laboratory, researchers can induce stress in participants in a variety of ways (see Fig. 2). First, the physiological aspects of a stress response can be targeted via the administration of synthetic hormones. For example, yohimbine (a compound that increases catecholamine concentrations; Powers et al. 2009; Smits et al. 2014) can be used to target the SAM axis, whereas synthetic GCs (e.g., hydrocortisone; Merz et al. 2014; Tabbert et al. 2010; de Quervain et al. 2011) can be used specifically for the HPA axis. Second, stress can be induced in participants using a physical stressor such as the Cold Pressor Test (CPT; Hines and Brown 1936). The CPT requires participants to submerge one of their hands in ice water for a variable short period of time. This test induces an aversive sensation that generates a stress response (Bentz et al. 2013). A

Methods used to administer stress	Advantages	Disadvantages
Synthetic hormone administration	1. Easily target a specific stress axis 2. Activates stress axis regardless of individual interpretation of the situation	1. Lack of subjective (i.e., psychological) stress component 2. More invasive than psychological protocols
Cold Pressor Test (CPT)	1. Easy administration 2. Participant is passive	1. Unable to target specific stress axis 2. Induces physical pain 3. Lack of subjective (i.e., psychological) stress component
Trier Social Stress Test (TSST)	1. Can be used to induce subjective (i.e., psychological) and objective (i.e., physiological) stress	1. Unable to target specific stress axis 2. Not universally stressful for all (individual interpretation)
Socially Evaluated Cold Pressor Test (SECPT)	1. Can be used to induce subjective (i.e., psychological) and objective (i.e., physiological) stress	1. Unable to target specific stress axis 2. Induces physical pain

Fig. 2 Advantages and disadvantages of methods used to administer stress in laboratory settings

third way to induce stress is via exposure to a psychosocial stressor such as the Trier Social Stress Test (TSST; Kirschbaum et al. 1993). The TSST consists of an oral presentation and mental arithmetic task (total of 10 min), all while the participants believe they are being filmed and evaluated by a committee of experts (Kirschbaum et al. 1993). Finally, a stressor consisting of both a physical and psychosocial component (e.g., Socially Evaluated Cold Pressor Test; SECPT; Schwabe et al. 2008) can be used to induce a stress response. Although this method is quite similar to the CPT, participants undergoing the SECPT are filmed while their hand is submerged in ice water. For the social component of this task, participants are told that their reactions are being observed by an experimenter (Meir Drexler et al. 2017). Importantly, the latter stressor type and psychosocial stressor types involve an important psychological and anticipation component (i.e., judgment from others) that can foster negative mood (Kirschbaum et al. 1993; Schwabe et al. 2008). In contrast, the administration of synthetic stress hormones does not involve these components. Moreover, the CPT and SECPT also induce physical pain in participants (Hines and Brown 1936; Schwabe et al. 2008). Therefore, pain is an important confound in these studies. For this reason, it can be difficult to generalize the effects found in CPT studies to those involving painless stressors (Antov et al. 2013). Finally, a portion of the aforementioned methods used to induce stress act on the SAM axis (Hines and Brown 1936; Smits et al. 2014), others on the HPA axis (de Quervain et al. 2011; Merz et al. 2014; Tabbert et al. 2010), whereas other methods can act on both axes (Hines and Brown 1936; Kirschbaum et al. 1993; Schwabe et al. 2008).

3 Impact of Stress on Fear Extinction

3.1 Impact of Stress-Related Synthetic Hormones in Men

As mentioned above, many stress-related synthetic hormones can be used to evoke a stress response in experimental settings. Several studies utilizing the administration of hydrocortisone before fear extinction have produced varying findings (Merz et al. 2014, 2018). Merz et al. (2014) first showed an alteration in extinction, with higher differential skin conductance responses (SCR) and lower differential activation of key fear circuitry regions (amygdala, nucleus accumbens, and medial prefrontal cortex), especially during the late extinction phase. Another study reported that hydrocortisone administration had an enhancing effect on fear extinction, as evidenced at the beginning of extinction learning by reduced SCR, lower activation of the amygdala-hippocampal complex, and enhanced functional connectivity between the vmPFC and anterior parahippocampal gyrus (Merz et al. 2018). Furthermore, when the latter study tested extinction recall 1 week later, the cortisol group still exhibited greater hippocampal-vmPFC connectivity to the extinguished cue and lower activation of the insular cortex to the non-extinguished cue compared to the placebo group.

Importantly, the main difference between these two studies lies in the delay between conditioning and extinction. In the first study, fear conditioning and extinction occurred on the same day and hydrocortisone was administered immediately after fear conditioning (45 min before fear extinction; Merz et al. 2014). In the second study, fear extinction occurred 24 h after fear conditioning. The delay between conditioning and extinction phases could modulate the direction of the effect that cortisol has on fear extinction. Based on the consolidation hypothesis, stress hormones (such as cortisol and NA) act on the amygdala during memory consolidation processes (for a review, see McGaugh 2000). That said, hydrocortisone administration immediately after the conditioning phase (and thus, during the consolidation of fear memory) may enhance fear memory consolidation and lead to higher resistance to extinction learning. With this, the enhancing effect of stress on extinction learning may require a consolidation period between conditioning and extinction learning (i.e., absence of any experimental manipulation; Merz et al. 2018). However, the time window needed to assess this effect is not known yet. Another hypothesis could be that when both threat- and safety-related information are presented on the same day (and thus, both memories are modifiable), cortisol may promote the consolidation of threat-related information by default, making the fear memory stronger and preventing any enhancing effect on fear extinction. This would explain findings on the enhancing effects of cortisol on extinction learning (safety learning) only when fear conditioning (threat learning) was already consolidated (occurring 24 h before).

3.2 *Impact of Stress-Related Synthetic Hormones in Women*

Most of the studies involving stress and fear extinction were performed solely in men (Merz et al. 2014, 2018). However, Tabbert et al. (2010) investigated the phenomenon in women taking oral contraceptives using hydrocortisone administration 45 min before conditioning and immediately followed by fear extinction. During conditioning (CS+ vs. CS− contrast), the cortisol group exhibited greater activation in the left ACC and left hippocampus, as well as trends in the left insula, right ACC, and right thalamus responses compared to the placebo group. The cortisol group also showed higher SCR during the entire extinction learning process (but not during conditioning) compared to the placebo group and this, irrespective of CS type. Interestingly, a group difference was found in the CS+ vs CS− contrast during extinction in the following regions: right hippocampus and right thalamus, with trends in the bilateral amygdala, right ACC, and right insula. This difference was driven by higher activation in these brain regions for the cortisol group toward the CS− compared to the CS+. According to previous research, this neural differentiation (greater activation to CS− than the CS+) in the hippocampus and amygdala is known to be involved in fear extinction learning (Knight et al. 2004; Phelps et al. 2004). Indeed, Phelps et al. (2004) reported a positive correlation between the neural differentiation in the amygdala and success in fear extinction (as indexed with SCR).

Therefore, increases in cortisol levels may enhance neural differentiation between the two stimuli and this could contribute to the enhancement of extinction learning (Tabbert et al. 2010). Greater responses to the CS− (compared to the CS+) during extinction have also been observed in other studies using fear-potentiated startle responses (FPS; Norrholm et al. 2006, 2008). To make sense of these results, the authors suggested either a generalization effect between the CSs at the beginning of extinction learning (Norrholm et al. 2006) or the anticipation of an additional conditioning phase involving reversed CS-US contingency (Norrholm et al. 2008). Given that few studies have been conducted in women and the small sample size of oral contraceptive users in the study by Tabbert et al. (2010), more studies are needed to understand the mechanisms by which synthetic stress hormones impact fear extinction in women.

3.3 Impact of SAM Axis Modulation

Finally, few studies in humans have explored the impact of catecholamines (particularly A and NA) on fear extinction. In a study conducted by Bos et al. (2012), participants received either propranolol (a β-adrenergic receptor antagonist) or a placebo pill before extinction learning (day 2), which was performed 24 h after fear conditioning (day 1). Their results showed that propranolol administration led to impaired extinction learning (as assessed by subjective ratings of US expectancy) that persisted during the test of extinction recall and reinstatement phase. As no impact of propranolol administration was observed on the SCR and FPS during extinction, the authors of the study concluded that the effects of propranolol on fear extinction seem to be limited to a cognitive level, as physiological measures of fear were unaffected. In another study, Lonsdorf et al. (2014) administered either reboxetine (a selective NA reuptake inhibitor that increases NA levels in the frontal cortex and hippocampus in an animal model; Hajós et al. 2004) or a placebo pill to men immediately after extinction learning (occurring 24 h after the fear conditioning phase). Based on fear ratings and SCR, results showed no impact of reboxetine administration on the consolidation of fear extinction learning. However, following reinstatement (1 week after extinction learning), the reboxetine group exhibited greater activation in the amygdala and posterior hippocampus, but lower activation in the vmPFC and anterior hippocampus relative to the placebo group. These results suggest that reboxetine-treated participants had greater activation in brain regions known to promote fear expression compared to placebo participants (who displayed greater activation in regions known to promote fear inhibition). Taken together, the results of these studies suggest that enhancement or blockade of NA may not be a promising avenue to facilitate fear extinction, at least at the physiological and neural levels.

3.4 Impact of Physical Stressors

Beyond the use of pharmacological protocols, researchers have also administered different types of stressors (physical and psychological) to study the effects of stress hormones on fear learning and extinction mechanisms. To assess the effects of the first hormonal cascade (SAM axis) on fear extinction, Antov et al. (2013, 2015) exposed male participants to either the CPT or a control condition (warm water) before fear conditioning (which occurred only 18 min before extinction learning; Antov et al. 2013) or immediately after fear conditioning, but right before fear extinction (Antov et al. 2015). In both studies, although the CPT group showed increased activity of the SAM axis, salivary cortisol concentrations remained stable (while they decreased in the control group). For this reason, we cannot rule out the possibility that the task also influenced HPA axis activity. During both the early and late phases of fear extinction learning in the first study, the CPT group exhibited a greater fear response to the CS+ compared to the CS− (when differentiation between CS+ and CS− should be minimal or absent). Again, exposure to a stressor before the conditioning phase seemed to enhance fear extinction resistance by promoting the consolidation of the fear memory (Antov et al. 2013). In the second study, no Group x Time interaction for the differential SCR during extinction was found. This suggests that no reduction of fear was observed in the latter. Of note, however, separate analyses for each group revealed a significant fear reduction in the CPT group during extinction (a result that was not found in the control group). In the latter study only, extinction recall was tested 24 h after extinction learning and led to a significantly better extinction retention index in the CPT group compared to controls. This points to a delayed beneficial effect of stress on extinction (Antov et al. 2015).

Finally, Bentz et al. (2013) exposed participants to a conditioning phase (day 1) and administered CPT (or a control condition) 20 min before fear extinction learning (day 2). On day 3, extinction memory recall was tested. When relying on the subjective measure of US expectancy, the authors observed a successful fear acquisition, but no apparent extinction of fear for both groups. However, at the beginning of days 2 and 3, men exposed to the CPT showed lower US expectancy than the control group. This suggests that stress induces lower explicit fear memory. This effect was not found in women. The authors noted that the observed sex differences may be due to lower cortisol responses in women in the CPT group compared to men in the same group (Bentz et al. 2013).

These three studies reached different conclusions while using diverse measures of fear (neural, SCR, subjective US expectancy) and sex distributions. Moreover, these studies utilized various methodologies regarding the timing of CPT administration and had certain limitations (e.g., non-successful conditioning and/or extinction in the control group). It is also important to remember that these three studies reported a potential impact of the CPT on the HPA axis, which ultimately prevented the researchers from isolating the effects of the SAM axis on fear extinction processes. Taken together, these differences and limitations across studies rendered it difficult to draw definitive conclusions about the effects of physical stressors on extinction learning.

3.5 Impact of Psychosocial and Combined Stressors

The studies summarized thus far have had mixed results. Studies employing psychological stressors are no exception regarding the impact of stress on fear extinction (Antov and Stockhorst 2014; Jackson et al. 2006; Simon-Kutscher et al. 2019). In these studies, participants were exposed to a variant of the TSST or control condition before the conditioning phase, which was immediately followed by extinction learning. However, only Antov and Stockhorst (2014) tested extinction memory recall.

To begin, Simon-Kutscher et al. (2019) reported a resistance effect of stress on fear extinction as seen by no reduction of SCR to the CS+ during extinction. The authors argue that the results obtained during extinction learning may be explained by an increased fear memory during acquisition due to TSST exposure, which could have promoted extinction resistance in that group. Given that cortisol levels were similar for both groups (i.e., stressed and controls) during extinction (but that SCR differed), cortisol may have promoted fear memory and in turn led to deficits in extinction through enhanced extinction resistance mechanisms. Similarly, Jackson et al. (2006) reported larger SCR to the CS+ during fear acquisition in stressed men only, where this effect was maintained during early and late trials of extinction learning. However, Antov and Stockhorst (2014) failed to find an impact of psychosocial stress on fear extinction learning in both men and women. In their study, the effects of stress were found during extinction recall as a function of participants' estradiol profiles. Specifically, stressed women in the early follicular phase of their menstrual cycle (low estradiol levels) showed greater fear recovery than their non-stressed counterparts. The opposite was found in mid-cycle women (high estradiol levels). Thus, although extinction retention in women seems to be modulated by stress when administered before fear conditioning, this effect varies according to estradiol levels. As the other studies did not test extinction recall, no study so far has replicated the results obtained by Antov and Stockhorst (2014).

Finally, only one study in the literature utilized the SECPT (Schwabe et al. 2008) to explore the impact of stress on fear extinction (Riggenbach et al. 2019). Here, stress was induced before the fear conditioning phase, 24 h before extinction learning, and 48 h before extinction memory testing. Despite a lack of evidence found for an effect of stress on extinction learning assessed by FPS, results showed that the reduction of threat expectancy for the CS+ was slower in stressed participants compared to controls. These findings reflect a resistance effect of fear extinction after exposure to the SECPT. In addition, extinction recall showed that stressed participants expressed greater FPS for both the CS+ and CS−. The authors also reported that cortisol responses to the SECPT were positively correlated with CS discrimination during extinction learning and recall. Again, stress exposure before fear conditioning seems to promote fear memory and alter fear extinction processes (Riggenbach et al. 2019).

As summarized above, the literature contains mixed findings regarding the impact of stress on fear extinction in a laboratory setting, such that stress can alter, enhance, or have no impact on fear extinction (for a summary of the studies conducted in laboratory settings discussed in this chapter, see Fig. 3). First, individual differences (e.g., personality traits, sex, and early life events) may influence how a person responds to stress on neuroendocrine, emotional, and physiological levels (Bangasser and Wicks 2017; Childs et al. 2014; Meaney et al. 1993; Orem et al. 2019). These inter-individual differences may contribute to the mixed findings regarding the impact of stress on fear extinction. Second, beyond the fact that these differences can also be partially explained by statistical power, stressor types, or population characteristics (such as sex), many studies have noted the importance of considering when the stressor is presented and the delay between conditioning and extinction phases (timing; Henckens et al. 2010, 2011, 2012; Meir Drexler et al. 2019). To have a better understanding of the effect of stress on fear extinction, it would be relevant to isolate each phase of the fear conditioning protocol by separating the phases by at least 24 h. This would reduce the chances of contamination between phases or the stressor acting on multiple phases. Together, this would allow researchers to isolate the effects of stress on the different phases and ultimately reach clearer conclusions.

4 Impact of Stress Hormones on Exposure Therapy

As mentioned in the opening of this chapter, studies examining the impact of stress hormones on fear extinction can be conducted in clinical settings. Several of these studies use the synthetic hormone administration techniques explained above to modulate stress hormone levels in patients and assess their impact on exposure-based therapy.

4.1 In Populations with Anxiety Disorders

Studies have exposed individuals with phobic disorder to exposure-based therapy sessions. In these studies, half of the participants were administered cortisol or yohimbine (to evaluate the influence of the HPA or SAM axis, respectively) 1 h before each session, whereas the remainder of the samples received a placebo. Three studies reported a greater decrease in subjective fear levels or phobic symptoms in the cortisol group over the course of exposure therapy compared to the placebo group (Soravia et al. 2006; de Quervain et al. 2011; Smits et al. 2014). For the SAM axis, a similar significant reduction in phobic fear in both groups (placebo vs. yohimbine) was found after therapy (Powers et al. 2009). Importantly,

Protocols and methods to administer stress		Types of measures and effects on fear extinction/recall							
		Skin conductance response		Subjective fear		Neural signals		Fear-potentiated startle response	
		Extinction	Recall	Extinction	Recall	Extinction	Recall	Extinction	Recall
Stress + COND + EXT	Synthetic hormone administration	↓	-	-	-	↑	-	-	-
	Cold Pressor Test	↓	-	-	-	-	-	-	-
	Trier Social Stress Test	↑↓	↑↓	-	-	-	-	-	-
Stress + COND + 24h + EXT	Socially Evaluated Cold Pressor Test	-	-	↓	-	-	-	Ø	↓
COND + Stress + EXT	Synthetic hormone administration	↓	-	-	-	↓	-	-	-
	Cold Pressor Test	↑	↑	-	-	-	-	-	-
COND + 24h + Stress + EXT	Synthetic hormone administration	↑↓	-	↓	↓	↑	↑	Ø	-
	Cold Pressor Test	Ø	-	Ø	↑	-	-	-	-
COND + 24h + EXT + Stress	Synthetic hormone administration	Ø	-	Ø	-	-	↓	-	-

Fig. 3 Effects of different protocols and methods of administering stress on fear extinction and recall in laboratory settings as quantified by physiological, subjective, neural, and fear-potentiated startle response measures. *COND* fear conditioning, *EXT* fear extinction, *24 h* delay of 24 h, ↑ enhanced effect, ↓ hindered effect, ↑↓ mixed results, Ø null effect, - no studies available

all studies showed a long-lasting effect of stress on exposure therapy as seen by lower fear levels at the last session and follow-ups (2-week and 1-month). With this, as the effect of stress remained at the follow-up session (where no drugs were administrated), it seems that modulation of the stress systems not only acts by blocking fear memory retrieval during therapy but most likely facilitates the consolidation of the safety-memory trace formed during exposure-based therapy (Soravia et al. 2006; Powers et al. 2009).

Collectively, these results converge to suggest an enhancing role of synthetic stress hormones (that act on either the SAM or HPA axes) in exposure-based therapy in phobic individuals. Importantly, one of these studies showed that endogenous cortisol released during exposure therapy by participants in the placebo group negatively correlated with fear ratings, suggesting that even endogenous cortisol may contribute to reduced fear symptoms in individuals suffering from specific phobia (Soravia et al. 2006). Similarly, higher endogenous cortisol elevation during exposure therapy also led to greater symptom improvement, as measured by subjective ratings, in individuals suffering from agoraphobia or panic disorder (Meuret et al. 2015).

Most of these effects are based on self-reports. Although de Quervain et al. (2011) also found beneficial effects of cortisol administration on physiological fear levels, more studies are needed to confirm whether the positive effects of synthetic and endogenous stress hormones on self-reported fear translate to physiological fear levels.

4.2 *In Populations with Post-Traumatic Stress Disorder (PTSD)*

To the best of our knowledge, only two studies explored the combination of stress hormone administration with exposure-based therapy in PTSD patients. A case study in two veterans with PTSD showed that the veteran who received a combination of hydrocortisone and exposure therapy exhibited a greater and faster decline in PTSD symptoms compared to the veteran who was given a placebo (Yehuda et al. 2010). Due to their intriguing findings, the research group conducted the same experiment in 24 veterans with PTSD. Half of the sample received hydrocortisone before exposure therapy, whereas the other half was given a placebo. As they were able to replicate the findings of their original study, hydrocortisone may enhance retention during exposure-based therapy and in consequence lead to a greater reduction in PTSD symptoms (Yehuda et al. 2015).

In addition, another study showed that higher endogenous cortisol levels may be an important biomarker for treatment responsiveness in populations with PTSD (van Gelderen et al. 2020). Here, salivary cortisol levels were measured before and after each therapy session, as well as in the afternoon following each session. Results showed that responders to treatment had higher pre- and post-session salivary cortisol levels relative to non-responders, which predicted greater improvements in PTSD symptoms. As cortisol levels in the hours following the treatment session did not differ between the responders and non-responders, the authors suggested that the beneficial effects of cortisol on exposure-based therapy are not attributable to higher basal cortisol levels, but rather to a higher anticipatory response before the therapy session (van Gelderen et al. 2020). Thus, endogenous cortisol responses to the therapy may have beneficial treatment outcomes.

In sum, the literature reviewed in this chapter suggests that fear extinction processes can be influenced by both exogenous and endogenous stress hormone modulation. The direction of the effects depends on various methodological factors. Overall, conclusions resulting from studies using laboratory-based protocols are mixed. However, stress hormone administration and endogenous stress hormone levels both seem to facilitate exposure-based therapy in clinical populations (for a review, see McGuire et al. 2014). This highlights the need for further well-controlled experimental studies to identify the mechanisms that promote such an effect.

As with stress, exercise can have an impact on numerous physiological and biological systems in humans (Cotman and Berchtold 2002; Nowakowski Sims 2015; for a review, see Tanner et al. 2018). Thus, it is relevant to explore the modulatory effects of exercise on fear extinction in both experimental and clinical settings.

5 The Influence of Physical Exercise

Exercise is known to improve physical health, sleep, mood, energy levels, and emotions, among others (Nowakowski Sims 2015). Exercise may also prevent, promote, and maintain good mental health (Saxena et al. 2005) and lead to various immediate physical changes including faster heart rate and respiration. Moreover, it has been shown that exercise can alter neurochemical processes, such as modulation of brain-derived neurotrophic factor (BDNF) levels (Cotman and Berchtold 2002) and anandamide (AEA; Crombie et al. 2018, 2019). On the one hand, BDNF is essential for the development and differentiation of neurons in several brain regions (for a review, see Notaras and van den Buuse 2020) that form the fear circuitry, including the hippocampus (Cotman and Berchtold 2002; Greenwood et al. 2009; Neeper et al. 1995), amygdala (Greenwood et al. 2009), and prefrontal cortex (Aksu et al. 2012; for a review, see Notaras and van den Buuse 2020). In addition, BDNF plays a role in adaptive plasticity, where the latter is important for learning, memory, and fear processes (for a review, see Notaras and van den Buuse 2020). Interestingly, animal studies have also reported that a lower natural BDNF concentration was associated with greater difficulty in extinguishing fear, whereas an increase in BDNF concentration appeared to prevent extinction failure (Peters et al. 2010). On the other hand, AEA is one of the important endocannabinoid (eCB) system biomarkers (Katona and Freund 2012) that modulate fear and extinction processes. This biomarker acts on the amygdala, hippocampus, and frontal cortex, among others

(Katona and Freund 2012; Lafenêtre et al. 2007; Lutz et al. 2015). Again, these three brain regions are highly important in fear extinction mechanisms (for a review, see Shin and Liberzon 2010). Therefore, given that exercise modulates biomarkers that are involved in both neural plasticity and fear-related circuitry, researchers have investigated the impact of exercise on fear extinction.

5.1 Impact of Exercise on Fear Extinction

The few human studies that have investigated this topic were initiated following results obtained from animal research. In general, animal studies have reported that aerobic exercise (e.g., running or walking) may enhance fear extinction when exercise is done after fear extinction learning, but not before (Jacquart et al. 2017; Roquet and Monfils 2018). The human literature has built on these findings to assess the impact of exercise on fear extinction.

In an experimental study, Keyan and Bryant (2019) exposed healthy young adults to a fear conditioning protocol and thereafter, randomly assigned them to either an exercise or control condition immediately after the fear extinction phase. Results showed that participants in the exercise group exhibited lower FPS to the CS+ than the control group during extinction recall, suggesting an enhancing effect of exercise on fear extinction processes. These effects may be due to the modulation of BDNF levels, although the latter was not quantified in this study. However, it is well-documented that the influence of BDNF could be modulated by genetic factors, such as the Val66Met polymorphism (Egan et al. 2003). For this reason, the authors tested whether the effect of exercise on fear extinction memory was moderated by this polymorphism. Although fear recovery and exercise condition were not associated with the Val homozygote carriers, the Met allele carriers showed greater fear recovery 24 h after fear extinction learning in the exercise group. The authors suggest that these individuals may have experienced higher exercise-induced stress without the beneficial effects of BDNF, therefore compromising fear extinction (Keyan and Bryant 2019). Due to the limited number of studies exploring this question, this remains speculative and requires further experimental studies to test this hypothesis.

Further, Crombie et al. (2021a, b) explored the impact of exercise on fear extinction in samples of women with PTSD. In their study, they sought to establish the nature of the influence of exercise on fear extinction (Crombie et al. 2021b) and better understand the underlying mechanisms (Crombie et al. 2021a). Although this study was conducted in a clinical sample, the researchers used a laboratory protocol in which the feared stimulus was unrelated to the participants' PTSD (Crombie et al. 2021a, b). Participants were randomly assigned to either the exercise or control condition immediately after the extinction learning phase (which occurred 24 h after conditioning). To assess AEA and BDNF concentrations, a subset of the participants provided blood samples before and after the exercise/control condition (Crombie et al. 2021a).

First, compared to the control group, the exercise group reported less threat expectancy to both CS+ and CS− after reinstatement. As no effect of exercise was identified during the initial fear recall test, researchers proposed that the potentiating effect of exercise on fear extinction consolidation may only be observed when later confronted with the feared stimulus. As such, greater extinction consolidation would protect the individual from fear reinstatement (Crombie et al. 2021b). Second, compared to the control condition, the exercise group showed a statistical trend toward increased AEA concentrations after the exercise protocol (Crombie et al. 2021a). Moreover, a trend revealed that the increases in AEA were negatively associated with lower threat expectancy after reinstatement. Compared to controls, participants in the exercise condition also showed the same negative association between BDNF concentrations and threat expectancy after reinstatement. Therefore, exercise may improve the consolidation of fear extinction learning in women with PTSD, where this effect seems to be mediated by AEA and BDNF concentrations (Crombie et al. 2021a).

6 Impact of Physical Exercise in Clinical Samples

As they rely on fear extinction mechanisms, a few studies have begun to explore the impact of exercise on exposure therapies that are commonly used in the treatment of PTSD and anxiety disorders (Jacquart et al. 2017; McInerney et al. 2021; Powers et al. 2015).

Powers et al. (2015) conducted the first pilot study in individuals diagnosed with PTSD to investigate the effect of BDNF concentrations on the influence of exercise on exposure therapy. While participants in the exercise condition exhibited higher BDNF concentrations and fewer PTSD symptoms at the end of the last session than participants in the control condition (Powers et al. 2015), two similar studies using virtual reality exposure therapy with participants with acrophobia did not observe an effect of exercise on therapy outcomes (Jacquart et al. 2017; McInerney et al. 2021). First, one noted that their divergent results could be explained by the nature of the participants' diagnoses in each study (PTSD vs. acrophobia) and the small sample used in the pilot study (Jacquart et al. 2017). Second, though only the participants in the exercise condition demonstrated a high heart rate during therapy, both groups demonstrated considerable fear reduction following exposure therapy, with no added value of exercise in the study conducted by McInerney et al. (2021). From this, the authors inferred that virtual reality exposure therapy alone may be sufficient to treat height phobias and that increased physical arousal through exercise appeared to have no additional effect.

In summary, this chapter aimed to explore the influence of stress and exercise on fear extinction. First, we highlighted the mixed nature of the literature regarding the effect of stress on fear extinction in experimental settings, where stress can reduce, enhance, or have no effect on fear extinction learning and recall. One major consideration for future studies pertains to the timing at which the modulation of the stress systems occurs with regard to the various phases of the protocol (i.e., conditioning, extinction). The limited number of studies targeting this question, individual differences in stress responses, different methods, diverse study populations, and small sample sizes may also account for some of these diverging results in the studies conducted thus far. Therefore, further studies are required to better understand how stress mechanisms impact extinction learning, consolidation, and retention. However, the studies performed in clinical populations tend to suggest a beneficial impact of stress hormone administration, and to some degree endogenous stress hormone levels, on exposure therapy assessed principally by subjective measures. More studies using physiological fear responses and neuroimaging techniques are still needed. Second, human studies investigating the impact of exercise on fear extinction (or exposure therapies) remain scarce. Though, overall results indicate an effect of exercise on enhancing fear extinction consolidation. These positive results were obtained in both experimental and clinical settings. In light of the limitations reported in the literature, future research is required to better understand the mechanisms underlying the impact of exercise on fear extinction. Important considerations for future studies include recruiting larger samples, targeting varying populations, examining genetic predispositions, considering the timing of exercise, measuring physiological fear responses, and using neuroimaging techniques. Overall, the modest literature suggests beneficial effects of both exercise and stress hormone modulation on exposure-based therapy in clinical populations. More fundamental and translational research is however necessary to tackle the mechanisms by which these effects occur and maximize its utility in clinical settings.

References

Aksu I, Baykara B, Ozbal S, Cetin F, Sisman AR, Dayi A, Gencoglu C, Tas A, Büyük E, Gonenc-Arda S, Uysal N (2012) Maternal treadmill exercise during pregnancy decreases anxiety and increases prefrontal cortex VEGF and BDNF levels of rat pups in early and late periods of life. Neurosci Lett 516(2):221–225. https://doi.org/10.1016/j.neulet.2012.03.091

Antov MI, Stockhorst U (2014) Stress exposure prior to fear acquisition interacts with estradiol status to alter recall of fear extinction in humans. Psychoneuroendocrinology 49:106–118. https://doi.org/10.1016/j.psyneuen.2014.06.022

Antov MI, Wölk C, Stockhorst U (2013) Differential impact of the first and second wave of a stress response on subsequent fear conditioning in healthy men. Biol Psychol 94(2):456–468. https://doi.org/10.1016/j.biopsycho.2013.08.007

Antov MI, Melicherová U, Stockhorst U (2015) Cold pressor test improves fear extinction in healthy men. Psychoneuroendocrinology 54:54–59. https://doi.org/10.1016/j.psyneuen.2015.01.009

Bangasser DA, Wicks B (2017) Sex-specific mechanisms for responding to stress. J Neurosci Res 95(1–2):75–82. https://doi.org/10.1002/jnr.23812

Beckner VE, Tucker DM, Delville Y, Mohr DC (2006) Stress facilitates consolidation of verbal memory for a film but does not affect retrieval. Behav Neurosci 120(3):518–527. https://doi.org/10.1037/0735-7044.120.3.518

Bentz D, Michael T, Wilhelm FH, Hartmann FR, Kunz S, von Rohr IRR, de Quervain DJ-F (2013) Influence of stress on fear memory processes in an aversive differential conditioning paradigm in humans. Psychoneuroendocrinology 38(7):1186–1197. https://doi.org/10.1016/j.psyneuen.2012.12.018

Bos MGN, Beckers T, Kindt M (2012) The effects of noradrenergic blockade on extinction in humans. Biol Psychol 89(3):598–605. https://doi.org/10.1016/j.biopsycho.2012.01.007

Buchanan TW, Tranel D (2008) Stress and emotional memory retrieval: effects of sex and cortisol response. Neurobiol Learn Mem 89(2):134–141. https://doi.org/10.1016/j.nlm.2007.07.003

Buchanan TW, Tranel D, Adolphs R (2006) Impaired memory retrieval correlates with individual differences in cortisol response but not autonomic response. Learn Mem 13(3):382–387. https://doi.org/10.1101/lm.206306

Cahill L, Gorski L, Le K (2003) Enhanced human memory consolidation with post-learning stress: interaction with the degree of arousal at encoding. Learn Mem 10(4):270–274. https://doi.org/10.1101/lm.62403

Cannon WB (1915) Bodily changes in pain, hunger, fear and rage. An account of recent researches into the function of emotional excitement. J Am Med Assoc LXIV(23):1933. https://doi.org/10.1001/jama.1915.02570490049033

Cannon WB, Lissák K (1939) Evidence for adrenaline in adrenergic neurones. Am J Physiol-Leg Cont 125(4):765–777. https://doi.org/10.1152/ajplegacy.1939.125.4.765

Childs E, White TL, de Wit H (2014) Personality traits modulate emotional and physiological responses to stress. Behav Pharmacol 25(5 and 6):493–502. https://doi.org/10.1097/FBP.0000000000000064

Cotman CW, Berchtold NC (2002) Exercise: a behavioral intervention to enhance brain health and plasticity. Trends Neurosci 25(6):7

Crombie KM, Brellenthin AG, Hillard CJ, Koltyn KF (2018) Psychobiological responses to aerobic exercise in individuals with posttraumatic stress disorder: psychobiological responses to exercise in PTSD. J Trauma Stress 31(1):134–145. https://doi.org/10.1002/jts.22253

Crombie KM, Leitzelar BN, Brellenthin AG, Hillard CJ, Koltyn KF (2019) Loss of exercise- and stress-induced increases in circulating 2-arachidonoylglycerol concentrations in adults with chronic PTSD. Biol Psychol 145:1–7. https://doi.org/10.1016/j.biopsycho.2019.04.002

Crombie KM, Sartin-Tarm A, Sellnow K, Ahrenholtz R, Lee S, Matalamaki M, Almassi NE, Hillard CJ, Koltyn KF, Adams TG, Cisler JM (2021a) Exercise-induced increases in anandamide and BDNF during extinction consolidation contribute to reduced threat following reinstatement: preliminary evidence from a randomized controlled trial. Psychoneuroendocrinology 132:105355. https://doi.org/10.1016/j.psyneuen.2021.105355

Crombie KM, Sartin-Tarm A, Sellnow K, Ahrenholtz R, Lee S, Matalamaki M, Adams TG, Cisler JM (2021b) Aerobic exercise and consolidation of fear extinction learning among women with posttraumatic stress disorder. Behav Res Ther 142:103867. https://doi.org/10.1016/j.brat.2021.103867

de Kloet ER, Karst H, Joëls M (2008) Corticosteroid hormones in the central stress response: quick-and-slow. Front Neuroendocrinol 29(2):268–272. https://doi.org/10.1016/j.yfrne.2007.10.002

de Quervain DJ-F, Roozendaal B, McGaugh JL (1998) Stress and glucocorticoids impair retrieval of long-term spatial memory. Nature 394(6695):787–790. https://doi.org/10.1038/29542

de Quervain DJ-F, Bentz D, Michael T, Bolt OC, Wiederhold BK, Margraf J, Wilhelm FH (2011) Glucocorticoids enhance extinction-based psychotherapy. Proc Natl Acad Sci 108(16):6621–6625. https://doi.org/10.1073/pnas.1018214108

de Quervain D, Schwabe L, Roozendaal B (2017) Stress, glucocorticoids and memory: implications for treating fear-related disorders. Nat Rev Neurosci 18(1):7–19. https://doi.org/10.1038/nrn.2016.155

Egan MF, Kojima M, Callicott JH, Goldberg TE, Kolachana BS, Bertolino A, Zaitsev E, Gold B, Goldman D, Dean M, Lu B, Weinberger DR (2003) The BDNF val66met polymorphism affects activity-dependent secretion of BDNF and human memory and hippocampal function. Cell 112(2):257–269. https://doi.org/10.1016/S0092-8674(03)00035-7

Greenwood BN, Strong PV, Foley TE, Fleshner M (2009) A behavioral analysis of the impact of voluntary physical activity on hippocampus-dependent contextual conditioning. Hippocampus 19(10):988–1001. https://doi.org/10.1002/hipo.20534

Gunnar MR, Vazquez D (2015) Stress neurobiology and developmental psychopathology. In: Cicchetti D, Cohen DJ (eds) Developmental psychopathology. Wiley, pp 533–577. https://doi.org/10.1002/9780470939390.ch13

Hajós M, Fleishaker JC, Filipiak-Reisner JK, Brown MT, Wong EHF (2004) The selective norepinephrine reuptake inhibitor antidepressant reboxetine: pharmacological and clinical profile. CNS Drug Rev 10(1):23–44. https://doi.org/10.1111/j.1527-3458.2004.tb00002.x

Heim C, Owens MJ, Plotsky PM, Nemeroff CB (1997) The role of early adverse life events in the etiology of depression and posttraumatic stress disorder: focus on corticotropin-releasing factor. Ann N Y Acad Sci 821(1 Psychobiology):194–207. https://doi.org/10.1111/j.1749-6632.1997.tb48279.x

Henckens MJAG, van Wingen GA, Joëls M, Fernández G (2010) Time-dependent effects of corticosteroids on human amygdala processing. J Neurosci 30(38):12725–12732. https://doi.org/10.1523/JNEUROSCI.3112-10.2010

Henckens MJAG, van Wingen GA, Joëls M, Fernández G (2011) Time-dependent corticosteroid modulation of prefrontal working memory processing. Proc Natl Acad Sci U S A 108(14):5801–5806. https://doi.org/10.1073/pnas.1019128108

Henckens MJAG, van Wingen GA, Joëls M, Fernández G (2012) Time-dependent effects of cortisol on selective attention and emotional interference: a functional MRI study. Front Integr Neurosci 6:66. https://doi.org/10.3389/fnint.2012.00066

Het S, Ramlow G, Wolf O (2005) A meta-analytic review of the effects of acute cortisol administration on human memory. Psychoneuroendocrinology 30(8):771–784. https://doi.org/10.1016/j.psyneuen.2005.03.005

Hines EA, Brown GE (1936) The cold pressor test for measuring the reactibility of the blood pressure: data concerning 571 normal and hypertensive subjects. Am Heart J 11(1):1–9. https://doi.org/10.1016/S0002-8703(36)90370-8

Jackson ED, Payne JD, Nadel L, Jacobs WJ (2006) Stress differentially modulates fear conditioning in healthy men and women. Biol Psychiatry 59(6):516–522. https://doi.org/10.1016/j.biopsych.2005.08.002

Jacquart J, Roquet RF, Papini S, Powers MB, Rosenfield D, Smits JAJ, Monfils M-H (2017) Effects of acute exercise on fear extinction in rats and exposure therapy in humans: null findings from five experiments. J Anxiety Disord 50:76–86. https://doi.org/10.1016/j.janxdis.2017.05.010

Joëls M, Baram TZ (2009) The neuro-symphony of stress. Nat Rev Neurosci 10(6):459–466. https://doi.org/10.1038/nrn2632

Katona I, Freund TF (2012) Multiple functions of endocannabinoid signaling in the brain. Annu Rev Neurosci 35(1):529–558. https://doi.org/10.1146/annurev-neuro-062111-150420

Keyan D, Bryant RA (2019) Acute exercise-induced enhancement of fear inhibition is moderated by BDNF Val66Met polymorphism. Transl Psychiatry 9(1):131. https://doi.org/10.1038/s41398-019-0464-z

Kirschbaum C, Pirke K-M, Hellhammer DH (1993) The "Trier Social Stress Test" – a tool for investigating psychobiological stress responses in a laboratory setting. Neuropsychobiology 28:76–81

Knight DC, Smith CN, Cheng DT, Stein EA, Helmstetter FJ (2004) Amygdala and hippocampal activity during acquisition and extinction of human fear conditioning. Cogn Affect Behav Neurosci 4(3):317–325. https://doi.org/10.3758/CABN.4.3.317

Lafenêtre P, Chaouloff F, Marsicano G (2007) The endocannabinoid system in the processing of anxiety and fear and how CB1 receptors may modulate fear extinction. Pharmacol Res 56(5):367–381. https://doi.org/10.1016/j.phrs.2007.09.006

Lonsdorf TB, Haaker J, Fadai T, Kalisch R (2014) No evidence for enhanced extinction memory consolidation through noradrenergic reuptake inhibition – delayed memory test and reinstatement in human fMRI. Psychopharmacology 231(9):1949–1962. https://doi.org/10.1007/s00213-013-3338-8

López JF, Akil H, Watson SJ (1999) Neural circuits mediating stress. Biol Psychiatry 46(11): 1461–1471. https://doi.org/10.1016/S0006-3223(99)00266-8

Lupien SJ, McEwen BS, Gunnar MR, Heim C (2009) Effects of stress throughout the lifespan on the brain, behaviour and cognition. Nat Rev Neurosci 10(6):434–445. https://doi.org/10.1038/nrn2639

Lutz B, Marsicano G, Maldonado R, Hillard CJ (2015) The endocannabinoid system in guarding against fear, anxiety and stress. Nat Rev Neurosci 16(12):705–718. https://doi.org/10.1038/nrn4036

Maheu FS, Joober R, Beaulieu S, Lupien SJ (2004) Differential effects of adrenergic and corticosteroid hormonal systems on human short- and long-term declarative memory for emotionally arousing material. Behav Neurosci 118(2):420–428. https://doi.org/10.1037/0735-7044.118.2.420

Marin M-F, Raymond C, Lupien SJ (2019) Memory and stress. In: Stress: physiology, biochemistry, and pathology. Elsevier, pp 69–78. https://doi.org/10.1016/B978-0-12-813146-6.00006-0

Mason JW (1975) A historical view of the stress field. J Hum Stress 1(1):6–12. https://doi.org/10.1080/0097840X.1975.9940399

Mcewen BS, Weiss JM, Schwartz LS (1968) Selective retention of corticosterone by limbic structures in rat brain. Nature 220(5170):911–912. https://doi.org/10.1038/220911a0

McEwen BS, Nasca C, Gray JD (2016) Stress effects on neuronal structure: hippocampus, amygdala, and prefrontal cortex. Neuropsychopharmacology 41(1):3–23. https://doi.org/10.1038/npp.2015.171

McGaugh JL (2000) Memory – a century of consolidation. Science 287(5451):248–251. https://doi.org/10.1126/science.287.5451.248

McGuire JF, Lewin AB, Storch EA (2014) Enhancing exposure therapy for anxiety disorders, obsessive-compulsive disorder and post-traumatic stress disorder. Expert Rev Neurother 14(8): 893–910. https://doi.org/10.1586/14737175.2014.934677

McInerney J, Brown P, Bird JC, Nickless A, Brown G, Freeman D (2021) Does raising heart rate prior to a behavioural test enhance learning in cognitive therapy for anxiety? An experimental test for the treatment of fear of heights using virtual reality. Behav Res Ther 144:103928. https://doi.org/10.1016/j.brat.2021.103928

Meaney MJ, Bhatnagar S, Larocque S, McCORMICK C, Shanks N, Sharma S, Smythe J, Viau V, Plotsky PM (1993) Individual differences in the hypothalamic-pituitary-adrenal stress response and the hypothalamic CRF system. Ann N Y Acad Sci 697(1 Corticotropin):70–85. https://doi.org/10.1111/j.1749-6632.1993.tb49924.x

Meir Drexler S, Hamacher-Dang TC, Wolf OT (2017) Stress before extinction learning enhances and generalizes extinction memory in a predictive learning task. Neurobiol Learn Mem 141: 143–149. https://doi.org/10.1016/j.nlm.2017.04.002

Meir Drexler S, Merz CJ, Jentsch VL, Wolf OT (2019) How stress and glucocorticoids timing-dependently affect extinction and relapse. Neurosci Biobehav Rev 98:145–153. https://doi.org/10.1016/j.neubiorev.2018.12.029

Merz CJ, Hermann A, Stark R, Wolf OT (2014) Cortisol modifies extinction learning of recently acquired fear in men. Soc Cogn Affect Neurosci 9(9):1426–1434. https://doi.org/10.1093/scan/nst137

Merz CJ, Hamacher-Dang TC, Stark R, Wolf OT, Hermann A (2018) Neural underpinnings of cortisol effects on fear extinction. Neuropsychopharmacology 43(2):384–392. https://doi.org/10.1038/npp.2017.227

Meuret AE, Trueba AF, Abelson JL, Liberzon I, Auchus R, Bhaskara L, Ritz T, Rosenfield D (2015) High cortisol awakening response and cortisol levels moderate exposure-based psychotherapy success. Psychoneuroendocrinology 51:331–340. https://doi.org/10.1016/j.psyneuen.2014.10.008

Neeper S, Gomez-Pinilla F, Choi J, Cotman CW (1995) Exercise and brain neurotrophins. Nature 373:109

Norrholm SD, Jovanovic T, Vervliet B, Myers KM, Davis M, Rothbaum BO, Duncan EJ (2006) Conditioned fear extinction and reinstatement in a human fear-potentiated startle paradigm. Learn Mem 13(6):681–685. https://doi.org/10.1101/lm.393906

Norrholm SD, Vervliet B, Jovanovic T, Boshoven W, Myers KM, Davis M, Rothbaum B, Duncan EJ (2008) Timing of extinction relative to acquisition: a parametric analysis of fear extinction in humans. Behav Neurosci 122(5):1016–1030. https://doi.org/10.1037/a0012604

Notaras M, van den Buuse M (2020) Neurobiology of BDNF in fear memory, sensitivity to stress, and stress-related disorders. Mol Psychiatry 25(10):2251–2274. https://doi.org/10.1038/s41380-019-0639-2

Nowakowski Sims E (2015) Group exercise as an adjunct treatment for trauma. J Trauma Treat s2. https://doi.org/10.4172/2167-1222.S2-014

Orem TR, Wheelock MD, Goodman AM, Harnett NG, Wood KH, Gossett EW, Granger DA, Mrug S, Knight DC (2019) Amygdala and prefrontal cortex activity varies with individual differences in the emotional response to psychosocial stress. Behav Neurosci 133(2):203–211. https://doi.org/10.1037/bne0000305

Payne JD, Jackson ED, Hoscheidt S, Ryan L, Jacobs WJ, Nadel L (2007) Stress administered prior to encoding impairs neutral but enhances emotional long-term episodic memories. Learn Mem 14(12):861–868. https://doi.org/10.1101/lm.743507

Peters J, Dieppa-Perea LM, Melendez LM, Quirk GJ (2010) Induction of fear extinction with hippocampal-infralimbic BDNF. Science 328(5983):1288–1290. https://doi.org/10.1126/science.1186909

Phelps EA, Delgado MR, Nearing KI, LeDoux JE (2004) Extinction learning in humans: role of the amygdala and vmPFC. Neuron 43(6):897–905. https://doi.org/10.1016/j.neuron.2004.08.042

Powers MB, Smits JAJ, Otto MW, Sanders C, Emmelkamp PMG (2009) Facilitation of fear extinction in phobic participants with a novel cognitive enhancer: a randomized placebo controlled trial of yohimbine augmentation. J Anxiety Disord 23(3):350–356. https://doi.org/10.1016/j.janxdis.2009.01.001

Powers MB, Medina JL, Burns S, Kauffman BY, Monfils M, Asmundson GJG, Diamond A, McIntyre C, Smits JAJ (2015) Exercise augmentation of exposure therapy for PTSD: rationale and pilot efficacy data. Cogn Behav Ther 44(4):314–327. https://doi.org/10.1080/16506073.2015.1012740

Reznikoff GA, Manaker S, Rhodes CH, Winokur A, Rainbow TC (1986) Localization and quantification of beta-adrenergic receptors in human brain. Neurology 36(8):1067–1067. https://doi.org/10.1212/WNL.36.8.1067

Riggenbach MR, Weiser JN, Mosley BE, Hipskind JJ, Wireman LE, Hess KL, Duffy TJ, Handel JK, Kaschalk MG, Reneau KE, Rorabaugh BR, Norrholm SD, Jovanovic T, Zoladz PR (2019) Immediate pre-learning stress enhances baseline startle response and fear acquisition in a fear-potentiated startle paradigm. Behav Brain Res 371:111980. https://doi.org/10.1016/j.bbr.2019.111980

Roquet RF, Monfils M-H (2018) Does exercise augment operant and Pavlovian extinction: a meta-analysis. J Psychiatr Res 96:73–93. https://doi.org/10.1016/j.jpsychires.2017.09.018

Sapolsky RM, Romero LM, Munck AU (2000) How do glucocorticoids influence stress responses? Integrating permissive, suppressive, stimulatory, and preparative actions*. Endocr Rev 21(1):55–89. https://doi.org/10.1210/edrv.21.1.0389

Saxena S, Van Ommeren M, Tang KC, Armstrong TP (2005) Mental health benefits of physical activity. J Ment Health 14(5):445–451. https://doi.org/10.1080/09638230500270776

Schwabe L, Haddad L, Schachinger H (2008) HPA axis activation by a socially evaluated cold-pressor test. Psychoneuroendocrinology 33(6):890–895. https://doi.org/10.1016/j.psyneuen.2008.03.001

Schwabe L, Joëls M, Roozendaal B, Wolf OT, Oitzl MS (2012) Stress effects on memory: an update and integration. Neurosci Biobehav Rev 36(7):1740–1749. https://doi.org/10.1016/j.neubiorev.2011.07.002

Selye H (1950) Stress and the general adaptation syndrome. BMJ 1(4667):1383–1392. https://doi.org/10.1136/bmj.1.4667.1383

Shin LM, Liberzon I (2010) The neurocircuitry of fear, stress, and anxiety disorders. Neuropsychopharmacology 35(1):169–191. https://doi.org/10.1038/npp.2009.83

Simon-Kutscher K, Wanke N, Hiller C, Schwabe L (2019) Fear without context: acute stress modulates the balance of cue-dependent and contextual fear learning. Psychol Sci 30(8):1123–1135. https://doi.org/10.1177/0956797619852027

Smits JAJ, Rosenfield D, Davis ML, Julian K, Handelsman PR, Otto MW, Tuerk P, Shiekh M, Rosenfield B, Hofmann SG, Powers MB (2014) Yohimbine enhancement of exposure therapy for social anxiety disorder: a randomized controlled trial. Biol Psychiatry 75(11):840–846. https://doi.org/10.1016/j.biopsych.2013.10.008

Soravia LM, Heinrichs M, Aerni A, Maroni C, Schelling G, Ehlert U, Roozendaal B, de Quervain DJ-F (2006) Glucocorticoids reduce phobic fear in humans. Proc Natl Acad Sci U S A 103(14):5585–5590. https://doi.org/10.1073/pnas.0509184103

Tabbert K, Merz CJ, Klucken T, Schweckendiek J, Vaitl D, Wolf OT, Stark R (2010) Cortisol enhances neural differentiation during fear acquisition and extinction in contingency aware young women. Neurobiol Learn Mem 94(3):392–401. https://doi.org/10.1016/j.nlm.2010.08.006

Tanner MK, Hake HS, Bouchet CA, Greenwood BN (2018) Running from fear: exercise modulation of fear extinction. Neurobiol Learn Mem 151:28–34. https://doi.org/10.1016/j.nlm.2018.03.021

Tollenaar MS, Elzinga BM, Spinhoven P, Everaerd W (2009) Immediate and prolonged effects of cortisol, but not propranolol, on memory retrieval in healthy young men. Neurobiol Learn Mem 91(1):23–31. https://doi.org/10.1016/j.nlm.2008.08.002

van Gelderen MJ, Nijdam MJ, de Vries F, Meijer OC, Vermetten E (2020) Exposure-related cortisol predicts outcome of psychotherapy in veterans with treatment-resistant posttraumatic stress disorder. J Psychiatr Res 130:387–393. https://doi.org/10.1016/j.jpsychires.2020.08.011

Yehuda R, Bierer LM, Pratchett L, Malowney M (2010) Glucocorticoid augmentation of prolonged exposure therapy: rationale and case report. Eur J Psychotraumatol 1(1):5643. https://doi.org/10.3402/ejpt.v1i0.5643

Yehuda R, Bierer LM, Pratchett LC, Lehrner A, Koch EC, Van Manen JA, Flory JD, Makotkine I, Hildebrandt T (2015) Cortisol augmentation of a psychological treatment for warfighters with posttraumatic stress disorder: randomized trial showing improved treatment retention and outcome. Psychoneuroendocrinology 51:589–597. https://doi.org/10.1016/j.psyneuen.2014.08.004

How Different Factors in Combination Change Fear Extinction Learning: The Case of Sex and Stress Hormones

Christian J. Merz

Contents

Abstract Effects of a specific factor on fear extinction or exposure therapy have revealed promising results, for example how sex or stress hormones exert the capability to critically change extinction learning and consolidation processes. However, we must acknowledge that in real life these factors do not operate in isolation, they go hand in hand. In this chapter, the available evidence regarding interactions of sex and stress hormones on extinction processes and exposure therapy will be integrated and discussed. First hints exist that these factors in combination critically target extinction learning and consolidation processes, calling for more detailed research on the exact underlying mechanisms. In addition to experiments with high sample sizes, we must aim for a collaborative effort of laboratories across the whole world to be able to identify critical combinations of factors associated with improved, but also impaired extinction processes and exposure therapy success. We expect that the revelation of further relevant factors will not only be limited to the interplay between sex and stress hormones but will include factors such as sleep and exercise as well. In the long run, uncovering the most important interaction effects will give us critical hints for differential treatment options to be realized in the sense of a personalized medicine approach.

C. J. Merz (✉)
Department of Cognitive Psychology, Institute of Cognitive Neuroscience, Ruhr University Bochum, Bochum, Germany
e-mail: christian.j.merz@rub.de

Curr Topics Behav Neurosci (2023) 64: 179–192
https://doi.org/10.1007/7854_2023_427
Published Online: 17 July 2023

Keywords Cortisol · Exposure therapy · Fear conditioning · Menstrual cycle · Oral contraceptives

1 Introduction

Lifelong hormonal changes exert a tremendous impact on our cognition, emotion, and behavior as well as our (mental) health by promoting adaptation to changes in the environment (Been et al. 2022). Especially, fast-acting phasic as well as slower tonic changes in the release of stress hormones (in terms of acute vs. chronic stress) and sex hormones (largely in terms of female menstrual cycle) sparked interest in their potential to modulate health and disease.

In the following chapter, stress and sex hormones will be introduced and their general impact on extinction learning and exposure therapy will be explored. Critically, their joint influence will be subsequently elucidated together with potential underlying mechanisms before providing final remarks on possible future directions.

2 Regulation and Importance of Stress and Sex Hormones

A fine-tuned neurobiological stress response is orchestrated when individuals are confronted with a potential or even anticipated threat, an essential mix relevant for adaptation regarding challenges ranging from daily hassles up to major life-threatening situations, through the complex process called allostasis (McEwen 2004). On the one hand, the sympathetic nervous system triggers a rapid release of (nor)epinephrine from the adrenal medulla initiating a fast increase in blood pressure or heart rate, conceptualized as fight-or-flight response (Cannon 1932). On the other hand, the hypothalamus-pituitary-adrenocortical (HPA) axis controls a slower stress response with the consecutive release of corticotropin-releasing hormone (hypothalamus), adrenocorticotropic hormone (pituitary), and glucocorticoids (GCs, adrenal cortex, in humans mostly cortisol; Joëls and Baram 2009). Cortisol levels typically reach their peak about 30–35 min after stress onset (Dickerson and Kemeny 2004; Goodman et al. 2017). In addition, cortisol can easily pass the blood–brain barrier due to its lipophilic structure and bind to mineralocorticoid and glucocorticoid receptors prominently expressed in the amygdala, hippocampus, or prefrontal cortex (de Kloet et al. 2005; Handa and Weiser 2014; Ulrich-Lai and Herman 2009). GCs are critically involved in the modulation of many important learning and memory processes (Shields et al. 2017) relevant for clinical applications (Wolf et al. 2016) including fear conditioning and particularly extinction learning (de Quervain et al. 2017, 2019; Merz and Wolf 2022).

Another important hormonal system controls the regulation and release of sex hormones, the hypothalamic-pituitary-gonadal (HPG) axis. A cyclic interplay of hormones ensures a communication between the hypothalamus (gonadotropin-releasing hormone), pituitary (follicle stimulating hormone and luteinizing hormone) and the gonads (ovaries and testes; estrogens (e.g., estradiol), gestagens (e.g., progesterone), and androgens (e.g., testosterone)). Sex hormones have manyfold organizational and activational effects, in particular regarding the regulation of the menstrual cycle in women (Heck and Handa 2019). While high progesterone concentrations are only observed after ovulation during the luteal phase (or during pregnancy), estradiol levels are high during the luteal phase with a prior peak mid-cycle during ovulation. During the follicular phase, i.e., after menstrual bleeding until ovulation, sex hormone availability is low. Importantly, women in a certain age-range frequently take oral contraceptives (OCs), which contain synthetic estrogens and/or progestins reducing endogenous sex hormone availability, thus circulating estrogens and progesterone levels are low under OC usage (Fleischman et al. 2010; Taylor et al. 2021). In general, sex hormone (estrogen, progesterone, and androgen) receptors are expressed throughout the whole brain (Cover et al. 2014; Handa and Weiser 2014). Comparable to stress hormones, menstrual cycle related changes in sex hormones as well as OCs critically influence brain functions and behavior (Hsu et al. 2021; Montoya and Bos 2017; Sundstrom Poromaa and Gingnell 2014) including fear conditioning and, more specifically, extinction learning (Merz et al. 2018b; Velasco et al. 2019).

In principle, hormonal axes do not work in isolation, they communicate with each other all the time, this is also the case for the HPA and HPG axes (Handa and Weiser 2014; Heck and Handa 2019). Activation of the HPA axis leads to a change in HPG axis activity and vice versa and the complex interactions of both axes modulate emotional learning and memory processes (cf. Merz and Wolf 2017). The literature on the intersection of sex and stress hormones on fear learning has been extensively reviewed elsewhere (Merz and Wolf 2015, 2017; Peyrot et al. 2020), in the following, let's focus on their potential to modulate extinction learning processes and exposure therapy.

3 Stress Hormone Effects on Extinction Learning and Exposure Therapy

Overall, stress hormones enhance extinction learning (cf. Chapter "Impact of Stress and Exercise on Fear Extinction" in this book) as also outlined in the Stress Timing affects Relapse (StaR) model (Meir Drexler et al. 2019, 2020). More specifically, stress hormones seem to reduce the context-dependency of the extinction memory trace leading to less renewal and return of fear. The underlying mechanism is assumed to rely on stress hormones reducing fear retrieval at the beginning of extinction training (Merz et al. 2018a) and facilitating the consolidation of the inhibitory extinction memory trace (Bentz et al. 2010; de Quervain and Margraf 2008).

Stress hormone effects on extinction learning have also been successfully translated into exposure therapy for patients with different mental disorders such as phobia or posttraumatic stress disorder (PTSD; for reviews: de Quervain et al. 2017, 2019). However, most of these results were obtained in men, what about the other half of the worldwide population, what about women (Cover et al. 2014; Lebron-Milad and Milad 2012)?

4 Sex Hormone Effects on Extinction Learning and Exposure Therapy

Indeed, women differ in their ability to extinguish and how well extinction memories are consolidated dependent on the stage of the menstrual cycle and OC intake (for recent reviews: Hsu et al. 2021; Jentsch et al. 2022; Merz et al. 2018b; Tang and Graham 2020; Velasco et al. 2019). In more detail, high levels of estrogens (and progesterone) at the time of extinction training facilitate consolidation of the extinction memory trace leading to a reduced return of fear (cf. Chapter "Impact of Sex Differences and Sex Hormones" in this book).

This approach has also been successfully implemented in exposure therapy in spider phobia (Li and Graham 2016) or PTSD (Glover et al. 2012; Pineles et al. 2016). Importantly, OC intake impairs extinction learning and consolidation processes as well as exposure therapy success (Raeder et al. 2019), while estrogen administration can rescue this effect (Graham and Milad 2013; White and Graham 2016). Concluding, high estrogen and progesterone levels together exert beneficial effects on extinction consolidation. In particular, high levels co-occur in the mid-luteal phase and low levels in the early follicular phase of the menstrual cycle. Thus, extinction and extinction-based therapies should specifically target the mid-luteal phase of the menstrual cycle in order to reduce possible return of fear phenomena and relapse.

In the following sections, the interaction between sex and stress hormone effects will be illustrated separately for extinction learning and exposure therapy.

5 A Joint Impact of Stress and Sex Hormones on Extinction Learning

Let's turn to the few studies having had a first look at the possible intertwined effects of sex and stress hormones on extinction training. Specifically in men, stress induction before extinction training facilitated extinction learning and consolidation indexed by reduced unconditioned stimulus (US) expectancy ratings during

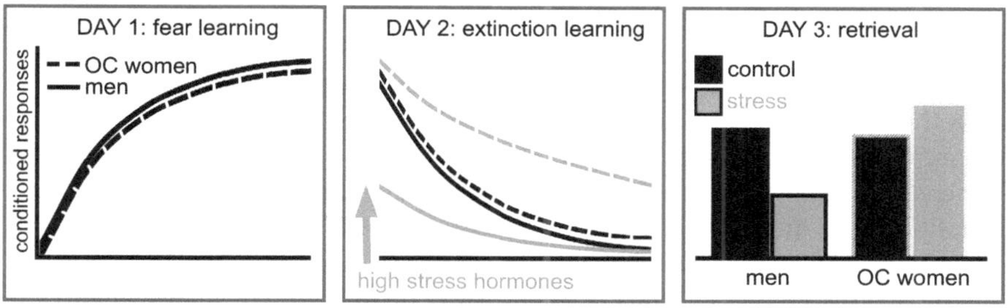

Fig. 1 Preliminary findings reveal that stress hormones interact with intake of oral contraceptives (OCs) to modulate extinction learning and retrieval. During fear learning on day 1, conditioned responses emerge over time in men and OC women alike. On day 2, high stress hormone concentrations exert beneficial effects on conditioned responding during extinction learning in men, but rather detrimental effects in OC women. High stress hormones might be due to cortisol administration, stress induction or the circadian rhythm (high cortisol levels in the morning compared to low cortisol levels in the evening). These results translate to retrieval on day 3 pointing to a further joint impact of stress hormones and OC intake on consolidation processes: High stress hormone concentrations before extinction learning leads to reduced retrieval of conditioned responses in men, but rather to increased conditioned responses in OC women. *Note:* Findings in free-cycling women seem to resemble those in men; thus, the observed effects seem to rely on OC intake per se and do not represent true sex differences

extinction training and a retrieval test 24 h later (Bentz et al. 2013). In OC women, this beneficial stress effect did not occur, rather the opposite picture emerged over time (Fig. 1; Bentz et al. 2013). Unfortunately, a direct comparison of obtained data in men and women was not reported, at least descriptively, an interaction effect could be assumed. A neural correlate of this favorable pre-extinction stress effect could be recently identified in the dorsal anterior cingulate cortex. GC administration before extinction training in healthy men and free-cycling women reduced activation of the dorsal anterior cingulate cortex during retrieval 24 later (Hagedorn et al. 2022). Importantly, no sex differences in the cortisol-related effect occurred which might indirectly indicate that the result pattern seems to converge at least in men and free-cycling women. In male and female patients with PTSD, GC administration before extinction training also reduced conditioned responding indexed by skin conductance responses, this result only emerged during extinction training, but not during retrieval 1 week later (Inslicht et al. 2022). In this study with a clinical sample, OC women as well as women tested in the follicular phase of their individual menstrual cycle were studied, but unfortunately sex hormone related effects were not reported. Thus, it remains to be shown if possible interactions between sex and stress hormones are restricted to OC women pointing to a special relevance of OCs in modulating extinction processes in the face of stress.

Surprisingly, while animal research revealed the joint impact of sex and stress hormones on fear learning (for reviews: Shors 2004, 2006), their interactive effects on fear extinction and consolidation remain largely unknown. Thus, it seems essential to also replicate and extend the relevant findings obtained in humans using available research designs in animals, keeping in mind that a direct translation of human experiments to animal experiments and vice versa bear its own challenges (comprehensively reviewed in Haaker et al. 2019). For example, typically used stressors in humans including a social evaluation component such as the Trier Social Stress Test (Kirschbaum et al. 1993) cannot be compared with stressors in animals such as restraint stress or foot shocks (Schöner et al. 2017), even though both lead to an increase in stress hormones. Additionally, rapidly changing sex hormone concentrations over the cycle in rodents (4 days cycle vs. 28 days cycle in women) makes it difficult to disentangle distinct effects of sex hormones on extinction learning and consolidation, since on each new experimental day a new cycle phase starts. Still, we need animal research to further improve our understanding of the exact mechanisms and causalities how stress and sex hormones jointly modulate extinction processes with experimental tools exclusively available for animal research. In particular, animal studies seem to be best suited to disentangle specific contributions of OCs on extinction processes (cf. Graham and Milad 2013), since OC intake can be experimentally manipulated and controlled, while OC intake remains a quasi-experimental factor in studies including women.

6 A Joint Impact of Stress and Sex Hormones on Exposure Therapy

Let's turn to reported interaction effects of sex and stress hormones on the application of extinction-based interventions in the clinical setting: exposure therapy. GC administration before exposure therapy led to less fear and anxiety in subjective reports as well as at the behavioral level (measured with the behavioral approach test; for recent reviews, see de Quervain et al. 2017, 2019). So far, this beneficial effect was observed in patients with spider phobia (Soravia et al. 2006, 2014), acrophobia (de Quervain et al. 2011), and social phobia (Soravia et al. 2006). While female and male patients were included in these studies, either sex differences were not considered or no sex differences were observed. However, a pharmacological intervention is not really applicable in the setting of licensed psychotherapists without support of medical staff. Thus, application of these stress hormone effects into clinical reality requires further attention.

High GC levels can also be achieved with stress induction, which was realized in a recent one-session exposure therapy study in women with spider phobia (Raeder et al. 2023). All patients entered the experimental protocol at a time of low availability of endogenous female sex hormones: OC women were included as well as women tested during the follicular phase of their individual menstrual cycle. OC contain exogenous female sex hormones reducing endogenous female sex hormone availability. Thus, the approach of directly comparing women in the follicular phase with OC women sheds light on the specific contribution of low female sex hormones vs. the impact of OCs per se on stress-hormone related exposure therapy success. Indeed, remarkable differences between groups emerged: exposure therapy success was impaired in stressed OC women, but not in women tested during the follicular phase. This first observation highlights the importance of considering sex differences and more crucially, sex hormones, menstrual cycle, and OC intake, when implementing stress-hormone associated effects in daily routine of clinical settings.

In addition to stress induction, the maybe most elegant and feasible way of establishing beneficial effects of high GC levels on exposure therapy in real life consists of the usage of the circadian cortisol rhythm. Typically, high cortisol levels occur after awakening in the morning slowly decreasing over the day to reach quite low cortisol levels in the evening. Interestingly, in a highly selective sample of patients with spider phobia, consisting only of OC women, exposure therapy was more effective when realized in morning compared to evening sessions (Lass-Hennemann and Michael 2014). In light of the previous report comparing spider-phobic women tested in the follicular phase with spider-phobic OC women (Raeder et al. 2023), it remains to be shown if the beneficial circadian rhythm-related stress-hormone effect is even more pronounced in men or women not taking OCs. Another approach confirmed this beneficial impact of high cortisol levels in a sample of male and female patients with panic disorder and agoraphobia (Meuret et al. 2015): Not only a higher cortisol awakening response, but also higher absolute cortisol levels during exposure sessions were related to a better exposure therapy outcome. Since no data on menstrual cycle and OCs were collected in this study, no further hints regarding interaction effects of sex and stress hormones emerged.

Taken together, high stress hormones (either due to the circadian rhythm, stress induction, or GC administration) before extinction training and exposure therapy enhance extinction learning and exposure therapy success leading to reduced return of fear during extinction retrieval and promoting a context-independent extinction memory trace (cf. Merz and Wolf 2022) as indicated in the STaR model (Meir Drexler et al. 2019, 2020). Possible interactions with sex hormones have not been consistently reported or investigated so far. If any, then OC usage might dampen or even reverse the observed stress-hormone effects (Fig. 1; cf. Jentsch et al. 2022), which underlines the importance of interactions between sex and stress hormones or more precisely between OC usage and stress hormones to modulate extinction processes.

7 Underlying Mechanisms

As introduced above, an extensive crosstalk between the HPA and HPG axis exists, which is also evident in a different cortisol release in response to stress with regard to the menstrual cycle in women: during the luteal phase, for example, cortisol increase is more pronounced relative to the follicular phase (Kajantie and Phillips 2006; Kudielka and Kirschbaum 2005). Moreover, OC intake leads to a rather blunted cortisol response in free (thus, biologically active) cortisol concentrations due to the ability of OCs to increase cortisol-binding globulin (capturing free cortisol molecules in the blood stream; Gervasio et al. 2022; Kirschbaum et al. 1999). Specifically in OC women, a direct and/or indirect reduction of a central sensitivity to GCs has been discussed before (Kuhlmann and Wolf 2005) as also considered for pro-inflammatory cytokine production to compensate for the blunted increase in free cortisol concentrations in response to stress (Rohleder et al. 2003). Additionally, continuous OC usage (containing synthetically estrogen and/or gestagen derivatives with androgenic or anti-androgenic properties; Pletzer and Kerschbaum 2014) could desensitize or downregulate estrogen and/or gestagen receptors localized in relevant brain structures for extinction learning such as the amygdala, hippocampus, and ventromedial prefrontal cortex (cf. Chapters "Molecular and Neural Circuits of Fear Extinction", "Neural Circuit of Fear Extinction: Psychophysiology", and "Neural Correlates of Fear Extinction in the Human Brain" in this book). Increases in stress hormones might disrupt this reduced excitability and eventually explain the reported divergent effects in OC women compared to men.

Keeping this in mind, the observed interactions between stress and sex hormones (and OCs) on extinction processes might result from different underlying cortisol responses to the same stressor (which could be ruled out by using a pharmacological approach, i.e., cortisol administration). Alternatively or additionally, these interactions might occur, because the female brain reacts differently to the same stress signature (structure- and/or function-wise) compared to the male brain. The underlying mechanism might encompass different binding capacities and/or affinities of mineralocorticoid and glucocorticoid receptors in women relative to men and even changing over the course of the menstrual cycle or due to OC intake (ter Horst et al. 2012; Turner 1997). Furthermore, progesterone (and also synthetic progestins as available in OCs) also binds to the mineralocorticoid receptor (Funder 2007) providing further evidence for a critical interaction between the HPA and HPG axes at the direct receptor level. Moreover, co-expression of sex and stress-hormone receptors might play a relevant rule for the observed interaction effects (Handa and Weiser 2014).

Taken together, the exact underlying mechanisms for sex-hormone dependent stress-hormone effects on extinction learning remain speculative and need critical evaluation in future studies. Particularly in light of the available literature so far, stress in combination with OC intake seems to exert a more critical influence rather than the combination with sex hormones per se (cf. Jentsch et al. 2022).

8 Where to Go from Here?

The reported beneficial effects for high stress hormones as well as high levels of estrogens and progesterone during extinction training and/or exposure therapy in isolation call for replication studies and further critical tests of their interaction with the following questions: what about combining these beneficial hormone influences and explore high sex and stress hormones to see if there is any additional positive effect? Or put the other way round: are low stress and sex hormone levels associated with impaired extinction learning and consolidation processes? Does this combination act as a vulnerability factor to develop an anxiety disorder or PTSD and can it limit treatment success? If yes, should women before entering exposure therapy discontinue OC intake? How long should OC discontinuation last in this case? Should exposure therapy sessions be best scheduled in the morning (high stress-hormone levels) and/or in the mid-luteal phase of the respective menstrual cycle? More generally, what about stress-hormone effects in transition periods with regard to sex hormones such as puberty, pregnancy and breastfeeding, or menopause (cf. Milligan-Saville and Graham 2016)?

These critical questions need to be urgently answered in the future in preclinical as well as clinical studies. Clearly, a closer look at the interplay of sex and stress hormones (and OCs) regarding extinction processes comprises the potential to disentangle critical combinations of factors associated with beneficial and adverse stress-hormone effects, especially regarding the clinical setting. We should keep in mind that in terms of a higher prevalence in women in comparison with men for anxiety disorders or PTSD (Cover et al. 2014; Kessler et al. 2005), this research is clearly needed to further understand sex differences in interaction with stress contributing to their development, maintenance, and treatment (e.g., with extinction-based therapies).

Apart from sex and stress hormones, interactions between critical factors involved in fear extinction as also outlined in other chapters of this book have rarely been investigated. Nevertheless, having a closer look particularly at the factors most promising in the modulation of extinction learning, on the one hand, and easiest to be incorporated in exposure therapy settings, on the other hand, seems to be essential to move the field forward. For example, stress hormones influence sleep architecture and sleep influence stress hormones in the context of fear extinction and PTSD (Pace-Schott et al. 2015). However, systematic sleep experiments also disentangling the contribution of sex differences and in general, sex hormone effects on sleep-related modulations of extinction learning are missing so far (for general sleep-related effects on extinction learning, cf. Chapter "Impact of Sleep on Fear Extinction" in this book). Thus, it is not clear if extinction memory consolidation processes during sleep equally occur in men and women and if circulating sex hormones over the course of the menstrual cycle or OC intake matter.

A very good example of further interaction effects comprises research regarding joint effects of sex and stress on fear extinction in interaction with exercise, which was recently investigated in rats (Drummond et al. 2021; cf. Chapter "Impact of Stress and Exercise on Fear Extinction" in this book). Nevertheless, combining two or even more factors in one experiment is still the exception rather than the rule, in animal studies and particularly in human studies (Cover et al. 2014; Lebron-Milad and Milad 2012).

Of course, the more between-subjects factors you include in an experiment, the larger your sample size gets, which can pose a significant challenge in terms of available resources such as required funding or realistic availability of participants in a certain amount of time. Still, the high sample size avenue needs to be pursued in the future, possibly not within a single laboratory, but more likely with collaborating laboratories all over the world working together with the identical fear conditioning paradigm, but different specializations. For this purpose, a shared database architecture needs to be established along the way with all relevant details concerning data protection, data usage, and joint analysis strategies. The goal of this avenue should consist of the identification of the most relevant factors linked to improved and impaired extinction learning and consolidation processes as well as their translation to exposure therapy (and other extinction-based interventions). When having reached the finish (admittedly, it will be a marathon rather than an 800-m race), a personalized medicine approach can be finally realized: Insights from individuals with a protective combination of factors might inspire specifically tailored treatment options or even prevention strategies for individuals at high risk to develop or already suffering from anxiety disorders or PTSD.

Funding and Disclosure

This work was supported by the Deutsche Forschungsgemeinschaft (DFG; German Research Foundation) within the SFB 1280 Extinction Learning (grant number 316803389 – SFB1280; project A09). The DFG had no role in writing of the manuscript or in the decision to submit this book chapter for publication.

Conflict of Interest Statement The author declares no conflict of interest.

References

Been LE, Sheppard PAS, Galea LAM, Glasper ER (2022) Hormones and neuroplasticity: a lifetime of adaptive responses. Neurosci Biobehav Rev 132:679–690

Bentz D, Michael T, de Quervain DJ-F, Wilhelm FH (2010) Enhancing exposure therapy for anxiety disorders with glucocorticoids: from basic mechanisms of emotional learning to clinical applications. J Anxiety Disord 24:223–230

Bentz D, Michael T, Wilhelm FH, Hartmann FR, Kunz S, von Rohr IRR, de Quervain DJ-F (2013) Influence of stress on fear memory processes in an aversive differential conditioning paradigm in humans. Psychoneuroendocrinology 38:1186–1197

Cannon WB (1932) The wisdom of the body. Norton, New York

Cover KK, Maeng LY, Lebron-Milad K, Milad MR (2014) Mechanisms of estradiol in fear circuitry: implications for sex differences in psychopathology. Transl Psychiatry 4:e422

de Kloet ER, Joëls M, Holsboer F (2005) Stress and the brain: from adaptation to disease. Nat Rev Neurosci 6:463–475

de Quervain DJ-F, Margraf J (2008) Glucocorticoids for the treatment of post-traumatic stress disorder and phobias: a novel therapeutic approach. Eur J Pharmacol 583:365–371

de Quervain DJ-F, Bentz D, Michael T, Bolt OC, Wiederhold BK, Margraf J, Wilhelm FH (2011) Glucocorticoids enhance extinction-based psychotherapy. Proc Natl Acad Sci U S A 108:6621–6625

de Quervain DJ-F, Schwabe L, Roozendaal B (2017) Stress, glucocorticoids and memory: implications for treating fear-related disorders. Nat Rev Neurosci 18:7–19

de Quervain DJ-F, Wolf OT, Roozendaal B (2019) Glucocorticoid-induced enhancement of extinction-from animal models to clinical trials. Psychopharmacology 236:183–199

Dickerson SS, Kemeny ME (2004) Acute stressors and cortisol responses: a theoretical integration and synthesis of laboratory research. Psychol Bull 130:355–391

Drummond KD, Waring ML, Faulkner GJ, Blewitt ME, Perry CJ, Kim JH (2021) Hippocampal neurogenesis mediates sex-specific effects of social isolation and exercise on fear extinction in adolescence. Neurobiol Stress 15

Fleischman DS, Navarrete CD, Fessler DMT (2010) Oral contraceptives suppress ovarian hormone production. Psychol Sci 21:750–752

Funder JW (2007) Why are mineralocorticoid receptors so nonselective? Curr Hypertens Rep 9: 112–116

Gervasio J, Zheng S, Skrotzki C, Pachete A (2022) The effect of oral contraceptive use on cortisol reactivity to the Trier Social Stress Test: a meta-analysis. Psychoneuroendocrinology 136: 105626

Glover EM, Jovanovic T, Mercer KB, Kerley K, Bradley B, Ressler KJ, Norrholm SD (2012) Estrogen levels are associated with extinction deficits in women with posttraumatic stress disorder. Biol Psychiatry 72:19–24

Goodman WK, Janson J, Wolf JM (2017) Meta-analytical assessment of the effects of protocol variations on cortisol responses to the Trier Social Stress Test. Psychoneuroendocrinology 80: 26–35

Graham BM, Milad MR (2013) Blockade of estrogen by hormonal contraceptives impairs fear extinction in female rats and women. Biol Psychiatry 73:371–378

Haaker J, Maren S, Marta A, Merz CJ, Richter J, Helene RS, Shira MD, Lange MD, Jungling K, Nees F, Seidenbecher T, Fullana MA, Carsten WT, Lonsdorf TB (2019) Making translation work: harmonizing cross-species methodology in the behavioural neuroscience of Pavlovian fear conditioning. Neurosci Biobehav Rev 107:329–345

Hagedorn B, Wolf OT, Merz CJ (2022) Cortisol before extinction generalization alters its neural correlates during retrieval. Psychoneuroendocrinology 136:105607

Handa RJ, Weiser MJ (2014) Gonadal steroid hormones and the hypothalamo-pituitary-adrenal axis. Front Neuroendocrinol 35:197–220

Heck AL, Handa RJ (2019) Sex differences in the hypothalamic-pituitary-adrenal axis' response to stress: an important role for gonadal hormones. Neuropsychopharmacology 44:45–58

Hsu C-MK, Ney LJ, Honan C, Felmingham KL (2021) Gonadal steroid hormones and emotional memory consolidation: a systematic review and meta-analysis. Neurosci Biobehav Rev 130: 529–542

Inslicht SS, Niles AN, Metzler TJ, Lipshitz SL, Otte C, Milad MR, Orr SP, Marmar CR, Neylan TC (2022) Randomized controlled experimental study of hydrocortisone and D-cycloserine effects on fear extinction in PTSD. Neuropsychopharmacology 47:1945–1952

Jentsch VL, Pötzl L, Wolf OT, Merz CJ (2022) Hormonal contraceptive usage influences stress hormone effects on cognition and emotion. Front Neuroendocrinol:101012

Joëls M, Baram TZ (2009) The neuro-symphony of stress. Nat Rev Neurosci 10:459–466

Kajantie E, Phillips DIW (2006) The effects of sex and hormonal status on the physiological response to acute psychosocial stress. Psychoneuroendocrinology 31:151–178

Kessler RC, Berglund P, Demler O, Jin R, Merikangas KR, Walters EE (2005) Lifetime prevalence and age-of-onset distributions of DSM-IV disorders in the national comorbidity survey replication. Arch Gen Psychiatry 62:593–602

Kirschbaum C, Pirke K-M, Hellhammer DH (1993) The 'Trier Social Stress Test' – a tool for investigating psychobiological stress responses in a laboratory setting. Neuropsychobiology 28: 76–81

Kirschbaum C, Kudielka BM, Gaab J, Schommer NC, Hellhammer DH (1999) Impact of gender, menstrual cycle phase, and oral contraceptives on the activity of the hypothalamus-pituitary-adrenal axis. Psychosom Med 61:154–162

Kudielka BM, Kirschbaum C (2005) Sex differences in HPA axis responses to stress: a review. Biol Psychol 69:113–132

Kuhlmann S, Wolf OT (2005) Cortisol and memory retrieval in women: influence of menstrual cycle and oral contraceptives. Psychopharmacology 183:65–71

Lass-Hennemann J, Michael T (2014) Endogenous cortisol levels influence exposure therapy in spider phobia. Behav Res Ther 60:39–45

Lebron-Milad K, Milad MR (2012) Sex differences, gonadal hormones and the fear extinction network: implications for anxiety disorders. Bio Moo Anx Dis 2:3

Li S, Graham BM (2016) Estradiol is associated with altered cognitive and physiological responses during fear conditioning and extinction in healthy and spider phobic women. Behav Neurosci 130:614–623

McEwen BS (2004) Protection and damage from acute and chronic stress – allostasis and allostatic overload and relevance to the pathophysiology of psychiatric disorders. Ann N Y Acad Sci 1032:1–7

Meir Drexler S, Merz CJ, Jentsch VL, Wolf OT (2019) How stress and glucocorticoids timing-dependently affect extinction and relapse. Neurosci Biobehav Rev 98:145–153

Meir Drexler S, Merz CJ, Jentsch VL, Wolf OT (2020) Stress modulation of fear and extinction in psychopathology and treatment. Neuroforum 26:133–141

Merz CJ, Wolf OT (2015) Stress and emotional learning in humans: evidence for sex differences. In: Shansky RM (ed) Sex differences in the central nervous system. Elsevier, Amsterdam, pp 149–170

Merz CJ, Wolf OT (2017) Sex differences in stress effects on emotional learning. J Neurosci Res 95:93–105

Merz CJ, Wolf OT (2022) How stress hormones shape memories of fear and anxiety in humans. Neurosci Biobehav Rev 142:104901

Merz CJ, Hamacher-Dang TC, Stark R, Wolf OT, Hermann A (2018a) Neural underpinnings of cortisol effects on fear extinction. Neuropsychopharmacology 43:384–392

Merz CJ, Kinner VL, Wolf OT (2018b) Let's talk about sex… differences in human fear conditioning. Curr Opin Behav Sci 23:7–12

Meuret AE, Trueba AF, Abelson JL, Liberzon I, Auchus R, Bhaskara L, Ritz T, Rosenfield D (2015) High cortisol awakening response and cortisol levels moderate exposure-based psychotherapy success. Psychoneuroendocrinology 51:331–340

Milligan-Saville JS, Graham BM (2016) Mothers do it differently: reproductive experience alters fear extinction in female rats and women. Transl Psychiatry 6:e928

Montoya ER, Bos PA (2017) How oral contraceptives impact social-emotional behavior and brain function. Trends Cogn Sci 21:125–136

Pace-Schott EF, Germain A, Milad MR (2015) Sleep and REM sleep disturbance in the pathophysiology of PTSD: the role of extinction memory. Biol Mood Anxiety Disord 5:3

Peyrot C, Brouillard A, Morand-Beaulieu S, Marin M-F (2020) A review on how stress modulates fear conditioning: let's not forget the role of sex and sex hormones. Behav Res Ther 129:103615

Pineles SL, Nillni YI, King MW, Patton SC, Bauer MR, Mostoufi SM, Gerber MR, Hauger R, Resick PA, Rasmusson AM, Orr SP (2016) Extinction retention and the menstrual cycle: different associations for women with posttraumatic stress disorder. J Abnorm Psychol 125: 349–355

Pletzer BA, Kerschbaum HH (2014) 50 years of hormonal contraception-time to find out, what it does to our brain. Front Neurosci 8:256

Raeder F, Heidemann F, Schedlowski M, Margraf J, Zlomuzica A (2019) No pills, more skills: the adverse effect of hormonal contraceptive use on exposure therapy benefit. J Psychiatry Res 119: 95–101

Raeder F, Merz CJ, Tegenthoff M, Dere E, Wolf OT, Margraf J, Schneider S, Zlomuzica A (2023) Do oral contraceptives modulate the effects of stress-induction on one-exposure session efficacy and generalization in women? Psychopharmacology 240:1075–1089

Rohleder N, Wolf JM, Piel M, Kirschbaum C (2003) Impact of oral contraceptive use on glucocorticoid sensitivity of pro-inflammatory cytokine production after psychosocial stress. Psychoneuroendocrinology 28:261–273

Schöner J, Heinz A, Endres M, Gertz K, Kronenberg G (2017) Post-traumatic stress disorder and beyond: an overview of rodent stress models. J Cell Mol Med 21:2248–2256

Shields GS, Sazma MA, McCullough AM, Yonelinas AP (2017) The effects of acute stress on episodic memory: a meta-analysis and integrative review. Psychol Bull 143:636–675

Shors TJ (2004) Learning during stressful times. Learn Mem 11:137–144

Shors TJ (2006) Stressful experience and learning across the lifespan. Annu Rev Psychol 57:55

Soravia LM, Heinrichs M, Aerni A, Maroni C, Schelling G, Ehlert U, Roozendaal B, de Quervain DJ-F (2006) Glucocorticoids reduce phobic fear in humans. Proc Natl Acad Sci U S A 103: 5585–5590

Soravia LM, Heinrichs M, Winzeler L, Fisler M, Schmitt W, Horn H, Dierks T, Strik W, Hofmann SG, de Quervain DJ-F (2014) Glucocorticoids enhance in vivo exposure-based therapy of spider phobia. Depress Anxiety 31:429–435

Sundstrom Poromaa I, Gingnell M (2014) Menstrual cycle influence on cognitive function and emotion processing-from a reproductive perspective. Front Neurosci 8:380

Tang S, Graham BM (2020) The role of hormonal and reproductive status in the treatment of anxiety disorders in women. In: Kim Y-K (ed) Anxiety disorders: rethinking and understanding recent discoveries. Springer, Singapore, pp 523–541

Taylor CM, Pritschet L, Jacobs EG (2021) The scientific body of knowledge – whose body does it serve? A spotlight on oral contraceptives and women's health factors in neuroimaging. Front Neuroendocrinol 60:100874

ter Horst JP, de Kloet ER, Schächinger H, Oitzl MS (2012) Relevance of stress and female sex hormones for emotion and cognition. Cell Mol Neurobiol 32:725–735

Turner BB (1997) Influence of gonadal steroids on brain corticosteroid receptors: a minireview. Neurochem Res 22:1375–1385

Ulrich-Lai YM, Herman JP (2009) Neural regulation of endocrine and autonomic stress responses. Nat Rev Neurosci 10:397–409

Velasco ER, Florido A, Milad MR, Andero R (2019) Sex differences in fear extinction. Neurosci Biobehav Rev 103:81–108

White EC, Graham BM (2016) Estradiol levels in women predict skin conductance response but not valence and expectancy ratings in conditioned fear extinction. Neurobiol Learn Mem 134:339–348

Wolf OT, Atsak P, de Quervain DJ-F, Roozendaal B, Wingenfeld K (2016) Stress and memory: a selective review on recent developments in the understanding of stress hormone effects on memory and their clinical relevance. J Neuroendocrinol 28:8

Part III
Psychopathology

Is Fear Extinction Impairment Central to Psychopathology?

Richard A. Bryant

Contents

Abstract As discussed in this chapter, there have been enormous advances in our understanding of how anxiety disorders develop, are maintained, and can be treated. Many of these advances have been the result of translational studies using fear conditioning and extinction models. Despite these successes, we recognize, as a field, that there are important limitations in the extent to which extinction can explain how anxiety disorders and behaviors remit. Clinically speaking, the outstanding challenge for treatment of anxiety disorders is to improve the current suboptimal success rates. Over the past 30 years, we have not improved our treatment success rates despite employing many pharmacological and pharmacological strategies. While extinction and related fear circuitry mechanisms most certainly appear to play a role in treatment of anxiety disorders, they are also apparently insufficient to fully accommodate the varied responses individuals exhibit with this treatment approach. Increasingly diverse and innovative approaches are needed that accommodate the multitude of change mechanisms involved in treating anxiety. However, this is not to suggest ignoring the key role that extinction and memory updating processes play in overcoming anxiety.

R. A. Bryant (✉)
School of Psychology, University of New South Wales, Sydney, NSW, Australia
e-mail: r.bryant@unsw.edu.au

Curr Topics Behav Neurosci (2023) 64: 195–212
https://doi.org/10.1007/7854_2023_439
Published Online: 6 September 2023

Keywords Extinction · Fear conditioning · Learning · Memory · Posttraumatic stress · Threat

1 Introduction

Principles and paradigms emerging from early models of fear conditioning, associative learning, and extinction processes have been seminal in how we have understood the etiology and treatment of psychopathological disorders. These models have been particularly seminal in conceptualizations and management of anxiety and stress-related disorders. These models have permitted enormous translational advances, from fundamental animal research to preclinical human studies, and eventually to clinical applications. This chapter provides an overview of the role of extinction in anxiety and stress-related disorders, and specifically addresses the question of the evidence for its traditional central role in the development and treatment of anxiety disorders. For the purpose of this review, this chapter includes under the umbrella term of anxiety and stress-related disorders, posttraumatic stress disorder (PTSD), specific phobia, social phobia, panic disorder, and obsessive-compulsive disorder. It initially reviews the principles of conditioning models and their relevance to anxiety, the evidence supporting the role of extinction in treatment of anxiety disorders, and finally it considers the evidence regarding the emphasis we place on extinction. This review considers that whereas there is little doubt that extinction is highly relevant for treatment of anxiety disorders and stress-related disorders, there are also good reasons to suggest that extinction does not adequately explain all aspects of remission of anxiety and stress.

2 The Rationale for Fear Conditioning and Extinction in Anxiety

Persistent anxiety has traditionally been understood through the lens of fear conditioning models. The historical roots of this tradition can be traced back to the seminal (and ethically dubious) work of Watson and Raynor, who subjected "young Albert" to the pairing of a rat with a very loud unexpected noise (an unconditioned stimulus); not surprisingly, the boy developed a strong fear of the rat as well as other furry animals, which is understood as his conditioned response, which generalized to related stimuli (Watson and Rayner 1920). Mowrer subsequently extended this conceptualization by explaining the role of avoidance and phobic behaviors with his two-factor model, which involves a combination of classical and operant conditioning processes (Mowrer 1960). In this account, anxiety develops when the initially acquired fear (which occurs via classical conditioning pairing of conditioned and unconditioned stimuli) is responded to with avoidance of the stimulus, which in turn leads to a reduction in anxiety. This operant process reinforces the avoidance, leading to persistent avoidance that is characteristic of anxiety disorders. In fear conditioning models, adaptive response to an acquired fear involves inhibitory learning that the conditioned stimuli, which typically involves internal or external cues that are associated with the fear response, are actually benign and do not signal threat (Bouton 1993). This is termed extinction learning, and it proposed that impairments in this process results in anxiety disorders because the individual has not been able to effectively learn that these otherwise benign cues are indicative of safety rather than threat.

Arguably the disorder that most clearly demonstrates this model is posttraumatic stress disorder (PTSD). In this disorder, it is understood that a traumatic event (the unconditioned stimulus) triggers extreme fear (the unconditioned response), and subsequent reminders of the trauma (the conditioned stimuli) elicit fear reactions that include intrusive memories, flashbacks, and intense anxiety (the conditioned responses; Pitman et al. 2012). A firefighter who develops PTSD may have strong anxiety reactions when he smells burning steak or sees flames at a barbeque because these are conditioned stimuli that trigger threats and elicit unwanted memories of the trauma. In PTSD models, it is proposed that most people successfully adapt after a traumatic experience because in the weeks and months after trauma exposure they engage in successful extinction learning as they are repeatedly exposed to conditioned stimuli that do not result in further trauma, and so there is inhibitory learning these cues signal safety rather than threat (Rauch et al. 2006). In this sense, PTSD is often understood as failed extinction retention insofar as people with PTSD purportedly have difficulty in consolidating the learning that these cues now signal safety (Davis and Myers 2002; Zuj and Norrholm 2019).

3 Current Evidence About Aberrant Extinction in Anxiety

There is much support for the role of fear conditioning models in anxiety disorders. Consistent with fear conditioning models, one meta-analysis of 44 fear conditioning studies that encompassed a range of anxiety disorders and included 963 patients and 1,222 healthy controls found that anxiety patients had increased responses to a CS – during fear acquisition ($d = 0.30$; Duits et al. 2015), suggesting that people with anxiety may have difficulty discriminating safety from threat (Lissek et al. 2014). It is possible that people with anxiety disorders may be more conditionable because of their deficits in discriminating between safety and threat, and their greater reactivity to threats in their environment. Supporting this possibility is evidence that PTSD individuals displayed higher levels of skin conductance and heart rate during conditioning than those without PTSD (Orr et al. 2000). The tendency to overgeneralize fear responses to safety and threat signals has also been documented in panic disorder (Lissek et al. 2010) and generalized anxiety disorder (Lissek et al. 2014). Indirect support for this hypothesis also comes from evidence that people who eventually develop PTSD display more arousal in the immediate phase after trauma exposure relative to those who do not develop PTSD. Numerous studies have reported higher resting heart rates in the acute phase in people who subsequently develop PTSD (Bryant et al. 2008; Shalev et al. 1998). There are also reports of lower gamma-aminobutyric acid (GABA) levels, a major inhibitory neurotransmitter, in those who develop PTSD (Vaiva et al. 2006). A study of patients in an Emergency Department found that higher skin conductance response during an interview regarding the trauma was predictive of later PTSD (Hinrichs et al. 2019). This finding has been observed in other studies that used trauma narratives in the acute trauma phase as a provocation stimulus, and these have noted increased heart rate and skin conductance in those that go on to develop PTSD (Goodman and Griffin 2018). Overall, there is convergent evidence that people with PTSD display distinct patterns of fear acquisition in fear conditioning paradigms (Blechert et al. 2007; Milad et al. 2009). It should be noted that the stronger fear conditioning evident in PTSD has been shown to result in greater fear load that a person needs to extinguish for successful adaptation to the trauma (Norrholm et al. 2011).

There is more consistent evidence that people with anxiety are characterized by impairments in extinction learning, which has been observed in PTSD (Blechert et al. 2007; Orr et al. 2000), panic disorder (Otto et al. 2014), social phobia (Hermann et al. 2002), and obsessive-compulsive disorder (Cooper and Dunsmoor 2021). Meta-analysis indicates that anxiety patients have stronger reactivity to CS+ trials during extinction relative to healthy controls ($d = 0.35$; Duits et al. 2015). In terms of PTSD, this deficit in learning safety signals has been demonstrated across different types of trauma-exposed populations (Jovanovic et al. 2010a) and appears to be associated with PTSD rather than depression after trauma (Jovanovic et al. 2010b). One study found that the difficulty in fear inhibition was only evident in individuals with more severe PTSD symptomatology (Jovanovic et al. 2009). There is also much evidence that people with PTSD having greater response, often measured with skin conductance, heart rate, or frontalis electromyogram, in response to trauma reminders (Orr et al. 1993, 1995), indicating that the persistent response to the conditioned stimuli is a common biomarker in PTSD.

The extent to which greater conditionability or impaired extinction in PTSD is a risk factor or a function of PTSD is complex. An earlier study that recruited monozygotic twins who either did or did not serve in the Vietnam war observed slower habituating skin conductance startle responses in veterans with PTSD and their non-combat-exposed co-twins relative to those without PTSD and also their non-combat exposed co-twin; this pattern points to habituating responses to startle (which may reflect a predisposition to have a conditioned response) possibly reflecting a pretrauma vulnerability factor (Pitman et al. 2006). In a similar vein, one study of new firefighter recruits found that elevated startle response prior to commencing active duty predicted severity of subsequent acute stress reactions after trauma exposure (Guthrie and Bryant 2005). Several studies have administered fear conditioning/extinction paradigms to groups who are high risk for subsequently being exposed to trauma, such as military prior to deployment and first responders during their recruitment training (Guthrie and Bryant 2006; Lommen et al. 2013; Orr et al. 2012). Each of these studies found that poorer extinction learning prior to trauma exposure predicted subsequent PTSD severity, suggesting extinction deficits pose a risk for PTSD development. Using a different approach, Milad and colleagues studied monozygotic twins who had either been exposed to combat and developed PTSD and the others had not (Milad et al. 2008). This study found that those twins with PTSD had higher skin conductance during extinction recall relative to their twins without PTSD, suggesting that the impairment could not be attributed simply to genetic risk factors. Overall, the evidence suggests that the primary deficit in fear learning processes in PTSD occurs in extinction retention rather than fear acquisition (Dvir et al. 2019; Pöhlchen et al. 2020).

4 Neural Evidence of Extinction in Anxiety

Prevailing neural models of extinction learning holds that key networks in the prefrontal regions of the brain are strongly implicated in extinction insofar as they down-regulate more limbic regions. Specifically, during conditioning there is sensory information projected to the amygdala, where the pairing of the neutral input with the aversive consequence occurs in the basolateral nucleus of the amygdala. These inputs subsequently result in long-term potentiation, resulting in the persistent conditioned response. This information is projected to the central nucleus of the amygdala, which is largely responsible for projections to the hypothalamus, periaqueductal gray, and associated regions involved in the increased release of stress hormones and increased sympathetic arousal that characterizes the fear response (LeDoux 2003). There is much evidence complementing the neural circuitry that underlies fear learning when recall of an extinction cue occurs in the same context in which extinction occurred as there is recruitment of the infralimbic regions of the medial prefrontal cortex, basolateral nucleus of the amygdala, and the hippocampus. Evidence suggests that the hippocampus activates the infralimbic network, which activates inhibitory interneurons in the amygdala, and this down-regulates the fear response (Quirk and Mueller 2008). In contrast, if the extinction stimulus occurs in a different context, then the hippocampus does not activate the infralimbic cortex and can result in a conditioned fear response (Milad et al. 2014). These circuits have been demonstrated in numerous functional magnetic resonance imaging (fMRI) studies during extinction procedures when extinction recall is occurring in and out of the initial extinction context (Milad et al. 2007; Milad and Quirk 2012).

Consistent with findings that anxiety disorders involve impoverished extinction learning, there is convergent evidence that the appropriate networks are not optimally recruited during extinction in patients with anxiety. Studies have shown that patients with PTSD show decreased activity in the ventral medial prefrontal cortex (vmPFC) and hippocampus during extinction recall (Garfinkel et al. 2014; Marin et al. 2016; Milad et al. 2009). The pattern over impaired inhibitory networks is replicated across other anxiety disorders, which reflect dysfunctional recruitment of extinction-related neural circuitry (Marin et al. 2017, 2020; Sylvester et al. 2012; Wen et al. 2021).

5 The Role of Extinction in Treatment of Anxiety Disorders

The major treatment for many anxiety disorders is based on extinction learning. Exposure therapy involves repeated and safe confrontations with the feared object (purportedly the conditioned stimulus), and it is postulated that exposure results in extinction learning that inhibits the initial conditioned fear association (Moscovitch et al. 2009). There is also evidence that capacity for extinction learning on a laboratory extinction paradigm prior to exposure therapy predicts more successful response to treatment in obsessive-compulsive disorder (Geller et al. 2019).

Many studies have investigated the extent to which habituation of distress between-(over days) and within-exposure (over minutes) therapy sessions reflects the extent to which similar reductions in fear are achieved during the overall course of exposure (over weeks), and predicts remission of PTSD symptoms. There have been numerous reports of between-session habituation predicting symptom reduction (Rauch et al. 2004; van Minnen and Hagenaars 2002), although other studies have reported symptom reduction is not associated with remission (Bluett et al. 2014). Fewer studies have noted that within-session habituation predicts treatment success (van Minnen and Hagenaars 2002), although most studies indicate that this pattern is not associated with remission (Bluett et al. 2014; Craske et al. 2008).

The notion that extinction plays a role in the treatment of anxiety disorders is further indicated by findings that better recruitment of extinction-related networks is predictive of more successful response to exposure therapies across a range of anxiety disorders (for a review, see Colvonen et al. 2017). There is also neuroimaging evidence that in people with an anxiety disorder the fear circuitry networks, and specifically those involving inhibitory extinction-related regions, are normalized after more successfully exposure therapy (for a review, see Manthey et al. 2021). Further, degree of treatment response during intense exposure therapy display has been shown to be associated with improved extinction learning after treatment relative to prior to therapy commencing (Maples-Keller et al. 2022). Taken together, these studies indicate that the neural networks implicated in extinction are both predictive of exposure therapy success and are also normalized to an extent following therapy.

6 Does Extinction Encompass the Totality of Anxiety Mechanisms?

This review has so far pointed to the supportive evidence of extinction models accounting for much of what we know about how anxiety disorders develop, are maintained, and can be treated. It needs to be recognized, however, that extinction models may not explain all facets of anxiety. Over the years there has been a range of criticisms of conditioning models of anxiety disorders that have been offered as a challenge to the centrality of this perspective. Arguably the most fundamental reason to question the role of fear conditioning and extinction as cornerstone processes in anxiety disorders is the apparent lack of a triggering event that precedes the onset of many anxiety disorders and that serves as the unconditioned stimulus. Whereas PTSD is clearly an exemplar of how anxiety can develop after exposure to a threatening event, other anxiety disorders are not so clear cut. Whereas some people with phobias can recall a traumatic experience occurring prior to the onset of their phobia (Davey 1989), others report no aversive experiences (Menzies and Clarke 1993a, b). Further, many people experience aversive events and do not develop an anxiety disorder (Breslau et al. 1991; Menzies and Clarke 1993a, b). Relatedly, there is often not a direct temporal contingency between the occurrence of a traumatic event and the onset of the anxiety disorder, which is most obviously seen in delayed-onset PTSD in which the disorder can occur many months after the event (Andrews et al. 2007); this delay could be interpreted as contrary to a conditioning response to the triggering aversive stimulus. It has also been noted that fear conditioning models would predict that there should be an equal probability of developing an anxiety disorder following all aversive experiences. Interestingly, however, there is a greater likelihood of developing a phobia of snakes, spiders, dogs, heights, thunder, or fire than more prevalent events such as knives, guns, or electrical sockets that have greater probability of causing harm (Poulton and Menzies 2002). Supporting this trend is evidence that people acquire fear more quickly to purportedly "hard-wired" stimuli such as snakes and spiders than other stimuli such as rabbits or flowers (Ohman and Mineka 2001).

In response to these criticisms, there have been many commentaries that have countered these criticisms by highlighting that the arguments put forward to challenge conditioning models are based on simplistic models of conditioning that do not accommodate subsequent refinements of conditioning theory (Field 2006). There is not a simple linear relationship between the occurrence of an aversive experience and a conditioned response. There are numerous contextual factors that can moderate both conditioning and extinction, including the person's amount of exposure to the conditioned stimulus prior to the conditioning (latent inhibition) or presence of an inhibitory stimulus at the time of an aversive experience that becomes associated with the conditioned stimuli (super learning). There are also numerous iterations of initial conditioning theories that have taken into account the variability in which fear is acquired and extinguished; full discussion of these theories is beyond the scope of the current review but are reviewed fully elsewhere including in multiple chapters of this book (Dunsmoor et al. 2015).

One of the key issues that is often ignored in the debate about the role of conditioning and extinction in relation to anxiety disorders is the role of individual differences (Lonsdorf et al. 2017). In the earlier years of research in conditioning, there was a focus on the average response of rodents or humans on conditioning/extinction paradigms. Of course this ignores the enormous heterogeneity that is seen in how individuals acquire fear learning and subsequently engage in extinction. More recent research in basic research has analyzed fear responses beyond the average focus and highlighted the importance of subgroups of responding (Galatzer-Levy et al. 2017). This development more closely mimics the real-world patterns of anxiety following exposure to a traumatic event which typically involves multiple trajectories of response (Galatzer-Levy et al. 2018). There are numerous moderating individual differences that can impact conditioning and extinction. One of the major moderating factors in anxiety disorders is sex, insofar as women are twice as likely to develop an anxiety disorder than men (Graham and Milad 2013). Extinction learning occurs variably in women across the menstrual cycle (Graham and Milad 2011) and hormonal contraceptive impedes extinction in women (Graham and Milad 2013). There is increasing evidence that low estrogen plays a role in impaired extinction, and hence may pose a risk factor for development for anxiety disorders (Glover et al. 2012, 2015). These observations highlight that hormonal factors can account for variation in extinction both between people, and also within a person over time.

Another fundamental individual difference that can moderate extinction processes is genetic predisposition. Twin studies have indicated that genetic factors can account for 30–40% of the risk for developing an anxiety disorder (Hettema et al. 2001), which is supported by evidence that genetic factors can contribute to fear conditioning (Lonsdorf and Kalisch 2011). The most studied genetic candidate in terms of fear conditioning is the serotonin transporter gene 5-HTTLPR (Bauer 2015), with the s-allele facilitating fear conditioning (Lonsdorf et al. 2009). As is expected, this impact does appear to be impacted by a person's life history (Hermann et al. 2012). Of relevance to treatment of anxiety disorders, this allele of the 5-HTTLPR is predictive of response to exposure therapy for PTSD (Bryant et al. 2010). Another polymorphism that is key for extinction learning is the gene that moderates expression of brain-derived neurotrophic factor (BDNF), a neurotrophin which has an important role in synaptic plasticity and long-term potentiation, and is involved in learning and memory processes (Andero and Ressler 2012). The BDNF val66met is implicated in fear learning (Lonsdorf et al. 2010) and is associated with impaired extinction learning (Soliman et al. 2010). Notably, the BDNF val66met allele is predictive of response to exposure therapy for PTSD (Felmingham et al. 2013). These genetic variants highlight that not all people will acquire fear and extinguish the initial conditioning to the same extent, and that one's genetic variant will interact with environmental factors in moderating how one can inhibit learnt fear (Nees et al. 2018).

One of the challenges for the application of extinction models to some anxiety and trauma-related stress conditions is that these disorders can be very heterogeneous and often involve a range of emotional states that are distinct from fear circuity processes. The most salient example of this is PTSD, which has been defined in DSM-5 in such a way that it can potentially manifest in 636,120 different combinations of symptoms (Galatzer-Levy and Bryant 2013), and empirical studies comprising very large samples have shown that most people diagnosed with PTSD present with a combination of symptoms not shared with any other person in the sample (Bryant et al. 2023a–c). Compounding this problem of heterogeneity is the explicit recognition that PTSD can encompass a range of negative emotional states, including anger, shame, guilt, and depression. The presence of these emotional states is in fact very common in PTSD (Contractor et al. 2018). Whereas extinction models can function well in explaining recovery from fear-based disorders, they do not operate comparably with other emotional states, and this can represent a limitation of their applicability to treatments of non-fear-based emotional states. Exemplifying this point is one study that found that exposure therapy for PTSD and prolonged grief functioned quite distinctly, and whereas habituation of distress occurred as expected in PTSD, this did not occur in treating prolonged grief (Bryant et al. 2023a–c).

7 Role of Reconsolidation

As noted above, the main rationale for interest in extinction processes in anxiety disorders is that they offer key explanatory value for exposure therapy. Although exposure therapy is the treatment of choice across the anxiety disorders, the evidence tells us that approximately one-half to one-third of patients do not respond optimally to this treatment approach (Loerinc et al. 2015). A further challenge for clinicians is that many patients who do respond to treatment subsequently relapse and their anxiety returns. This return of fear is explained in extinction principles via a number of processes, including fear reinstatement, fear renewal, and spontaneous recovery (Bouton 2004). The underlying mechanism for each of these processes is that in extinction learning the original fear memory trace is inhibited but not erased, and therefore the memory can be reactivated.

One mechanism that has been proposed as a relevant paradigm for conceptualizing optimal treatment of anxiety disorders involves fear memory reconsolidation (Monfils and Holmes 2018). This process involves previously consolidated memories becoming destabilized during retrieval, and in the destabilization window protein synthesis permits longer-term storage; this allows potential blockade of fear responses to the original memory if the memory is reconsolidated without the fear response (Nader et al. 2000). Subsequent studies have shown that fear memories can be modified by different pharmacological interventions during the period of memory destabilization (Kindt et al. 2009). There has been debate about the overlapping and distinctive mechanisms involved in extinction and reconsolidation, however there is evidence that different mechanisms exist (Monfils et al. 2009). This is further elaborated in an included chapter in this book.

Building on this paradigm, several studies have attempted to treat anxiety disorders by utilizing reconsolidation processes. One pilot study of treating PTSD found that administering propranolol, a beta-blocker that reduces noradrenergic response, after retrieval of trauma memories resulted in reduced physiological responding to trauma memories relative to a placebo (Brunet et al. 2008). A larger trial using the same paradigm found that propranolol resulted in greater PTSD symptom reduction relative to placebo (Brunet et al. 2018). Using a similar approach, another study found that administering propranolol to spider phobics after exposure therapy led to more reduced avoidance of spiders a year after treatment relative to administering placebo or propranolol without prior memory reactivation (Soeter and Kindt 2015).

Other approaches have used psychological interventions during the reactivation period to update the fear memory. Several studies have shown in analog studies of trauma memories that intrusive memories of a trauma film can be reduced by providing a visual interference task (James et al. 2015) or thinking of attachment figures (Bryant and Datta 2019) after retrieval of the initial memory. Extending this to the clinical domain, one study found that interfering with visual processing after recalling a trauma while still in an emergency department after traumatic injury resulted in reduced subsequent intrusive memories (Iyadurai et al. 2018).

8 Optimizing Extinction to Enhance Exposure Therapy

The pattern for exposure therapy to only have moderate success in treating anxiety disorders (Loerinc et al. 2015) has led to many attempts to augment exposure treatments. These attempts have typically focused on pharmacological strategies to boost extinction during exposure. In short, these strategies have focused on targeting neurotransmitters such as glutamate, dopamine, serotonin, norepinephrine, as well as hormones such as glucocorticoids that are implicated in neural regions and are pivotal for extinction learning (Lebois et al. 2019). For example, several studies have used d-cycloserine (DCS), which is a partial glutamatergic agonist that has been shown to augment extinction learning in animals (Richardson et al. 2004). Researchers were initially excited by the promise of DCS to enhance the effects of exposure therapy because of a series of trials indicating benefits of DCS relative to placebo in combination with exposure therapy for a range of anxiety disorders (Hofmann et al. 2006; Ressler et al. 2004). There are mixed findings in terms of DCS enhancing the benefits of exposure therapy, and comprehensive reviews conclude that the effects of DCS are actually quite modest (Mataix-Cols et al. 2017). Other candidates have been tried to augment extinction processes in the context of exposure therapy. On the basis that glucocorticoid administration facilitates extinction in animals (Cai et al. 2006), one pilot study reported that hydrocortisone can augment exposure therapy (Yehuda et al. 2015); this has yet to be replicated in a full trial. On the basis that yohimbine can elevate norepinephrine levels, which may augment engagement with trauma memories and subsequent extinction, it has been used in one study to augment trauma focused (T-F) psychotherapy (Tuerk et al. 2018); this study found no effect for yohimbine at follow-up assessment. Recent studies have attempted to augment psychotherapy for PTSD by combining it with MDMA (3,4-methylenedioxymethamphetamine) administration with the expectation that it facilitates engagement with trauma memories and the therapist; pilot studies show promising results but have yet to be replicated in full trials, and importantly this approach has not combined MDMA with exposure therapy (Mithoefer et al. 2018). Importantly, these trials have combined MDMA with generic psychotherapy rather than exposure therapy. Other augmentation attempts have used brain stimulation techniques, including transcranial direct current stimulation and transcranial magnetic stimulation, but they currently lack adequate evidence (Lebois et al. 2019). Finally, there is evidence that behavioral strategies such as brief intensive exercise can augment exposure therapy for PTSD (Bryant et al. 2023a–c), possibly because it augments consolidation of extinction learning (Keyan and Bryant 2019). However, this has yet to be proven in a fully powered trial.

9 Extinction and Predictors of Treatment Response

In light of the modest gains made by attempts to augment exposure therapy via modulating extinction processes strong (Lebois et al. 2019), it is worth considering the extent to which extinction-related factors actually predict treatment response. Supporting the proposal that extinction is a key mechanism underpinning exposure therapy, there is considerable evidence that greater treatment response is associated with pre-treatment elevated arousal (Norrholm et al. 2016), glucocorticoid receptor sensitivity (Yehuda et al. 2015), and recruitment of neural network implicated in fear extinction (Colvonen et al. 2017). More importantly, there is evidence that performance on laboratory-based conditioning and extinction paradigms can be predictive of successful response to exposure therapy for anxiety disorders (Forcadell et al. 2017; Geller et al. 2019; Maples-Keller et al. 2022; Waters and Pine 2016).

It is important to note, however, that the evidence for these extinction-based factors predicting treatment response is very mixed. For example, there are studies indicating that treatment responders and non-responders do not differ on conditioning or extinction performance (Raeder et al. 2020; Wannemueller et al. 2018). Taken together, the extent to which paradigms hypothesized to underpin exposure therapy, primarily relying on extinction processes, has limited support in terms of predicting exposure therapy outcomes (Scheveneels et al. 2021). One of the reasons for this modest prediction is the accumulating evidence that many other factors appear to predict treatment outcome that are apparently unrelated to extinction or fear circuitry factors. For example, poorer treatment outcome can be associated with childhood trauma exposure (Bosch et al. 2020), comorbid psychiatric conditions (Stein et al. 2012), older age (Litz et al. 2019), and lower socioeconomic status (Wachen et al. 2014). Treatment response can also be improved by positive support from others (Thrasher et al. 2010), while negative support impedes treatment outcome (Tarrier et al. 2000). There is much evidence that cognitive style impacts poor treatment outcomes, including catastrophic thinking (Holliday et al. 2018; Zalta et al. 2014) and anxiety sensitivity (Zandberg et al. 2016). Non-fear emotions at baseline also predict poor treatment response, including shame and guilt (Oktedalen et al. 2015) and anger (Lloyd et al. 2014). The observation that factors seemingly unrelated to fear or extinction processes are predictive of exposure therapy response suggests that there may be other factors in addition to extinction that are involved in remission of anxiety.

10 Concluding Comments

This review highlights that fear conditioning and extinction models have made enormous advances in our understanding of how anxiety disorders develop, are maintained, and can be treated. There are important limitations, however, in the extent to which extinction can explain how anxiety remits. From a clinical perspective, the outstanding challenge for treatment of anxiety disorders is to improve the current limited success rates. The available evidence suggests that attempts to augment exposure therapy have met with limited success and despite the many different strategies employed, we have not improved our treatment success rates over the past 30 years. While extinction and related fear circuitry mechanisms definitely appear to play a role in treatment of anxiety disorders, they are also apparently insufficient to fully accommodate the variable responses people have to this treatment approach. More diverse approaches are needed that accommodate the diverse change mechanisms that are involved in treating anxiety. In proposing this, however, it is important to not neglect the key role that extinction and memory updating processes play in overcoming anxiety.

References

Andero R, Ressler KJ (2012) Fear extinction and BDNF: translating animal models of PTSD to the clinic. Genes Brain Behav 11:503–512

Andrews B, Brewin CR, Philpott R, Stewart L (2007) Delayed-onset posttraumatic stress disorder: a systematic review of the evidence. Am J Psychiatr 164:1319–1326

Bauer EP (2015) Serotonin in fear conditioning processes. Behav Brain Res 277:68–77

Blechert J, Michael T, Vriends N, Margraf J, Wilhelm FH (2007) Fear conditioning in posttraumatic stress disorder: evidence for delayed extinction of autonomic, experiential, and behavioural responses. Behav Res Ther 45:2019–2033

Bluett EJ, Zoellner LA, Feeny NC (2014) Does change in distress matter? Mechanisms of change in prolonged exposure for PTSD. J Behav Ther Exp Psychiatry 45:97–104

Bosch J, Mackintosh MA, Wells SY, Wickramasinghe I, Glassman LH, Morland LA (2020) PTSD treatment response and quality of life in women with childhood trauma histories. Psychol Trauma 12:55–63

Bouton ME (1993) Context, time, and memory retrieval in the interference paradigms of Pavlovian learning. Psychol Bull 114:80–99

Bouton ME (2004) Context and behavioral processes in extinction. Learn Mem 11:485–494

Breslau N, Davis G, Andreski P, Peterson E (1991) Traumatic events and posttraumatic stress disorder in an urban population of young adults. Arch Gen Psychiatry 48:216–222

Brunet A, Orr SP, Tremblay J, Robertson K, Nader K, Pitman RK (2008) Effect of post-retrieval propranolol on psychophysiologic responding during subsequent script-driven traumatic imagery in post-traumatic stress disorder. J Psychiatr Res 42:503–506

Brunet A, Saumier D, Liu A, Streiner DL, Tremblay J, Pitman RK (2018) Reduction of PTSD symptoms with pre-reactivation propranolol therapy: a randomized controlled trial. Am J Psychiatr 175:427–433

Bryant RA, Datta S (2019) Reconsolidating distressing memories by thinking of attachment figures. Clin Psychol Sci 7:1249–1256

Bryant RA, Creamer M, O'Donnell M, Silove D, McFarlane AC (2008) A multisite study of initial respiration rate and heart rate as predictors of posttraumatic stress disorder. J Clin Psychiatry 69: 1694–1701

Bryant RA, Felmingham KL, Falconer EM, Pe Benito L, Dobson-Stone C, Pierce KD, Schofield PR (2010, Jun 15) Preliminary evidence of the short allele of the serotonin transporter gene predicting poor response to cognitive behavior therapy in posttraumatic stress disorder. Biol Psychiatry 67(12):1217–1219

Bryant RA, Azevedo S, Yadav S, Keyan D, Rawson N, Dawson K, Tockar J, Garber B, Hadzi-Pavlovic D (2023a) Habituation of distress during exposure and its relationship to treatment outcome in post-traumatic stress disorder and prolonged grief disorder. Eur J Psychotraumatol 14(2):2193525

Bryant RA, Dawson KS, Azevedo S, Yadav S, Cahill C, Kenny L, Maccallum F, Tran J, Rawson N, Tockar J, Garber B, Keyan D (2023b) Augmenting trauma-focused psychotherapy for post-traumatic stress disorder with brief aerobic exercise in Australia: a randomised clinical trial. Lancet Psychiatry 10(1):21–29

Bryant RA, Galatzer-Levy I, Hadzi-Pavlovic D (2023c) The heterogeneity of posttraumatic stress disorder in DSM-5. JAMA Psychiatry 80(2):189–191

Cai WH, Blundell J, Han J, Greene RW, Powell CM (2006) Postreactivation glucocorticoids impair recall of established fear memory. J Neurosci 26:9560–9566

Colvonen PJ, Glassman LH, Crocker LD, Buttner MM, Orff H, Schiehser DM, Norman SB, Afari N (2017) Pretreatment biomarkers predicting PTSD psychotherapy outcomes: a systematic review. Neurosci Biobehav Rev 75:140–156

Contractor AA, Greene T, Dolan M, Elhai JD (2018) Relations between PTSD and depression symptom clusters in samples differentiated by PTSD diagnostic status. J Anxiety Disord 59:17–26

Cooper SE, Dunsmoor JE (2021) Fear conditioning and extinction in obsessive-compulsive disorder: a systematic review. Neurosci Biobehav Rev 129:75–94

Craske MG, Kircanski K, Zelikowsky M, Mystkowski J, Chowdhury N, Baker A (2008) Optimizing inhibitory learning during exposure therapy. Behav Res Ther 46:5–27

Davey GC (1989) Dental phobias and anxieties: evidence for conditioning processes in the acquisition and modulation of a learned fear. Behav Res Ther 27(1):51–58. https://doi.org/10.1016/0005-7967(89)90119-8

Davis M, Myers KM (2002) The role of glutamate and gamma-aminobutyric acid in fear extinction: clinical implications for exposure therapy. Biol Psychiatry 52:998–1007

Duits P, Cath DC, Lissek S, Hox JJ, Hamm AO, Engelhard IM, van den Hout MA, Baas JM (2015) Updated meta-analysis of classical fear conditioning in the anxiety disorders. Depress Anxiety 32:239–253

Dunsmoor JE, Niv Y, Daw N, Phelps EA (2015) Rethinking extinction. Neuron 88:47–63

Dvir M, Horovitz O, Aderka IM, Shechner T (2019) Fear conditioning and extinction in anxious and nonanxious youth: a meta-analysis. Behav Res Ther 120:103431. https://doi.org/10.1016/j.brat.2019.103431

Felmingham KL, Dobson-Stone C, Schofield PR, Quirk GJ, Bryant RA (2013) The brain-derived neurotrophic factor Val66Met polymorphism predicts response to exposure therapy in posttraumatic stress disorder. Biol Psychiatry 73:1059–1063

Field AP (2006) Is conditioning a useful framework for understanding the development and treatment of phobias? Clin Psychol Rev 26:857–875

Forcadell E, Torrents-Rodas D, Vervliet B, Leiva D, Tortella-Feliu M, Fullana MA (2017) Does fear extinction in the laboratory predict outcomes of exposure therapy? A treatment analog study. Int J Psychophysiol 121:63–71

Galatzer-Levy I, Bryant RA (2013) 636,120 ways to have posttraumatic stress disorder: the relative merits of categorical and dimensional approaches to posttraumatic stress. Perspect Psychol Sci 8:651–662

Galatzer-Levy IR, Andero R, Sawamura T, Jovanovic T, Papini S, Ressler KJ et al (2017) A cross species study of heterogeneity in fear extinction learning in relation to FKBP5 variation and expression: implications for the acute treatment of posttraumatic stress disorder. Neuropharmacology 116:188–195

Galatzer-Levy IR, Huang SH, Bonanno GA (2018) Trajectories of resilience and dysfunction following potential trauma: a review and statistical evaluation. Clin Psychol Rev 63:41–55

Garfinkel SN, Abelson JL, King AP, Sripada RK, Wang X, Gaines LM et al (2014) Impaired contextual modulation of memories in PTSD: an fMRI and psychophysiological study of extinction retention and fear renewal. J Neurosci 34:13435–13443

Geller DA, McGuire JF, Orr SP, Small BJ, Murphy TK, Trainor K et al (2019) Fear extinction learning as a predictor of response to cognitive behavioral therapy for pediatric obsessive compulsive disorder. J Anxiety Disord 64:1–8

Glover EM, Jovanovic T, Mercer KB, Kerley K, Bradley B, Ressler KJ, Norrholm SD (2012) Estrogen levels are associated with extinction deficits in women with posttraumatic stress disorder. Biol Psychiatry 72(1):19–24

Glover EM, Jovanovic T, Norrholm SD (2015) Estrogen and extinction of fear memories: implications for posttraumatic stress disorder treatment. Biol Psychiatry 78(3):178–185

Goodman BF, Griffin MG (2018) Prospectively predicting PTSD status with heart rate reactivity and recovery in interpersonal violence survivors. Psychiatry Res 259:270–276

Graham BM, Milad MR (2011) The study of fear extinction: implications for anxiety disorders. Am J Psychiatr 168:1255–1265

Graham BM, Milad MR (2013) Blockade of estrogen by hormonal contraceptives impairs fear extinction in female rats and women. Biol Psychiatry 73:371–378

Guthrie RM, Bryant RA (2005) Auditory startle response in firefighters before and after trauma exposure. Am J Psychiatr 162:283–290

Guthrie RM, Bryant RA (2006) Extinction learning before trauma and subsequent posttraumatic stress. Psychosom Med 68:307–311

Hermann C, Ziegler S, Birbaumer N, Flor H (2002) Psychophysiological and subjective indicators of aversive Pavlovian conditioning in generalized social phobia. Biol Psychiatry 52:328–337

Hermann A, Kupper Y, Schmitz A, Walter B, Vaitl D, Hennig J et al (2012) Functional gene polymorphisms in the serotonin system and traumatic life events modulate the neural basis of fear acquisition and extinction. PLoS One 7(9):e44352

Hettema JM, Neale MC, Kendler KS (2001) A review and meta-analysis of the genetic epidemiology of anxiety disorders. Am J Psychiatr 158:1568–1578

Hinrichs R, van Rooij SJ, Michopoulos V, Schultebraucks K, Winters S, Maples-Keller J et al (2019) Increased skin conductance response in the immediate aftermath of trauma predicts PTSD risk. Chronic Stress 3:2470547019844441

Hofmann SG, Meuret AE, Smits JA, Simon NM, Pollack MH, Eisenmenger K et al (2006) Augmentation of exposure therapy with D-cycloserine for social anxiety disorder. Arch Gen Psychiatry 63:298–304

Holliday R, Holder N, Surís A (2018) Reductions in self-blame cognitions predict PTSD improvements with cognitive processing therapy for military sexual trauma-related PTSD. Psychiatry Res 263:181–184

Iyadurai L, Blackwell SE, Meiser-Stedman R, Watson PC, Bonsall MB, Geddes JR et al (2018) Preventing intrusive memories after trauma via a brief intervention involving Tetris computer game play in the emergency department: a proof-of-concept randomized controlled trial. Mol Psychiatry 23:674–682

James EL, Bonsall MB, Hoppitt L, Tunbridge EM, Geddes JR, Milton AL et al (2015) Computer game play reduces intrusive memories of experimental trauma via reconsolidation-update mechanisms. Psychol Sci 26:1201–1215

Jovanovic T, Norrholm SD, Fennell JE, Keyes M, Fiallos AM, Myers KM et al (2009) Posttraumatic stress disorder may be associated with impaired fear inhibition: relation to symptom severity. Psychiatry Res 167:151–160

Jovanovic T, Norrholm SD, Blanding NQ, Phifer JE, Weiss T, Davis M et al (2010a) Fear potentiation is associated with hypothalamic-pituitary-adrenal axis function in PTSD. Psychoneuroendocrinology 35:846–857

Jovanovic T, Norrholm SD, Blanding NQ, Davis M, Duncan E, Bradley B et al (2010b) Impaired fear inhibition is a biomarker of PTSD but not depression. Depress Anxiety 27:244–251

Keyan D, Bryant RA (2019) The capacity for acute exercise to modulate emotional memories: a review of findings and mechanisms. Neurosci Biobehav Rev 107:438–449

Kindt M, Soeter M, Vervliet B (2009) Beyond extinction: erasing human fear responses and preventing the return of fear. Nat Neurosci 12:256–258

Lebois LAM, Seligowski AV, Wolff JD, Hill SB, Ressler KJ (2019) Augmentation of extinction and inhibitory learning in anxiety and trauma-related disorders. Annu Rev Clin Psychol 15:257–284

LeDoux J (2003) The emotional brain, fear, and the amygdala. Cell Mol Neurobiol 23:727–738

Lissek S, Rabin S, Heller RE, Lukenbaugh D, Geraci M, Pine DS et al (2010) Overgeneralization of conditioned fear as a pathogenic marker of panic disorder. Am J Psychiatr 167:47–55

Lissek S, Kaczkurkin AN, Rabin S, Geraci M, Pine DS, Grillon C (2014) Generalized anxiety disorder is associated with overgeneralization of classically conditioned fear. Biol Psychiatry 75:909–915

Litz BT, Berke DS, Kline NK, Grimm K, Rusowicz-Orazem L, Resick PA et al (2019) Patterns and predictors of change in trauma-focused treatments for war-related posttraumatic stress disorder. J Consult Clin Psychol 87:1019–1029

Lloyd D, Nixon RD, Varker T, Elliott P, Perry D, Bryant RA et al (2014) Comorbidity in the prediction of cognitive processing therapy treatment outcomes for combat-related posttraumatic stress disorder. J Anxiety Disord 28:237–240

Loerinc AG, Meuret AE, Twohig MP, Rosenfield D, Bluett EJ et al (2015) Response rates for CBT for anxiety disorders: need for standardized criteria. Clin Psychol Rev 42:72–82

Lommen MJ, Engelhard IM, Sijbrandij M, van den Hout MA, Hermans D (2013) Pre-trauma individual differences in extinction learning predict posttraumatic stress. Behav Res Ther 51: 63–67

Lonsdorf TB, Kalisch R (2011) A review on experimental and clinical genetic associations studies on fear conditioning, extinction and cognitive-behavioral treatment. Transl Psychiatry 1:e41

Lonsdorf TB, Weike AI, Nikamo P, Schalling M, Hamm AO, Ohman A (2009) Genetic gating of human fear learning and extinction: possible implications for gene-environment interaction in anxiety disorder. Psychol Sci 20:198–206

Lonsdorf TB, Weike AI, Golkar A, Schalling M, Hamm AO, Ohman A (2010) Amygdala-dependent fear conditioning in humans is modulated by the BDNFval66met polymorphism. Behav Neurosci 124:9–15

Lonsdorf TB, Menz MM, Andreatta M, Fullana MA, Golkar A, Haaker J et al (2017) Don't fear 'fear conditioning': methodological considerations for the design and analysis of studies on human fear acquisition, extinction, and return of fear. Neurosci Biobehav Rev 77:247–285

Manthey A, Sierk A, Brakemeier EL, Walter H, Daniels JK (2021) Does trauma-focused psychotherapy change the brain? A systematic review of neural correlates of therapeutic gains in PTSD. Eur J Psychotraumatol 12:1929025

Maples-Keller J, Watkins LE, Nylocks KM, Yasinski C, Coghlan C, Black K, Jovanovic T, Rauch SAM, Rothbaum BO, Norrholm SD (2022) Acquisition, extinction, and return of fear in veterans in intensive outpatient prolonged exposure therapy: a fear-potentiated startle study. Behav Res Ther 154:104124

Marin MF, Song H, VanElzakker MB, Staples-Bradley LK, Linnman C, Pace-Schott EF et al (2016) Association of resting metabolism in the fear neural network with extinction recall activations and clinical measures in trauma-exposed individuals. Am J Psychiatr 173:930–938

Marin MF, Zsido RG, Song H, Lasko NB, Killgore WD, Rauch SL, Simon NM, Milad MR (2017) Skin conductance responses and neural activations during fear conditioning and extinction recall across anxiety disorders. JAMA Psychiatry 74(6):622–631

Marin MF, Hammoud MZ, Klumpp H, Simon NM, Milad MR (2020) Multimodal categorical and dimensional approaches to understanding threat conditioning and its extinction in individuals with anxiety disorders. JAMA Psychiatry 77(6):618–627. https://doi.org/10.1001/jamapsychiatry.2019.4833

Mataix-Cols D, Fernandez de la Cruz L, Monzani B, Rosenfield D, Andersson E, Perez-Vigil A et al (2017) D-cycloserine augmentation of exposure-based cognitive behavior therapy for anxiety, obsessive-compulsive, and posttraumatic stress disorders: a systematic review and meta-analysis of individual participant data. JAMA Psychiatry 74:501–510

Menzies RG, Clarke JC (1993a) The etiology of childhood water phobia. Behav Res Ther 31:499–501

Menzies RG, Clarke JC (1993b) The etiology of fear of heights and its relationship to severity and individual response patterns. Behav Res Ther 31:355–365

Milad MR, Quirk GJ (2012) Fear extinction as a model for translational neuroscience: ten years of progress. Annu Rev Psychol 63(1):129–151

Milad MR, Wright CI, Orr SP, Pitman RK, Quirk GJ, Rauch SL (2007) Recall of fear extinction in humans activates the ventromedial prefrontal cortex and hippocampus in concert. Biol Psychiatry 62:446–454

Milad MR, Orr SP, Lasko NB, Chang Y, Rauch SL, Pitman RK (2008) Presence and acquired origin of reduced recall for fear extinction in PTSD: results of a twin study. J Psychiatr Res 42:515–520

Milad MR, Pitman RK, Ellis CB, Gold AL, Shin LM, Lasko NB et al (2009) Neurobiological basis of failure to recall extinction memory in posttraumatic stress disorder. Biol Psychiatry 66:1075–1082

Milad MR, Rosenbaum BL, Simon NM (2014) Neuroscience of fear extinction: implications for assessment and treatment of fear-based and anxiety related disorders. Behav Res Ther 62:17–23

Mithoefer MC, Mithoefer AT, Feduccia AA, Jerome L, Wagner M, Wymer J et al (2018) 3,4-methylenedioxymethamphetamine (MDMA)-assisted psychotherapy for post-traumatic stress disorder in military veterans, firefighters, and police officers: a randomised, double-blind, dose-response, phase 2 clinical trial. Lancet Psychiatry 5:486–497

Monfils MH, Holmes EA (2018) Memory boundaries: opening a window inspired by reconsolidation to treat anxiety, trauma-related, and addiction disorders. Lancet Psychiatry 5:1032–1042

Monfils MH, Cowansage KK, Klann E, LeDoux JE (2009, May 15) Extinction-reconsolidation boundaries: key to persistent attenuation of fear memories. Science 324(5929):951–955

Moscovitch DA, Antony MM, Swinson RP (2009) Exposure-based treatments for anxiety disorders: theory and process. In: Antony MM, Stein MB (eds) Oxford handbook of anxiety and related disorders. Oxford University Press, pp 461–475

Mowrer OH (1960) Learning theory and behavior. Wiley, New York

Nader K, Schafe GE, Le Doux JE (2000) Fear memories require protein synthesis in the amygdala for reconsolidation after retrieval. Nature 406:722–726

Nees F, Witt SH, Flor H (2018) Neurogenetic approaches to stress and fear in humans as pathophysiological mechanisms for posttraumatic stress disorder. Biol Psychiatry 83(10):810–820

Norrholm SD, Jovanovic T, Olin IW, Sands LA, Karapanou I, Bradley B, Ressler KJ (2011) Fear extinction in traumatized civilians with posttraumatic stress disorder: relation to symptom severity. Biol Psychiatry 69(6):556–563

Norrholm SD, Jovanovic T, Gerardi M, Breazeale KG, Price M, Davis M et al (2016) Baseline psychophysiological and cortisol reactivity as a predictor of PTSD treatment outcome in virtual reality exposure therapy. Behav Res Ther 82:28–37

Ohman A, Mineka S (2001) Fears, phobias, and preparedness: toward an evolved module of fear and fear learning. Psychol Rev 108:483–522

Oktedalen T, Hoffart A, Langkaas TF (2015) Trauma-related shame and guilt as time-varying predictors of posttraumatic stress disorder symptoms during imagery exposure and imagery rescripting – a randomized controlled trial. Psychother Res 25:518–532

Orr SP, Pitman RK, Lasko NB, Herz LR (1993) Psychophysiological assessment of posttraumatic stress disorder imagery in World War II and Korean combat veterans. J Abnorm Psychol 102: 152–159

Orr SP, Lasko NB, Shalev AY, Pitman RK (1995) Physiologic responses to loud tones in Vietnam veterans with posttraumatic stress disorder. J Abnorm Psychol 104:75–82

Orr SP, Metzger LJ, Lasko NB, Macklin ML, Peri T, Pitman RK (2000) De novo conditioning in trauma-exposed individuals with and without posttraumatic stress disorder. J Abnorm Psychol 109:290–298

Orr SP, Lasko NB, Macklin ML, Pineles SL, Chang Y, Pitman RK (2012) Predicting post-trauma stress symptoms from pre-trauma psychophysiologic reactivity, personality traits and measures of psychopathology. Biol Mood Anxiety Disorder 2(1):8

Otto MW, Moshier SJ, Kinner DG, Simon NM, Pollack MH, Orr SP (2014) De novo fear conditioning across diagnostic groups in the affective disorders: evidence for learning impairments. Behav Ther 45:619–629

Pitman RK, Gilbertson MW, Gurvits TV, May FS, Lasko NB, Metzger LJ, Shenton ME, Yehuda R, Orr SP (2006) Clarifying the origin of biological abnormalities in PTSD through the study of identical twins discordant for combat exposure. Ann N Y Acad Sci 1071:242–254

Pitman RK, Rasmusson AM, Koenen KC, Shin LM, Orr SP, Gilbertson MW et al (2012) Biological studies of post-traumatic stress disorder. Nat Rev Neurosci 13:769–787

Pöhlchen D, Leuchs L, Binder FP, Blaskovich B, Nantawisarakul T, Topalidis P, Brückl TM, Norrholm SD, Jovanovic T, BeCOME working group, Spoormaker VI (2020) No robust differences in fear conditioning between patients with fear-related disorders and healthy controls. Behav Res Ther 129:103610. https://doi.org/10.1016/j.brat.2020.103610

Poulton R, Menzies RG (2002) Non-associative fear acquisition: a review of the evidence from retrospective and longitudinal research. Behav Res Ther 40:127–149

Quirk GJ, Mueller D (2008) Neural mechanisms of extinction learning and retrieval. Neuropsychopharmacology 33:56–72

Raeder F, Merz CJ, Margraf J, Zlomuzica A (2020) The association between fear extinction, the ability to accomplish exposure and exposure therapy outcome in specific phobia. Sci Rep 10: 4288

Rauch SAM, Foa EB, Furr JM, Filip JC (2004) Imagery vividness and perceived anxious arousal in prolonged exposure treatment for PTSD. J Trauma Stress 17:461–465

Rauch SL, Shin LM, Phelps EA (2006) Neurocircuitry models of posttraumatic stress disorder and extinction: human neuroimaging research-past, present, and future. Biol Psychiatry 60:376–382

Ressler KJ, Rothbaum BO, Tannenbaum L, Anderson P, Graap K, Zimand E et al (2004) Cognitive enhancers as adjuncts to psychotherapy: use of d-cycloserine in phobic individuals to facilitate extinction of fear. Arch Gen Psychiatry 61:1136–1144

Richardson R, Ledgerwood L, Cranney J (2004) Facilitation of fear extinction by D-cycloserine: theoretical and clinical implications. Learn Mem 11:510–516

Scheveneels S, Boddez Y, Hermans D (2021) Predicting clinical outcomes via human fear conditioning: a narrative review. Behav Res Ther 142:103870

Shalev AY, Sahar T, Freedman S, Peri T, Glick N, Brandes D et al (1998) A prospective study of heart rate response following trauma and the subsequent development of posttraumatic stress disorder. Arch Gen Psychiatry 55:553–559

Soeter M, Kindt M (2015) An abrupt transformation of phobic behavior after a post-retrieval amnesic agent. Biol Psychiatry 78:880–886

Soliman F, Glatt CE, Bath KG, Levita L, Jones RM, Pattwell SS et al (2010) A genetic variant BDNF polymorphism alters extinction learning in both mouse and human. Science 327:863–866

Stein NR, Dickstein BD, Schuster J, Litz BT, Resick PA (2012) Trajectories of response to treatment for posttraumatic stress disorder. Behav Ther 43(4):790–800

Sylvester CM, Corbetta M, Raichle ME, Rodebaugh TL, Schlaggar BL, Sheline YI et al (2012) Functional network dysfunction in anxiety and anxiety disorders. Trends Neurosci 35:527–535

Tarrier N, Sommerfield C, Pilgrim H, Faragher B (2000) Factors associated with outcome of cognitive-behavioural treatment of chronic post-traumatic stress disorder. Behav Res Ther 38: 191–202

Thrasher S, Power M, Morant N, Marks I, Dalgleish T (2010) Social support moderates outcome in a randomized controlled trial of exposure therapy and (or) cognitive restructuring for chronic posttraumatic stress disorder. Can J Psychiatr 55:187–190

Tuerk PW, Wangelin BC, Powers MB, Smits JAJ, Acierno R, Myers US et al (2018) Augmenting treatment efficiency in exposure therapy for PTSD: a randomized double-blind placebo-controlled trial of yohimbine HCl. Cogn Behav Ther 47:351–371

Vaiva G, Boss V, Ducrocq F, Fontaine M, Devos P, Brunet A et al (2006) Relationship between posttrauma GABA plasma levels and PTSD at 1-year follow-up. Am J Psychiatr 163:1446–1448

van Minnen A, Hagenaars M (2002) Fear activation and habituation patterns as early process predictors of response to prolonged exposure treatment in PTSD. J Trauma Stress 15:359–367

Wachen JS, Jimenez S, Smith K (2014) Long-term functional outcomes of women receiving cognitive processing therapy and prolonged exposure. Psychol Trauma 6:S58–S65

Wannemueller A, Moser D, Kumsta R, Johren HP, Adolph D, Margraf J (2018) Mechanisms, genes and treatment: experimental fear conditioning, the serotonin transporter gene, and the outcome of a highly standardized exposure-based fear treatment. Behav Res Ther 107:117–126

Waters AM, Pine DS (2016) Evaluating differences in Pavlovian fear acquisition and extinction as predictors of outcome from cognitive behavioural therapy for anxious children. J Child Psychol Psychiatry 57:869–876

Watson JB, Rayner R (1920) Conditioned emotional reactions. J Exp Psychol Gen 3:1–14

Wen Z, Chen ZS, Milad MR (2021) Fear extinction learning modulates large-scale brain connectivity. NeuroImage 238:118261

Yehuda R, Bierer LM, Pratchett LC, Lehrner A, Koch EC, Van Manen JA et al (2015) Cortisol augmentation of a psychological treatment for warfighters with posttraumatic stress disorder: randomized trial showing improved treatment retention and outcome. Psychoneuroendocrinology 51:589–597

Zalta AK, Gillihan SJ, Fisher AJ, Mintz J, McLean CP, Yehuda R, Foa EB (2014) Change in negative cognitions associated with PTSD predicts symptom reduction in prolonged exposure. J Consult Clin Psychol 82:171–175

Zandberg LJ, Rosenfield D, McLean CP, Powers MB, Asnaani A, Foa EB (2016) Concurrent treatment of posttraumatic stress disorder and alcohol dependence: predictors and moderators of outcome. J Consult Clin Psychol 84:43–56

Zuj DV, Norrholm SD (2019) The clinical applications and practical relevance of human conditioning paradigms for posttraumatic stress disorder. Prog Neuro-Psychopharmacol Biol Psychiatry 88:339–351

Getting Better with Age? A Review of Psychophysiological Studies of Fear Extinction Learning Across Development

Anaïs F. Stenson, John M. France, and Tanja Jovanovic

Contents

Abstract A critical developmental task is learning what constitutes reliable threat and safety signals in the environment. In humans, atypical fear learning processes are implicated in many mental health conditions, particularly fear and anxiety disorders, pointing to the potential for laboratory measures of fear learning to facilitate early identification of at-risk individuals. This chapter reviews studies of fear learning and extinction learning that incorporate peripheral measures of psychophysiological response and include a developmental sample. Broadly, these studies indicate substantial consistency in differential learning and extinction across development, as assessed with multiple paradigms, across physiological indices. Importantly, though, response coherence across measures (e.g., physiological, neural, and behavioral) was inconsistent across studies. There was also less consistency in results from studies that probed associations between anxiety and fear learning processes. These mixed findings highlight the need for additional examination of when and why there

A. F. Stenson (✉), J. M. France, and T. Jovanovic
Department of Psychiatry and Behavioral Neurosciences, Wayne State University School of Medicine, Detroit, MI, USA
e-mail: astenson@wayne.edu

Curr Topics Behav Neurosci (2023) 64: 213–236
https://doi.org/10.1007/7854_2023_441
Published Online: 1 September 2023

is variability, both across development and in relation to individual differences factors, including mental health, childhood adversity, and sex. In addition, there remains a need for studies that test for developmental change in extinction recall learning and whether stimulus type impacts learning across development. Longitudinal studies designed to address these questions could provide novel insight into the developmental trajectory of fear learning and extinction.

Keywords Fear learning · Development · Extinction learning · Anxiety

1 Introduction

A critical developmental task is learning what constitutes reliable threat and safety signals in the environment. This ability is necessary for survival and is highly conserved across species (LaBar and LeDoux 2003; LeDoux 1993). However, because the significance of a signal may change across time or contexts, the ability to learn new associations that supersede prior learning is also essential. In other words, individuals need to learn what signals predict danger or safety, but must also be able to adapt when those associations change. In humans, atypical fear learning processes are implicated in many mental health conditions, particularly fear and anxiety disorders (e.g., Lissek et al. (2005), but see also Duits et al. (2015); Pöhlchen et al. (2020)). Historically, most human studies of fear learning processes have been conducted in adults, but a more recent and growing literature focuses on these processes across development.

In humans, the ability to acquire conditioned fear responses is normative and emerges early in development (Gao et al. 2010; Ingram and Fitzgerald 1974). In this review, we will use the term "fear conditioning" to refer to this type of aversive Pavlovian conditioning, but we note that this process is also referred to as fear learning and fear acquisition. Other chapters in this volume provide substantial information about the fundamentals of conditioning and extinction, so we will not provide further information about that here. This chapter reviews and synthesizes the current understanding of fear extinction learning across development and suggests directions for future research. We focus on studies that evaluate developmental change, either by comparing fear conditioning and extinction learning across an age range or between age groups, because these can provide insight into both continuity and change in this key cognitive process. Table 1 summarizes key details from each study reviewed here.

Fear conditioning and extinction have been studied across both ontogeny and phylogeny, however, there is less research on fear extinction than fear conditioning across human development. Extinction learning is a critical process to examine during different developmental stages, because it may be impaired in a range of mental health disorders (e.g., Jovanovic and Ressler (2010); Lissek et al. (2005); but see also Duits et al. (2015); Pöhlchen et al. (2020)). Fear extinction is an influential

Table 1 Paradigm details for the screaming lady, bells, or shapes tasks

Paradigm	Papers	Conditioned Stimulus (CS)	Unconditioned Stimulus (US)	ACQ trials	Reinforcement rate	ACQ to EXT delay	EXT trials
Screaming lady	Abend et al. (2020), Britton et al. (2013), Chauret et al. (2019), Gold et al. (2020), Shechner et al. (2015)	Photographs of female faces with neutral expressions	Scream and female face with fearful expression	20 (10 per CS)	80%	5–10 min	16 (8 per CS)
Bells	Michalska et al. (2016), McLaughlin et al. (2016), Shechner et al. (2015), Susman et al. (2021), Jenness et al. (2019)	Colored cartoon bells	Alarm tone	20 (10 per CS)	80%	Same session[a]	16 (8 per CS)
Shapes 1	Craske et al. (2008), Waters and Pine (2016)	Geometric shapes	Aversive tone	16 (8 per CS)	100%	Same session[a]	8 (4 per CS)
Shapes 2	Jovanovic et al. (2014)	Geometric shapes	Air blast to larynx	27 (9 each CS+, CS−, and noise alone)	100%	10 min	36 (12 each CS+, CS−, and noise alone)
Shapes 3	Pattwell et al. (2012)	Geometric shapes	Aversive tone	48 (24 per CS)	50%	24 h	48 (24 per CS)

Note: *ACQ* acquisition, *EXT* extinction
[a] Duration unspecified

translational model that has yielded "bench-to-bedside" insights into treatment for fear- and anxiety-based disorders (Craske et al. 2018; Maples-Keller et al. 2022; Maples-Keller et al. 2017; Michopoulos et al. 2015). Principles of fear learning and extinction inform exposure therapy, a widely used treatment for both post-traumatic stress disorder (PTSD) and many anxiety disorders (Rothbaum et al. 2012). In addition, recent evidence suggests that individual differences in sympathetic responsivity during fear extinction learning may predict response to cognitive behavioral therapy (CBT) for pediatric obsessive-compulsive disorder (Geller et al. 2019). See the chapter by Kitt, Odriozola, and Gee for additional discussion. The clinical relevance of extinction learning merits an in-depth developmental account, both in order to understand the mechanisms implicated during periods of increased vulnerability to mental health problems and to inform interventions.

1.1 Psychophysiological Measurement of Fear

The measurement of fear is central to assessment of both fear conditioning and extinction, since both processes are indexed by changes in fear expression. There is a vigorous scientific and philosophical debate about what constitutes fear, and whether fear conditioning and extinction paradigms measure threat response or fear (LeDoux and Pine 2016). In keeping with a large body of literature, we use the term fear to refer to multiple indices of fear expression, including increased eyeblink startle response, sympathetic activation, and self-reported feelings of fear.

The *eyeblink startle response* is a reflex that is amplified in aversive contexts, including differential fear conditioning (Bradley et al. 2001; Davis et al. 1993; Lang et al. 1990) and reflects a defensive motivational state. The eyeblink is generated by a fast contraction of the *orbicularis oculi* within 20–100 ms after an acoustic stimulus that is measured via electrodes placed around the eye (Lang et al. 1990). Acoustic stimuli activate the startle response via a tri-synaptic circuit that includes the cochlear root neurons, the nucleus reticularis pontis caudalis, and motor neurons in the spinal cord (Davis et al. 1993; Lee et al. 1996). For further discussion of psychophysiology and human extinction learning, please see the chapter by Woodford and Norrholm in this volume.

Blink reflexes can be elicited by auditory and visual stimuli in infants as young as 8 weeks old, and the magnitude of the reflex can be modulated by attentional processes (Richards 2000). Eyeblink startle response magnitude generally increases across childhood (Quevedo et al. 2010); however, Schmitz et al. (2014) found that across pubertal development baseline startle decreased, but fear-potentiated startle increased. Stenson et al. (2021) also reported that children's fear-potentiated startle magnitude, but not startle to a safety cue, increased across pubertal development.

Most studies of fear conditioning that measure eyeblink startle assess responses to acoustic startle probes presented independently, called "noise-alone" trials, as well as during presentation of the CS+ and CS− (Abend et al. 2020; Jovanovic et al. 2014; Norrholm et al. 2011). Inclusion of a noise-alone condition enables quantification of individual differences in startle response (i.e., startle to the noise-alone trials), which can then be used as a baseline against which the CSs responses are compared. Typically, studies that use eyeblink startle to assess fear response subtract the average startle magnitude to these "noise-alone" trials from the response to the CS+ and CS−. This approach yields a "fear-potentiated startle" (FPS) measure. The use of FPS provides the opportunity to separate out learning effects from differences across individuals that could be linked to factors like age, anxiety, or gender. FPS has been associated with amygdala activity in animal research (Davis 1992) and human neuroimaging studies (Kuhn et al. 2020).

Skin conductance responses (SCRs) index autonomic arousal, specifically, activation of the sympathetic nervous system, which is triggered by emotional engagement and orienting responses (Lang et al. 1998). Sympathetic nerves innervate sweat glands in the skin, causing them to produce more sweat; this change is measured with electrodes placed on the palm or fingers (Andreassi 2010). Neuroimaging studies indicate that SCRs are mediated primarily by amygdala activation (Wood et al. 2014), as well as by activity in the right orbitofrontal cortex, right anterior insula, left lingual gyrus, right fusiform gyrus, and left cerebellum (Critchley et al. 2002).

Skin conductance levels and responsivity have been found to be generally stable across childhood, both within individuals and cross-sectionally (El-Sheikh 2007; Leclair-Visonneau et al. 2016). Adults and children have similar SCRs to arousing, negatively valenced stimuli (McManis et al. 2002) and habituate to those more slowly than to neutral stimuli, but adults' SCR may habituate to an aversive stimulus more quickly than adolescents' (Miller and Shields 1980). Although skin conductance is understood to be developmentally stable from infancy into early adulthood, it may differ across demographic groups. For instance, some evidence suggests that individuals of African ancestry may have lower SC levels, and for this reason are more likely to have their SCR data excluded from fear conditioning and extinction studies (Kredlow et al. 2017).

Measures of cardiac function and response, including heart rate and heart rate variability, are less commonly used in fear conditioning and extinction studies than either eyeblink startle or skin conductance response. However, two recent studies suggest that resting respiratory sinus arrhythmia (RSA), a measure of fluctuation in heart rate across respiratory cycles modulated by vagal innervation of the sinoatrial node (Beauchaine 2015), may be an important predictor of longitudinal associations between fear extinction learning and later mental health, particularly in individuals exposed to violence during development (Jenness et al. 2019; Susman et al. 2021).

2 The Neurobiological Bases of Fear Learning and Extinction

The neurobiology of fear learning and extinction are covered elsewhere this volume, so here we provide a brief overview for context. (Please see chapter in this volume by Woodford and colleagues for further detail on these processes). Animal and human studies have identified the amygdala, a small nucleus located deep within the temporal lobes, as necessary for the acquisition of fear (Davis et al. 1993; LeDoux 1993). Following activation of the amygdala by a threat cue, projections from the lateral amygdala to the hypothalamus and brainstem activate these regions, which in turn mediate a suite of fear-induced physiological responses (Davis et al. 1993; LeDoux et al. 1988). In humans, the neural bases of fear conditioning and extinction have been widely studied in adults using functional magnetic resonance imaging (fMRI). These studies have identified two primary networks that support fear conditioning and extinction. A meta-analysis found that threat cues (CS+) differentially activate the amygdala, anterior insula, and dorsal anterior cingulate cortex (Fullana et al. 2016). In contrast, safety signals (CS−) preferentially activate the hippocampus, posterior cingulate cortex, and the ventromedial prefrontal cortex (vmPFC) (Fullana et al. 2016). The vmPFC plays a critical role in safety learning during extinction and over time (Milad et al. 2007).

There is evidence for age-related change in the neural correlates of fear extinction learning, which is also reviewed in the chapter by Kitt, Oriozola, and Gee. One study found that during fear conditioning adolescents engaged subcortical regions, including the amygdala and hippocampus, more than adults (Lau et al. 2011). During extinction, adults engaged the dorsolateral prefrontal cortex (dlPFC) more than adolescents when the safety signal (CS−) appeared. In adults, but not adolescents, self-reported fear of the CS− was positively correlated with this dlPFC activation (Lau et al. 2011). Subsequent studies comparing adolescents and adults have reported that adolescents activate the vmPFC, dlPFC, temporoparietal junction (TPJ), and posterior cingulate cortex (PCC) less than adults (Ganella et al. 2017), and that adolescents exhibited greater amygdala activation and delayed medial PFC (mPFC) engagement to the CS+ relative to the CS− during late extinction (Morriss et al. 2019). Two seminal cross-sectional studies have also identified developmental shifts in amygdala-prefrontal connectivity during the transition from middle childhood to adolescence (Gabard-Durnam et al. 2014; Gee et al. 2013a). Amygdala-prefrontal cortex functional connectivity was positive in children <10 years, whereas after age 10 amygdala-prefrontal connectivity was negative (Gee et al. 2013a). In contrast, functional connectivity between the amygdala and prefrontal cortex increased with age (from 4 to 23 years; Gabard-Durnam et al. (2014)). Together, these results suggest that the PFC plays a role in modulating fear response during extinction in adults, but that children and adolescents may engage these regions differently than adults, potentially due to ongoing development of structural or functional connectivity between the amygdala and PFC.

Concordance between physiology, behavior, and neural signatures of extinction may provide important insight into potential developmental change in fear extinction learning. One study found that SCRs during extinction learning were negatively correlated with both vmPFC and PCC activation in adults, but not adolescents (Ganella et al. 2017). Another reported that greater dlPFC activation during extinction predicted greater subjective fear ratings to the threat cue in adults, while lower dlPFC activation predicted fear ratings in adolescents (Lau et al. 2011). These results suggest age may moderate the relationship between prefrontal activity and fear response, as measured by both psychophysiology and self-report. The generalizability of these findings is somewhat limited, however, by relatively small sample sizes and mixed patterns of results across these studies. More robust inferences about the coherence of neural and behavioral indices of fear learning and extinction across development will require larger, longitudinal samples that utilize multiple measures.

3 Evidence for Age Effects on Fear Extinction Learning

Substantial evidence indicates that differential fear conditioning is consistently observed across self-report and psychophysiological measures from middle childhood onward (e.g.,(Abend et al. (2020); Gold et al. (2020); Jovanovic et al. (2014); Michalska et al. (2016); Neumann et al. (2008)). A small handful of studies have examined fear conditioning with psychophysiological measures in younger children. One tested differential conditioning in 12 infants using skin potential responses measured from the foot; approximately half of the infants demonstrated differential conditioning (Ingram and Fitzgerald 1974). Another study assessed differential conditioning, as measured by SCRs, in a longitudinal cohort of children from age 3 to age 8 (Gao et al. 2010). Results indicated that children as young as 3 may show conditioning, but extinction learning was not tested. To date, no studies have tested both conditioning and extinction using measures of psychophysiological response in children younger than 5 years (see Table 2 for a list of studies that utilize psychophysiological measures).

Most developmental studies of fear conditioning and extinction that include psychophysiological measures utilize one of three different paradigms. See Table 1 for details of each. The chapter in this volume by Kitt and co-authors focuses on the neural circuitry believed to underlie pediatric and developmental aspects of extinction learning, and so here we focus on the psychophysiological evidence provided by results from these widely used paradigms.

The first is the “Screaming Lady” task, which presents black and white photographs of women’s faces displaying neutral expressions as the CSs. During fear conditioning, the CS+ is followed by one woman’s face displaying a fearful expression while a scream sound is played. The second is the “Bells” task, which presents cartoon images of colored bells as the CSs and a noxious sound as the US during fear conditioning, and then presents the CSs without USs during extinction. The third is the “Shapes” task, which uses colored geometric shapes for the CSs but includes a wider variety of paradigm parameters and USs than the Screaming Lady and Bells tasks.

3.1 The Screaming Lady Paradigm

Abend et al. (2020) conducted a well-powered, cross-sectional analysis of differential fear conditioning and extinction learning in 8- to 50-year-olds ($N = 351$) with

Table 2 List of studies included in this review. All test extinction learning using at least one psychophysiological measure and assess age-related change

Task	Publication	Psychophysiology	Ages	Groups and/or individual difference variable	*N*
Screaming lady	Abend et al. (2020)	FPS, SCR	8–50 years	Anxiety: yes or no; age (continuous)	351
	Britton et al. (2013)	FPS, SCR	8–19 and 21–45 years	Anxiety: yes or no; age: adolescent or adult	114
	Chauret et al. (2014)	SCR	10–17 years	Sex: female or male	120
	Chauret et al. (2019)	SCR	10–17 years	Anxiety: diagnosis, family history, or neither	77
	Ganella et al. (2017)	SCR	14–16 and 25–35 years	Age: adolescent or adult	31
	Ganella et al. (2018)	SCR	14–16 and 25–35 years	Age: adolescent or adult	31
	Gold et al. (2020)	SCR	8–50 years	Anxiety: yes or no; age (continuous)	80
	Lau et al. (2011)	SCR	10–17 and 18–50 years	Age: adolescent or adult	77
	Shechner et al. (2015)[a]	FPS, SCR	Youth and adults[b]	Anxiety: diagnosis or no diagnosis	84
Bells	Jenness et al. (2019)	RSA, SCR	6–18 years	Variable abuse exposure	94
	McLaughlin et al. (2016)	SCR	6–18 years	Variable abuse exposure	94
	Michalska et al. (2016)	SCR	5–10 years	Age: 5–6, 7–8, and 9–10-year-olds	54
	Michalska et al. (2019)	SCR	11–15 years	Social reticence: high or low in childhood	59
	Shechner et al. (2015)[a]	FPS, SCR	Youth and adults[b]	Anxiety: diagnosis or no diagnosis	84
	Susman et al. (2021)	RSA, SCR	9–17 years	Violence exposure: yes or no	165
Shapes 1	Craske et al. (2008)	SCR	7–12 years	Anxiety: diagnosis, family history, or neither	49
	Waters and Pine (2016)	SCR	7–13 years	Anxiety: diagnosis or no diagnosis. Within anxiety group: CBT responder or non-responder	76
Shapes 2	Jovanovic et al. (2014)	SCR and FPS	8–13 years	Age: younger or older than 10; anxiety: median split into lower vs. higher symptom groups	60

(continued)

Table 2 (continued)

Task	Publication	Psychophysiology	Ages	Groups and/or individual difference variable	*N*
Shapes 3	Pattwell et al. (2012)	SCR	5–11, 12–17, and 18–28 years	Age: child, adolescent, or adult	83
Other	Marusak et al. (2021)	SCR	6–11 years	Trauma exposure: yes or no	44

Notes: CBT cognitive behavioral therapy. *FPS* fear potentiated startle. *RSA* respiratory sinus arrythmia

[a] Shechner et al. (2015) is listed twice, once each in the Bells and Screaming Lady task sections, because both tasks were used in the study

[b] The participants' exact age range was not provided

and without anxiety (see Sect. 4 for anxiety results). FPS and SCRs were measured throughout conditioning and extinction, and participants provided subjective fear ratings for each CS before conditioning, after conditioning, and after extinction. Most (250) participants also completed a separate structural brain imaging session. Participants of all ages had similar SCRs and comparable levels of self-reported fear across fear conditioning and extinction. Both children and adults demonstrated fear conditioning and extinction learning, as measured by FPS and self-reported fear of the CSs. SCRs to the US were larger in younger participants, but were unrelated to anxiety, brain structure, or subjective fear ratings. There were no significant associations between brain structure and eyeblink startle response. In contrast, SCR magnitudes were associated with both thicker cortex and greater gray matter volumes in an age-dependent manner, such that these associations were positive in younger participants but became negative in older participants. Interpretation of this age difference is, however, somewhat complicated by normative age differences in gray matter volume, which peaks in childhood and then declines across this lifespan (Mills et al. 2016). Viewed together, these findings indicate comparable fear extinction learning, as measured by self-report of fear, SCRs, and FPS, across a wide swath of development, and provide evidence for age-dependent relationships between SCR magnitude, cortical thickness, and gray matter volume.

The "Screaming Lady" task was also used in a study of age- and anxiety-linked differences in fear learning, extinction, and extinction recall assessed SCRs, neural activity, and subjective fear indices in 8- to 48-year-olds ($N = 114$; (Britton et al. 2013). Outcomes were compared between four groups: anxious and non-anxious children (8–19 years) and adults (23–48 years) defined in the same manner. Acquisition and extinction were conducted during one session, followed by an extinction recall test in the MRI scanner approximately 3 weeks later that assessed participants' threat perception and explicit memory for the CS-US contingency. Successful differential conditioning was observed across SCR, FPS, and subjective fear measures for all age and anxiety groups. In contrast, FPS data indicated that adults displayed more robust extinction than children ($p < 0.02$) until the last two trials, in which children and adults alike displayed comparable extinction ($p > 0.20$). Children also had larger SCRs than adults to both CSs across acquisition and extinction, a result consistent with greater sympathetic response in younger vs. older participants. All groups reported significantly more fear of the CS+ than the CS− except youth with anxiety, indicating an interaction of age and anxiety. This result, as well as the associations between anxiety and brain activity, is discussed in Sect. 4.

3.2 The Bells Paradigm

Michalska et al. (2016) examined fear conditioning ($N = 54$) and extinction ($N = 48$) in 5- to 10-year-olds. Participants completed acquisition and extinction in a single session while SCRs were recorded. Children also rated their subjective fear for each CS twice: after fear conditioning and again after extinction. During acquisition, children had larger SCRs to the CS+ than the CS−, indicating that they acquired a conditioned fear response. During extinction, SCRs were not significantly different between CS+ and CS− trials, congruent with successful extinction learning. In contrast, self-reported fear was significantly higher to the CS+ vs. the CS− during extinction. These patterns were consistent across the participants' age range and may indicate that physiological responses (SCRs) and subjective fear diverged during the extinction task. Together, these results suggest that some degree of fear extinction learning may be present in school-aged children as young as 5 years, as indicated by SCRs, but that they continue to perceive the CS+ as more threatening than the CS− after extinction. These results may indicate that physiological and self-report measures tap different processes in children vs. adults, for whom self-report and physiological responsivity generally converge when operationally defined as psychiatrically healthy (Mauss et al. 2005).

3.3 The Shapes Paradigm

Jovanovic et al. (2014) examined fear conditioning and extinction in 8- to 13-year-olds ($N = 60$) who were assessed for anxiety symptoms. Multiple measures, including self-report of contingency awareness, SCR, and eyeblink startle were used to assess age and anxiety effects. The CSs were colored shapes and the US was an air blast to the larynx. Conditioning and extinction were separated by a 10-min delay. Additional paradigm details are provided in Table 1 and anxiety-specific findings are described in Sect. 4. Participants' self-reported contingency awareness reflected effective differential conditioning and extinction, regardless of age or anxiety status. During conditioning, SCRs and FPS responses were compared between younger (<10) and older (>10) children. Both age groups evidenced successful fear conditioning and extinction, as measured by FPS and SCRs, although SCR magnitudes tended to be larger in younger vs. older children ($p = 0.06$). These results suggest comparable fear conditioning and extinction in 8- to 13-year-olds as measured by self-report, EMG, and SCR measures.

Pattwell et al. (2012) compared fear conditioning and extinction learning between children ($n = 30$, 5–11 years), adolescents ($n = 28$, 12–17 years), and adults ($n = 25$, 18–28 years) assessed for anxiety. Pubertal status was assessed in adolescents only. Conditioning occurred on Day 1 and extinction was tested on Day 2. Notably, this is one of few studies to test extinction learning in different age groups after $a > 1$ h delay from conditioning. All groups acquired differential fear learning, measured via SCRs; there were no age-linked differences in response to either the CS+ or CS−. In contrast, diminished extinction was observed in adolescents vs. children ($p = 0.02$) and adults ($p = 0.08$). Children and adults displayed comparable extinction learning ($p = 0.70$). Anxiety status, sex, and pubertal status (adolescents only) were not associated with conditioning or extinction. Their results indicate possible adolescent-specific deficits in extinction learning, relative to children and adults, that are unrelated to anxiety, age, pubertal development, or sex in this cohort.

3.4 Multiple Paradigms

Citing high attrition and potential confounds introduced by the use of social stimuli in the Screaming Lady task, Shechner et al. (2015) compared child and adult performance on a fear conditioning and extinction across the Screaming Lady and Bells tasks. Children ($n = 15$ with anxiety, $n = 17$ without anxiety) and adults ($n = 22$ with anxiety, $n = 30$ without anxiety) completed both tasks while FPS and SCRs were recorded, enabling comparison of fear responsivity across age groups, anxiety status, and task. Self-reported fear of each CS was assessed before and after conditioning, and again after extinction learning. Attrition was not significantly different between tasks for participants with or without anxiety, regardless of age group. Within-subjects comparisons of self-reported fear, SCRs, and FPS between the tasks indicated significant correlations, with small to medium effect sizes (0.34–0.65). Direct comparisons between the tasks revealed higher self-reported fear to the Screaming Lady vs. the Bells task, however, differential conditioning (CS+ > CS−) was greater in the Bells task. Participants with anxiety reported more fear across tasks than non-anxious participants, and this result was driven largely by greater fear during the Screaming Lady task. SCRs indicated successful conditioning and extinction for both tasks in children, but only for the Bells task in adults. Results of FPS analyses indicated similar responses across tasks, age, and anxiety groups. Broadly, results were consistent with comparable conditioning and extinction performance across both age and anxiety groups, but also highlight that task features may impact performance differently between these groups.

Viewed together, these studies suggest similar acquisition of conditioned fear, per FPS and SCR measures, in children and adults. In contrast, some evidence suggests that extinction learning may be less efficient in children than adults. Britton et al. (2013) reported that children acquire extinction learning more slowly than adults,

Pattwell et al. (2012) observed attenuated extinction in adolescents relative to children and adults, and Michalska et al. (2016) reported that children continue to report higher levels of fear to the CS+ vs. the CS− even after completing extinction. These results suggest that extinction learning may be less robust in children than adults, but that this pattern may only be apparent when both psychophysiological and self-report measures are utilized. Notably, however, the well-powered study conducted by Abend et al. (2020) reported comparable fear extinction learning across both psychophysiological and self-reported fear in participants aged 8–50 years. Clarification of the extent to which there may be developmental change in extinction learning performance may require additional studies that test participants across a wide age range using multiple paradigms (e.g., Shechner et al. (2015)) and measures of fear responsivity.

4 Effects of Anxiety on Fear Extinction Learning Across Development

Many studies of differential conditioning and extinction learning across development have also examined the impact of anxiety. The prevalence of anxiety disorders increases substantially from childhood into adolescence and early adulthood (Merikangas et al. 2010; Paus et al. 2008), highlighting the value of identifying neurobiological mechanisms linked to this age-related increase in risk, including fear responsivity, learning, and inhibition. In this section we highlight results from studies that have examined the impact of anxiety on conditioning and extinction during different developmental windows using the Screaming Lady (Abend et al. 2020; Britton et al. 2013; Chauret et al. 2019; Gold et al. 2020) and Shapes (Craske et al. 2008; Jovanovic et al. 2014; Waters and Pine 2016) tasks.

Multiple studies that used the Screaming Lady task to examine fear conditioning and extinction learning across development also tested for effects of anxiety. Results from Britton et al. (2013), Abend et al. (2020), and Gold et al. (2020) all indicated that children and adults with and without anxiety demonstrated successful differential conditioning. Britton et al. (2013) found that self-report of fear during extinction learning, however, indicated that participants with anxiety did not extinguish as effectively as those without anxiety, regardless of age. Notably, though, both children and participants with anxiety discontinued participation at significantly higher rates (odds ratio: 4.2) than those without anxiety. This differential attrition suggests that the paradigm was not invariant to age or diagnostic group, thus complicating interpretation of the study results. In contrast, both Abend et al. (2020) and Gold et al. (2020) reported that participants demonstrated differential fear conditioning and extinction learning, regardless of age or anxiety. Abend et al. (2020), however, reported that participants with anxiety had larger SCRs overall and reported more fear of both CSs across both tasks. These results are consistent with the hypotheses that (1) anxiety is associated with greater sympathetic activation in anticipation of a threat but not acute threat response (i.e., US response) and (2) is not linked to aberrant fear acquisition or extinction. Gold et al. (2020) also reported that participants with anxiety reported greater fear and had larger SCRs than participants without anxiety across both tasks. In addition, brain function and connectivity during the extinction recall task differed according to anxiety status. This association was moderated by age: participants without anxiety had greater AMY-vmPFC connectivity during CS+ trials than participants with anxiety, and this pattern emerged gradually across the tested age range (8–50 years). These results are suggestive of developmental differences in brain function during fear recall in pediatric vs. adult anxiety and highlight the importance of examining the neurobiological underpinnings of mental health risk across a wide age range.

Both Jovanovic et al. (2014) and Waters and Pine (2016) used the Shapes task to examine age and anxiety effects on differential conditioning and extinction. Jovanovic et al. (2014) did not find age effects on conditioning or extinction between children younger and older than 10 years, however, when these age groups were further divided into lower- and higher-anxiety groups, interactions between age group and anxiety emerged for FPS during conditioning. Specifically, older children discriminated between the CS+ and CS− more effectively than younger children, and effect was driven by older children in the high anxiety group. In contrast, age and anxiety did not interact to impact FPS during extinction learning. For SCRs, there were no significant interactions between age and anxiety during conditioning or extinction. These results suggest that although fear conditioning and extinction are generally similar across the tested age range (8–13 years), there may be interactions between age and anxiety that yield a more complex pattern of results.

Waters and Pine (2016) tested whether individual differences in sympathetic responsivity (SCRs) and perceptions of the CSs during fear conditioning and extinction learning predicted response to cognitive behavioral therapy (CBT) in 7-to-13-year-olds with ($n = 44$) and without anxiety ($n = 32$). All participants completed conditioning and extinction, and then anxious participants completed CBT. Both SCRs and subjective ratings across the tasks differentiated the children whose symptoms improved following CBT from non-responders. Specifically, the CBT responders had significant reductions in SCRs across extinction and perceived the CSs to be less aversive than the non-responders. Overall, the CBT responders had comparable SCR and subjective evaluations as the non-anxious participants. Although replication in a larger sample is needed, this result suggests that sympathetic responses across extinction may predict which children with anxiety could benefit from CBT.

Two studies have examined the impact of intergenerational anxiety risk (i.e., family history of anxiety) on fear conditioning and extinction learning. Chauret et al. (2019) used the Screaming Lady paradigm with $N = 76$ 10- to 17-year-olds who had either (1) a current anxiety diagnosis, (2) a parent with an anxiety disorder, or (3) no anxiety history. Both conditioning and extinction learning were completed in an MRI while SCRs were recorded. All groups indicated successful conditioning per SCRs, however, group differences emerged in neural responsivity. Both at-risk and healthy children acquired conditioned fear, as measured by activation within fronto-amygdala circuitry, whereas anxious children did not. In addition, during acquisition children without anxiety risk showed increased activation in fronto-amygdala circuitry to the threat cue vs. the safety signal (CS+ > CS−), whereas the at-risk group responded more to the safety cue (CS− > CS+) in the amygdala and dlPFC and the anxiety group had diminished responsivity (CS− > CS+) in the left dorsal anterior cingulate cortex (dACC) and ventrolateral PFC (vlPFC). During extinction there were also significant group differences in SCRs, which were greater for the CS− relative to the CS+ in the at-risk and anxious groups, but not for the no anxiety group. Amygdala responsivity during extinction was also greater in anxious children vs. the at-risk and no anxiety groups.

Craske et al. (2008) used the Shapes paradigm to test intergenerational effects of anxiety on fear conditioning and extinction learning in a single session. Extinction recall was tested 2 weeks later. Participants were 7- to 12-year-old children ($N = 48$) classified, as in Chauret et al. (2019), as either (1) current anxiety ($n = 23$), (2) a familial history of anxiety ($n = 15$), or (3) no anxiety history ($n = 11$). All groups demonstrated differential conditioning per SCRs (CS+ > CS−), however, both the anxious and at-risk children had larger SCRs across all tasks than children in the control group. SCRs indicated resistance to within-session extinction and extinction recall for the anxiety vs. no anxiety groups. At-risk children initially had larger SCRs to both CSs than control children during extinction, but the two groups were comparable by late extinction. The at-risk group also had larger SCRs across extinction recall, regardless of block or CS type, relative to the no anxiety group. Anxious children rated the CS+ as more unpleasant than children in the at-risk and control groups; the at-risk and control group ratings were not significantly different from each other. In contrast, there were no group differences in awareness of the CS-US contingency. Together with the results from Chauret et al. (2019), these findings indicated that extinction learning may be impaired in both children with current anxiety and children with a familial history of anxiety, even when the child does not have a current anxiety diagnosis. Importantly, these studies suggest that both sympathetic and neural responsivity during extinction learning may enable identification of individuals at increased with for anxiety prior to the development of an anxiety disorder.

Many of the studies reviewed here report that anxiety was not associated with physiological indices of differential conditioning and extinction learning, but did impact participants' subjective fear (Abend et al. 2020; Gold et al. 2020; Lau et al. 2008; Shechner et al. 2015). In contrast, Craske et al. (2008) reported diminished extinction learning and recall per SCRs, as well as more negative perceptions of the CS+, in anxious children relative to both children at-risk for anxiety and the control group; the anxious and at-risk groups both had elevated SCRs across tasks, compared to the control group. Chauret et al. (2019) also observed group differences in SCRs and neural responses during both conditioning and extinction learning in children with anxiety, at-risk for anxiety, and without increased risk. On balance, these studies provide somewhat more evidence for elevated sympathetic responsivity across CSs and for anxiety effects on self-reported fear during conditioning and/or extinction (Abend et al. 2020; Britton et al. 2013; Craske et al. 2008; Gold et al. 2020). Importantly, however, the findings from Craske et al. (2008) and

Chauret et al. (2019) indicate that the two-group approach (current anxiety diagnosis or not) may obscure group differences by including participants at elevated risk for anxiety, due to familial history or other factors, in the no anxiety group. Relatedly, Waters and Pine (2016) observed that among children with an anxiety diagnosis, differences in sympathetic response across extinction predicted future response to CBT. Like Craske et al. (2008) and Chauret et al. (2019), this finding highlights the need for assessment of individual difference factors that may contribute to variability within anxious and non-anxious groups. Ideally future studies will screen participants without current anxiety for factors that may put them at risk, as well as those that may predict treatment response.

5 Effects of Childhood Adversity on Fear Conditioning and Extinction

Several recent studies of fear conditioning and extinction have examined how early exposure to high levels of threat (e.g., violence) impacts differential conditioning and extinction learning processes (Jenness et al. 2019; Jovanovic et al. 2014; Jovanovic et al. 2022; Marusak et al. 2021; McLaughlin et al. 2016; Stenson et al. 2021). Childhood adversity is associated with greater risk for fear and anxiety disorders, but the mechanisms through which this vulnerability is conferred are not fully understood. Effects of adversity on fear conditioning and extinction learning may be one such mechanism.

McLaughlin et al. (2016) tested the impact of maltreatment exposure on conditioning and extinction in 6- to 18-year-old children ($N = 94$) with and without maltreatment exposure using the Bells task. Both self-reported fear and SCRs were measured in all participants, and structural imaging data were acquired ($n = 60$). Across the age range, both groups acquired conditioning and extinction, per fear and SCRs, however, the maltreatment group acquired conditioning, as measured by SCRs, more slowly than children without maltreatment. In contrast, self-reported fear was not different between the two groups. The imaging data indicated that maltreatment was associated with smaller amygdala and hippocampal volumes and that the volume of these structures was negatively associated with SCRs to the CS+ during early conditioning. Amygdala volume was also negatively associated with self-reported fear to the CS− following conditioning. These findings suggest that reduced volume of brain structures involved in fear conditioning may mediate the relationships between maltreatment exposure and a) differential conditioning, and b) fear inhibition in the presence of a safety signal (CS−). In spite of these maltreatment-linked differences in conditioning, no group differences emerged for extinction learning. These results suggest that maltreatment exposure is associated with deficits in differential conditioning that may be mediated by the impact of maltreatment on relevant brain structures.

Marusak et al. (2021) used a novel virtual reality paradigm to test fear conditioning, extinction learning, and recall in ($N = 44$) 6- to 11-year-olds with and without trauma exposure. The CSs were avatars of men and the UCS was an aversive noise. Conditioning and extinction were completed on 1 day, but in different virtual contexts. Recall was tested in a subset of participants ($n = 35$) in the scanner 1 day later. Both groups displayed comparable differential conditioning and extinction learning per SCRs. During extinction recall, however, trauma-exposed children were more avoidant of both CSs and displayed more activation in the dACC and anterior insula to the no longer reinforced CS+. These results suggest poorer extinction recall in trauma-exposed children relative to unexposed children and raise the possibility that this may be one of the mechanisms that account for the link between childhood trauma and increased mental health risk (Merrick et al. 2019).

Three recent studies have examined associations between violence or trauma exposure, differential fear conditioning and extinction learning, and mental health in children. Jovanovic et al. (2022) assessed fear conditioning and extinction learning via FPS in 9-year-old children ($N = 48$) with violence exposure, and incorporated an assessment of their awareness of the contingency shift (i.e., from reinforced to no longer reinforced CS+) during extinction by presenting prompts for the children to indicate whether they thought the CS+ would be followed by the US. Awareness of the contingency shift was associated with successful conditioning but not with extinction learning. In children without awareness of this shift there was a strong positive association between trauma exposure and post-traumatic stress disorder (PTSD) symptoms, whereas for children who were aware of this shift this association was not significant.

Both Jenness et al. (2019) and Susman et al. (2021) examined longitudinal relationships between violence or maltreatment exposure, respiratory sinus arrhythmia (RSA), differential conditioning and extinction learning, and mental health using the Bells task. Jenness et al. (2019) assessed 6- to 18-year-olds ($N = 94$) with varying levels of child abuse exposure. Participants' RSA was measured at rest, and they self-reported PTSD symptomatology and child abuse exposure. They then completed conditioning and extinction tasks while SCRs were recorded. Among the participants with abuse histories, (1) high RSA predicted lower PTSD symptoms and enhanced extinction learning, and (2) extinction learning mediated the association of abuse with PTSD symptoms only in participants with high RSA. In contrast, participants without abuse histories and with low RSA had better extinction learning than those with high RSA. Together, these findings indicate that extinction learning may be an important protective mechanism in maltreated children. Susman et al. (2021) examined these associations in 9- to 17-year-olds ($N = 134$), almost half of whom had prior violence exposure. Participants completed measurement of resting-state RSA, as well as conditioning and extinction tasks; Mental health was assessed 2 years later. During extinction, SCRs to CS+ and CS− were positively associated with internalizing symptoms at follow-up; SCRs to the CS− were also linked to higher PTSD and externalizing symptoms. Neither vagal tone, violence exposure, nor their interaction was associated with fear extinction learning. The absence of an association between violence exposure and extinction learning contrasts with the Jenness et al. (2019) finding that high RSA enhanced extinction learning in maltreated children. However, greater RSA was associated with attenuation of the association between violence exposure and both (1) PTSD and (2) externalizing symptoms.

Together these studies indicate multiple possible pathways through which childhood adversity may contribute to increased mental health risk via effects on the physiological and neurobiological processes that modulate differential fear conditioning and extinction learning. McLaughlin et al. (2016) reported negative associations between childhood maltreatment and brain structure that may be linked to diminished differentiation of threat and safety signals in the environment. Marusak et al. (2021) observed differential activation of the dACC and anterior insula in trauma-exposed children during fear extinction recall. The findings reported by Jovanovic et al. (2022) suggest that individual differences in both environmental exposures, such as violence, and cognitive processes, like awareness of contingency shifts, may impact psychopathology risk in violence-exposed children. Jenness et al. (2019) observed that high RSA may buffer children with abuse histories against both poor extinction learning and later PTSD and the relationship between extinction learning and PTSD symptoms may depend on having high RSA. Susman et al. (2021) found that elevated sympathetic responsivity to CSs during extinction may predict later symptomatology and that higher RSA may buffer violence-exposed children against poor mental health outcomes following violence exposure. These studies provide intriguing evidence for contributions of childhood adversity to individual differences in fear conditioning and extinction learning, but there remains a need for additional studies to clarify both (1) the extent to which different types of adversity (e.g., childhood maltreatment vs. community violence exposure) may modulate neural, physiological, and cognitive processes to impact fear conditioning and extinction learning, and (2) whether any such effects mediate mental health outcomes.

6 Future Directions

Overall, the studies we have reviewed suggest developmental continuity in differential fear conditioning or extinction from middle childhood through adulthood, as measured by SCRs and FPS (Abend et al. 2020; Adolph et al. 2022; Gold et al. 2020; Jovanovic et al. 2014; Michalska et al. 2016). Many studies have examined extinction learning across development using psychophysiology, however, there are outstanding gaps and questions to be addressed. Below, we highlight key next steps for the study of differential conditioning and extinction learning across development.

All of the studies discussed used paradigms that were not designed to test clinically-relevant aspects of extinction learning. First, few tested extinction recall, therefore, it is unclear if developmental differences in extinction learning might present as diminished recall in younger participants, relative to adults. The ability to retain extinction learning is critical to the treatment of many fear- and anxiety-based disorders, including PTSD and specific phobias. Second, only one of these studies tested extinction across contexts (Marusak et al. 2021). The protracted development of medial temporal lobe structures (Bachevalier and Vargha-Khadem 2005; Lavenex and Banta Lavenex 2013) that support contextual memory (e.g., the hippocampus) suggest an extended developmental trajectory for that the impact of context on conditioning and extinction. Findings that exposure to adversity may impact hippocampal volume (van Rooij et al. 2020) and activation to CS+ (McLaughlin et al. 2016) suggest the need to examine how these effects impact extinction learning across contexts.

Several developmental factors suggest the importance of examining of sex as a biological variable in relation to fear learning and extinction processes, but few studies have tested for sex differences. Chauret et al. (2014) randomly assigned ($N = 120$) 10- to 17-year-olds to conditioning and extinction tasks that used either male or female faces as the CS+ and CS−. Regardless of sex, participants reported more fear of male vs. female faces during conditioning and extinction. During conditioning, boys' SCRs were amplified when viewing male faces, whereas girls had higher SCRs following female faces. During extinction, boys continued to have larger SCRs when viewing male faces. These results suggest that tasks that include socially salient CSs may be processed differently depending on both stimulus (i.e., CS) sex and participant sex. Because anxiety and PTSD risk is higher for girls than boys beginning in adolescence (Merikangas et al. 2010), future studies that test for impacts of participant and stimulus sex on fear conditioning and extinction may provide insight into sex-specific risk factors (Garza and Jovanovic 2017).

Although several studies include adolescents, few have directly addressed puberty. Pubertal development triggers sex-specific changes in the neuroendocrine milieu, which are linked to broad changes in brain structure (Herting and Sowell 2017). Puberty drives significant increases in fear response, measured as FPS, during differential conditioning (Stenson et al. 2021). Schmitz et al. (2014) also observed increases in FPS across puberty within a longitudinal sample, but this effect was specific to girls. Pattwell et al. (2012) did not find effects of puberty on extinction learning, but the small sample ($n = 28$ adolescents) included both sexes and may not have been adequately powered to detect effects of puberty. Studies that examine how puberty impacts differential conditioning and extinction learning in both sexes may provide insight into sex-specific developmental changes that contribute to increased vulnerability to fear- and anxiety-based mental health challenges in girls and women.

In addition to puberty, during adolescence there is development within circuits that facilitate fear conditioning and extinction learning (Gabard-Durnam et al. 2014; Gee et al. 2013b), elevated autonomic responses to emotionally salient stimuli (Silk et al. 2009), shifts in emotion regulation abilities (Zimmermann and Iwanski 2014), and preferential processing of socially-relevant cues (e.g., faces; Blakemore and Mills (2014); Passarotti et al. (2009)). Age-related changes in emotion regulation, which develops and becomes more efficient between early adolescence (11 years) and adulthood (Zimmermann and Iwanski 2014), may be particularly relevant to extinction recall: prior research indicates that when children successfully regulate their emotional arousal, memory for emotionally salient stimuli is diminished (Leventon and Bauer 2016). Many studies that include adolescents have not been designed to assess how changes in these processes may impact conditioning and extinction learning. Importantly, if aspects of the task are developmentally sensitive, then developmental shifts in fear conditioning and extinction learning may be obscured. For instance, the Screaming Lady task uses social stimuli that may be processed differently by adolescents vs. children or adults. The use of paradigms designed to systematically manipulate developmentally relevant factors could clarify the extent of developmental change in fear conditioning, extinction learning, and extinction recall.

7 Conclusions

Many of these studies report discordant subjective and physiological responses in individuals with anxiety, particularly children. Recent theoretical papers have proposed that aversive Pavlovian conditioning processes are importantly different in humans vs. non-human animals, because humans experience a higher-order feeling of fear that is both phenomenologically and neurobiologically distinct from the process engaged by threat learning and extinction in non-human animals (LeDoux and Hofmann 2018; LeDoux and Pine 2016). This theoretical framework, along with the findings that anxiety may be associated with elevated feelings of fear or expectancy of an aversive event, but not differential physiological responses to the CS+ and CS−, indicates a need for more research into how individuals with and without anxiety engage cognitive processes, such as emotion regulation, in ways that may modulate their subjective experience during differential fear conditioning and extinction learning. Because many of these processes develop slowly, a developmental approach to this research may shed light on the relationships between extinction learning, extinction recall, and mental health risk across the lifespan.

A growing body of evidence demonstrates that differential conditioning and extinction learning can be studied with physiological, neuroimaging, and self-report measures across a wide swath of development. Broadly, there is evidence for substantial consistency in differential learning and extinction across development, as assessed with a set of widely used paradigms, across physiological indices. Importantly, though, response coherence across measures (e.g., physiological, neural, and behavioral) was inconsistent across studies. This highlights a need for additional examination of when and why there is variability across these measures, both across development and in relation to individual differences factors, including mental health, childhood adversity, and sex. Moving forward, the use of comparable paradigms and measures in studies of fear extinction across development would facilitate comparison of results across samples and age groups, in turn advancing understanding of fear extinction across development (Ryan et al. 2019). Finally, longitudinal studies that assess within-person change using multiple measures of learning and emotional responsivity will provide the most robust insights into fear conditioning and extinction across development.

References

Abend R, Gold AL, Britton JC, Michalska KJ, Shechner T, Sachs JF, Winkler AM, Leibenluft E, Averbeck BB, Pine DS (2020) Anticipatory threat responding: associations with anxiety, development, and brain structure. Biol Psychiatry 87(10):916–925. https://doi.org/10.1016/j.biopsych.2019.11.006

Adolph D, Flasinski T, Lippert MW, Pflug V, Hamm AO, Richter J, Margraf J, Schneider S (2022) Measuring extinction learning across the lifespan – adaptation of an optimized paradigm to closely match exposure treatment procedures. Biol Psychol 170:108311. https://doi.org/10.1016/j.biopsycho.2022.108311

Andreassi JL (2010) Psychophysiology: human behavior and physiological response. Psychology Press

Bachevalier J, Vargha-Khadem F (2005) The primate hippocampus: ontogeny, early insult and memory. Curr Opin Neurobiol 15(2):168–174. https://doi.org/10.1016/j.conb.2005.03.015

Beauchaine TP (2015) Respiratory sinus arrhythmia: a transdiagnostic biomarker of emotion dysregulation and psychopathology. Curr Opin Psychol 3:43–47. https://doi.org/10.1016/j.copsyc.2015.01.017

Blakemore SJ, Mills KL (2014) Is adolescence a sensitive period for sociocultural processing? Annu Rev Psychol 65:187–207. https://doi.org/10.1146/annurev-psych-010213-115202

Bradley MM, Codispoti M, Cuthbert BN, Lang PJ (2001) Emotion and motivation I: defensive and appetitive reactions in picture processing. Emotion 1(3):276–298. https://doi.org/10.1037//1528-3542.1.3.276

Britton JC, Grillon C, Lissek S, Norcross MA, Szuhany KL, Chen G, Ernst M, Nelson EE, Leibenluft E, Shechner T, Pine DS (2013) Response to learned threat: an FMRI study in adolescent and adult anxiety. Am J Psychiatry 170(10):1195–1204. https://doi.org/10.1176/appi.ajp.2013.12050651

Chauret M, La Buissonniere-Ariza V, Lamoureux Tremblay V, Suffren S, Servonnet A, Pine DS, Maheu FS (2014) The conditioning and extinction of fear in youths: what's sex got to do with it? Biol Psychol 100:97–105. https://doi.org/10.1016/j.biopsycho.2014.06.001

Chauret M, Suffren S, Pine DS, Nassim M, Saint-Amour D, Maheu FS (2019) Fear conditioning and extinction in anxious youth, offspring at-risk for anxiety and healthy comparisons: an fMRI study. Biol Psychol 148:107744. https://doi.org/10.1016/j.biopsycho.2019.107744

Craske MG, Waters AM, Lindsey Bergman R, Naliboff B, Lipp OV, Negoro H, Ornitz EM (2008) Is aversive learning a marker of risk for anxiety disorders in children? Behav Res Ther 46(8):954–967. https://doi.org/10.1016/j.brat.2008.04.011

Craske MG, Hermans D, Vervliet B (2018) State-of-the-art and future directions for extinction as a translational model for fear and anxiety. Philos Trans R Soc Lond B Biol Sci 373(1742). https://doi.org/10.1098/rstb.2017.0025

Critchley HD, Melmed RN, Featherstone E, Mathias CJ, Dolan RJ (2002) Volitional control of autonomic arousal: a functional magnetic resonance study. Neuroimage 16(4):909–919. https://doi.org/10.1006/nimg.2002.1147

Davis M (1992) The role of the amygdala in fear-potentiated startle: implications for animal models of anxiety. Trends Pharmacol Sci 13(1):35–41. https://doi.org/10.1016/0165-6147(92)90014-w

Davis M, Falls WA, Campeau S, Kim M (1993) Fear-potentiated startle: a neural and pharmacological analysis. Behav Brain Res 58(1–2):175–198. https://doi.org/10.1016/0166-4328(93)90102-v

Duits P, Cath DC, Lissek S, Hox JJ, Hamm AO, Engelhard IM, van den Hout MA, Baas JM (2015) Updated meta-analysis of classical fear conditioning in the anxiety disorders. Depress Anxiety 32(4):239–253. https://doi.org/10.1002/da.22353

El-Sheikh M (2007) Children's skin conductance level and reactivity: are these measures stable over time and across tasks? Dev Psychobiol 49(2):180–186. https://doi.org/10.1002/dev.20171

Fullana MA, Harrison BJ, Soriano-Mas C, Vervliet B, Cardoner N, Avila-Parcet A, Radua J (2016) Neural signatures of human fear conditioning: an updated and extended meta-analysis of fMRI studies. Mol Psychiatry 21(4):500–508. https://doi.org/10.1038/mp.2015.88

Gabard-Durnam LJ, Flannery J, Goff B, Gee DG, Humphreys KL, Telzer E, Hare T, Tottenham N (2014) The development of human amygdala functional connectivity at rest from 4 to 23 years: a cross-sectional study. Neuroimage 95:193–207. https://doi.org/10.1016/j.neuroimage.2014.03.038

Ganella DE, Drummond KD, Ganella EP, Whittle S, Kim JH (2017) Extinction of conditioned fear in adolescents and adults: a human fMRI study. Front Hum Neurosci 11:647. https://doi.org/10.3389/fnhum.2017.00647

Gao Y, Raine A, Venables PH, Dawson ME, Mednick SA (2010) The development of skin conductance fear conditioning in children from ages 3 to 8 years. Dev Sci 13(1):201–212. https://doi.org/10.1111/j.1467-7687.2009.00874.x

Garza K, Jovanovic T (2017) Impact of gender on child and adolescent PTSD. Curr Psychiatry Rep 19(11):87. https://doi.org/10.1007/s11920-017-0830-6

Gee DG, Gabard-Durnam LJ, Flannery J, Goff B, Humphreys KL, Telzer EH, Hare TA, Bookheimer SY, Tottenham N (2013a) Early developmental emergence of human amygdala-prefrontal connectivity after maternal deprivation. Proc Natl Acad Sci U S A 110(39): 15638–15643. https://doi.org/10.1073/pnas.1307893110

Gee DG, Humphreys KL, Flannery J, Goff B, Telzer EH, Shapiro M, Hare TA, Bookheimer SY, Tottenham N (2013b) A developmental shift from positive to negative connectivity in human amygdala-prefrontal circuitry. J Neurosci 33(10):4584–4593. https://doi.org/10.1523/JNEUROSCI.3446-12.2013

Geller DA, McGuire JF, Orr SP, Small BJ, Murphy TK, Trainor K, Porth R, Wilhelm S, Storch EA (2019) Fear extinction learning as a predictor of response to cognitive behavioral therapy for pediatric obsessive compulsive disorder. J Anxiety Disord 64:1–8. https://doi.org/10.1016/j.janxdis.2019.02.005

Gold AL, Abend R, Britton JC, Behrens B, Farber M, Ronkin E, Chen G, Leibenluft E, Pine DS (2020) Age differences in the neural correlates of anxiety disorders: an fMRI study of response to learned threat. Am J Psychiatry 177(5):454–463. https://doi.org/10.1176/appi.ajp.2019.19060650

Herting MM, Sowell ER (2017) Puberty and structural brain development in humans. Front Neuroendocrinol 44:122–137. https://doi.org/10.1016/j.yfrne.2016.12.003

Ingram E, Fitzgerald HE (1974) Individual differences in infant orienting and autonomic conditioning. Dev Psychobiol 7(4):359–367. https://doi.org/10.1002/dev.420070413

Jenness JL, Miller AB, Rosen ML, McLaughlin KA (2019) Extinction learning as a potential mechanism linking high vagal tone with lower PTSD symptoms among abused youth. J Abnorm Child Psychol 47(4):659–670. https://doi.org/10.1007/s10802-018-0464-0

Jovanovic T, Ressler KJ (2010) How the neurocircuitry and genetics of fear inhibition may inform our understanding of PTSD. Am J Psychiatry 167(6):648–662. https://doi.org/10.1176/appi.ajp.2009.09071074

Jovanovic T, Nylocks KM, Gamwell KL, Smith A, Davis TA, Norrholm SD, Bradley B (2014) Development of fear acquisition and extinction in children: effects of age and anxiety. Neurobiol Learn Mem 113:135–142. https://doi.org/10.1016/j.nlm.2013.10.016

Jovanovic T, Wiltshire CN, Reda MH, France J, Wanna CP, Minton ST, Davie W, Grasser LR, Winters S, Schacter H, Marusak HA, Stenson AF (2022) Uncertain in the face of change: lack of contingency shift awareness during extinction is associated with higher fear-potentiated startle and PTSD symptoms in children. Int J Psychophysiol 178:90–98. https://doi.org/10.1016/j.ijpsycho.2022.06.008

Kredlow MA, Pineles SL, Inslicht SS, Marin MF, Milad MR, Otto MW, Orr SP (2017) Assessment of skin conductance in African American and non-African American participants in studies of conditioned fear. Psychophysiology 54(11):1741–1754. https://doi.org/10.1111/psyp.12909

Kuhn M, Wendt J, Sjouwerman R, Buchel C, Hamm A, Lonsdorf TB (2020) The neurofunctional basis of affective startle modulation in humans: evidence from combined facial electromyography and functional magnetic resonance imaging. Biol Psychiatry 87(6):548–558. https://doi.org/10.1016/j.biopsych.2019.07.028

LaBar KS, LeDoux JE (2003) Emotional learning circuits in animals and humans. In: Davidson RJ, Scherer KR, Goldsmith HH (eds) Handbook of affective sciences. Oxford University Press, pp 52–65. https://login.proxy.library.emory.edu/login?url=https://search.ebscohost.com/login.aspx?direct=trueanddb=psyhandAN=2009-07773-004andsite=ehost-live

Lang PJ, Bradley MM, Cuthbert B (1990) Emotion, attention, and the startle reflex. Psychol Rev 97(3):377–395. https://doi.org/10.1037/0033-295X.97.3.377

Lang PJ, Bradley MM, Cuthbert BN (1998) Emotion, motivation, and anxiety: brain mechanisms and psychophysiology. Biol Psychiatry 44(12):1248–1263. https://doi.org/10.1016/s0006-3223(98)00275-3

Lau JY, Britton JC, Nelson EE, Angold A, Ernst M, Goldwin M, Grillon C, Leibenluft E, Lissek S, Norcross M, Shiffrin N, Pine DS (2011) Distinct neural signatures of threat learning in adolescents and adults. Proc Natl Acad Sci U S A 108(11):4500–4505. https://doi.org/10.1073/pnas.1005494108

Lau JYF, Lissek S, Nelson EE, Lee Y, Roberson-Nay R, Poeth K, Jenness J, Ernst M, Grillon C, Pine DS (2008) Fear conditioning in adolescents with anxiety disorders: results from a novel experimental paradigm. J Am Acad Child Adolesc Psychiatry 47(1):94–102. https://doi.org/10.1097/chi.0b01e31815a5f01

Lavenex P, Banta Lavenex P (2013) Building hippocampal circuits to learn and remember: insights into the development of human memory. Behav Brain Res 254:8–21. https://doi.org/10.1016/j.bbr.2013.02.007

Leclair-Visonneau L, Bosquet T, Magot A, Fayet G, Gras-Le Guen C, Hamel A, Pereon Y (2016) Electrochemical skin conductance for quantitative assessment of sweat function: normative values in children. Clin Neurophysiol Pract 1:43–45. https://doi.org/10.1016/j.cnp.2016.07.001

LeDoux JE (1993) Emotional memory systems in the brain. Behav Brain Res 58(1–2):69–79. https://doi.org/10.1016/0166-4328(93)90091-4

LeDoux JE, Hofmann SG (2018) The subjective experience of emotion: a fearful view. Curr Opin Behav Sci 19:67–72. https://doi.org/10.1016/j.cobeha.2017.09.011

LeDoux JE, Pine DS (2016) Using neuroscience to help understand fear and anxiety: a two-system framework. Am J Psychiatry 173(11):1083–1093. https://doi.org/10.1176/appi.ajp.2016.16030353

LeDoux JE, Iwata J, Cicchetti P, Reis DJ (1988) Different projections of the central amygdaloid nucleus mediate autonomic and behavioral correlates of conditioned fear. J Neurosci 8(7):2517–2529. https://doi.org/10.1523/jneurosci.08-07-02517.1988

Lee Y, Lopez DE, Meloni EG, Davis M (1996) A primary acoustic startle pathway: obligatory role of cochlear root neurons and the nucleus reticularis pontis caudalis. J Neurosci 16(11):3775–3789. https://www.ncbi.nlm.nih.gov/pubmed/8642420

Leventon JS, Bauer PJ (2016) Emotion regulation during the encoding of emotional stimuli: effects on subsequent memory. J Exp Child Psychol 142:312–333. https://doi.org/10.1016/j.jecp.2015.09.024

Lissek S, Powers AS, McClure EB, Phelps EA, Woldehawariat G, Grillon C, Pine DS (2005) Classical fear conditioning in the anxiety disorders: a meta-analysis. Behav Res Ther 43(11):1391–1424. https://doi.org/10.1016/j.brat.2004.10.007

Maples-Keller JL, Price M, Jovanovic T, Norrholm SD, Odenat L, Post L, Zwiebach L, Breazeale K, Gross R, Kim SJ, Rothbaum BO (2017) Targeting memory reconsolidation to prevent the return of fear in patients with fear of flying. Depress Anxiety 34(7):610–620. https://doi.org/10.1002/da.22626

Maples-Keller J, Watkins LE, Nylocks KM, Yasinski C, Coghlan C, Black K, Jovanovic T, Rauch SAM, Rothbaum BO, Norrholm SD (2022) Acquisition, extinction, and return of fear in veterans in intensive outpatient prolonged exposure therapy: a fear-potentiated startle study. Behav Res Ther 154:104124. https://doi.org/10.1016/j.brat.2022.104124

Marusak HA, Hehr A, Bhogal A, Peters C, Iadipaolo A, Rabinak CA (2021) Alterations in fear extinction neural circuitry and fear-related behavior linked to trauma exposure in children. Behav Brain Res 398:112958. https://doi.org/10.1016/j.bbr.2020.112958

Mauss IB, Levenson RW, McCarter L, Wilhelm FH, Gross JJ (2005) The tie that binds? Coherence among emotion experience, behavior, and physiology. Emotion 5(2):175–190. https://doi.org/10.1037/1528-3542.5.2.175

McLaughlin KA, Sheridan MA, Gold AL, Duys A, Lambert HK, Peverill M, Heleniak C, Shechner T, Wojcieszak Z, Pine DS (2016) Maltreatment exposure, brain structure, and fear conditioning in children and adolescents. Neuropsychopharmacology 41(8):1956–1964. https://doi.org/10.1038/npp.2015.365

McManis MH, Kagan J, Snidman NC, Woodward SA (2002) EEG asymmetry, power, and temperament in children. Dev Psychobiol 41(2):169–177. https://doi.org/10.1002/dev.10053

Merikangas KR, He J-P, Burstein M, Swanson SA, Avenevoli S, Cui L, Benjet C, Georgiades K, Swendsen J (2010) Lifetime prevalence of mental disorders in US adolescents: results from the National Comorbidity Survey Replication–Adolescent Supplement (NCS-A). J Am Acad Child Adolesc Psychiatry 49(10):980–989

Merrick MT, Ford DC, Ports KA, Guinn AS, Chen J, Klevens J, Metzler M, Jones CM, Simon TR, Daniel VM, Ottley P, Mercy JA (2019) Vital signs: estimated proportion of adult health problems attributable to adverse childhood experiences and implications for prevention – 25 states, 2015-2017. MMWR Morb Mortal Wkly Rep 68(44):999–1005. https://doi.org/10.15585/mmwr.mm6844e1

Michalska KJ, Shechner T, Hong M, Britton JC, Leibenluft E, Pine DS, Fox NA (2016) A developmental analysis of threat/safety learning and extinction recall during middle childhood. J Exp Child Psychol 146:95–105. https://doi.org/10.1016/j.jecp.2016.01.008

Michopoulos V, Norrholm SD, Jovanovic T (2015) Diagnostic biomarkers for posttraumatic stress disorder: promising horizons from translational neuroscience research. Biol Psychiatry 78(5): 344–353. https://doi.org/10.1016/j.biopsych.2015.01.005

Milad MR, Wright CI, Orr SP, Pitman RK, Quirk GJ, Rauch SL (2007) Recall of fear extinction in humans activates the ventromedial prefrontal cortex and hippocampus in concert. Biol Psychiatry 62(5):446–454. https://doi.org/10.1016/j.biopsych.2006.10.011

Miller EM, Shields SA (1980) Skin-conductance response as a measure of adolescents emotional reactivity. Psychol Rep 46(2):587–590. https://doi.org/10.2466/pr0.1980.46.2.587

Mills KL, Goddings AL, Herting MM, Meuwese R, Blakemore SJ, Crone EA, Dahl RE, Guroglu B, Raznahan A, Sowell ER, Tamnes CK (2016) Structural brain development between childhood and adulthood: convergence across four longitudinal samples. Neuroimage 141:273–281. https://doi.org/10.1016/j.neuroimage.2016.07.044

Morriss J, Christakou A, van Reekum CM (2019) Multimodal evidence for delayed threat extinction learning in adolescence and young adulthood. Sci Rep 9(1):7748. https://doi.org/10.1038/s41598-019-44150-1

Neumann DL, Waters AM, Westbury HR, Henry J (2008) The use of an unpleasant sound unconditional stimulus in an aversive conditioning procedure with 8- to 11-year-old children. Biol Psychol 79(3):337–342. https://doi.org/10.1016/j.biopsycho.2008.08.005

Norrholm SD, Anderson KM, Olin IW, Jovanovic T, Kwon C, Warren VT, McCarthy A, Bosshardt L, Sabree J, Duncan EJ, Rothbaum BO, Bradley B (2011) Versatility of fear-potentiated startle paradigms for assessing human conditioned fear extinction and return of fear. Front Behav Neurosci 5:77. https://doi.org/10.3389/fnbeh.2011.00077

Passarotti AM, Sweeney JA, Pavuluri MN (2009) Neural correlates of incidental and directed facial emotion processing in adolescents and adults. Soc Cogn Affect Neurosci 4(4):387–398. https://doi.org/10.1093/scan/nsp029

Pattwell SS, Duhoux S, Hartley CA, Johnson DC, Jing D, Elliott MD, Ruberry EJ, Powers A, Mehta N, Yang RR, Soliman F, Glatt CE, Casey BJ, Ninan I, Lee FS (2012) Altered fear learning across development in both mouse and human. Proc Natl Acad Sci U S A 109(40): 16318–16323. https://doi.org/10.1073/pnas.1206834109

Paus T, Keshavan M, Giedd JN (2008) Why do many psychiatric disorders emerge during adolescence? Nat Rev Neurosci 9(12):947–957. https://doi.org/10.1038/nrn2513

Pöhlchen D, Leuchs L, Binder FP, Blaskovich B, Nantawisarakul T, Topalidis P, Bruckl TM, Norrholm SD, Jovanovic T, Be C, w. g., and Spoormaker, V. I. (2020) No robust differences in fear conditioning between patients with fear-related disorders and healthy controls. Behav Res Ther 129:103610. https://doi.org/10.1016/j.brat.2020.103610

Quevedo K, Smith T, Donzella B, Schunk E, Gunnar M (2010) The startle response: developmental effects and a paradigm for children and adults. Dev Psychobiol 52(1):78–89. https://doi.org/10.1002/dev.20415

Richards JE (2000) Development of multimodal attention in young infants: modification of the startle reflex by attention. Psychophysiology 37(1):65–75. https://www.ncbi.nlm.nih.gov/pubmed/10705768

Rothbaum BO, Kearns MC, Price M, Malcoun E, Davis M, Ressler KJ, Lang D, Houry D (2012) Early intervention may prevent the development of posttraumatic stress disorder: a randomized pilot civilian study with modified prolonged exposure. Biol Psychiatry 72(11):957–963. https://doi.org/10.1016/j.biopsych.2012.06.002

Ryan KM, Zimmer-Gembeck MJ, Neumann DL, Waters AM (2019) The need for standards in the design of differential fear conditioning and extinction experiments in youth: a systematic review and recommendations for research on anxiety. Behav Res Ther 112:42–62. https://doi.org/10.1016/j.brat.2018.11.009

Schmitz A, Grillon C, Avenevoli S, Cui L, Merikangas KR (2014) Developmental investigation of fear-potentiated startle across puberty. Biol Psychol 97:15–21. https://doi.org/10.1016/j.biopsycho.2013.12.002

Shechner T, Britton JC, Ronkin EG, Jarcho JM, Mash JA, Michalska KJ, Leibenluft E, Pine DS (2015) Fear conditioning and extinction in anxious and nonanxious youth and adults: examining a novel developmentally appropriate fear-conditioning task. Depress Anxiety 32(4):277–288. https://doi.org/10.1002/da.22318

Silk JS, Siegle GJ, Whalen DJ, Ostapenko LJ, Ladouceur CD, Dahl RE (2009) Pubertal changes in emotional information processing: pupillary, behavioral, and subjective evidence during emotional word identification. Dev Psychopathol 21(1):7–26. https://doi.org/10.1017/S0954579409000029

Stenson AF, Nugent NR, van Rooij SJH, Minton ST, Compton AB, Hinrichs R, Jovanovic T (2021) Puberty drives fear learning during adolescence. Dev Sci 24(1):e13000. https://doi.org/10.1111/desc.13000

Susman ES, Weissman DG, Sheridan MA, McLaughlin KA (2021) High vagal tone and rapid extinction learning as potential transdiagnostic protective factors following childhood violence exposure. Dev Psychobiol 63(6):e22176. https://doi.org/10.1002/dev.22176

van Rooij SJH, Smith RD, Stenson AF, Ely TD, Yang X, Tottenham N, Stevens JS, Jovanovic T (2020) Increased activation of the fear neurocircuitry in children exposed to violence. Depress Anxiety 37(4):303–312. https://doi.org/10.1002/da.22994

Waters AM, Pine DS (2016) Evaluating differences in Pavlovian fear acquisition and extinction as predictors of outcome from cognitive behavioural therapy for anxious children. J Child Psychol Psychiatry 57(7):869–876. https://doi.org/10.1111/jcpp.12522

Wood KH, Ver Hoef LW, Knight DC (2014) The amygdala mediates the emotional modulation of threat-elicited skin conductance response. Emotion 14(4):693–700. https://doi.org/10.1037/a0036636

Zimmermann P, Iwanski A (2014) Emotion regulation from early adolescence to emerging adulthood and middle adulthood age differences, gender differences, and emotion-specific developmental variations. Int J Behav Dev 38(2):182–194. https://doi.org/10.1177/0165025413515405

Extinction Learning Across Development: Neurodevelopmental Changes and Implications for Pediatric Anxiety Disorders

Elizabeth R. Kitt, Paola Odriozola, and Dylan G. Gee

Contents

Abstract Alterations in extinction learning relate to the development and maintenance of anxiety disorders across the lifespan. While exposure therapy, based on principles of extinction, can be highly effective for treating anxiety, many patients do not show sufficient improvement following treatment. In particular, evidence suggests that exposure therapy does not work sufficiently for up to 40% of children who receive this evidence-based treatment.

Importantly, fear learning and extinction, as well as the neural circuitry supporting these processes, undergo dynamic changes across development. An improved understanding of developmental changes in extinction learning and the associated neural circuitry may help to identify targets to improve treatment response in clinically anxious children and adolescents. In this chapter, we provide a brief overview of methods used to study fear learning and extinction in developmental populations. We then review what is currently known about the developmental

E. R. Kitt, P. Odriozola, and D. G. Gee (✉)
Department of Psychology, Yale University, New Haven, CT, USA
e-mail: dylan.gee@yale.edu

Curr Topics Behav Neurosci (2023) 64: 237–256
https://doi.org/10.1007/7854_2023_430
Published Online: 3 August 2023

changes that occur in extinction learning and related neural circuitry. We end this chapter with a discussion of the implications of these neurodevelopmental changes for the characterization and treatment of pediatric anxiety disorders.

Keywords Extinction learning · Anxiety · Development · fMRI

1 Introduction

Across the lifespan, the ability to detect threats in the environment and learn appropriate fear associations is an essential adaptive process. In the context of normative stressors, fear is an adaptive emotional response to immediate threats (both real and perceived). However, excessive or disproportionate fear can result in clinically significant levels of distress and can ultimately lead to the development of anxiety disorders. A growing literature suggests that alterations in extinction learning relate to the development of anxiety disorders across the lifespan (Britton et al. 2011; Graham and Milad 2011; Sevenster et al. 2018).

Based on evidence of altered fear learning and extinction in individuals with anxiety disorders, exposure therapy, which is based on principles of extinction learning, has become the gold-standard treatment for anxiety disorders (McNally 2007). Exposure therapy is effective for many children and adolescents (Kendall et al. 2008). Nevertheless, evidence suggests that it does not work sufficiently for up to 40% of youth with anxiety disorders (Creswell et al. 2020; Hudson et al. 2015). Importantly, a surge of translational research has emphasized that fear learning and extinction, as well as the neural circuitry supporting these processes, undergo dynamic changes across development (Casey et al. 2015). An improved understanding of developmental changes in extinction learning and the associated neural circuitry may help to identify targets to improve treatment response in clinically anxious children and adolescents.

In the following chapter, we first provide a brief overview of methods used to study fear learning and extinction in developmental populations. We then review what is currently known about the developmental changes that occur in extinction learning and related neural circuitry. We end this chapter with a discussion of the implications of these neurodevelopmental changes for the characterization and treatment of pediatric anxiety disorders.

2 The Basics of Operationalizing Fear Learning and Extinction Across Development

Classic fear conditioning and extinction paradigms have provided important tools to study the acquisition and extinction of fear across species and across the lifespan. Prior chapters have described these methods as used with adults in detail. Here, we highlight specific methods used with children and adolescents. While similar paradigms are used across the lifespan, it is important to note several methodological and ethical challenges involved in the study of fear learning and extinction in developmental samples. For example, effective fear conditioning requires the use of a potent unconditioned stimulus (UCS). In adult human samples, this UCS is typically an electric shock (Shechner et al. 2014). However, electric shocks pose the potential risks of causing pain or heightened anxiety in youth. Thus, researchers often select an alternative UCS that is developmentally appropriate yet still potent enough to contribute to the acquisition of fear associations. Commonly used UCSs include loud, aversive noises, sometimes coupled with aversive images (e.g., a "screaming lady" paradigm coupling a fearful human face and a loud recorded scream; Britton et al. 2013; Glenn et al. 2012b; Neumann and Waters 2006; Shechner et al. 2014) or airblast to the larynx (e.g., Jovanovic et al. 2005, 2014; Monk et al. 2003). Recently, some researchers have used shock in vicarious learning paradigms, in which the child observes an adult (e.g., a parent) undergoing a fear learning paradigm using an electric shock as the UCS. After watching this, children are shown the same stimuli and told that they may receive a shock; however, no shocks are administered to the child (e.g., Bilodeau-Houle et al. 2020; Marin et al. 2020).

Another important methodological consideration involves the measurement of the conditioned response (CR). Past research has employed a range of methods to measure the expression of fear. In rodents, defensive behaviors indicative of preparation for an expected aversive event are often used as an indicator of fear response. For example, rodents may show increased startle response or freezing (Curzon et al. 2009). In humans, common measures include participant self-report of fear, autonomic reactivity (e.g., skin conductance, heart rate), endocrine responses (e.g., cortisol levels), and behavioral responses (e.g., fear-potentiated startle; Lonsdorf et al. 2017; Shechner et al. 2014). While multiple measures, including skin conductance response and self-report, have been found to be reliable measures of fear response in children (Britton et al. 2013; Neumann et al. 2008; Shechner et al. 2014), many studies choose to combine both physiological and self-reported measures to mitigate the risk that young children may have difficulty reliably reporting on their fear responses (Shechner et al. 2014).

Cross-species research examining fear learning and extinction has provided valuable insights into the development of the neural circuitry supporting fear learning and extinction. Initially, the majority of studies of fear learning and extinction relied on rodent models. This has provided a strong foundation for our understanding of fear learning in humans, as the neural circuitry supporting fear learning is highly conserved across species (Birn et al. 2014). Advancements in human neuroimaging have provided an instrumental tool for furthering the study of these processes in humans; however, it is important to consider several key challenges when interpreting the results of human neuroimaging studies, especially with developmental samples. In particular, neuroimaging studies often suffer from low statistical power and small sample sizes (Button et al. 2013; Zacharek et al. 2021). Flexibility in possible analysis methods has further contributed to low reproducibility of neuroimaging results (Poldrack et al. 2017). Moreover, children may be less likely to tolerate neuroimaging, particularly with longer scans leading to greater motion artifacts and attrition of younger participants (Zacharek et al. 2021). While continued studies will be necessary to replicate findings from human neuroimaging, we present existing cross-species data, which provides critical insights into the development of fear learning and extinction and the neural circuitry underlying these processes.

3 Fear Learning and Extinction Across Development

Behavioral and physiological evidence suggests that fear learning and extinction emerge early in development. From a young age, children demonstrate patterns of fear acquisition (Gao et al. 2010). As children age, they show gradual, linear improvement in their ability to discriminate between neutral stimuli and aversively conditioned stimuli (Gao et al. 2010; Glenn et al. 2012a). Children also show attenuated fear responses following repeated presentations of a previously aversively conditioned stimulus without the aversive UCS, suggesting that extinction learning begins early in development (Michalska et al. 2016; Waters et al. 2017). However, unlike the linear development of fear learning, evidence suggests that fear extinction undergoes non-linear changes across normative development. Although findings are mixed (Waters et al. 2017), cross-species evidence suggests that extinction learning is diminished during adolescence, relative to childhood and adulthood (Baker et al. 2014; McCallum et al. 2010; Pattwell et al. 2012). In other words, adolescents (approximately ages 12–17 years) show a diminished capacity to extinguish previously conditioned fears compared to both children (age 11 years and younger) and adults (age 18 years and older). Furthermore, juvenile rodents do not show the same forms of relapse following extinction that are typically seen in adults (i.e., spontaneous recovery, reinstatement, and renewal; Gogolla et al. 2009; Kim and Richardson 2008; Yap and Richardson 2007). While further research in humans is necessary, this may suggest that extinction learning in the juvenile stage may be associated with true erasure of the original fear memory trace, unlike in adulthood (Kim and Richardson 2008). Taken together, these findings indicate important changes in fear learning and extinction throughout development.

4 Neurodevelopmental Changes Underlying Fear Learning and Extinction

These behavioral and physiological changes in fear learning and extinction are paralleled by structural and functional changes in neural circuitry across development. Convergent evidence across humans and rodents indicates that interactions among the amygdala, prefrontal cortex (PFC), and hippocampus – regions that develop in childhood and adolescence – play a critical role in fear learning and extinction (e.g., Milad and Quirk 2002; Phelps et al. 2004; Sotres-Bayon et al. 2006). In this section, we review the dynamic neurodevelopmental changes in these regions that may underlie developmental shifts in fear learning and extinction processes.

4.1 Developmental Changes in Neural Circuitry Supporting Fear Learning and Extinction

The neural circuitry implicated in fear learning and extinction, most notably the amygdala, vmPFC, and hippocampus, undergoes dynamic changes across development. The amygdala shows early maturation both structurally (Giedd et al. 1996; Humphrey 1968; Ulfig et al. 2003) and functionally (Gabard-Durnam et al. 2018; Graham et al. 2016; Rogers et al. 2017), reaching peak volume by pre-adolescence (ages 9–11 years old; Uematsu et al. 2012). In line with findings of early development of fear learning, the amygdala exhibits differential responses to threat cues in humans as young as infants (Graham et al. 2013).

As opposed to the earlier maturation of the amygdala, the hippocampus and PFC show more protracted development (Akers et al. 2014; Casey et al. 2008; Raineki et al. 2010). Prefrontal regions continue to develop well into late adolescence, with gray matter volume decreasing throughout adolescence (Casey et al. 2008; Sturman and Moghaddam 2011). Similarly, connections between the vmPFC, hippocampus, and amygdala show continued structural (Gee et al. 2016; Lebel et al. 2012; Swartz et al. 2014) and functional (Casey et al. 2019; Gabard-Durnam et al. 2014; Gee et al. 2013; Jalbrzikowski et al. 2017; Perlman and Pelphrey 2011) development over the course of adolescence.

This protracted maturation may contribute to a limited capacity for prefrontal regulation of the amygdala and hippocampus in childhood and adolescence and could underlie the changes observed in extinction learning across development. Early in life, fear learning relies primarily on the amygdala (Adolphs et al. 1994; LaBar et al. 1998). However, later in development, fear learning appears to depend on the broader circuit involving the hippocampus and PFC (Britton et al. 2011; Lau et al. 2011). Of note, while projections from the vmPFC (or infralimbic cortex in rodents) undergo protracted development, projections from the dACC (or prelimbic cortex in rodents) are augmented in rodents during the adolescent period compared to adulthood (Pattwell et al. 2016). Given the role of the vmPFC in fear extinction and the role of the dACC in the maintenance of fear, these changes in neural circuitry may reflect a period of vulnerability for fear-related disorders in adolescence (Casey et al. 2015; Lee et al. 2014).

5 Fear Learning and Extinction in Pediatric Anxiety Disorders

Anxiety disorders are the most prevalent psychiatric disorders in youth (Costello et al. 2005; Kessler et al. 2005; Merikangas et al. 2010). Childhood and adolescence represent times of particular risk for anxiety disorders, with the majority of lifetime anxiety disorders emerging during this period (Beesdo et al. 2009; Kessler et al. 2005). When not effectively addressed, anxiety disorders have the potential for a chronic course, persisting across the lifespan (Gregory et al. 2007). Importantly, disruptions in fear learning and extinction are considered central to the etiology and maintenance of anxiety disorders (Mineka and Oehlberg 2008). Exposure therapy, based on principles of extinction, is the first-line psychosocial treatment for pediatric anxiety disorders (Abramowitz 2013; Britton et al. 2011). Exposure therapy involves repeatedly confronting anxiety-provoking stimuli, such that patients begin to form new memories of the previously phobic stimulus without the feared outcome. Over time, patients learn to reduce fear responses to these stimuli as the original fear memory trace competes with the new extinction memory (Abramowitz 2013, 2019; Britton et al. 2011). While exposure therapy can be highly effective for treating anxiety, many patients do not show sufficient improvement following treatment (McNally 2007). In particular, evidence suggests that exposure therapy does not work sufficiently for up to 40% of children who receive this evidence-based treatment (Creswell et al. 2020; Hudson et al. 2015). A closer look at the dynamic changes occurring in the neural circuitry implicated in fear learning and extinction during childhood and adolescence could pave the way for the development and optimization of developmentally tailored interventions for pediatric anxiety disorders (Casey et al. 2015; Gee and Kribakaran 2020; Lee et al. 2014; Odriozola and Gee 2021).

In this section, we first review the existing literature on alterations in fear learning and extinction in pediatric anxiety disorders. We then provide a brief overview of anxiety-related differences in neural regions linked to fear learning and extinction. Finally, we end this section by discussing potential implications of this research for the treatment of pediatric anxiety disorders.

5.1 *Alterations in Fear Learning and Extinction in Pediatric Anxiety*

To date, relatively few studies have examined patterns of fear learning and extinction in pediatric anxiety; however, the literature on this subject has begun to shed light on how these processes may function in youth with anxiety disorders. This burgeoning body of research emphasizes the importance of taking both a clinical and developmental approach to examine how development and anxiety may interact in dynamic ways to affect processes of fear learning and extinction.

Across the lifespan, individuals with anxiety disorders show elevated fear responses to previously conditioned stimuli, including those that were never presented with the UCS (i.e., a safety cue [CS−]) through the process of stimulus generalization (Abend et al. 2020; Craske et al. 2008; Dvir et al. 2019; Harrewijn et al. 2020; Jovanovic et al. 2014; Lissek et al. 2005). Of note, a recent study found that patients with anxiety showed elevated physiological fear responses to the conditioned stimulus [CS+] and the CS− but *not* the UCS, suggesting that anxiety-related differences may lie in responding to anticipatory rather than acute threat (Abend et al. 2020). Preliminary evidence suggests that individual differences in extinction learning may be linked with treatment response, with non-responders showing no reduction in skin conductance response during extinction (Treanor et al. 2021; Waters and Pine 2016).

A recent meta-analysis found that youth with anxiety disorders show similar patterns of differential fear acquisition and extinction compared to those without anxiety disorders (Dvir et al. 2019). Although the effect was smaller for extinction learning, youth across diagnostic groups showed modest differential learning effects between the CS+ and CS− during both fear learning and extinction, suggesting successful fear conditioning and extinction (Dvir et al. 2019). Of note, this meta-analysis was based on a small number of studies, and there are significant inconsistencies between individual studies, with several suggesting that anxious youth may show reduced differential learning during fear conditioning and following extinction (for a review, see Grasser and Jovanovic 2021). Interestingly, a recent study found an interaction between contingency awareness and anxiety on children's fear responses, such that children with greater anxiety who were unaware of contingency changes during extinction showed increased fear responses to safety cues and did not show discrimination between the CS+ and CS− during fear conditioning. This pattern of findings may indicate a potential role for contingency awareness in anxiety-related differences in differential learning (Jovanovic et al. 2022).

Taken together, these results suggest that anxious youth show elevated fear responses during conditioning across stimuli. However, results are mixed with regard to anxiety-related differences in differential learning. Further research will be necessary to examine whether youth with anxiety disorders show alterations in extinction learning.

5.2 Anxiety-Related Alterations in Relevant Neural Circuitry in Youth

Anxiety disorders have been linked with both structural and functional alterations in key neural circuitry underlying fear learning and extinction, including the PFC, amygdala, and hippocampus (Gee and Casey 2017; Martin et al. 2009; Zacharek et al. 2021). While research on anxiety-related structural changes has primarily been conducted in adults, growing research in youth indicates similar patterns of structural differences in pediatric anxiety disorders in these key regions (Strawn et al. 2021). Preliminary evidence suggests that children with anxiety disorders have increased cortical volumes in the left vmPFC and reduced right hippocampal volumes compared to non-anxious youth (Gold et al. 2017). Several studies have also identified structural alterations in the amygdala in youth with anxiety disorders. However, the direction of these anxiety-related differences has been mixed, with some studies reporting smaller amygdala volumes in youth with anxiety disorders (Milham et al. 2005; Strawn et al. 2015) and others reporting larger amygdala volumes in anxious youth (e.g., De Bellis et al. 2000; Qin et al. 2014). While future work will be necessary to resolve the discrepancies across studies, this literature suggests that youth with anxiety disorders may exhibit structural alterations in key neural circuitry underlying fear learning and extinction.

Youth with anxiety disorders also show altered functioning in the neural circuitry underpinning fear learning and extinction. When viewing threatening emotional stimuli, such as fearful or angry faces, anxious youth show elevated activation in the amygdala, ACC, and ventral PFC (e.g., McClure et al. 2007; Monk et al. 2008; Thomas et al. 2001). Furthermore, youth with anxiety disorders exhibit altered PFC–amygdala functional connectivity in response to emotional stimuli, as opposed to children without anxiety disorders (Monk et al. 2008). Importantly, differences in prefrontal activation to emotional stimuli have been found to predict improved response to both exposure-based behavioral treatments and medication-based treatments for pediatric anxiety, with greater prefrontal activation to threatening faces predicting better treatment response (Kujawa et al. 2016). Taken together, these findings indicate that children with anxiety disorders may have diminished prefrontal control of amygdala reactivity, relative to children without anxiety disorders.

Anxious youth have also been found to show altered patterns of neural activation during fear conditioning paradigms. A recent study found that youth with anxiety disorders show reduced activation in the left dACC and left ventrolateral PFC to the threat cue during fear conditioning, relative to both non-anxious and “at-risk” youth (defined as having a parent with current or past anxiety; Chauret et al. 2019). Adolescents with anxiety disorders have also been found to exhibit reduced medial PFC (mPFC) activation to the safety cue during fear conditioning, relative to adolescents without anxiety disorders (Haddad et al. 2015). This pattern of results may provide further support for excessive fear expression to safety in youth with anxiety disorders. Furthermore, both adolescents and adults with anxiety disorders have been found to show reduced subgenual ACC activation compared to their non-anxious, age-matched counterparts when asked to report on their current level of subjective fear during extinction recall (Britton et al. 2013). This growing literature underscores important anxiety-related alterations in the neural circuitry supporting fear learning and extinction in youth with anxiety disorders.

5.3 Age-Modulated Patterns of Diagnostic Differences in Relevant Neural Circuitry

Building upon these diagnostic and age-related differences in neural circuitry underlying fear learning and extinction, preliminary evidence suggests that some of these anxiety-related alterations may vary across development. In a sample of both anxious and non-anxious youth and adults, an extinction recall task was used to examine participants' responses to morphed stimuli of varying resemblance to a previously conditioned CS+ and CS−. Adults with anxiety disorders were found to exhibit reduced vmPFC activation during explicit threat appraisal compared to non-anxious adults (Britton et al. 2013). By contrast, adolescents with anxiety disorders showed a U-shaped pattern of vmPFC activation, such that activation was strongest in response to the most extreme CS+ and CS− (Britton et al. 2013). The authors interpreted these findings as further suggesting a unique pattern of elevated sensitivity to threat and safety in adolescents with anxiety disorders.

In addition to these age-related diagnostic differences in activation, recent findings point to age-modulated diagnostic alterations in functional connectivity. Patterns of left amygdala functional connectivity with the vmPFC during the same extinction recall task were found to vary depending on both age and diagnosis. Specifically, while making explicit judgments about current fear or previous threat associations (versus when making judgments about physical aspects of stimuli), anxious adults showed more positive amygdala–vmPFC connectivity, whereas anxious children showed negative amygdala–vmPFC coupling (Gold et al. 2016). While this initial study compared patterns of connectivity during different attentional states, a recent study from the same group examined differences in connectivity in response to different morphed stimuli. Again, results indicated that age moderated anxiety-related differences in amygdala–vmPFC connectivity. Specifically, adults with anxiety disorders, as compared to non-anxious adults, showed more negative amygdala–vmPFC coupling in response to stimuli that increasingly resembled the previously conditioned safety cue; however, this age-related pattern was reversed in youth (Fig. 1; Gold et al. 2020). Taken together, these findings suggest that youth with anxiety disorders may show alterations in amygdala–vmPFC connectivity during extinction recall, and these alterations may vary across development.

In summary, these findings indicate potential age-specific vulnerabilities related to threat learning and extinction in children and adolescents with anxiety disorders. These age-related differences further underscore the importance of extending findings of fear learning and extinction in adults into younger samples, as youth may show unique patterns across development.

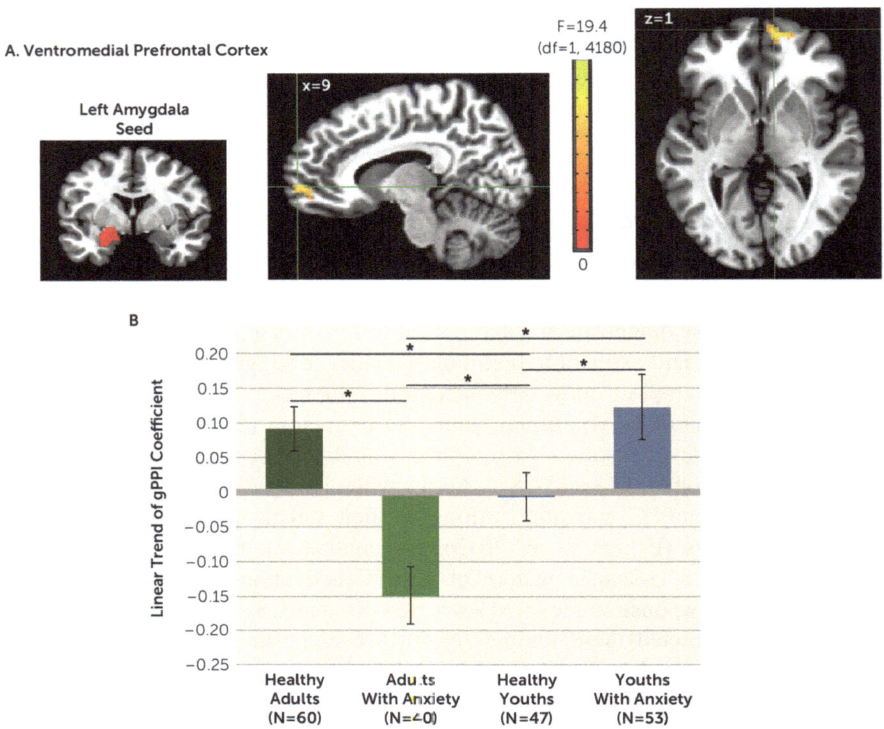

Fig. 1 In a study of extinction recall in both anxious and non-anxious youth and adults, diagnostic differences in left amygdala–vmPFC functional connectivity were found to vary by age. Approximately 3 weeks after fear conditioning and extinction, participants completed an extinction recall task with morphed stimuli of varying similarity to the previously conditioned CS+ and CS−. Relative to healthy adults, adults with anxiety disorders exhibited more negative left amygdala–vmPFC connectivity to stimuli that more closely resembled the previously conditioned CS−. Conversely, the opposite pattern was observed in youth, such that youth with anxiety disorders exhibited more positive left amygdala–vmPFC connectivity. Reprinted with permission from "Age differences in the neural correlates of anxiety disorders: an fMRI study of response to learned threat," by A. L. Gold et al. 2020, *American Journal of Psychiatry, 177*(5), 454–463. (Copyright © 2020). All Rights Reserved

5.4 Implications for Treatment for Pediatric Anxiety

The current literature delineates developmental changes in key neural circuitry supporting extinction learning. Importantly, findings described above suggest that youth with anxiety disorders may exhibit different patterns of neural activation and connectivity during fear learning and extinction compared to both their non-anxious peers and anxious adults (e.g., Britton et al. 2013; Gold et al. 2016, 2020). Building upon this growing understanding of age- and anxiety-related differences in fear learning and extinction could provide important opportunities to optimize treatments for pediatric anxiety based on the developing brain.

5.5 *Safety Cue Learning*

While translational evidence suggests that neural circuitry supporting extinction learning may be diminished in adolescence, other approaches to fear reduction that depend less heavily on the PFC may result in longer-term reduction of conditioned fear responding during this period of development. For example, extinction learning that occurs during memory reconsolidation, which relies more heavily on the amygdala than the PFC, has been shown to reduce fear expression in adolescents and adults (Agren et al. 2012; Johnson and Casey 2015; Schiller et al. 2013). Cross-species evidence also suggests that fear reduction via safety cue learning targets an alternative pathway that may be augmented during adolescence (Fig. 2).

Conditioned inhibition via safety cue learning involves reducing fear in the presence of a threat using a stimulus trained to represent the absence of threat (the CS−, or "safety cue") (Christianson et al. 2012; Kazama et al. 2013; Myers and Davis 2004; Odriozola and Gee 2021). During conditioning, the CS+ is paired with the UCS, whereas the CS− is never paired with the UCS, thus indicating the absence of threat. Importantly, the CS+ and CS− must be combined to test for conditioned inhibition. If the CS− is functioning as a conditioned inhibitor, when the CS+ and CS− are combined, the CS− should inhibit the CR associated with the CS+.

Whereas extinction learning involves the creation of a competing, parallel association between the CS+ and threat, safety cue learning reduces threat via a separate association between a distinct stimulus (the CS−) and the absence of threat (Christianson et al. 2012). Extinction learning is supported by prefrontal control of the amygdala, which undergoes protracted development during adolescence (e.g., Pattwell et al. 2012). By contrast, recent cross-species findings suggest that safety cue learning involves projections between the hippocampus and the dACC (prelimbic cortex in rodents), which may be strengthened during adolescence (Meyer et al. 2019). An emerging body of translational research indicates that safety cue learning may be effective for reducing responses to threat in rodents, nonhuman primates, and healthy adult humans (Kazama et al. 2013; Meyer et al. 2019; Myers and Davis 2004). Furthermore, a recent pilot study provided initial evidence that youth with anxiety disorders may show unique patterns of activation during conditioned inhibition via safety cue learning, relative to youth without anxiety disorders (Harrewijn et al. 2020).

While more work is necessary to understand how safety cue learning in the laboratory could inform interventions in the clinic, this preliminary evidence suggests that safety cue learning could present a pathway for fear reduction that targets the unique state of the developing brain in youth with anxiety disorders. Prevailing

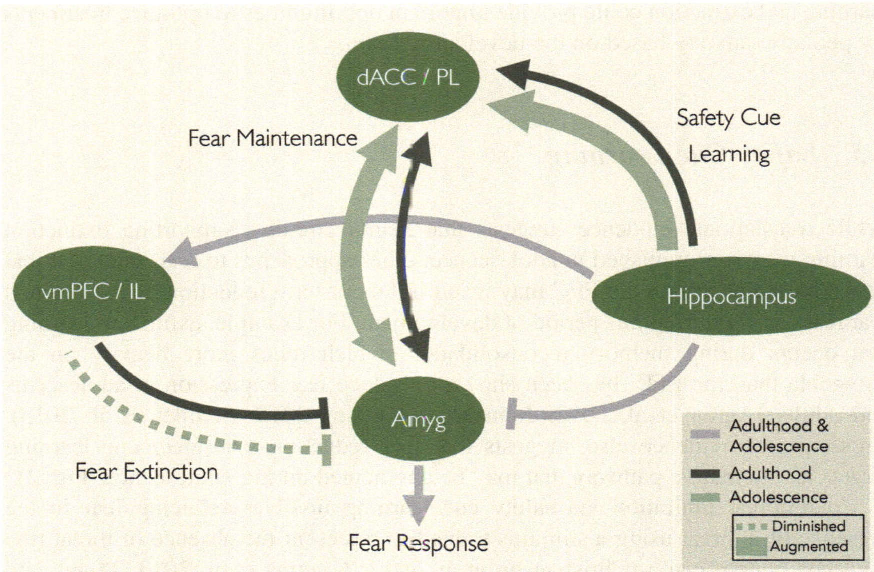

Fig. 2 Schematic of developmental changes in neural circuitry supporting fear maintenance, fear extinction, and safety cue learning, a proposed developmentally tailored method of fear reduction during adolescence. Rodent models suggest that connections between the dACC (or PL in rodents) and the amygdala, which support the maintenance of fear, may be augmented during adolescence. Projections between the vmPFC (or IL in rodents) and the amygdala, which play a key role in fear extinction, are weaker during adolescence; however, connections between the hippocampus and the dACC, which are proposed to support safety cue learning, may be strengthened in adolescence. Targeting this alternative pathway during adolescence may represent a developmentally optimized means of fear reduction. *Amyg* amygdala, *dACC* dorsal anterior cingulate cortex, *vmPFC* ventromedial prefrontal cortex, *PL* prelimbic cortex, *IL* infralimbic cortex. Adapted with permission from "Learning about Safety: Conditioned Inhibition as a Novel Approach to Fear Reduction Targeting the Developmental Brain," by P. Odriozola and D. G. Gee, 2021, *American Journal of Psychiatry, 178*(2), 136–155. (Copyright © 2021). All Rights Reserved

theories of exposure therapy suggest that, while safety cues reduce fear in the short term, they may contribute to the maintenance of anxiety in the long term (Craske et al. 2014). However, judicious incorporation of safety cues into exposure therapy, particularly during adolescence, may enhance treatment acceptability, improve adherence to treatment, and ultimately lead to improved treatment outcomes for youth (Odriozola and Gee 2021). Future studies will play an important role in elucidating how safety cue learning could inform a novel approach to augmenting traditional exposure-based treatments based on neurodevelopment changes in anxious youth.

5.6 Parental Involvement in Treatment

The role of caregivers in supporting youths' regulation of fear also shifts dramatically across development. Across species, young children show approach behaviors toward stimuli that have been conditioned in their parent's presence, even with stimuli that have been aversively conditioned (Moriceau and Sullivan 2006; Tottenham et al. 2019). In other words, when conditioning occurs in the presence of their parent, young children exhibit behaviors that may contradict traditional fear learning, approaching rather than avoiding aversively conditioned stimuli. Furthermore, children have been found to show increased discrimination between a CS+ and CS− (i.e., between threat and safety cues) when in the presence of their parent but not when their parent is absent (van Rooij et al. 2017). Parents may also influence their child's fear learning through vicarious learning. In a recent study, children watched videos of their parents undergoing a differential conditioning paradigm. When viewing the stimuli themselves, children showed greater fear responses to the stimulus that was reinforced with shock for their parent, despite the fact that none of the stimuli were reinforced for the child (Marin et al. 2020). This supports the idea that observation of parents may contribute to children's acquisition of fear.

This parental modulation of children's affective learning is paralleled by changes in parental buffering of physiological and neurological processes. Parental presence is associated with reduced amygdala reactivity, increased mPFC–amygdala functional connectivity, diminished endocrine responses to threat, and improved regulatory behavior in childhood (Feldman et al. 2010; Gee et al. 2014; Gunnar and Donzella 2002; Hostinar et al. 2014). Over the course of normative development, the effect of parents on their child's fear reduction weakens, in parallel with the continued development of amygdala–mPFC circuitry. The effects of caregivers on stimulus discrimination, endocrine responses to threat, amygdala–mPFC connectivity, and regulatory behavior are typically seen in childhood but not adolescence (Gee et al. 2014; Hostinar et al. 2015). Thus, while the neural circuitry underlying fear learning and extinction is still developing, caregivers may play an important regulatory role in modulating affective learning and reactivity to threat (Callaghan and Tottenham 2016; Gee 2016, 2020).

Parents also play an important role in supporting fear reduction in anxious children. Children with animal phobias have been found to show more approach behaviors toward phobia-related animals when in the presence of a parent (Ollendick et al. 2012). Critically, research suggests that children with anxiety disorders may show increased reliance on their parents to reduce their fear. A growing body of research indicates that family accommodation (i.e., changes that family members make to their own behavior or schedule to reduce their child's distress) is common among families with anxious children (Lebowitz et al. 2013; Norman et al. 2015; Storch et al. 2015). Moreover, higher levels of family accommodation are associated with worse youth symptom severity, and evidence suggests that family accommodation may contribute to the maintenance or exacerbation of pediatric anxiety disorders (Iniesta-Sepúlveda et al. 2021; La Buissonnière-Ariza et al. 2018), perhaps in part by reducing opportunities for youth to engage in extinction learning. By removing potential opportunities for youth to experience feared situations or stimuli, family accommodation may hinder youth's ability to develop new associations.

Consistent with this role of parents in anxious children's fear reduction, parent-based treatments for pediatric anxiety have been found to be highly effective. For example, a parent-based psychosocial intervention that specifically targets the reduction of family accommodation has been found to be an acceptable and efficacious treatment (Lebowitz et al. 2014). A recent noninferiority trial found that such parent-based treatment, in which the child does not interact with the therapist, was noninferior to traditional child-focused, exposure-based treatment (Lebowitz et al. 2020). Further examination of the role of parents in regulating fear in anxious youth, as well as continued exploration of how this could be leveraged in treatment, will be valuable to further enhance treatment efficacy for pediatric anxiety disorders.

6 Conclusion

In summary, a growing body of translational developmental research emphasizes key alterations in fear learning and extinction, as well as the neural circuitry that supports these processes, over the course of development. While both fear learning and extinction emerge at a young age, they develop at different rates. Specifically, fear learning shows a linear pattern of progression across development, whereas extinction learning appears to exhibit a non-linear trajectory over the course of development, with extinction learning diminished in adolescence. These developmental changes in fear learning and extinction are paralleled by dynamic structural and functional changes in neural circuitry, most notably the amygdala, vmPFC, dACC, and hippocampus. Converging theory and empirical evidence underscore the role of altered fear learning and extinction in the etiology and maintenance of pediatric anxiety disorders. Given that exposure therapy, the first-line psychosocial treatment for pediatric anxiety disorders, is based on principles of extinction learning, building on our growing understanding of neurodevelopmental changes associated with extinction learning may improve our ability to optimize interventions for youth with anxiety disorders.

References

Abend R, Gold AL, Britton JC, Michalska KJ, Shechner T, Sachs JF, Winkler AM, Leibenluft E, Averbeck BB, Pine DS (2020) Anticipatory threat responding: associations with anxiety, development, and brain structure. Biol Psychiatry 87(10):916–925

Abramowitz JS (2013) The practice of exposure therapy: relevance of cognitive-behavioral theory and extinction theory. Behav Ther 44(4):548–558

Abramowitz JS, Deacon BJ, Whiteside SP (2019) Exposure therapy for anxiety: principles and practice. Guilford Publications

Adolphs R, Tranel D, Damasio H, Damasio A (1994) Impaired recognition of emotion in facial expressions following bilateral damage to the human amygdala. Nature 372(6507):669–672

Agren T, Engman J, Frick A, Björkstrand J, Larsson E-M, Furmark T, Fredrikson M (2012) Disruption of reconsolidation erases a fear memory trace in the human amygdala. Science 337(6101):1550–1552

Akers KG, Martinez-Canabal A, Restivo L, Yiu AP, De Cristofaro A, Hsiang H-L, Wheeler AL, Guskjolen A, Niibori Y, Shoji H (2014) Hippocampal neurogenesis regulates forgetting during adulthood and infancy. Science 344(6184):598–602

Baker KD, Den ML, Graham BM, Richardson R (2014) A window of vulnerability: impaired fear extinction in adolescence. Neurobiol Learn Mem 113:90–100

Beesdo K, Knappe S, Pine DS (2009) Anxiety and anxiety disorders in children and adolescents: developmental issues and implications for DSM-V. Psychiatr Clin 32(3):483–524

Bilodeau-Houle A, Bouchard V, Morand-Beaulieu S, Herringa RJ, Milad MR, Marin M-F (2020) Anxiety sensitivity moderates the association between father-child relationship security and fear transmission. Front Psychol 11:579514

Birn RM, Shackman AJ, Oler JA, Williams LE, McFarlin DR, Rogers GM, Shelton SE, Alexander AL, Pine DS, Slattery MJ (2014) Evolutionarily conserved prefrontal-amygdalar dysfunction in early-life anxiety. Mol Psychiatry 19(8):915–922

Britton JC, Lissek S, Grillon C, Norcross MA, Pine DS (2011) Development of anxiety: the role of threat appraisal and fear learning. Depress Anxiety 28(1):5–17

Britton JC, Grillon C, Lissek S, Norcross MA, Szuhany KL, Chen G, Ernst M, Nelson EE, Leibenluft E, Shechner T (2013) Response to learned threat: an fMRI study in adolescent and adult anxiety. Am J Psychiatr 170(10):1195–1204

Button KS, Ioannidis J, Mokrysz C, Nosek BA, Flint J, Robinson ES, Munafò MR (2013) Power failure: why small sample size undermines the reliability of neuroscience. Nat Rev Neurosci 14(5):365–376

Callaghan BL, Tottenham N (2016) The neuro-environmental loop of plasticity: a cross-species analysis of parental effects on emotion circuitry development following typical and adverse caregiving. Neuropsychopharmacology 41(1):163–176

Casey BJ, Getz S, Galvan A (2008) The adolescent brain. Dev Rev 28(1):62–77

Casey B, Glatt CE, Lee FS (2015) Treating the developing versus developed brain: translating preclinical mouse and human studies. Neuron 86(6):1358–1368

Casey B, Heller AS, Gee DG, Cohen AO (2019) Development of the emotional brain. Neurosci Lett 693:29–34

Chauret M, Suffren S, Pine DS, Nassim M, Saint-Amour D, Maheu FS (2019) Fear conditioning and extinction in anxious youth, offspring at-risk for anxiety and healthy comparisons: an fMRI study. Biol Psychol 148:107744

Christianson JP, Fernando AB, Kazama AM, Jovanovic T, Ostroff LE, Sangha S (2012) Inhibition of fear by learned safety signals: a mini-symposium review. J Neurosci 32(41):14118–14124

Costello EJ, Egger HL, Angold A (2005) The developmental epidemiology of anxiety disorders: phenomenology, prevalence, and comorbidity. Child Adolesc Psychiatr Clin 14(4):631–648

Craske MG, Waters AM, Bergman RL, Naliboff B, Lipp OV, Negoro H, Ornitz EM (2008) Is aversive learning a marker of risk for anxiety disorders in children? Behav Res Ther 46(8): 954–967

Craske MG, Treanor M, Conway CC, Zbozinek T, Vervliet B (2014) Maximizing exposure therapy: an inhibitory learning approach. Behav Res Ther 58:10–23

Creswell C, Waite P, Hudson J (2020) Practitioner review: anxiety disorders in children and young people–assessment and treatment. J Child Psychol Psychiatry 61(6):628–643

Curzon P, Rustay NR, Browman KE (2009) Cued and contextual fear conditioning for rodents. In: Buccafusco JJ (ed) Methods of behavior analysis in neuroscience, 2nd edn. CRC Press/Taylor & Francis

De Bellis MD, Casey B, Dahl RE, Birmaher B, Williamson DE, Thomas KM, Axelson DA, Frustaci K, Boring AM, Hall J (2000) A pilot study of amygdala volumes in pediatric generalized anxiety disorder. Biol Psychiatry 48(1):51–57

Dvir M, Horovitz O, Aderka IM, Shechner T (2019) Fear conditioning and extinction in anxious and non-anxious youth: a meta-analysis. Behav Res Ther 120:103431

Feldman R, Singer M, Zagoory O (2010) Touch attenuates infants' physiological reactivity to stress. Dev Sci 13(2):271–278

Gabard-Durnam LJ, Flannery J, Goff B, Gee DG, Humphreys KL, Telzer E, Hare T, Tottenham N (2014) The development of human amygdala functional connectivity at rest from 4 to 23 years: a cross-sectional study. NeuroImage 95:193–207

Gabard-Durnam L, O'Muircheartaigh J, Dirks H, Dean D III, Tottenham N, Deoni S (2018) Human amygdala functional network development: a cross-sectional study from 3 months to 5 years of age. Dev Cogn Neurosci 34:63–74

Gao Y, Raine A, Venables PH, Dawson ME, Mednick SA (2010) The development of skin conductance fear conditioning in children from ages 3 to 8 years. Dev Sci 13(1):201–212

Gee DG (2016) Sensitive periods of emotion regulation: influences of parental care on frontoamygdala circuitry and plasticity. New Dir Child Adolesc Dev 153:87–110

Gee DG (2020) Caregiving influences on emotional learning and regulation: applying a sensitive period model. Curr Opin Behav Sci 36:177–184

Gee D, Casey B (2017) Neuroimaging and the neuroanatomical circuits implicated in anxiety, fear, and stress-related disorders. In: Sadock B, Sadock V, Ruiz P (eds) Comprehensive textbook of psychiatry, pp 4490–4530

Gee DG, Kribakaran S (2020) Developmental differences in neural responding to threat and safety: implications for treating youths with anxiety. Am J Psychiatr 177(5):378–380

Gee DG, Humphreys KL, Flannery J, Goff B, Telzer EH, Shapiro M, Hare TA, Bookheimer SY, Tottenham N (2013) A developmental shift from positive to negative connectivity in human amygdala–prefrontal circuitry. J Neurosci 33(10):4584–4593

Gee DG, Gabard-Durnam L, Telzer EH, Humphreys KL, Goff B, Shapiro M, Flannery J, Lumian DS, Fareri DS, Caldera C (2014) Maternal buffering of human amygdala-prefrontal circuitry during childhood but not during adolescence. Psychol Sci 25(11):2067–2078

Gee DG, Fetcho RN, Jing D, Li A, Glatt CE, Drysdale AT, Cohen AO, Dellarco DV, Yang RR, Dale AM (2016) Individual differences in frontolimbic circuitry and anxiety emerge with adolescent changes in endocannabinoid signaling across species. Proc Natl Acad Sci 113(16): 4500–4505

Giedd JN, Vaituzis AC, Hamburger SD, Lange N, Rajapakse JC, Kaysen D, Vauss YC, Rapoport JL (1996) Quantitative MRI of the temporal lobe, amygdala, and hippocampus in normal human development: ages 4–18 years. J Comp Neurol 366(2):223–230

Glenn CR, Klein DN, Lissek S, Britton JC, Pine DS, Hajcak G (2012a) The development of fear learning and generalization in 8–13 year-olds. Dev Psychobiol 54(7):675–684

Glenn CR, Lieberman L, Hajcak G (2012b) Comparing electric shock and a fearful screaming face as unconditioned stimuli for fear learning. Int J Psychophysiol 86(3):214–219

Gogolla N, Caroni P, Lüthi A, Herry C (2009) Perineuronal nets protect fear memories from erasure. Science 325(5945):1258–1261

Gold AL, Shechner T, Farber MJ, Spiro CN, Leibenluft E, Pine DS, Britton JC (2016) Amygdala–cortical connectivity: associations with anxiety, development, and threat. Depress Anxiety 33(10):917–926

Gold AL, Steuber ER, White LK, Pacheco J, Sachs JF, Pagliaccio D, Berman E, Leibenluft E, Pine DS (2017) Cortical thickness and subcortical gray matter volume in pediatric anxiety disorders. Neuropsychopharmacology 42(12):2423–2433

Gold AL, Abend R, Britton JC, Behrens B, Farber M, Ronkin E, Chen G, Leibenluft E, Pine DS (2020) Age differences in the neural correlates of anxiety disorders: an fMRI study of response to learned threat. Am J Psychiatr 177(5):454–463

Graham BM, Milad MR (2011) The study of fear extinction: implications for anxiety disorders. Am J Psychiatr 168(12):1255–1265

Graham AM, Fisher PA, Pfeifer JH (2013) What sleeping babies hear: a functional MRI study of interparental conflict and infants' emotion processing. Psychol Sci 24(5):782–789

Graham AM, Buss C, Rasmussen JM, Rudolph MD, Demeter DV, Gilmore JH, Styner M, Entringer S, Wadhwa PD, Fair DA (2016) Implications of newborn amygdala connectivity for fear and cognitive development at 6-months-of-age. Dev Cogn Neurosci 18:12–25

Grasser LR, Jovanovic T (2021) Safety learning during development: implications for development of psychopathology. Behav Brain Res 408:113297

Gregory AM, Caspi A, Moffitt TE, Koenen K, Eley TC, Poulton R (2007) Juvenile mental health histories of adults with anxiety disorders. Am J Psychiatr 164(2):301–308

Gunnar MR, Donzella B (2002) Social regulation of the cortisol levels in early human development. Psychoneuroendocrinology 27(1–2):199–220

Haddad AD, Bilderbeck A, James AC, Lau JY (2015) Fear responses to safety cues in anxious adolescents: preliminary evidence for atypical age-associated trajectories of functional neural circuits. J Psychiatr Res 68:301–308

Harrewijn A, Kitt ER, Abend R, Matsumoto C, Odriozola P, Winkler AM, Leibenluft E, Pine DS, Gee DG (2020) Comparing neural correlates of conditioned inhibition between children with and without anxiety disorders – a preliminary study. Behav Brain Res:112994

Hostinar CE, Sullivan RM, Gunnar MR (2014) Psychobiological mechanisms underlying the social buffering of the hypothalamic–pituitary–adrenocortical axis: a review of animal models and human studies across development. Psychol Bull 140(1):256

Hostinar CE, Johnson AE, Gunnar MR (2015) Parent support is less effective in buffering cortisol stress reactivity for adolescents compared to children. Dev Sci 18(2):281–297

Hudson JL, Rapee RM, Lyneham HJ, McLellan LF, Wuthrich VM, Schniering CA (2015) Comparing outcomes for children with different anxiety disorders following cognitive behavioural therapy. Behav Res Ther 72:30–37

Humphrey T (1968) The development of the human amygdala during early embryonic life. J Comp Neurol 132(1):135–165

Iniesta-Sepúlveda M, Rodriguez-Jimenez T, Lebowitz ER, Goodman WK, Storch EA (2021) The relationship of family accommodation with pediatric anxiety severity: meta-analytic findings and child, family and methodological moderators. Child Psychiatry Hum Dev 52(1):1–14

Jalbrzikowski M, Larsen B, Hallquist MN, Foran W, Calabro F, Luna B (2017) Development of white matter microstructure and intrinsic functional connectivity between the amygdala and ventromedial prefrontal cortex: associations with anxiety and depression. Biol Psychiatry 82(7): 511–521

Johnson D, Casey B (2015) Extinction during memory reconsolidation blocks recovery of fear in adolescents. Sci Rep 5(1):1–5

Jovanovic T, Keyes M, Fiallos A, Myers KM, Davis M, Duncan EJ (2005) Fear potentiation and fear inhibition in a human fear-potentiated startle paradigm. Biol Psychiatry 57(12):1559–1564

Jovanovic T, Nylocks KM, Gamwell KL, Smith A, Davis TA, Norrholm SD, Bradley B (2014) Development of fear acquisition and extinction in children: effects of age and anxiety. Neurobiol Learn Mem 113:135–142

Jovanovic T, Wiltshire C, Reda M, France J, Wanna C, Minton S, Davie W, Grasser L, Winters S, Schacter H (2022) Uncertain in the face of change: lack of contingency shift awareness during extinction is associated with higher fear-potentiated startle and PTSD symptoms in children. Int J Psychophysiol 178:90–98

Kazama AM, Schauder KB, McKinnon M, Bachevalier J, Davis M (2013) A novel AX+/BX– paradigm to assess fear learning and safety-signal processing with repeated-measure designs. J Neurosci Methods 214(2):177–183

Kendall PC, Hudson JL, Gosch E, Flannery-Schroeder E, Suveg C (2008) Cognitive-behavioral therapy for anxiety disordered youth: a randomized clinical trial evaluating child and family modalities. J Consult Clin Psychol 76(2):282

Kessler RC, Berglund P, Demler O, Jin R, Merikangas KR, Walters EE (2005) Lifetime prevalence and age-of-onset distributions of DSM-IV disorders in the National Comorbidity Survey Replication. Arch Gen Psychiatry 62(6):593–602

Kim JH, Richardson R (2008) The effect of temporary amygdala inactivation on extinction and reextinction of fear in the developing rat: unlearning as a potential mechanism for extinction early in development. J Neurosci 28(6):1282–1290

Kujawa A, Swain JE, Hanna GL, Koschmann E, Simpson D, Connolly S, Fitzgerald KD, Monk CS, Phan KL (2016) Prefrontal reactivity to social signals of threat as a predictor of treatment response in anxious youth. Neuropsychopharmacology 41(8):1983–1990

La Buissonnière-Ariza V, Schneider SC, Højgaard D, Kay BC, Riemann BC, Eken SC, Lake P, Nadeau JM, Storch EA (2018) Family accommodation of anxiety symptoms in youth undergoing intensive multimodal treatment for anxiety disorders and obsessive-compulsive disorder: nature, clinical correlates, and treatment response. Compr Psychiatry 80:1–13

LaBar KS, Gatenby JC, Gore JC, LeDoux JE, Phelps EA (1998) Human amygdala activation during conditioned fear acquisition and extinction: a mixed-trial fMRI study. Neuron 20(5):937–945

Lau JY, Britton JC, Nelson EE, Angold A, Ernst M, Goldwin M, Grillon C, Leibenluft E, Lissek S, Norcross M (2011) Distinct neural signatures of threat learning in adolescents and adults. Proc Natl Acad Sci 108(11):4500–4505

Lebel C, Gee M, Camicioli R, Wieler M, Martin W, Beaulieu C (2012) Diffusion tensor imaging of white matter tract evolution over the lifespan. NeuroImage 60(1):340–352

Lebowitz ER, Woolston J, Bar-Haim Y, Calvocoressi L, Dauser C, Warnick E, Scahill L, Chakir AR, Shechner T, Hermes H (2013) Family accommodation in pediatric anxiety disorders. Depress Anxiety 30(1):47–54

Lebowitz ER, Omer H, Hermes H, Scahill L (2014) Parent training for childhood anxiety disorders: the SPACE program. Cogn Behav Pract 21(4):456–469

Lebowitz ER, Marin C, Martino A, Shimshoni Y, Silverman WK (2020) Parent-based treatment as efficacious as cognitive-behavioral therapy for childhood anxiety: a randomized noninferiority study of supportive parenting for anxious childhood emotions. J Am Acad Child Adolesc Psychiatry 59(3):362–372

Lee FS, Heimer H, Giedd JN, Lein ES, Šestan N, Weinberger DR, Casey B (2014) Adolescent mental health – opportunity and obligation. Science 346(6209):547–549

Lissek S, Powers AS, McClure EB, Phelps EA, Woldehawariat G, Grillon C, Pine DS (2005) Classical fear conditioning in the anxiety disorders: a meta-analysis. Behav Res Ther 43(11): 1391–1424

Lonsdorf TB, Menz MM, Andreatta M, Fullana MA, Golkar A, Haaker J, Heitland I, Hermann A, Kuhn M, Kruse O (2017) Don't fear 'fear conditioning': methodological considerations for the design and analysis of studies on human fear acquisition, extinction, and return of fear. Neurosci Biobehav Rev 77:247–285

Marin M-F, Bilodeau-Houle A, Morand-Beaulieu S, Brouillard A, Herringa RJ, Milad MR (2020) Vicarious conditioned fear acquisition and extinction in child–parent dyads. Sci Rep 10(1):1–10

Martin EI, Ressler KJ, Binder E, Nemeroff CB (2009) The neurobiology of anxiety disorders: brain imaging, genetics, and psychoneuroendocrinology. Psychiatr Clin 32(3):549–575

McCallum J, Kim JH, Richardson R (2010) Impaired extinction retention in adolescent rats: effects of D-cycloserine. Neuropsychopharmacology 35(10):2134–2142

McClure EB, Monk CS, Nelson EE, Parrish JM, Adler A, Blair RJR, Fromm S, Charney DS, Leibenluft E, Ernst M (2007) Abnormal attention modulation of fear circuit function in pediatric generalized anxiety disorder. Arch Gen Psychiatry 64(1):97–106

McNally RJ (2007) Mechanisms of exposure therapy: how neuroscience can improve psychological treatments for anxiety disorders. Clin Psychol Rev 27(6):750–759

Merikangas KR, He J-P, Burstein M, Swanson SA, Avenevoli S, Cui L, Benjet C, Georgiades K, Swendsen J (2010) Lifetime prevalence of mental disorders in US adolescents: results from the National Comorbidity Study-Adolescent Supplement (NCS-A). J Am Acad Child Adolesc Psychiatry 49(10):980–989. https://doi.org/10.1016/j.jaac.2010.05.017

Meyer HC, Odriozola P, Cohodes EM, Mandell JD, Li A, Yang R, Hall BS, Haberman JT, Zacharek SJ, Liston C, Lee FS, Gee DG (2019) Ventral hippocampus interacts with prelimbic cortex during inhibition of threat response via learned safety in both mice and humans. Proc Natl Acad Sci 116(52):26970–26979

Michalska KJ, Shechner T, Hong M, Britton JC, Leibenluft E, Pine DS, Fox NA (2016) A developmental analysis of threat/safety learning and extinction recall during middle childhood. J Exp Child Psychol 146:95–105

Milad MR, Quirk GJ (2002) Neurons in medial prefrontal cortex signal memory for fear extinction. Nature 420(6911):70–74

Milham MP, Nugent AC, Drevets WC, Dickstein DS, Leibenluft E, Ernst M, Charney D, Pine DS (2005) Selective reduction in amygdala volume in pediatric anxiety disorders: a voxel-based morphometry investigation. Biol Psychiatry 57(9):961–966

Mineka S, Oehlberg K (2008) The relevance of recent developments in classical conditioning to understanding the etiology and maintenance of anxiety disorders. Acta Psychol 127(3):567–580

Monk CS, Grillon C, Baas JM, McClure EB, Nelson EE, Zarahn E, Charney DS, Ernst M, Pine DS (2003) A neuroimaging method for the study of threat in adolescents. Dev Psychobiol 43(4): 359–366

Monk CS, Telzer EH, Mogg K, Bradley BP, Mai X, Louro HM, Chen G, McClure-Tone EB, Ernst M, Pine DS (2008) Amygdala and ventrolateral prefrontal cortex activation to masked angry faces in children and adolescents with generalized anxiety disorder. Arch Gen Psychiatry 65(5):568–576

Moriceau S, Sullivan RM (2006) Maternal presence serves as a switch between learning fear and attraction in infancy. Nat Neurosci 9(8):1004–1006

Myers KM, Davis M (2004) AX+, BX-discrimination learning in the fear-potentiated startle paradigm: possible relevance to inhibitory fear learning in extinction. Learn Mem 11(4): 464–475

Neumann DL, Waters AM (2006) The use of an unpleasant sound as an unconditional stimulus in a human aversive Pavlovian conditioning procedure. Biol Psychol 73(2):175–185

Neumann DL, Waters AM, Westbury HR, Henry J (2008) The use of an unpleasant sound unconditional stimulus in an aversive conditioning procedure with 8-to 11-year-old children. Biol Psychol 79(3):337–342

Norman KR, Silverman WK, Lebowitz ER (2015) Family accommodation of child and adolescent anxiety: mechanisms, assessment, and treatment. J Child Adolesc Psychiatr Nurs 28(3): 131–140

Odriozola P, Gee DG (2021) Learning about safety: conditioned inhibition as a novel approach to fear reduction targeting the developing brain. Am J Psychiatr 178(2):136–155

Ollendick TH, Lewis KM, Cowart MJ, Davis T III (2012) Prediction of child performance on a parent–child behavioral approach test with animal phobic children. Behav Modif 36(4):509–524

Pattwell SS, Duhoux S, Hartley CA, Johnson DC, Jing D, Elliott MD, Ruberry EJ, Powers A, Mehta N, Yang RR (2012) Altered fear learning across development in both mouse and human. Proc Natl Acad Sci 109(40):16318–16323

Pattwell SS, Liston C, Jing D, Ninan I, Yang RR, Witztum J, Murdock MH, Dincheva I, Bath KG, Casey B (2016) Dynamic changes in neural circuitry during adolescence are associated with persistent attenuation of fear memories. Nat Commun 7(1):1–9

Perlman SB, Pelphrey KA (2011) Developing connections for affective regulation: age-related changes in emotional brain connectivity. J Exp Child Psychol 108(3):607–620

Phelps EA, Delgado MR, Nearing KI, LeDoux JE (2004) Extinction learning in humans: role of the amygdala and vmPFC. Neuron 43(6):897–905

Poldrack RA, Baker CI, Durnez J, Gorgolewski KJ, Matthews PM, Munafò MR, Nichols TE, Poline J-B, Vul E, Yarkoni T (2017) Scanning the horizon: towards transparent and reproducible neuroimaging research. Nat Rev Neurosci 18(2):115–126

Qin S, Young CB, Duan X, Chen T, Supekar K, Menon V (2014) Amygdala subregional structure and intrinsic functional connectivity predicts individual differences in anxiety during early childhood. Biol Psychiatry 75(11):892–900

Raineki C, Holman PJ, Debiec J, Bugg M, Beasley A, Sullivan RM (2010) Functional emergence of the hippocampus in context fear learning in infant rats. Hippocampus 20(9):1037–1046

Rogers CE, Sylvester CM, Mintz C, Kenley JK, Shimony JS, Barch DM, Smyser CD (2017) Neonatal amygdala functional connectivity at rest in healthy and preterm infants and early internalizing symptoms. J Am Acad Child Adolesc Psychiatry 56(2):157–166

Schiller D, Kanen JW, LeDoux JE, Monfils M-H, Phelps EA (2013) Extinction during reconsolidation of threat memory diminishes prefrontal cortex involvement. Proc Natl Acad Sci 110(50):20040–20045

Sevenster D, Visser RM, D'Hooge R (2018) A translational perspective on neural circuits of fear extinction: current promises and challenges. Neurobiol Learn Mem 155:113–126

Shechner T, Hong M, Britton JC, Pine DS, Fox NA (2014) Fear conditioning and extinction across development: evidence from human studies and animal models. Biol Psychol 100:1–12

Sotres-Bayon F, Cain CK, LeDoux JE (2006) Brain mechanisms of fear extinction: historical perspectives on the contribution of prefrontal cortex. Biol Psychiatry 60(4):329–336

Storch EA, Salloum A, Johnco C, Dane BF, Crawford EA, King MA, McBride NM, Lewin AB (2015) Phenomenology and clinical correlates of family accommodation in pediatric anxiety disorders. J Anxiety Disord 35:75–81

Strawn JR, Hamm L, Fitzgerald DA, Fitzgerald KD, Monk CS, Phan KL (2015) Neurostructural abnormalities in pediatric anxiety disorders. J Anxiety Disord 32:81–88

Strawn JR, Lu L, Peris TS, Levine A, Walkup JT (2021) Research review: pediatric anxiety disorders–what have we learnt in the last 10 years? J Child Psychol Psychiatry 62(2):114–139

Sturman DA, Moghaddam B (2011) The neurobiology of adolescence: changes in brain architecture, functional dynamics, and behavioral tendencies. Neurosci Biobehav Rev 35(8):1704–1712

Swartz JR, Carrasco M, Wiggins JL, Thomason ME, Monk CS (2014) Age-related changes in the structure and function of prefrontal cortex–amygdala circuitry in children and adolescents: a multi-modal imaging approach. NeuroImage 86:212–220

Thomas KM, Drevets WC, Dahl RE, Ryan ND, Birmaher B, Eccard CH, Axelson D, Whalen PJ, Casey B (2001) Amygdala response to fearful faces in anxious and depressed children. Arch Gen Psychiatry 58(11):1057–1063

Tottenham N, Shapiro M, Flannery J, Caldera C, Sullivan RM (2019) Parental presence switches avoidance to attraction learning in children. Nat Hum Behav 3(10):1070–1077

Treanor M, Rosenberg BM, Craske MG (2021) Pavlovian learning processes in pediatric anxiety disorders: a critical review. Biol Psychiatry 89(7):690–696

Uematsu A, Matsui M, Tanaka C, Takahashi T, Noguchi K, Suzuki M, Nishijo H (2012) Developmental trajectories of amygdala and hippocampus from infancy to early adulthood in healthy individuals

Ulfig N, Setzer M, Bohl J (2003) Ontogeny of the human amygdala. Ann N Y Acad Sci 985(1): 22–33

van Rooij SJ, Cross D, Stevens JS, Vance LA, Kim YJ, Bradley B, Tottenham N, Jovanovic T (2017) Maternal buffering of fear-potentiated startle in children and adolescents with trauma exposure. Soc Neurosci 12(1):22–31

Waters AM, Pine DS (2016) Evaluating differences in Pavlovian fear acquisition and extinction as predictors of outcome from cognitive behavioural therapy for anxious children. J Child Psychol Psychiatry 57(7):869–876

Waters AM, Theresiana C, Neumann DL, Craske MG (2017) Developmental differences in aversive conditioning, extinction, and reinstatement: a study with children, adolescents, and adults. J Exp Child Psychol 159:263–278

Yap CS, Richardson R (2007) Extinction in the developing rat: an examination of renewal effects. Dev Psychobiol 49(6):565–575

Zacharek SJ, Kribakaran S, Kitt ER, Gee DG (2021) Leveraging big data to map neurodevelopmental trajectories in pediatric anxiety. Dev Cogn Neurosci 50:100974

Fear Extinction Learning in Posttraumatic Stress Disorder

Yana Lokshina, Jony Sheynin, Gregory S. Vogt, and Israel Liberzon

Contents

Abstract Impairments in fear extinction processes have been implicated in the genesis and maintenance of debilitating psychopathologies, including Posttraumatic stress disorder (PTSD). PTSD, classified as a trauma- and stressor-related disorder, is characterized by four symptom clusters: intrusive recollections of trauma, avoidance of trauma-related stimuli, alterations in cognition and mood, and hyperarousal. One

---

Y. Lokshina and I. Liberzon (✉)
Department of Psychiatry and Behavioral Science, Texas A&M University Health Science Center, Bryan, TX, USA

Department of Psychological and Brain Sciences, Texas A&M University, College Station, TX, USA

Texas A&M Institute for Neuroscience, Texas A&M University, College Station, TX, USA
e-mail: liberzon@tamu.edu

J. Sheynin
Department of Psychiatry and Behavioral Science, Texas A&M University Health Science Center, Bryan, TX, USA

G. S. Vogt
Department of Psychiatry and Behavioral Science, Texas A&M University Health Science Center, Bryan, TX, USA

Department of Psychological and Brain Sciences, Texas A&M University, College Station, TX, USA

Curr Topics Behav Neurosci (2023) 64: 257–270
https://doi.org/10.1007/7854_2023_436
Published Online: 4 August 2023

of the key pathological feature associated with the persistence of these symptoms is impaired fear extinction, as delineated in multiple studies employing Pavlovian fear-conditioning paradigms. These paradigms, comprising fear acquisition, extinction, extinction recall, and fear renewal phases, have illuminated the neurobiological substrates of PTSD. Dysfunctions in the neural circuits that mediate these fear learning and extinction processes can result in failure to extinguish fear responses and retain extinction memory, giving rise to enduring experience of fear and anxiety. The protective avoidance behaviors observed in individuals with PTSD further exacerbate intrusive symptoms and pose challenges to effective treatment strategies. A comprehensive analysis of fear conditioning and extinction processes, along with the underlying neurobiology, could significantly enhance our understanding of PTSD pathophysiology. This chapter delineates the role of fear extinction processes in PTSD, investigates the underlying neurobiological substrates, and underscores the therapeutic implications, while also identifying future research directions.

Deficits in fear extinction processes may have significant implications for the development and maintenance of debilitating forms of psychopathology. Some of the most salient examples include the links between altered fear processing/extinction learning and posttraumatic stress disorder (PTSD), a psychiatric condition that develops in a subset of individuals following exposure to severe trauma. From the clinical perspective, PTSD is defined as a trauma- and stressor-related disorder with four clusters of symptoms, including experience of intrusive thoughts (i.e., trauma-related memories, dreams, flashbacks), avoidance of internal or external reminders of trauma, negative change in cognition and mood, and heightened physiological arousal (American Psychiatric Association 2013). Importantly, previous research on PTSD has widely utilized Pavlovian (classical) fear-conditioning paradigms to study the neurobiological correlates of PTSD symptomology, and impaired fear extinction has been proposed to be an integral mechanism underlying persistence of symptoms in individuals with PTSD (Briscione et al. 2014; Rothbaum and Davis 2003; VanElzakker et al. 2014; Zuj et al. 2015).

Pavlovian fear-conditioning paradigms incorporate fear conditioning (acquisition) and extinction phases and often also include extinction recall and fear renewal phases. These distinct fear-learning processes are governed by multiple, sometimes overlapping neural circuits. Dysfunctions within these neural circuits are associated with failure to appropriately extinguish fear responses and/or to retain extinction (safety) memory, potentially leading to the persistent or repeating experiences of fear and anxiety often present in PTSD patients. Furthermore, individuals with PTSD may also engage in protective avoidant behaviors, which can contribute to severe avoidance symptoms, including isolation, further exacerbate intrusive symptoms, and interfere with implementation of effective PTSD treatments, such as exposure-based therapy (Milad et al. 2014). A comprehensive investigation of fear conditioning, extinction, extinction retention/recall, and fear renewal processes, along with underlying neurobiological mechanisms, is thus likely to substantially advance our understanding of the pathophysiology of PTSD. In this chapter, we outline the experimental paradigms commonly used to study fear/safety learning in PTSD, followed by discussion of the underlying neurobiological mechanisms and neural circuits involved. We proceed with discussing the implications of these fear-learning processes for therapeutic approaches and conclude with remaining gaps and directions for future research.

1 Experimental Paradigm of Fear Extinction in PTSD

A classical Pavlovian fear-conditioning paradigm has been commonly used as a translational model to study fear extinction alterations in PTSD. This paradigm incorporates a fear-conditioning phase, during which subjects are exposed to a neutral signal (conditioned stimulus (CS), e.g. visual or audio cue), which is followed by an aversive stimulus (unconditioned stimulus (US), e.g. loud noise or electric shock) intended to trigger a fear response. During this phase, the subjects learn to associate the CS with the US and, and as a result, demonstrate fear response during the appearance of the CS alone. The fear response can be measured across different domains, including activation of relevant brain circuitry (assessed with functional MRI), activation of autonomic response (assessed by muscle activity, heart rate, or skin conductance response (SCR)), as well as a cognitive learning of this association (assessed as self-report of expectancy scores).

After this conditioning phase, subjects undergo fear extinction, during which the CS is repeatedly presented but no longer followed by the aversive US. Subjects normally show a reduction of the fear response over time, as a result of learned fear extinction processes. In some variations of this paradigm, the experiment proceeds on the following day when subjects' retention of the extinction memory is tested (i.e., extinction recall phase), by undergoing similar trials to extinction phase, i.e. repeated CS's without US. It has been shown that while autonomic responses during conditioning and extinction phases in PTSD patients are usually intact, patients demonstrate great SCR during extinction recall phase (Garfinkel et al. 2014; Milad et al. 2008), suggesting a failure in PTSD patients to effectively retain or recall the extinction memory.

Extinction recall and retention are dependent on contextual processing (Maren et al. 2013), and it has been proposed by our group that the deficits in contextual processing (CP) are central to PTSD pathophysiology and symptom development (Liberzon and Abelson 2016). If this is the case, it will be important to test the processing of contextual information and its effects on fear and extinction learning in PTSD patients. This can be readily accomplished in paradigms using visual cues (CSs), by introducing distinct on-screen backgrounds during the conditioning ("dangerous context"), extinction, and extinction recall phases ("safe context") (Marusak et al. 2021; Milad et al. 2009). To further study effect of contextual modulation of fear memory, one can implement an additional phase (i.e., fear renewal), after fear conditioning and fear extinction, where the CS is presented in the dangerous context and renewal of the fear response is expected. Indeed, the report of both greater SCR in PTSD patients during extinction recall, and of lower SCR during fear renewal, suggests ineffective retrieval of the fear and the extinction memory traces by dangerous and safe contexts, respectively (Garfinkel et al. 2014; Milad et al. 2009). Of note, since control groups are often composed of trauma-exposed individuals, reported alterations are mostly associated with PTSD symptoms rather than the traumatic exposure. In the following paragraphs, we describe the brain regions that are involved in the altered processing of fear memories demonstrated by PTSD patients. Further discussion of extinction and the return of fear using psychophysiological methods can be found in the Woodford and colleagues chapter found in this text.

2 Neurobiology of PTSD-Related Alterations in Fear Responses

2.1 Alterations During Fear Conditioning and Fear Extinction

Over the past few decades, a substantial body of evidence derived from rodent studies indicates that the amygdala, hippocampus, as well as prelimbic and infralimbic regions of medial prefrontal cortex (mPFC) are the key brain regions involved in fear conditioning and extinction processes (Sierra-Mercado et al. 2011; VanElzakker et al. 2014). Specifically, the amygdala nuclei are critical for fear conditioning, and it is also believed to regulate learning and storage of extinction memories (Zhang et al. 2020); the mPFC is implicated in consolidation of fear extinction memories (Bisby et al. 2018); and the hippocampus is involved in encoding and establishing contextual representations in general and fear extinction in particular (Senn et al. 2014; Sotres-Bayon and Quirk 2010). Human neuroimaging studies on fear conditioning and extinction have produced findings that replicated findings from rodents' research, showing that the amygdala, hippocampus, ventromedial PFC, and dorsal anterior cingulate cortex (dACC) (functional homologs of infralimbic and prelimbic cortex in rodents, respectively) play a key role in these processes (Feng et al. 2014; Zuj et al. 2016). Importantly, while prior research on underlying neural mechanisms of extinction processes in PTSD has reported disorder-related alterations in activity of the amygdala, hippocampus, and cortical regions (Harnett et al. 2020), evidence of altered neural circuits activations in PTSD patients, during fear conditioning and fear extinction phase, is inconsistent and awaits further investigation. Additionally, in study of cellular and molecular level changes occurred during fear extinction processes, one cannot avoid using the animal models of PTSD. However, some of existing animal models of PTSD rely on face validity only and the findings are not necessarily pertinent for PTSD. On the other hand, models that have construct validity, including single prolonged stress model (SPS; Liberzon et al. 1997), provide important data regarding cellular and molecular processes that may take place within the involved neural circuits.

Amygdala: Increased activity of the amygdala in PTSD during fear conditioning and extinction has been reported in some (Fani et al. 2015; Linnman et al. 2011; Milad et al. 2009; Suarez-Jimenez et al. 2020), but not all functional MRI studies (Shvil et al. 2014). While the specific molecular mechanisms of fear extinction deficits in PTSD remain to be determined, emerging evidence has indicated that amygdaloid *N*-methyl-D-aspartate (NMDA) receptors play a critical role in fear extinction (Rodrigues et al. 2001; Santini et al. 2001; Walker et al. 2002), and intra-amygdala administration of NMDA receptor agonists, such as D-cycloserine, has been shown to enhance fear extinction (Walker et al. 2002), see chapter by the Ressler author group in this text for further discussion of D-cycloserine. It has also been proposed that glucocorticoid receptors (GR) regulate glutamatergic activity in the amygdala, and GR down-regulation seen after stress exposure could cause anandamide depletion in the (basolateral) amygdala. This could, in turn, lead to poorer engagement of noradrenergic transmission, resulting in deficits in fear extinction (Maren and Holmes 2016).

vmPFC: Decreased activity of vmPFC in PTSD during fear extinction has been similarly reported in some (Milad et al. 2009; Rougemont-Bucking et al. 2011; Shin and Liberzon 2010), but not all studies (Shvil et al. 2014). Rodent studies have shown decreased levels of glutamate and glutamine (the precursor metabolite for neuronal glutamate) (Knox et al. 2010), and loss of N-methyl-d-aspartate (NMDA) and α-amino-3-hydroxy-5-methyl-4-isoxazole propionic acid (AMPA) receptors in the mPFC due to stress exposure (Moench and Wellman 2015). Additionally, upregulation of GR in mPFC has been reported in PTSD (Wen et al. 2015), but to date, no comprehensive analysis of this upregulation in fear extinction deficits in PTSD has been conducted.

Hippocampus: While the role of the hippocampus in mediating context specificity of fear extinction is well-established (Corcoran et al. 2005), the findings on PTSD-related alterations in hippocampal activity during this process are inconsistent. Specifically, while decreased activity of hippocampus during extinction has been reported (Knight et al. 2004), other groups failed to find any PTSD-related hippocampal alterations during this stage (Hoffman et al. 2014; Wicking et al. 2016).

dACC: Similar to the regions discussed above, there are no clear PTSD-related alterations in dACC activity detected during fear conditioning or extinction. While there is a report of increased activity of dACC during these two phases (Rougemont-Bucking et al. 2011), another report found no alterations (Shvil et al. 2014).

2.2 Alterations During Extinction Recall and Fear Renewal

While the literature on extinction retention/recall and fear renewal in PTSD is more limited than on fear conditioning and extinction, it is generally composed of more consistent reports. Specifically, hypoactivation of vmPFC and hippocampus, and hyperactivation of dACC, have been reported during extinction recall in PTSD patients (Milad et al. 2009; Rougemont-Bucking et al. 2011). Increased amygdala activity during extinction recall has also been reported in PTSD (Garfinkel et al. 2014), although not in all studies (Shvil et al. 2014). During fear renewal, Garfinkel et al. (2014) reported lower hippocampal activity in PTSD patients. On a cellular level, Jin and Maren (2015) reported elevated Fos activity in a small population of hippocampal neurons that project to both the mPFC and basolateral complex of the amygdala (BLA) during fear renewal. Such activity is disrupted as a result of stress exposure and is potentially related to the alterations seen on MRI.

2.3 *Altered Contextual Processing*

As the processing of fear extinction, extinction recall, and fear renewal is context specific (Maren et al. 2013), attention should be given to contextual processing (CP) and the underlying neural circuitry that modulate these processes. Effective CP is dependent on the ability to identify contextual representations as "new" or dissimilar and nonoverlapping, when compared to previously encountered context, a process known as pattern separation (PS), and the ability to recognize pattern as "old" or previously encountered, based on partial information, a process known as pattern completion (PC) (Ngo et al. 2021). CP circuitry is therefore dependent on effective and balanced PC and PS functions performed by subfields of hippocampus, and its interconnection with mPFC and thalamic nuclei, such as the nucleus reuniens. Specifically, the hippocampal subfields cornu ammonis 3 (CA3) and dentate gyrus are involved in PS (McHugh et al. 2007), and the CA1 subfield is thought to be implicated in PC (Leutgeb and Leutgeb 2007). PTSD has been associated with CP deficits, and it has been suggested that impairments in PS and PC processes could lead to dominance of fear over safety memories, inappropriate cue-related behavioral responsivity, and hypervigilance during extinction recall and fear renewal (Liberzon and Abelson 2016). In terms of PS and PC balance, this could represent dominance of PC processes identifying multiple situations as previously encountered "dangerous" context. Hypoactivation of hippocampus and vmPFC, and reduced connectivity between these regions, have been reported in PTSD patients during extinction recall and fear renewal, supporting the role of CP deficits in these processes (Wang et al. 2016). While this indirect evidence is compelling, additional work examining PC and PS function and activation of specific subfields of the hippocampus, in PTSD, will be needed to confirm these hypotheses regarding PTSD pathophysiology.

2.4 *Relation to Other Fear-Related Psychopathologies*

Deficits in extinction learning and extinction recall have been implicated in other fear-related psychopathologies, like certain anxiety disorders (e.g., generalized anxiety disorder, panic disorder, and specific phobia) and obsessive-compulsive disorder (OCD). One of the most widely studied neural circuits in these anxiety disorders is the cortico-limbic circuit, which plays a key role in fear processing. It has been proposed that dysregulation of this circuit is implicated in the development and maintenance of anxiety disorders by some (Swartz and Monk 2014). Specifically, amygdala, insula, and dACC hyperactivity might underly heightened fear responses observed in anxiety disorder, whereas vmPFC hypoactivity might be linked to diminished regulation of threat responses in these disorders (Marin et al. 2017, 2020). In concert, a recent study that followed a transdiagnostic dimensional approach across several anxiety disorders found that activation of dACC and amygdala is associated with arousal levels in response to threat, supporting the notion of common neurobiological processes in a number of anxiety-/fear-related conditions, irrespective of the specific diagnosis (Marin et al. 2020). On the other hand, growing literature on hippocampal alterations, and hippocampus-based contextual processing deficits in PTSD, suggesting that the hippocampus plays a more critical role in PTSD symptoms development (Liberzon and Abelson 2016), indicating that there might be both common and distinct processes that contribute to the development of specific phenotypes.

Similarly in OCD, while much of the research is dominated by the cortico-striatal-thalamo-cortical model, Milad and Rauch (2012) proposed a framework that incorporates neural regions associated with fear conditioning and extinction into the understanding of OCD pathophysiology. Specifically, they suggest that dACC and amygdala are linked to the expression and regulation of fear in OCD and that impaired fear extinction may be a factor contributing to the maintenance, but not necessarily the etiology of OCD. Indeed, evidence suggests abnormalities in extinction learning (i.e., elevated conditioned responses) or recall (i.e., poor discrimination between fear and safety signals) in those with OCD (see review by Cooper and Dunsmoor 2021). Though sparse, initial neuroimaging studies have implicated abnormal vmPFC activity in extinction learning and recall deficits in OCD (Apergis-Schoute et al. 2017; Milad et al. 2013).

Taken together, fear-based psychopathologies, including PTSD, anxiety disorders, and OCD, likely share some neural alterations that are evident in fear conditioning and extinction paradigms. This is not surprising, considering the relevance of these paradigms to the excessive fear and anxiety symptoms, in these disorders, as well as the therapeutic efficacy of exposure-based cognitive behavioral therapies, shared by these disorders. Approaches to study neural deficits underlying psychopathological processes, irrespective of specific diagnosis, should thus be considered and are expected to promote the emerging field of precision psychiatry. Lastly, even when similar neural alterations are identified across disorders, the specific cognitive/emotional or behavioral dysfunction in the disorder should be verified, e.g. reported alterations in vmPFC and dACC could also be implicated in impaired decision-making and error-detection in OCD (Cooper and Dunsmoor 2021; Milad et al. 2013).

3 Fear Extinction in Exposure-Based Therapy

Research on fear conditioning and extinction also provides insight into candidate mechanisms underlying treatment approaches for PTSD. Prolonged exposure therapy (PE) is considered the gold-standard psychotherapeutic treatment for PTSD and has demonstrated efficacy compared to other psychological treatments (Cusack et al. 2016; Foa and McLean 2016; Rauch et al. 2012). One component of PE is *in vivo* exposures to trauma reminders that a patient completes between sessions, which target behavioral patterns of PTSD patients that react to various cues as life-threatening and dangerous, and subsequently avoid them. Continued avoidance of feared stimuli negatively reinforces this behavior and can prevent extinction learning from occurring. A second component of PE includes imaginal exposures that further facilitate remediating erroneous and negative cognitions a patient has by revisiting and processing the most distressing traumatic memory within a session. Such exposures help patients to organize the memory of the trauma, distinguish between thinking of and re-experiencing the trauma, and promote the remembrance of the trauma without causing undue distress and anxiety (Foa and McLean 2016).

While within-session exposures and habituation to activated fear appear to aid in extinction learning, Foa and McLean (2016) suggest it is the between-session habituation component of PE that appears to be linked to the consolidation and retention of fear extinction, as well as positive treatment outcomes. As previously mentioned, PTSD patients demonstrate abnormalities in their ability to retain and recall extinction memories (Garfinkel et al. 2014; Greco and Liberzon 2016; Milad et al. 2008) – a deficit that PE appears to specifically target through between-session *in vivo* exposures. Foa and McLean (2016) hypothesize that this aspect of learning and memory is likely governed by prefrontal cortical regions, such as the mPFC. In support, research has shown that greater activation and larger volumes of prefrontal cortical areas are associated with better outcomes after treatment with cognitive behavioral therapy (Milad et al. 2014) Additionally, it has been suggested that the extinction learning prompted through exposure therapy induces plasticity in the mPFC (through increased activation and protein synthesis), which may account for the resolution of inflexible thinking seen in PTSD (Paredes and Morilak 2019). In addition, it was reported that repeated exposure therapy was associated with strengthened connectivity between key regions involved in fear extinction, specifically, between amygdala and hippocampus, mPFC and insula, striatum and hippocampus (Cisler et al. 2014).

Importantly, effectiveness of exposure-based therapy could be limited due to the context specificity of extinction learning, as a CS no-US association learned in the therapy office might not efficiently extend to the CS–US association in other contexts. The use of virtual reality to mimic multiple contexts, combined with exposure therapy, has been examined to further improve PE efficacy (Difede et al. 2007). A second approach under investigation for enhancing extinction-based therapy for PTSD is pharmacological. Specifically, D-cycloserine, a NMDA-receptor agonist, in combination with exposure-based therapies has been studied in several anxiety- and stress-related disorders, including PTSD, with mixed results (de Kleine et al. 2012; Otto et al. 2010). Combining exposure therapy with cannabinoids, which can also facilitate formation of extinction memory (Morena et al. 2018), could be investigated in future. In the same vein, CP circuitry deficit (specifically, hypoactivation of mPFC) could be amenable to glutamatergic transmission modulation (for instance, via ketamine administration) in combination with CP "retraining."

Interestingly, apart from psychotherapy and pharmacological treatment, it was also proposed that direct neuromodulation of mPFC and amygdala may promote extinction learning and therefore represents a potential treatment approach to PTSD (Marin et al. 2014). Specifically, rodent studies have shown that deep brain stimulation (electrical stimulation via implanted electrodes) of the rat analogs of vmPFC and dACC can modulate extinction recall and that such stimulation was associated with improvement in a rat model of PTSD (Rodriguez-Romaguera et al. 2012; Stidd et al. 2013). Further, transcranial magnetic stimulation (non-invasive electromagnetic induction) can be used to target the dorsolateral prefrontal cortex (dlPFC), a region that is coupled with vmPFC and thus may also be involved in fear extinction (Marin et al. 2014). Indeed, a meta-analysis reported improvements in PTSD symptoms with transcranial magnetic stimulation (Karsen et al. 2014). Lastly, a very recent study by Maples-Keller and colleagues showed a link between improved laboratory extinction learning and clinical prolonged exposure treatment outcomes in veteran patients with PTSD receiving intensive outpatient treatment (Maples-Keller et al. 2022).

4 Future Directions

First, as we had argued above, further investigation of both extinction recall and fear renewal is important for understanding of contextual modulation of fear memories. Second, considering a report of altered recall of extinguished fear after trauma exposure (Marusak et al. 2021), it would be important to include both trauma-exposed and trauma-unexposed control groups to dissociate effects of trauma and PTSD on fear-learning processes (Lokshina et al. 2021). Third, as exaggerated avoidance symptoms are prevalent in PTSD, it would be interesting to further study the relationship between altered avoidance and fear learning in this population (Glotzbach et al. 2012; Sripada et al. 2013). Fourth, additional biological targets, such as the contribution of specific neuronal populations (Jin and Maren 2015), could be investigated to better understand the individual differences and inconsistencies that have been reported in the literature (Greco and Liberzon 2016).

Further, an examination of other brain regions to expand the traditional focus on the amygdala-hippocampal-vmPFC "fear network" in fear extinction is also needed. For instance, while functional connectivity was reported to increase from early to late extinction learning across a broad range of brain regions, including default-mode, frontoparietal and ventral-attention networks, such increased connectivity was not detected in PTSD patients (Wen et al. 2021, 2022). Finally, utilization of biologically based computational models could also prove effective, shedding further light on the specific functional contribution of the involved brain regions and subregions, the nature of their interactions, as well as the role of specific neuronal populations (Mattera et al. 2020; Moustafa et al. 2013; Radell et al. 2017). In sum, while this chapter summarized the growing literature on fear extinction in PTSD, it also presented additional research directions that seem promising in providing a more complete understanding of the alterations of fear learning in this population.

References

American Psychiatric Association (2013) Diagnostic and statistical manual of mental disorders, 5th edn. American Psychiatric Association, Washington

Apergis-Schoute AM, Gillan CM, Fineberg NA, Fernandez-Egea E, Sahakian BJ, Robbins TW (2017) Neural basis of impaired safety signaling in obsessive compulsive disorder. Proc Natl Acad Sci U S A 114(12):3216–3221. https://doi.org/10.1073/pnas.1609194114

Bisby MA, Baker KD, Richardson R (2018) Elucidating the mechanisms of fear extinction in developing animals: a special case of NMDA receptor-independent extinction in adolescent rats. Learn Mem 25(4):158–164. https://doi.org/10.1101/lm.047209.117

Briscione MA, Jovanovic T, Norrholm SD (2014) Conditioned fear associated phenotypes as robust, translational indices of trauma-, stressor-, and anxiety-related behaviors. Front Psych 5:88. https://doi.org/10.3389/fpsyt.2014.00088

Cisler JM, Steele JS, Lenow JK, Smitherman S, Everett B, Messias E, Kilts CD (2014) Functional reorganization of neural networks during repeated exposure to the traumatic memory in posttraumatic stress disorder: an exploratory fMRI study. J Psychiatr Res 48(1):47–55. https://doi.org/10.1016/j.jpsychires.2013.09.013

Cooper SE, Dunsmoor JE (2021) Fear conditioning and extinction in obsessive-compulsive disorder: a systematic review. Neurosci Biobehav Rev 129:75–94. https://doi.org/10.1016/j.neubiorev.2021.07.026

Corcoran KA, Desmond TJ, Frey KA, Maren S (2005) Hippocampal inactivation disrupts the acquisition and contextual encoding of fear extinction. J Neurosci 25(39):8978–8987. https://doi.org/10.1523/JNEUROSCI.2246-05.2005

Cusack K, Jonas DE, Forneris CA, Wines C, Sonis J, Middleton JC et al (2016) Psychological treatments for adults with posttraumatic stress disorder: a systematic review and meta-analysis. Clin Psychol Rev 43:128–141. https://doi.org/10.1016/j.cpr.2015.10.003

de Kleine RA, Hendriks GJ, Kusters WJ, Broekman TG, van Minnen A (2012) A randomized placebo-controlled trial of D-cycloserine to enhance exposure therapy for posttraumatic stress disorder. Biol Psychiatry 71(11):962–968. https://doi.org/10.1016/j.biopsych.2012.02.033

Difede J, Cukor J, Jayasinghe N, Patt I, Jedel S, Spielman L et al (2007) Virtual reality exposure therapy for the treatment of posttraumatic stress disorder following September 11, 2001. J Clin Psychiatry 68(11):1639–1647

Fani N, King TZ, Brewster R, Srivastava A, Stevens JS, Glover EM et al (2015) Fear-potentiated startle during extinction is associated with white matter microstructure and functional connectivity. Cortex 64:249–259. https://doi.org/10.1016/j.cortex.2014.11.006

Feng P, Feng T, Chen Z, Lei X (2014) Memory consolidation of fear conditioning: bi-stable amygdala connectivity with dorsal anterior cingulate and medial prefrontal cortex. Soc Cogn Affect Neurosci 9(11):1730–1737. https://doi.org/10.1093/scan/nst170

Foa EB, McLean CP (2016) The efficacy of exposure therapy for anxiety-related disorders and its underlying mechanisms: the case of OCD and PTSD. Annu Rev Clin Psychol 12:1–28. https://doi.org/10.1146/annurev-clinpsy-021815-093533

Garfinkel SN, Abelson JL, King AP, Sripada RK, Wang X, Gaines LM, Liberzon I (2014) Impaired contextual modulation of memories in PTSD: an fMRI and psychophysiological study of extinction retention and fear renewal. J Neurosci 34(40):13435–13443. https://doi.org/10.1523/JNEUROSCI.4287-13.2014

Glotzbach E, Ewald H, Andreatta M, Pauli P, Muhlberger A (2012) Contextual fear conditioning predicts subsequent avoidance behaviour in a virtual reality environment. Cognit Emot 26(7):1256–1272. https://doi.org/10.1080/02699931.2012.656581

Greco JA, Liberzon I (2016) Neuroimaging of fear-associated learning. Neuropsychopharmacology 41(1):320–334. https://doi.org/10.1038/npp.2015.255

Harnett NG, Goodman AM, Knight DC (2020) PTSD-related neuroimaging abnormalities in brain function, structure, and biochemistry. Exp Neurol 330:113331. https://doi.org/10.1016/j.expneurol.2020.113331

Hoffman AN, Lorson NG, Sanabria F, Foster Olive M, Conrad CD (2014) Chronic stress disrupts fear extinction and enhances amygdala and hippocampal Fos expression in an animal model of post-traumatic stress disorder. Neurobiol Learn Mem 112:139–147. https://doi.org/10.1016/j.nlm.2014.01.018

Jin J, Maren S (2015) Fear renewal preferentially activates ventral hippocampal neurons projecting to both amygdala and prefrontal cortex in rats. Sci Rep 5:8388. https://doi.org/10.1038/srep08388

Karsen EF, Watts BV, Holtzheimer PE (2014) Review of the effectiveness of transcranial magnetic stimulation for post-traumatic stress disorder. Brain Stimul 7(2):151–157. https://doi.org/10.1016/j.brs.2013.10.006

Knight DC, Smith CN, Cheng DT, Stein EA, Helmstetter FJ (2004) Amygdala and hippocampal activity during acquisition and extinction of human fear conditioning. Cogn Affect Behav Neurosci 4(3):317–325. https://doi.org/10.3758/CABN.4.3.317

Knox D, Perrine SA, George SA, Galloway MP, Liberzon I (2010) Single prolonged stress decreases glutamate, glutamine, and creatine concentrations in the rat medial prefrontal cortex. Neurosci Lett 480(1):16–20. https://doi.org/10.1016/j.neulet.2010.05.052

Leutgeb S, Leutgeb JK (2007) Pattern separation, pattern completion, and new neuronal codes within a continuous CA3 map. Learn Mem 14(11):745–757. https://doi.org/10.1101/lm.703907

Liberzon I, Abelson JL (2016) Context processing and the neurobiology of post-traumatic stress disorder. Neuron 92(1):14–30. https://doi.org/10.1016/j.neuron.2016.09.039

Liberzon I, Krstov M, Young EA (1997) Stress-restress: effects on ACTH and fast feedback. Psychoneuroendocrinology 22(6):443–453. https://doi.org/10.1016/s0306-4530(97)00044-9

Linnman C, Zeffiro TA, Pitman RK, Milad MR (2011) An fMRI study of unconditioned responses in post-traumatic stress disorder. Biol Mood Anxiety Disord 1(1):8. https://doi.org/10.1186/2045-5380-1-8

Lokshina Y, Sheynin J, Liberzon I (2021) Post-trauma brain: a commentary on functional brain alterations after trauma and implications to posttraumatic stress disorder. Curr Res Psychiatry 1(3):44–47. https://doi.org/10.46439/Psychiatry.1.015

Maples-Keller J, Watkins LE, Nylocks KM, Yasinski C, Coghlan C, Black K et al (2022) Acquisition, extinction, and return of fear in veterans in intensive outpatient prolonged exposure therapy: a fear-potentiated startle study. Behav Res Ther 154:104124. https://doi.org/10.1016/j.brat.2022.104124

Maren S, Holmes A (2016) Stress and fear extinction. Neuropsychopharmacology 41(1):58–79. https://doi.org/10.1038/npp.2015.180

Maren S, Phan KL, Liberzon I (2013) The contextual brain: implications for fear conditioning, extinction and psychopathology. Nat Rev Neurosci 14(6):417–428. https://doi.org/10.1038/nrn3492

Marin MF, Camprodon JA, Dougherty DD, Milad MR (2014) Device-based brain stimulation to augment fear extinction: implications for PTSD treatment and beyond. Depress Anxiety 31(4): 269–278. https://doi.org/10.1002/da.22252

Marin MF, Zsido RG, Song H, Lasko NB, Killgore WDS, Rauch SL et al (2017) Skin conductance responses and neural activations during fear conditioning and extinction recall across anxiety disorders. JAMA Psychiatry 74(6):622–631. https://doi.org/10.1001/jamapsychiatry.2017.0329

Marin MF, Hammoud MZ, Klumpp H, Simon NM, Milad MR (2020) Multimodal categorical and dimensional approaches to understanding threat conditioning and its extinction in individuals with anxiety disorders. JAMA Psychiatry 77(6):618–627. https://doi.org/10.1001/jamapsychiatry.2019.4833

Marusak HA, Hehr A, Bhogal A, Peters C, Iadipaolo A, Rabinak CA (2021) Alterations in fear extinction neural circuitry and fear-related behavior linked to trauma exposure in children. Behav Brain Res 398:112958. https://doi.org/10.1016/j.bbr.2020.112958

Mattera A, Pagani M, Baldassarre G (2020) A computational model integrating multiple phenomena on cued fear conditioning, extinction, and reinstatement. Front Syst Neurosci 14:569108. https://doi.org/10.3389/fnsys.2020.569108

McHugh TJ, Jones MW, Quinn JJ, Balthasar N, Coppari R, Elmquist JK et al (2007) Dentate gyrus NMDA receptors mediate rapid pattern separation in the hippocampal network. Science 317(5834):94–99. https://doi.org/10.1126/science.1140263

Milad MR, Rauch SL (2012) Obsessive-compulsive disorder: beyond segregated cortico-striatal pathways. Trends Cogn Sci 16(1):43–51. https://doi.org/10.1016/j.tics.2011.11.003

Milad MR, Orr SP, Lasko NB, Chang Y, Rauch SL, Pitman RK (2008) Presence and acquired origin of reduced recall for fear extinction in PTSD: results of a twin study. J Psychiatr Res 42(7):515–520. https://doi.org/10.1016/j.jpsychires.2008.01.017

Milad MR, Pitman RK, Ellis CB, Gold AL, Shin LM, Lasko NB et al (2009) Neurobiological basis of failure to recall extinction memory in posttraumatic stress disorder. Biol Psychiatry 66(12): 1075–1082. https://doi.org/10.1016/j.biopsych.2009.06.026

Milad MR, Furtak SC, Greenberg JL, Keshaviah A, Im JJ, Falkenstein MJ et al (2013) Deficits in conditioned fear extinction in obsessive-compulsive disorder and neurobiological changes in the fear circuit. JAMA Psychiatry 70(6):608–618.; quiz 554. https://doi.org/10.1001/jamapsychiatry.2013.914

Milad MR, Rosenbaum BL, Simon NM (2014) Neuroscience of fear extinction: implications for assessment and treatment of fear-based and anxiety related disorders. Behav Res Ther 62:17–23. https://doi.org/10.1016/j.brat.2014.08.006

Moench KM, Wellman CL (2015) Stress-induced alterations in prefrontal dendritic spines: implications for post-traumatic stress disorder. Neurosci Lett 601:41–45. https://doi.org/10.1016/j.neulet.2014.12.035

Morena M, Berardi A, Colucci P, Palmery M, Trezza V, Hill MN, Campolongo P (2018) Enhancing endocannabinoid neurotransmission augments the efficacy of extinction training and ameliorates traumatic stress-induced behavioral alterations in rats. Neuropsychopharmacology 43(6): 1284–1296. https://doi.org/10.1038/npp.2017.305

Moustafa AA, Gilbertson MW, Orr SP, Herzallah MM, Servatius RJ, Myers CE (2013) A model of amygdala-hippocampal-prefrontal interaction in fear conditioning and extinction in animals. Brain Cogn 81(1):29–43. https://doi.org/10.1016/j.bandc.2012.10.005

Ngo CT, Michelmann S, Olson IR, Newcombe NS (2021) Pattern separation and pattern completion: behaviorally separable processes? Mem Cognit 49(1):193–205. https://doi.org/10.3758/s13421-020-01072-y

Otto MW, Tolin DF, Simon NM, Pearlson GD, Basden S, Meunier SA et al (2010) Efficacy of d-cycloserine for enhancing response to cognitive-behavior therapy for panic disorder. Biol Psychiatry 67(4):365–370. https://doi.org/10.1016/j.biopsych.2009.07.036

Paredes D, Morilak DA (2019) A rodent model of exposure therapy: the use of fear extinction as a therapeutic intervention for PTSD. Front Behav Neurosci 13:46. https://doi.org/10.3389/fnbeh.2019.00046

Radell ML, Myers CE, Sheynin J, Moustafa AA (2017) Computational models of post-traumatic stress disorder (PTSD). In: Moustafa AA (ed) Computational models of brain and behavior. Wiley, New York, pp 43–55. https://doi.org/10.1002/9781119159193.ch4

Rauch SA, Eftekhari A, Ruzek JI (2012) Review of exposure therapy: a gold standard for PTSD treatment. J Rehabil Res Dev 49(5):679–687. https://doi.org/10.1682/jrrd.2011.08.0152

Rodrigues SM, Schafe GE, LeDoux JE (2001) Intra-amygdala blockade of the NR2B subunit of the NMDA receptor disrupts the acquisition but not the expression of fear conditioning. J Neurosci 21(17):6889–6896. https://doi.org/10.1523/JNEUROSCI.21-17-06889.2001

Rodriguez-Romaguera J, Do Monte FH, Quirk GJ (2012) Deep brain stimulation of the ventral striatum enhances extinction of conditioned fear. Proc Natl Acad Sci U S A 109(22):8764–8769. https://doi.org/10.1073/pnas.1200782109

Rothbaum BO, Davis M (2003) Applying learning principles to the treatment of post-trauma reactions. Ann N Y Acad Sci 1008:112–121. https://doi.org/10.1196/annals.1301.012

Rougemont-Bucking A, Linnman C, Zeffiro TA, Zeidan MA, Lebron-Milad K, Rodriguez-Romaguera J et al (2011) Altered processing of contextual information during fear extinction in PTSD: an fMRI study. CNS Neurosci Ther 17(4):227–236. https://doi.org/10.1111/j.1755-5949.2010.00152.x

Santini E, Muller RU, Quirk GJ (2001) Consolidation of extinction learning involves transfer from NMDA-independent to NMDA-dependent memory. J Neurosci 21(22):9009–9017. https://doi.org/10.1523/JNEUROSCI.21-22-09009.2001

Senn V, Wolff SB, Herry C, Grenier F, Ehrlich I, Grundemann J et al (2014) Long-range connectivity defines behavioral specificity of amygdala neurons. Neuron 81(2):428–437. https://doi.org/10.1016/j.neuron.2013.11.006

Shin LM, Liberzon I (2010) The neurocircuitry of fear, stress, and anxiety disorders. Neuropsychopharmacology 35(1):169–191. https://doi.org/10.1038/npp.2009.83

Shvil E, Sullivan GM, Schafer S, Markowitz JC, Campeas M, Wager TD et al (2014) Sex differences in extinction recall in posttraumatic stress disorder: a pilot fMRI study. Neurobiol Learn Mem 113:101–108. https://doi.org/10.1016/j.nlm.2014.02.003

Sierra-Mercado D, Padilla-Coreano N, Quirk GJ (2011) Dissociable roles of prelimbic and infralimbic cortices, ventral hippocampus, and basolateral amygdala in the expression and extinction of conditioned fear. Neuropsychopharmacology 36(2):529–538. https://doi.org/10.1038/npp.2010.184

Sotres-Bayon F, Quirk GJ (2010) Prefrontal control of fear: more than just extinction. Curr Opin Neurobiol 20(2):231–235. S0959-4388(10)00027-9 [pii]. https://doi.org/10.1016/j.conb.2010.02.005

Sripada RK, Garfinkel SN, Liberzon I (2013) Avoidant symptoms in PTSD predict fear circuit activation during multimodal fear extinction. Front Hum Neurosci 7:672. https://doi.org/10.3389/fnhum.2013.00672

Stidd DA, Vogelsang K, Krahl SE, Langevin JP, Fellous JM (2013) Amygdala deep brain stimulation is superior to paroxetine treatment in a rat model of posttraumatic stress disorder. Brain Stimul 6(6):837–844. https://doi.org/10.1016/j.brs.2013.05.008

Suarez-Jimenez B, Albajes-Eizagirre A, Lazarov A, Zhu X, Harrison BJ, Radua J et al (2020) Neural signatures of conditioning, extinction learning, and extinction recall in posttraumatic stress disorder: a meta-analysis of functional magnetic resonance imaging studies. Psychol Med 50(9):1442–1451. https://doi.org/10.1017/S0033291719001387

Swartz JR, Monk CS (2014) The role of corticolimbic circuitry in the development of anxiety disorders in children and adolescents. Neurobiol Childhood:133–148. https://doi.org/10.1007/7854_2013_242

VanElzakker MB, Dahlgren MK, Davis FC, Dubois S, Shin LM (2014) From Pavlov to PTSD: the extinction of conditioned fear in rodents, humans, and anxiety disorders. Neurobiol Learn Mem 113:3–18. https://doi.org/10.1016/j.nlm.2013.11.014

Walker DL, Ressler KJ, Lu KT, Davis M (2002) Facilitation of conditioned fear extinction by systemic administration or intra-amygdala infusions of D-cycloserine as assessed with fear-potentiated startle in rats. J Neurosci 22(6):2343–2351. https://doi.org/10.1523/JNEUROSCI.22-06-02343.2002

Wang Q, Jin J, Maren S (2016) Renewal of extinguished fear activates ventral hippocampal neurons projecting to the prelimbic and infralimbic cortices in rats. Neurobiol Learn Mem 134 Pt A:38–43. https://doi.org/10.1016/j.nlm.2016.04.002

Wen L, Han F, Shi Y (2015) Changes in the glucocorticoid receptor and Ca(2)(+)/calreticulin-dependent signalling pathway in the medial prefrontal cortex of rats with post-traumatic stress disorder. J Mol Neurosci 56(1):24–34. https://doi.org/10.1007/s12031-014-0464-7

Wen Z, Chen ZS, Milad MR (2021) Fear extinction learning modulates large-scale brain connectivity. Neuroimage 238:118261. https://doi.org/10.1016/j.neuroimage.2021.118261

Wen Z, Seo J, Pace-Schott EF, Milad MR (2022) Abnormal dynamic functional connectivity during fear extinction learning in PTSD and anxiety disorders. Mol Psychiatry 27(4):2216–2224. https://doi.org/10.1038/s41380-022-01462-5

Wicking M, Steiger F, Nees F, Diener SJ, Grimm O, Ruttorf M et al (2016) Deficient fear extinction memory in posttraumatic stress disorder. Neurobiol Learn Mem 136:116–126. https://doi.org/10.1016/j.nlm.2016.09.016

Zhang X, Kim J, Tonegawa S (2020) Amygdala reward neurons form and store fear extinction memory. Neuron 105(6):1077–1093 e1077. https://doi.org/10.1016/j.neuron.2019.12.025

Zuj DV, Palmer MA, Lommen MJ, Felmingham KL (2016) The centrality of fear extinction in linking risk factors to PTSD: a narrative review. Neurosci Biobehav Rev 69:15–35. https://doi.org/10.1016/j.neubiorev.2016.07.014

Part IV
Therapeutics and Neuromodulations of Fear Extinction

Exposure Therapy and Its Mechanisms

Gabriella E. Hamlett, Edna B. Foa, and Lily A. Brown

Contents

Abstract Exposure therapy is the gold-standard treatment approach for pathological anxiety. This therapeutic approach builds on principles of extinction training from traditional fear conditioning and extinction protocols. In this chapter, we discuss principles of exposure therapy in the clinic and the laboratory experimental results that guide our decisions in the therapy. We discuss emotional processing theory and inhibitory learning principles, with a focus on expectation violation. We conclude with future research directions needed to improve exposure therapy outcomes among patients with anxiety-related disorders.

Keywords Anxiety · Emotional processing theory · Expectation violation · Exposure therapy

G. E. Hamlett, E. B. Foa, and L. A. Brown (✉)
Department of Psychiatry, Center for the Treatment and Study of Anxiety, University of Pennsylvania, Philadelphia, PA, USA
e-mail: lilybr@upenn.edu

Curr Topics Behav Neurosci (2023) 64: 273–288
https://doi.org/10.1007/7854_2023_428
Published Online: 3 August 2023

1 Introduction to Exposure Therapy

Fear conditioning and extinction is well supported as an experimental analog for exposure therapy. In human fear conditioning paradigms, a neutral cue (conditioned stimulus, CS) such as a red square is paired with the presentation of an aversive cue (unconditioned stimulus, US), such as a mild electric shock. The subject eventually expresses a conditioned fear response when the originally neutral cue is present. The subsequent presentation of the CS in the absence of the US leads to fear extinction learning (for review, see Milad and Quirk 2012). Exposure-based treatments for anxiety-related disorders have important parallels with fear extinction paradigms (Gillihan and Foa 2011). We provide further detail on modern fear conditioning and extinction paradigms later in this chapter.

Exposure therapy is an evidence-based treatment for anxiety-related disorders, including obsessive compulsive disorder (OCD), specific phobias, social anxiety disorder, and trauma-and stressor-related disorders (Hans and Hiller 2013; Ougrin 2011; Stewart and Chambless 2009). In general, exposure therapy is associated with medium to large effect-size reductions in anxiety and secondary symptoms, including depression (Ougrin 2011). Exposure therapy is also associated with significant improvements in quality of life, occupational outcomes, social support, and reductions in medical over-utilization (Hofmann 2014). For these reasons, clinical practice guidelines recommend exposure therapy as an evidence-based treatment for anxiety-related disorders (American Psychological Association 2017; Department of Veterans Affairs & Department of Defense 2017).

However, like other cognitive behavior therapies, exposure therapy is not universally efficacious. Approximately 25% of patients are not willing to begin exposure therapy (Garcia-Palacios et al. 2007). In addition, between 10 and 50% of individuals with an anxiety-related disorder who receive exposure therapy do not respond to treatment or do not meet criteria for remission (Brown et al. 2019; Craske 1999; Springer et al. 2018). Dropout from exposure therapy is also a concern, with 15–30% of individuals prematurely terminating therapy (Benbow and Anderson 2019; Eftekhari et al. 2020; Imel et al. 2013; Ong et al. 2016). Further, up to 14% of patients report return of fear following exposure therapy (van Dis et al. 2020). Thus, while exposure therapy is an excellent treatment option for anxiety, it is not always efficacious, well-tolerated, or sustained.

To improve the efficacy, tolerance, and maintenance of exposure therapy, it is essential to understand its core mechanisms of action. By understanding the mechanisms that underlie the change in exposure therapy, we can modify its delivery and tailor in order to improve outcomes. Improved understanding of mechanisms of action for exposure therapy may also provide opportunities to reduce treatment length, which may, in turn, reduce waitlists and improve access to care.

In this chapter, we begin by providing a general background on Fear Conditioning and Extinction. We then describe exposure therapy in the clinic and present two theories that were developed to understand the mechanisms underlying exposure therapy, emotional processing theory (EPT; Foa and Kozak 1986; Foa and McNally

1996) and inhibitory learning theory (ILT; Craske et al. 2012). We provide evidence for each of the theories as well as evidence from human studies that support or refute the mechanisms of action. We conclude with directions for future research to improve our understanding of the mechanisms of action of exposure therapy and extinction learning.

2 Exposure Therapy in the Clinic

Individuals with anxiety-related disorders commonly report vivid fear-conditioning memories. For example, an individual struggling with social anxiety disorder may report an instance of a social humiliation during a public speech. This person may subsequently fear and avoid social situations where they fear rejection or humiliation. As another example, a person with panic disorder my make an association between low-grade panic sensations (e.g., difficulty breathing, heart racing) and a full-blown panic attack. The person may then avoid low-grade panic sensations by not exercising, wearing loose clothes, and turning on a fan. As a final example, a person with posttraumatic stress disorder (PTSD) may form an association between the physical attributes of an assault perpetrator (e.g., associating blonde hair with the assault itself, such that the person may fear and avoid individuals with blonde hair). The goal of exposure therapy is to create opportunities for forming a new association about the relative safety to feared cues or creating new perceptions to compete with existing ones. In the case of PTSD described above, exposure therapy involves coaching an individual to approach objectively safe people, places, or situations that remind the trauma survivor of the trauma. People who have similar attributes to an assault perpetrator are feared and the individual with PTSD may then avoid individuals with similar attributes to the original perpetrator. The goal of exposure therapy is to modify existing associations (Foa and Kozak 1986) or to form a new associative memory that competes with the old association so that individuals who physically resemble the perpetrator do not evoke the expectation of assault (Foa and McNally 1996; Craske et al. 2012).

Exposure therapy utilizes three core presentations of exposures in practice. The first type is in vivo exposure, or real-world, confrontations with fear-associated stimuli or contexts. As such, these in vivo exposure practices involve approaching people, situations, or objects that are safe yet are fear evoking. The second type is imaginal exposure, which involves mentally revisiting a distressing memory of a past event (e.g., a trauma or an extreme social rejection) or exposure to an anxiety-evoking catastrophic image (e.g., an image of being fired, finding out that a family member is deceased, going bankrupt, etc.). The goal of imaginal exposure is to reduce the impact of distressing images or memories that the person is tempted to suppress. Imaginal exposure is then followed by processing aspects of the distressing memory. The third type is interoceptive exposure, or repeated clinical experiences with bodily sensations and cues. Due to the highly physiological nature of this disorder's features, interoceptive exposures are most useful for individuals fearing panic sensations.

2.1 Emotional Processing Theory

Exposure therapy has its origins in classical conditioning and habituation (Groves and Thompson 1970). As a result of failure of classical learning theory to account for several clinical observations, the mechanisms underlying exposure therapy were first introduced by Lang (1977) and Rachman (1980) and then further developed by Foa and Kozak (1986) into emotional processing theory (EPT), a comprehensive conceptualization of the processes involved in exposure therapy. EPT incorporates elements of learning theory (with an emphasis on extinction and habituation, information processing theory, and cognitive behavioral therapy theory [Foa and Kozak 1986]). The original EPT was later updated in several publications (e.g., Foa et al. 2006; Foa and McNally 1996; Gillihan and Foa 2011). The theory posits that anxiety disorders develop and are maintained through the generation of a "fear structure," a memory structure that primes people for escape and avoidance. It contains three types of information: (1) information about the feared stimulus, (2) information about the psychophysiological and psychological responses to the stimulus, and (3) information about the semantic meaning of the stimulus and its responses. EPT proposes that for successful treatment of fear-based disorders, the fear structure must be activated to allow for the incorporation of the new information which is provided through exposure to the feared stimuli and/or responses. According to the original EPT (Foa and Kozak 1986), exposure to a feared but safe situation was thought to lessen or diminish the association between stimuli and/or response elements and the associated meaning of "danger" in the fear structure. This reduction in associative strength was thought to result in a modified structure that eliminates the unrealistic fear. Foa and McNally (1996) proposed an updated model of EPT, in which the new information presented during exposure forms an alternate structure which is incompatible with the existing structure by changing the meaning of the stimuli and/or responses. The original structure and the alternate structure both remain in memory and compete with one another. According to this model, relapse occurs when certain external or internal factors activate the original structure. This organization is consistent with contemporary learning theory underlying the acquisition, extinction, and return of conditioned fear as well as theories of memory reconsolidation (see Myers and Davis 2002).

Emotional processing, conceptualized as a hypothetical construct, was proposed to be the mechanism underlying exposure therapy. As a hypothetical construct, *emotional processing* cannot be measured directly. Therefore, EPT suggested four indicators that can be used as a proxy to determine whether *emotional processing* has occurred. The four indicators are: fear activation, within-session habituation, between-session habituation, and change in negative cognitions (Fig. 1). According to both the original EPT (Foa and Kozak 1986) and the reformulation by Foa and

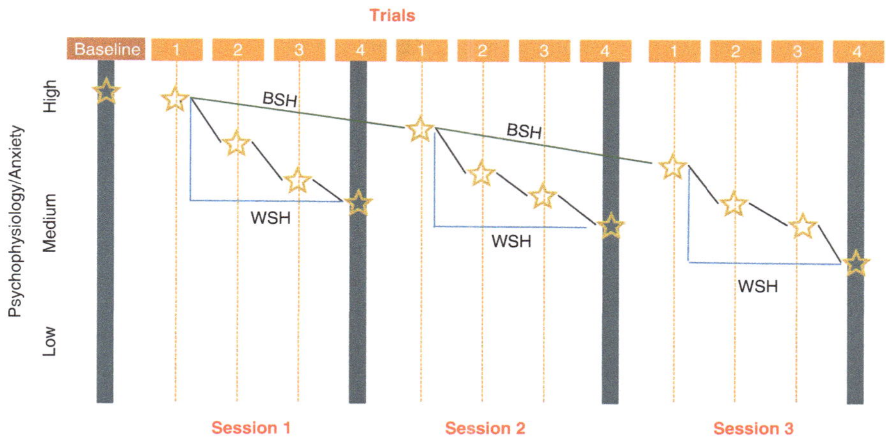

Fig. 1 Depiction of within- and between-session habituation over three sessions of exposure therapy/extinction training. Note: *BSH* between-session habituation, or a reduction in responding from one session to the next (green lines); *WSH* within-session habituation, or a reduction in responding within a single session (blue lines); Stars indicate level of response (which could reflect physiology or self-reported anxiety level or a behavioral indicator of anxiety)

McNally (1996), the modified structure and/or the alternate, non-fear structure are formed via the integration of the change in meaning that is incompatible with the original fear structure. Thus, according to EPT, change in negative cognitions is a core indicator of emotional processing.

Below we operationalize these four indicators of emotional processing and provide empirical evidence from the literature on the extent to which these indicators are associated with exposure therapy success.

Fear Activation EPT conceptualizes that fear activation is characterized by a marked increase in subjective distress that is necessary for effective exposure treatment (Foa and Kozak 1986; Gillihan and Foa 2011). To achieve fear activation, there must be a close match between information presented during exposure and specific salient elements in the fear structure (Foa and Kozak 1986). General information that matches any part of this fear structure may also activate the other parts of the fear structure (Foa and Kozak 1986).

Within-Session Habituation Habituation is defined as decreased responsiveness over repeated stimulation (Groves and Thompson 1970). Within-session habituation is the reduction in fear responding within a given exposure session or series of extinction trials (Foa and Kozak 1986). For example, if a person is completing an exposure where they approach a spider, within-session habituation could either be defined as the reduction in distress from the beginning to the end of the exposure practice, or from the highest distress level ("peak" distress) to the end of the exposure practice.

Between-Session Habituation In contrast, between-session habituation refers to the reduction in fear responding across multiple exposure or extinction training sessions (Foa and Kozak 1986). For example, a person may practice exposure to spiders this week and next week; between-session habituation would occur if their distress at either the beginning of next week's session or at the peak of next week's session is lower than this week's distress.

Change in Negative Cognitions The final indicator of emotional processing conceptualized by EPT is change in negative cognitions (Foa and Kozak 1986; Foa and McLean 2016). According to EPT, negative cognitions significantly contribute to the development and maintenance of anxiety-related disorders. Thus, recovery requires disconfirmation of these negative cognitions and beliefs (e.g., disconfirming an original negative belief that spiders are dangerous and can cause death). Failure to learn and then generalize from repeated evidence that exposure to a feared stimulus does not lead to the feared harm precludes modification of the belief, thus defies disconfirmation (Foa and Kozak 1986).

3 Evidence Supporting the Four Indicators of Emotional Processing from Human Studies

Fear Activation According to EPT and in studies of pharmacologic manipulation of fear responses during extinction, greater activation during exposure therapy is purported to be associated with greater reductions in anxiety (Cain et al. 2004; Morris and Bouton 2007). Several behavioral and neural studies using animal models provide supporting evidence that the level of fear activation during extinction learning predicts fear responding and is associated with better treatment outcomes (Barry et al. 1965; Bouton et al. 2001; Cain et al. 2004; Morris and Bouton 2007). Furthermore, several clinical studies have also supported this indicator, demonstrating an effect of initial fear activation on improved extinction (Riggs et al. 1992; Foa et al. 1995; Kozak et al. 1988; Pitman et al. 1996; van Minnen and Hagenaars 2002). Foa et al. (1995) explored the hypothesis that fear activation during prolonged exposure treatment (as measured by facial fear expression and subjective distress) promotes improvement in treatment outcomes. They found that individuals receiving prolonged exposure therapy (PE) for PTSD who expressed greater initial fear activation had better PTSD treatment outcomes compared to individuals expressing less initial fear activation. Similarly, Pitman et al. (1996) examined emotional processing and treatment outcomes (e.g., fear activation/arousal, within-session habituation, between-session habituation, and self-reported fear, sadness, arousal, anger, and guilt) in veterans with chronic PTSD receiving PE. They found that greater fear activation was related to greater improved PTSD treatment outcomes. In a study examining male combat veterans who received PE treatment, Drs. Wangelin and Tuerk (2015) found that certain physiological markers of PTSD (heart rate and skin conductance) diminished significantly between pre- and post-treatment. Importantly, those with higher heart rate reactivity in response to trauma stimuli pre-PE (i.e., greater pre-treatment fear activation) had better treatment outcomes post-treatment in terms of PTSD symptom reduction.

However, some studies did not find a relationship between fear activation and treatment outcome. Baker et al. (2010) examined fear during in vivo exposure therapy in students scoring high on acrophobia measures. The study did not find an association between fear activation and treatment outcome. Culver et al. (2012) investigated the effects of arousal during public speaking in individuals with fear of public speaking. The addition of excitatory stimuli (meant to increase fear activation) did not predict outcome at 1-week follow-up testing, as measured physiologically, subjectively, and behaviorally. In a meta-analysis testing the predictions of EPT, Rupp et al. (2017) found that initial fear activation was not associated with success of exposure therapy.

Within-Session Habituation A number of studies have tested whether within-session habituation would enhance subsequent fear extinction or treatment outcome, with mixed results (Cain et al. 2003; Kozak et al. 1988; Moody et al. 2006; Rescorla and Durlach 1987; Riley et al. 1995). Some studies examining PE found that within-session reductions in distress were associated with greater PTSD symptom reduction (e.g., Gallagher and Resick 2012; Norton et al. 2011; Rupp et al. 2017). Gallagher and Resick (2012) found that greater within-session decreases in mean subjective units of distress scale (SUDS) ratings for participants in PE were associated with a greater decrease in PTSD symptoms from pre- to post-treatment. Norton et al. (2011) found that among individuals with anxiety disorders, greater within-session habituation during the first exposure session was associated with lower risk of treatment dropout and lower anxiety symptoms at post-treatment. More recently, in a meta-analysis, Rupp et al. (2017) found that within-session habituation was associated with improved outcomes.

However, a number of studies have found that within-session habituation was not related to treatment outcome and therefore did not support the role of within-session habituation in promoting emotional processing (Kozak et al. 1988; Moody et al. 2006; Riley et al. 1995; van Minnen and Foa 2006). van Minnen and Foa (2006) compared the efficacy of 60 min of imaginal exposure versus 30 min of imaginal exposure; if within-session habituation is a predictor of improvement, the 60 min group should outperform the 30 min group. In contrast, there were no outcome differences between the groups; thus, within-session habituation may not be a necessary condition for successful treatment of PTSD. Similarly, Nacasch et al. (2015) explored the relationship of treatment outcome and within-session habituation in adult veterans with chronic PTSD who were randomly assigned to 90-min ($n = 19$) or 60-min ($n = 20$) sessions of PE. They found that participants receiving shorter sessions showed less within-session habituation but had comparable treatment outcomes to those with longer session. Thus, level of within-session SUDS was not related to treatment response. Sripada and Rauch (2015) found that within-session habituation showed an *increase rather than decrease* in outcome. In summary, the degree of within-session habituation is not consistently associated with outcome.

Between-Session Habituation Many studies found that greater between-session habituation was associated with improved treatment outcomes (e.g., Kozak et al. 1988; Rauch et al. 2004). Sripada and Rauch (2015) found that between-session habituation was strongly associated with improved treatment response. Baker et al. (2010) found that between-session habituation predicted short-term change, though it was less predictive of follow-up response rates. Rupp et al.'s (2017) meta-analysis found that between-session habituation was positively associated with treatment outcome.

Change in Negative Cognitions and Beliefs Changes in negative cognitions and beliefs are a critical mechanism of action of exposure therapy (Foa and McLean 2016; Kleim et al. 2007; McLean et al. 2015; Foa and Kozak 1986; Wilhelm et al. 2015; Köcher et al. 2022). Several studies have found that changes in negative cognitions and beliefs are related to exposure therapy treatment outcome across various anxiety disorders (McLean et al. 2015; Wilhelm et al. 2015; Köcher et al. 2022). Wilhelm et al. (2015) found that, among individuals with OCD, changes in maladaptive beliefs were a key mechanism of action and precede behavioral change and symptom severity reduction during exposure therapy. Zalta et al. (2014) found that session-to-session reduction in negative cognitions drove successive reductions in PTSD symptoms and not vice versa. This finding established that changes in negative PTSD-related cognitions (i.e., disconfirmation of negative expectations) are a central mechanism of PE treatment. Change in negative trauma-related cognitions mediated change in PTSD and depressive symptoms whereas change in PTSD and depressive symptoms did not mediate change in negative cognitions (McLean et al. 2015). Another longitudinal intervention study explored the role of positive and negative metacognitive beliefs after exposure-focused treatment among children with primary diagnoses of specific phobia, separation anxiety, or social phobia. Metacognitive beliefs and anxiety symptoms were measured pre- and post-treatment. Köcher et al. (2022) found that reduction in negative metacognitive beliefs (i.e., "If I keep worrying, I may go crazy") was correlated with greater reduction in anxiety symptoms (Köcher et al. 2022).

In summary, most studies supported fear activation, between-session habituation, and negative cognitions as indicators of *emotional processing*. Only few studies supported within-session habituation as an indicator of *emotional processing*. The majority did not.

4 Advances in Fear Conditioning and Extinction

As previously mentioned, fear conditioning and extinction is well supported as an experimental analog for the development of learned fear behaviors and subsequent exposure therapy, allowing researchers to measure the mechanisms underlying fear extinction. Since EPT was conceived, there were several developments in fear conditioning and extinction learning that require consideration. Modern fear-conditioning designs also include a safe conditioned stimulus (CS-), which is presented during fear conditioning but never paired with the US. For instance, a subject may be presented with a blue light that is never paired with a shock (CS-) and a green light that is paired with a shock (CS+) during the same fear-conditioning phase. The inclusion of a CS- and a CS+ allows for nuanced analyses to determine differential psychophysiological, expectancy, and subjective distress profiles based on the anticipation of shock or the expected absence of shock in the presence of the light. Some studies also include multiple CS+ stimuli types (CS+a, a green light, and CS+b, a yellow light) to allow for manipulation of effects following additional learning, as described below. Lastly, it is common to present a partial reinforcement schedule for conditioning to prevent a ceiling effect. Thus, if fear conditioning includes eight presentations of each CS, the CS+ will be paired with the US on six of the trials (75% reinforcement schedule).

In modern extinction learning designs, the CS+ is presented in the absence of the shock and the CS- continues to be presented in the absence of the shock. In some studies that included multiple CS+ types (CS+a and CS+b), only one type of CS+ is presented during extinction learning to determine the impact of generalization of extinction learning. For instance, if a subject is presented with the CS- and only CS +a during extinction learning, future responding to CS+b may provide some information about generalization of extinction learning. In other words, when a subject learns that a green light (CS+a) is no longer paired with a shock, one can evaluate whether the subject generalizes their learning to the yellow light (CS+b) which was never presented in the absence of a shock. Thus, during extinction learning, the subject is likely developing a new inhibitory association (CS → no US) to compete with the original fear association (CS → US; Rescorla and Wagner 1972). The experimental design allows for a determination of psychophysiological reactivity, US expectancy, and self-reported distress to a cue that was never paired with a shock (CS-), a cue that was previously paired with a shock but is no longer paired with a shock (CS+a) and a cue that was previously paired with a shock and which has not been encountered since the initial pairing (CS+b). Studies incorporating this type of experimental modifications are discussed throughout the present text.

5 Inhibitory Learning Theory (ILT) and Its Application to Exposure Therapy

Inhibitory learning theory (ILT) was developed to account for the significant percentage of patients that experience a return of fear following treatment (van Dis et al. 2020) and to update conceptualizations of underlying mechanisms of exposure therapy in an effort to improve therapeutic interventions (Craske et al. 2008). While both ILT and EPT suggest that exposure therapy does not erase or weaken the existing fear-conditioning association (CS → US), according to ILT, the original CS develops two meanings that both remain in memory and compete for retrieval (e.g., Bouton 1993; Bouton and King 1983; Rescorla 1996).

While ILT conceptualizes fear activation as a necessary condition for extinction, it does not necessarily conceptualize it as an indicator of exposure therapy success. Instead, ILT suggests that exposure therapy results in the generation of inhibitory associations (CS → no US) that can be acquired independently of fear levels and that competes with the excitatory association in the presence of the CS (Craske et al. 2008).

Expectation Violation Based, in part, on findings from the Rescorla–Wagner model of learning (1972), inhibitory learning suggests that expectation violation is the core mechanism underlying extinction learning. The Rescorla–Wagner model suggests that learning is the result of a difference between an expected outcome and an actual outcome (plus other variables that factor into the salience and association value of the variables). Indicators of expectation violation include subjective reports of US expectancies as well as variability in self-reported or objectively measured anxiety over trials which may indicate fluctuations in expectancies over time. According to ILT, variability in US expectancy leads to more opportunities to learn and form an inhibitory (CS → no US) association later in extinction training compared to a straightforward reduction in US expectancy over time (as occurs in within or between-session habituation). In a situation in which habituation of US expectancy occurred, an individual would no longer have a US expectancy during later extinction learning trials. That would mean that on those trials, no learning can occur. In contrast, in situations where individuals experience variability in US expectancy during training, they may continue to expect the US during later extinction trials, making ongoing learning possible. For instance, variability in skin conductance response to the CS+ over multiple trials during extinction training indicates that on some trials, the participant was distressed (possibly expecting a US) whereas on others, they were not (possibly not expecting a US) and then again, they later became distressed (US expectancy increased; Brown et al. 2017). Likewise, fear-potentiated startle and other psychophysiological and neural indicators lend themselves to this type of block-by-block analysis.

The persistence of expectations even when presented with contradicting experiences is a significant maintenance factor for fear-based disorders (Craske et al. 2012; Rief et al. 2015). Expectancy violation occurs during exposure when there is a mismatch between one's anticipated outcome ("I will receive the US") and the actual outcome (no US was presented; Rescorla and Wagner 1972). Whether expectation violations result in altered future expectations depends on three coping processes including accommodation, assimilation, and immunization (Gollwitzer et al. 2017; Pinquart et al. 2021). Accommodation refers to how individuals adjust their expectations to fit an unexpected outcome. Assimilation is the mechanism by which individuals avoid expectation-inconsistent outcomes or actively contribute to a higher likelihood of expectation consistent outcomes. Finally, the third coping process, self-immunization is the process through which individuals minimize the potential impact of expectation violating information as a means of maintaining the stability of their self-concepts (Gollwitzer et al. 2017). In accordance with inhibitory learning models, individuals are more likely to change their expectation if there is a greater difference between the disconfirming event and the original expectation; the more salient the disconfirmation is, the more likely the expectation is to change (Craske et al. 2012). Thus, exposure therapists should prioritize practices that allow for more salient contradictory experiences to increase expectation violation (D'Astolfo and Rief 2017).

6 Evidence Supporting Expectation Violation

One indicator of extinction learning is variability in expectancy of an aversive outcome (US). In fear conditioning and extinction studies in a non-clinical human sample, variability in US expectancy during a CS+ was associated with enhanced extinction performance on a test of extinction retention (Brown et al. 2017). In a clinical sample of individuals with contamination fears, Kircanski et al. (2012) randomly assigned participants to one of two groups: blocked and constant exposure or random and variable exposure. Both groups exhibited decreases in subjective fear from pre- to post-treatment and through 2-week follow-up, with no significant differences between groups. However, greater variability in subjective fear over repeated exposure trials predicted lower subjective fear at 2-week follow-up (Kircanski et al. 2012).

Similarly, Culver et al. (2012) found that emotional variability was an indicator of greater expectation violation in public speaking anxiety. Emotional variability was operationalized as the standard deviation of SUDS ratings across exposures and standard deviation of all heart rate scores throughout exposure. Higher emotional variability in fear responding during exposure facilitated long-term fear reduction (Culver et al. 2012). That is, greater emotion variability throughout exposure predicted lower mean-SUDS at time point two. These results also further support hypotheses that habituation during exposure is not necessary for long-term fear reduction (Culver et al. 2012). Furthermore, sustained arousal (defined as higher heart rate at the end of exposure and *less* within-session habituation of distress during exposure) as well as variability in subjective fear responding during exposure were better predictors of long-term anxiety outcomes than habituation of fear across exposure (Culver et al. 2012).

Nearly all ILT recommendations for enhancing exposure therapy ultimately attempt to enhance expectation violation. For instance, some studies have focused on removing safety signals, or cues that offer a discrimination in the perceived likelihood of US and that therefore activate an inhibitory (CS → No US association). According to ILT, the presence of a safety signal would reduce US expectancy during and extinction trial, therefore reducing the amount of learning that can occur on that trial. Among individuals with anxiety disorders, safety signals include the presence of another person, therapists, medications, food, or drink (Barlow and

Cerny 1988; Craske et al. 2008). Indeed, several studies have highlighted that the presence (or perception of the presence) of safety signals impairs extinction learning or exposure therapy outcome (Deacon and Maack 2008; Shin and Newman 2018).

Other studies have focused on strategies to promote deepened extinction by combining multiple feared cues in a single exposure session (typically with one extinguished first, and then presented in combination with a new CS+). According to ILT, presenting an extinguished CS+ (CS+a) with a non-extinguished CS+ (CS+b) should allow for greater US expectancy in the presence of their joint presentation. A common clinical example of this is to have a person listen to an imaginal exposure script while engaging in an in vivo exposure. In humans and animals, deepened extinction trials reduce return of fear to the originally extinguished cue (CS+a) and the newly extinguished cue (CS+b; Culver et al. 2015; Rescorla 2006). This effect might be driven by increased expectation of the US when both cues are presented, which allows for more opportunity for expectation violation. Some studies have failed to show an effect (Lovibond et al. 2000; Vervliet et al. 2007). It is possible that the lack of benefit of deepened extinction in these studies could be attributable to blocking effects (Soltysik et al. 1983). In other words, a cue that already elicits an inhibitory association (CS+a that was already extinguished) may reduce learning about a new cue (CS+b) when CS+b is presented without a US because the individual may attribute the lack of US to the presence of the CS+a (treating it like a conditioned inhibitor).

7 Summary and Directions for Future Research

In summary, exposure therapy is a gold-standard, evidence-based practice for anxiety-related disorders, including PTSD and OCD. Fear conditioning and extinction offers useful experimental homologs for the acquisition of learned fear behaviors and resulting clinical exposure therapy. Models of fear conditioning and extinction learning have informed theories of mechanisms of action for exposure therapy. According to EPT, fear activation, within-session habituation, between-session habituation, and change in negative cognitions may be critical indicators of emotional processing; however, there is mounting evidence suggesting that within-session habituation does not predict exposure outcome and thus it is at best a weak a week indicator. On the other hand, between-sessions habituation and change in negative cognitions have been gaining much support from clinical studies and the latter is now considered the strongest indicator of EPT mechanism of exposure therapy. Some studies have supported the utility of expectation violation, refers also as disconfirmation of negative cognitions, is a mechanism of action for exposure therapy, consistent with both ILT and EPT. Future research is needed to determine whether strategies to directly target these mechanisms of action improve response rates or reduce return of fear for patients with anxiety-related disorders.

References

American Psychiatric Association (2017) Clinical practice guidelines for the treatment of posttraumatic stress disorder (PTSD) in adults. https://doi.org/10.1037/e501872017-001

Baker A, Mystkowski J, Culver N, Yi R, Mortazavi A, Craske MG (2010) Does habituation matter? Emotional processing theory and exposure therapy for acrophobia. Behav Res Ther 48(11): 1139. https://doi.org/10.1016/j.brat.2010.07.009

Barlow DH, Cerny JA (1988) Psychological treatment of panic. Guilford Press

Barry H, Etheredge E, Miller N (1965) Counterconditioning and extinction of fear fail to transfer from amobarbital to nondrug state. Psychopharmacologia 8(2):150–156. https://doi.org/10.1007/BF00404176

Benbow AA, Anderson PL (2019) A meta-analytic examination of attrition in virtual reality exposure therapy for anxiety disorders. J Anxiety Disord 61:18–26. https://doi.org/10.1016/j.janxdis.2018.06.006

Bouton ME (1993) Context, time, and memory retrieval in the interference paradigms of Pavlovian learning. Psychol Bull 114(1):80

Bouton ME, King DA (1983) Contextual control of the extinction of conditioned fear: tests for the associative value of the context. J Exp Psychol Anim Behav Process 9(3):248

Bouton ME, Mineka S, Barlow DH (2001) A modern learning theory perspective on the etiology of panic disorder. Psychol Rev 108(1):4–32. https://doi.org/10.1037/0033-295x.108.1.4

Brown LA, LeBeau RT, Chat KY, Craske MG (2017) Associative learning versus fear habituation as predictors of long-term extinction retention. Cognit Emot. https://www.tandfonline.com/doi/abs/10.1080/02699931.2016.1158695

Brown LA, Clapp JD, Kemp JJ, Yarvis JS, Dondanville KA, Litz BT, Mintz J, Roache JD, Young-McCaughan S, Peterson AL, Foa EB, For the STRONG STAR Consortium (2019) The pattern of symptom change during prolonged exposure therapy and present-centered therapy for PTSD in active duty military personnel. Psychol Med 49(12):1980–1989. https://doi.org/10.1017/S0033291718002714

Cain CK, Blouin AM, Barad M (2003) Temporally massed CS presentations generate more fear extinction than spaced presentations. J Exp Psychol Anim Behav Process 29(4):323. https://doi.org/10.1037/0097-7403.29.4.323

Cain CK, Blouin AM, Barad M (2004) Adrenergic transmission facilitates extinction of conditional fear in mice. Learn Mem 11(2):179–187. https://doi.org/10.1101/lm.71504

Craske MG (1999) Anxiety disorders: psychological approaches to theory and treatment, vol 425. Westview Press, p xxi

Craske M, Kircanski K, Zelikowsky M, Mystkowski J, Chowdhury J, Baker A (2008) Optimizing inhibitory learning during exposure therapy. Behav Res Ther 46(1). https://doi.org/10.1016/j.brat.2007.10.003

Craske MG, Liao B, Brown LA, Vervliet B (2012) Role of inhibition in exposure therapy. J Exp Psychopathol 3(3):322–345. https://doi.org/10.5127/jep.026511

Culver NC, Stoyanova M, Craske MG (2012) Emotional variability and sustained arousal during exposure. J Behav Ther Exp Psychiatry 43(2):787–793. https://doi.org/10.1016/j.jbtep.2011.10.009

Culver NC, Vervliet B, Craske MG (2015) Compound extinction: using the Rescorla–Wagner model to maximize exposure therapy effects for anxiety disorders. Clin Psychol Sci 3(3): 335–348. https://doi.org/10.1177/2167702614542103

D'Astolfo L, Rief W (2017) Learning about expectation violation from prediction error paradigms – a meta-analysis on brain processes following a prediction error. Front Psychol. https://doi.org/10.3389/fpsyg.2017.01253

Deacon B, Maack DJ (2008) The effects of safety behaviors on the fear of contamination: an experimental investigation. Behav Res Ther 46(4):537–547. https://doi.org/10.1016/j.brat.2008.01.010

Department of Veterans Affairs, & Department of Defense (2017) VA/DOD clinical practice guideline for the management of posttraumatic stress disorder and acute stress disorder. Department of Veterans Affairs and Department of Defense, p 200

Eftekhari A, Crowley JJ, Mackintosh MA, Rosen CS (2020) Predicting treatment dropout among veterans receiving prolonged exposure therapy. Psychol Trauma Theory Res Pract Policy 12(4): 405–412. https://doi.org/10.1037/tra0000484

Foa EB, Kozak M (1986) Emotional processing of fear: exposure to corrective information. Psychol Bull 99:20–35. https://doi.org/10.1037/0033-2909.99.1.20

Foa EB, McLean CP (2016) The efficacy of exposure therapy for anxiety-related disorders and its underlying mechanisms: the case of OCD and PTSD. Ann Rev. https://doi.org/10.1146/annurev-clinpsy-021815-093533

Foa EB, McNally R (1996) Mechanisms of change in exposure therapy. In: Current controversies in the anxiety disorders. Guilford Press, pp 329–343

Foa EB, Riggs DS, Massie ED, Yarczower M (1995) The impact of fear activation and anger on the efficacy of exposure treatment for posttraumatic stress disorder. Behav Ther 26(3):487–499

Foa EB, Huppert JD, Cahill SP (2006) Emotional processing theory: an update. In: Rothbaum BO (ed) Pathological anxiety: emotional processing in etiology and treatment. The Guilford Press, pp 3–24

Gallagher MW, Resick PA (2012) Mechanisms of change in cognitive processing therapy and prolonged exposure therapy for PTSD: preliminary evidence for the differential effects of hopelessness and habituation. Cogn Ther Res 36(6):750–755. https://doi.org/10.1007/s10608-011-9423-6

Garcia-Palacios A, Botella C, Hoffman H, Fabregat S (2007) Comparing acceptance and refusal rates of virtual reality exposure vs. in vivo exposure by patients with specific phobias. Cyber Psychol Behav 10(5):722–724. https://doi.org/10.1089/cpb.2007.9962

Gillihan SJ, Foa EB (2011) Fear extinction and emotional processing theory. In: Associative learning and conditioning theory. Oxford University Press. https://oxford.universitypressscholarship.com/view/10.1093/acprof:oso/9780199735969

Gollwitzer M, Thorwart A, Meissner K (2017) Editorial: psychological responses to violations of expectations. Front Psychol 8. https://doi.org/10.3389/fpsyg.2017.02357

Groves PM, Thompson RF (1970) Habituation: a dual-process theory. Psychol Rev 77(5):419. https://doi.org/10.1037/h0029810

Hans E, Hiller W (2013) A meta-analysis of nonrandomized effectiveness studies on outpatient cognitive behavioral therapy for adult anxiety disorders. Clin Psychol Rev 33(8):954–964. https://doi.org/10.1016/j.cpr.2013.07.003

Hofmann SG (2014) Effect of cognitive-behavioral therapy for anxiety disorders on quality of life: a meta-analysis. J Consult Clin Psychol 82(3):375. https://doi.org/10.1037/a0035491

Imel ZE, Laska K, Jakupcak M, Simpson TL (2013) Meta-analysis of dropout in treatments for posttraumatic stress disorder. J Consult Clin Psychol 81(3):394–404. https://doi.org/10.1037/a0031474

Kircanski K, Mortazavi A, Castriotta N, Baker A, Mystkowski J, Yi R, Craske MG (2012) Challenges to the traditional exposure paradigm: variability in exposure therapy for contamination fears. J Behav Ther Exp Psychiatry 43(2). https://doi.org/10.1016/j.jbtep.2011.10.010

Kleim B, Ehlers A, Glucksman E (2007) Early predictors of chronic post-traumatic stress disorder in assault survivors. Psychol Med 37(10). https://doi.org/10.1017/S0033291707001006

Köcher LM, Pflug V, Schneider S, Christiansen H (2022) Do metacognitions of children and adolescents with anxiety disorders change after intensified exposure therapy? Children 9(2):168. https://doi.org/10.3390/children9020168

Kozak MJ, Foa EB, Steketee GS (1988) Process and outcome of exposure treatment with obsessive-compulsives: psychophysiological indicators of emotional processing. Behav Ther 19(2): 157–169. https://doi.org/10.1016/S0005-7894(88)80039-X

Lang PJ (1977) Imagery in therapy: an information processing analysis of fear. Behav Ther 8(5): 862–886. https://doi.org/10.1016/S0005-7894(77)80157-3

Lovibond P, Davis N, O'Flaherty A (2000) Protection from extinction in human fear conditioning. Behav Res Ther 38(10):967–983. https://doi.org/10.1016/s0005-7967(99)00121-7
McLean CP, Yeh R, Rosenfield D, Foa EB (2015) Changes in negative cognitions mediate PTSD symptom reductions during client-centered therapy and prolonged exposure for adolescents. Behav Res Ther 68:64–69. https://doi.org/10.1016/j.brat.2015.03.008
Milad MR, Quirk GJ (2012) Fear extinction as a model for translational neuroscience: ten years of progress. Annu Rev Psychol 63:129–151. https://doi.org/10.1146/annurev.psych.121208.131631
Moody EW, Sunsay C, Bouton ME (2006) Priming and trial spacing in extinction: effects on extinction performance, spontaneous recovery, and reinstatement in appetitive conditioning. Q J Exp Psychol 59:809–829. https://doi.org/10.1080/17470210500299045
Morris RW, Bouton ME (2007) The effect of yohimbine on the extinction of conditioned fear: a role for context. Behav Neurosci 121(3):501–514. https://doi.org/10.1037/0735-7044.121.3.501
Myers KM, Davis M (2002) Behavioral and neural analysis of extinction. Neuron 36(4):567–584. https://doi.org/10.1016/s0896-6273(02)01064-4. PMID: 12441048
Nacasch N, Huppert JD, Su Y-J, Kivity Y, Dinshtein Y, Yeh R, Foa EB (2015) Are 60-minute prolonged exposure sessions with 20-minute imaginal exposure to traumatic memories sufficient to successfully treat PTSD? A randomized noninferiority clinical trial. Behav Ther 46(3): 328–341. https://doi.org/10.1016/j.beth.2014.12.002
Norton P, Hayes-Skelton S, Klenck S (2011) What happens in session does not stay in session: changes within exposures predict subsequent improvement and dropout. J Anxiety Disord 25(5). https://doi.org/10.1016/j.janxdis.2011.02.006
Ong CW, Clyde JW, Bluett EJ, Levin ME, Twohig MP (2016) Dropout rates in exposure with response prevention for obsessive-compulsive disorder: what do the data really say? J Anxiety Disord 40:8–17. https://doi.org/10.1016/j.janxdis.2016.03.006
Ougrin D (2011) Efficacy of exposure versus cognitive therapy in anxiety disorders: systematic review and meta-analysis. BMC Psychiatry 11(1):200. https://doi.org/10.1186/1471-244X-11-200
Pinquart M, Endres D, Teige-Mocigemba S, Panitz C, Schütz A (2021) Why expectations do or do not change after expectation violation: a comparison of seven models. Conscious Cogn 89. https://doi.org/10.1016/j.concog.2021.103086
Pitman RK, Orr SP, Altman B, Longpre RE, Poiré RE, Macklin ML, Michaels MJ, Steketee GS (1996) Emotional processing and outcome of imaginal flooding therapy in Vietnam veterans with chronic posttraumatic stress disorder. Compr Psychiatry 37(6):409–418
Rachman S (1980) Emotional processing. Behav Res Ther 18(1):51–60. https://doi.org/10.1016/0005-7967(80)90069-8
Rauch S, Foa EB, Furr J, Filip J (2004) Imagery vividness and perceived anxious arousal in prolonged exposure treatment for PTSD. J Trauma Stress 17(6). https://doi.org/10.1007/s10960-004-5794-8
Rescorla RA (1996) Preservation of Pavlovian associations through extinction. Q J Exp Psychol: Sect B 49(3):245–258
Rescorla RA (2006) Deepened extinction from compound stimulus presentation. J Exp Psychol Anim Behav Process 32(2):135–144. https://doi.org/10.1037/0097-7403.32.2.135
Rescorla RA, Durlach PJ (1987) The role of context in intertrial interval effects in autoshaping. Q J Exp Psychol Sect B 39(1b):35–48. https://doi.org/10.1080/14640748708402250
Rescorla RA, Wagner A (1972) A theory of Pavlovian conditioning: variations in the effectiveness of reinforcement and nonreinforcement. Curr Res Theory 2. https://www.researchgate.net/publication/233820243_A_theory_of_Pavlovian_conditioning_Variations_in_the_effectiveness_of_reinforcement_and_nonreinforcement
Rief W, Glombiewski J, Gollwitzer M, Schubö A, Schwarting R, Thorwart A (2015) Expectancies as core features of mental disorders. Curr Opin Psychiatry 28. https://doi.org/10.1097/YCO.0000000000000184

Riggs DS, Dancu CV, Gershuny BS, Greenberg D, Foa EB (1992) Anger and post-traumatic stress disorder in female crime victims. J Trauma Stress 5(4):613–625

Riley WT, McCormick MG, Simon EM, Stack K, Pushkin Y, Overstreet MM, Carmona JJ, Magakian C (1995) Effects of alprazolam dose on the induction and habituation processes during behavioral panic induction treatment. J Anxiety Disord 9(3):217–227. https://doi.org/10.1016/0887-6185(95)00003-7

Rupp C, Doebler P, Ehring T, Vossbeck-Elsebusch A (2017) Emotional processing theory put to test: a meta-analysis on the association between process and outcome measures in exposure therapy. Clin Psychol Psychother 24(3). https://doi.org/10.1002/cpp.2039

Shin KE, Newman MG (2018) Using retrieval cues to attenuate return of fear in individuals with public speaking anxiety. Behav Ther 49(2):212–224. https://doi.org/10.1016/j.beth.2017.07.011

Soltysik SS, Wolfe GE, Nicholas T, Wilson WJ, Garcia-Sanchez JL (1983) Blocking of inhibitory conditioning within a serial conditioned stimulus-conditioned inhibitor compound: maintenance of acquired behavior without an unconditioned stimulus. Learn Motiv 14(1):1–29. https://doi.org/10.1016/0023-9690(83)90010-3

Springer KS, Levy HC, Tolin DF (2018) Remission in CBT for adult anxiety disorders: a meta-analysis. Clin Psychol Rev 61:1–8. https://doi.org/10.1016/j.cpr.2018.03.002

Sripada RK, Rauch SAM (2015) Between-session and within-session habituation in prolonged exposure therapy for posttraumatic stress disorder: a hierarchical linear modeling approach. J Anxiety Disord 30:81–87. https://doi.org/10.1016/j.janxdis.2015.01.002

Stewart RE, Chambless DL (2009) Cognitive-behavioral therapy for adult anxiety disorders in clinical practice: a meta-analysis of effectiveness studies. J Consult Clin Psychol 77(4):595–606. https://doi.org/10.1037/a0016032

van Dis EAM, van Veen SC, Hagenaars MA, Batelaan NM, Bockting CLH, van den Heuvel RM, Cuijpers P, Engelhard IM (2020) Long-term outcomes of cognitive behavioral therapy for anxiety-related disorders: a systematic review and meta-analysis. JAMA Psychiatry 77(3):265–273. https://doi.org/10.1001/jamapsychiatry.2019.3986

van Minnen A, Foa EB (2006) The effect of imaginal exposure length on outcome of treatment for PTSD. J Trauma Stress 19(4). https://doi.org/10.1002/jts.20146

van Minnen A, Hagenaars M (2002) Fear activation and habituation patterns as early process predictors of response to prolonged exposure treatment in PTSD. J Trauma Stress 15(5):359–367

Vervliet B, Vansteenwegen D, Hermans D, Eelen P (2007) Concurrent excitors limit the extinction of conditioned fear in humans. Behav Res Ther 45(2):375–383. https://doi.org/10.1016/j.brat.2006.01.009

Wangelin BC, Tuerk PW (2015) Taking the pulse of prolonged exposure therapy: physiological reactivity to trauma imagery as an objective measure of treatment response. Depress Anxiety 32(12):927–934

Wilhelm S, Berman NC, Keshaviah A, Schwartz RA, Steketee GS (2015) Mechanisms of change in cognitive therapy for obsessive compulsive disorder: role of maladaptive beliefs and schemas. Behav Res Ther 65:5. https://doi.org/10.1016/j.brat.2014.12.006

Zalta A, Gillihan S, Fisher A, Mintz J, McLean C, Yehuda R, Foa EB (2014) Change in negative cognitions associated with PTSD predicts symptom reduction in prolonged exposure. J Consult Clin Psychol 82(1). https://doi.org/10.1037/a0034735

Enhancing Fear Extinction: Pharmacological Approaches

Olga Y. Ponomareva, Robert J. Fenster, and Kerry J. Ressler

Contents

Abstract Extinction is the process by which the memory of a learned conditioned association decreases over time and with introduction of new associations. It is a vital part of fear learning, and it is critical to recovery in multiple fear-related disorders, including Specific and Social Phobias, Panic Disorder, Obsessive Compulsive Disorder (OCD), and Posttraumatic Stress Disorder (PTSD). The process of extinction is also the underlying mechanism for recovery in gold-standard therapies for PTSD, including prolonged exposure, cognitive processing therapy, eye movement desensitization and procession, as well as other empirically-based paradigms. Pharmacological modulators of extinction are thus promising targets for treatment of fear-related disorders. We focus here on emerging psychopharmacological treatments to facilitate extinction: D-cycloserine, scopolamine, losartan, ketamine, and 3,4-methylenedioxymethamphetamine. We also provide an overview

---

Olga Y. Ponomareva and Robert J. Fenster contributed equally to this work.

O. Y. Ponomareva, R. J. Fenster, and K. J. Ressler (✉)
McLean Hospital, Harvard Medical School, Belmont, MA, USA
e-mail: kressler@mclean.harvard.edu

Curr Topics Behav Neurosci (2023) 64: 289–306
https://doi.org/10.1007/7854_2023_443
Published Online: 17 August 2023

of recent advances in molecular pathways that show promise as targets for extincion and inhibitory learning, including pathways related to cannabinoid, brain-derived neurotrophic factor, hypothalamic-pituitary-adrenal signaling, and promising work in neurosteroid compounds.

Keywords Allopregnanolone · Angiotensin · BDNF · Cannabinoid · Cortisol · Cycloserine · Extinction · Fear · Ketamine · Learning · Losartan · MDMA · Medication · Mifepristone · Pharmacology · Plasticity · Scopolamine

1 Introduction

Extinction is the process by which the memory of a learned conditioned association decreases over time and with introduction of new associations. It is a vital part of fear learning, and it is critical to recovery in multiple fear-related disorders, including Specific and Social Phobias, Panic Disorder, Obsessive Compulsive Disorder (OCD), and Posttraumatic Stress Disorder (PTSD). The process of extinction is also the underlying mechanism for recovery in gold-standard therapies for PTSD – extinction is a critical process underlying prolonged exposure, cognitive processing therapy, eye movement desensitization and procession, as well as other empirically-based paradigms.

For almost two decades, beginning with the work of D-cycloserine (Ressler et al. 2004), translational researchers have attempted to enhance neural mechanisms of extinction with agents that target specific aspects of the emotional learning and memory process. Importantly, current medication treatments for anxiety- and trauma-related syndromes focus on general symptom reduction without targeting the core pathology related to specific memories, cue- or situation-driven triggers. Trauma-focused therapies, which often focus on the process of extinction, outperform pharmacotherapy for the treatment of PTSD and are more likely to lead to improved sustained remission over time when compared to treatment with medications only (Lee et al. 2016). Thus, if medications could enhance the process of forming new learning, they would effectively serve as "pathology reversing" compounds to disrupt the intractable, negative, ruminative memory traces underlying trauma- and stressor-related disorders.

2 Overview of Psychopharmacological Interventions

Although limited in current use and availability, psychopharmacological treatments can be used to facilitate extinction and inhibitory learning, often as an adjunct to ongoing trauma-focused therapy. We focus here on D-cycloserine, scopolamine, losartan, ketamine, and 3,4-methylenedioxymethamphetamine (MDMA), along with a mention of emerging treatments in the fields of cannabinoid and brain-derived neurotrophic factor (BDNF)-related pathways, antagonists of the hypothalamic–pituitary–adrenal (HPA) pathway, and the neurosteroid allopregnanolone and related compounds. There are a number of other drugs at various stages of development, but space limits our ability to discuss them in detail here (see Singewald et al. for a recent comprehensive review) (Singewald et al. 2015).

3 D-Cycloserine

D-cycloserine is an antibiotic used to treat tuberculosis (Rodrigues et al. 2014). It acts as a superagonist at GluN2C-containing N-methyl-D-aspartate (NMDA) receptors, but as a partial agonist at GluN2A-, GluN2B-, and GluN2D-containing NMDA receptors (Dravid et al. 2010). Initial work from rodent models suggested that D-cycloserine promotes extinction learning (Ledgerwood et al. 2003; Walker et al. 2002). Several human trials were also positive (Ressler et al. 2004; Difede et al. 2014; Inslicht et al. 2021; Zhu et al. 2001; Baker et al. 2017; Rosenfield et al. 2019; Mataix-Cols et al. 2017); however, several studies and reviews have been mixed (Rodrigues et al. 2014). Notably, the limited sample size, sample heterogeneity, and limited number of current studies do not permit drawing a definitive conclusion at this time.

Importantly, D-cycloserine reaches peak concentrations in the cerebral spinal fluid 2 hours after consumption (Baron et al. 1955), which can be slowed by fatty food intake (Zhu et al. 2001). To reach peak effect in the middle of the therapy session, a recent study requested participants to take the medication 90 minutes prior to exposure therapy onset, thus optimizing any beneficial effects of the drug (Difede et al. 2014).

One possible reason for the mixed results seen in clinical trials is that D-cycloserine binds to a range of NMDA receptors, combinations of which play a role in promoting either fear or extinction learning (Baker et al. 2017). It could thus simultaneously promote both extinction learning and reconsolidation of fear memory. Future work targeting drugs specific for GluN2B- and/or GluN2C-containing NMDA receptors may thus be more effective in promoting extinction when optimally timed to reach peak concentration during exposure. Alternatively, combining a general NMDA-dependent plasticity agent, such as D-cycloserine with a more valence-specific compound that preferentially leads to extinction and fear inhibition, may be another way to address the issue of potential increased fear reconsolidation/sensitization. It is clear that additional work on precise targeting of NMDA enhancement to improve extinction vs reconsolidation processes will be needed. Fortunately, the lessons learned over the past two decades of work with D-cycloserine in the augmentation of extinction learning will surely benefit future studies using different targets and approaches.

4 Scopolamine

Scopolamine is an antagonist that blocks muscarinic acetylcholine receptors, used to prevent nausea, vomiting, and motion sickness (Craske et al. 2018). Work in rodents suggests that scopolamine facilitates the retrieval of inhibitory memories and makes extinction learning more context independent (Chang and Liang 2012; Zelikowsky et al. 2013; Douchamps et al. 2013). Multiple clinical studies have demonstrated rapid antidepressant effects of scopolamine (Navarria et al. 2015), however trials in treating anxiety- or trauma-related disorders and in augmenting extinction specifically are relatively new. Two recent clinical trials suggested improvement in extinction in randomized clinical trials (Sewart et al. 2021; Craske et al. 2019). Authors of the latter studies found that scopolamine treatment impaired hippocampal processes and augmented extinction during exposure, with effects that persisted one month later.

While findings at context renewal were limited and further investigation with varying scopolamine dosages will be important, these early translational findings are exciting. Future studies must also consider that scopolamine is commonly used to pharmacologically induce amnesia in animals, and it is known to interfere with the ability of the hippocampal memory system to effectively bind sensory, especially spatial, information to episodic memory traces (Newman et al. 2017).

5 Losartan

Losartan is an antagonist of angiotensin II type 1 (AT_1) receptors, used to treat high blood pressure (Marvar et al. 2014). Multiple lines of evidence suggest individuals with PTSD are more likely to have hypertension and an increased risk for heart attack and stroke (Edmondson 2014), with stress-induced increase in angiotensin II as a major mechanism of elevating blood pressure (Saavedra and Benicky 2007). AT_1 binding also becomes elevated in the hypothalamus in response to stress, which can be countered by angiotensin receptor blockers like losartan (Armando et al. 2007; Bregonzio et al. 2008). Furthermore, there are AT1 receptors within the amygdala that facilitate fear learning (Marvar et al. 2014). Additional work in mice suggests that losartan facilitates extinction learning without an effect on fear learning, blood pressure, or other measures of neuroendocrine stress (Marvar et al. 2014; Hurt et al. 2015).

Clinical studies have likewise suggested that angiotensin receptor blockers may protect against stress-induced brain pathology in patients with vascular disease and hypertension (Amenta et al. 2002; Anderson et al. 2011). A retrospective observational study of traumatized individuals found that individuals on various angiotensin receptor blockers and angiotensin-converting enzyme inhibitors had fewer PTSD symptoms (Khoury et al. 2012). A follow-up replication and extension found a similar result in a larger (>100,000) medical cohort along with identifying supportive genetic evidence for the role of this receptor and pathway in PTSD (Seligowski et al. 2021). However, a recent randomized controlled clinical trial (RCT) found no improvement in PTSD symptoms on losartan vs placebo (Stein et al. 2021). As with many of the other examples of "failure to translate" from preclinical to clinical findings, further work needs to be done to determine if the lack of efficacy in this RCT is due to dosing, timing, inclusion/exclusion factors (e.g., would this target only be efficacious in patients with high AT1 binding?), or if this pathway is truly not useful for targeting extinction in humans.

6 Ketamine

Ketamine is a noncompetitive NMDA receptor antagonist traditionally used as an anesthetic at higher doses and an analgesic at lower doses (Hirota and Lambert 1996; Orser et al. 1997). Ketamine stops downstream signaling initiated by NMDA receptor activation (McGhee et al. 2008), which results in increased aminomethylphosphonic acid (AMPA)-dependent plasticity (Autry et al. 2011). There is also evidence that ketamine can modulate opioid and monoaminergic receptors (Hirota and Lambert 1996). Given its fast antidepressant effects, much work with ketamine has been performed in the field of depression, leading to its FDA approval and current clinical use for treatment-resistant depression (Browne and Lucki 2013) .

Ketamine may also be helpful for facilitating extinction learning and/or reducing fear responding in anxiety- and trauma-related disorders, with several clinical studies underway. A recent study in veterans demonstrated that repeated IV ketamine infusions led to improvement of PTSD symptoms in patients with comorbid major depressive disorder (MDD), however, this study was limited by open-label design and small sample size (Albott et al. 2018). In patients with chronic PTSD, ketamine appears to be well tolerated, with one study suggesting an acute improvement of PTSD symptoms 24 hours after ketamine infusion (Feder et al. 2014), and another study showing modest improvement of symptoms after 2 weeks of ketamine infusions (Feder et al. 2021). A small study in veterans demonstrated the feasibility of using ketamine as an adjunct treatment to prolonged exposure therapy for PTSD, although the efficacy of this treatment remains to be determined (Shiroma et al. 2020).

Additionally, a prospective study of comorbid PTSD and chronic pain suggests that subanesthetic doses of ketamine are comparable to treatment with the non-steroidal anti-inflammatory drug ketorolac in decreasing symptoms of PTSD and chronic pain (Dadabayev et al. 2020); however, clinical relevance of these results also remains to be seen. Further, a recent multicenter trial of veterans and service members failed to show a ketamine dose response on improvement of PTSD symptoms, while demonstrating improvement of depression symptoms at the standard dose (Abdallah et al. 2022). In summary, current studies of ketamine in PTSD demonstrate mixed results and are limited by trial design and subject number, however, several more clinical trials are underway to address some of these limitations.

One consideration is the potential effect duration, given lack of current evidence of any lasting changes from ketamine on PTSD symptoms. Given these mixed findings, and significant concerns that ketamine induces transient psychotic and dissociative states (Krystal et al. 1994), especially concerning for individuals with dissociative subtype of PTSD, more work is needed to examine ketamine's efficacy for augmenting extinction.

7 MDMA

3,4-Methylenedioxy methamphetamine (MDMA) is a monoamine transporter substrate that passes through dopamine, norepinephrine, and serotonin transporters (Verrico et al. 2007). It has been historically used in a psychotherapeutic context before being labeled an illegal drug ("Ecstasy" or "Molly") by the FDA in the 1980s. MDMA improves mood and facilitates feelings of well-being, often making individuals feel more extroverted but also inducing perceptual distortion in vision (Peroutka et al. 1988; Vollenweider et al. 1998).

On a cellular level, MDMA enters presynaptic nerve terminals to inhibit the vesicular monoamine transporter 2 (VMAT2) and activate the trace amine-associated receptor 1 (TAAR1) (Berry et al. 2017). As a result, due to the blocking of VMAT2, presynaptic nerve terminals fill with monoamines that can no longer be loaded into synaptic vesicles. In addition, TAAR1 activation leads monoamine transporters to disseminate monoamines instead of recycling them back into the presynaptic neuron. This floods synapses with monoamines and causes the observed behavioral effects of MDMA. Downstream signaling modulates levels of oxytocin, prolactin, and cortisol in the brain (Wolff et al. 2006; Harris et al. 2002; Dumont et al. 2009; Mas et al. 1999; Thompson et al. 2007). Rodent studies have shown a direct effect of MDMA on extinction processing (Young et al. 2015), while early human studies of MDMA and extinction learning have, in general, not replicated the enhancing effects on extinction seen in rodents (Vizeli et al. 2022; Maples-Keller et al. 2022). Ongoing studies are actively exploring the effect of MDMA in clinical settings.

So far, several trials have been conducted in PTSD subjects with promising preliminary findings. In a randomized controlled pilot study, individuals who received MDMA during two 8-hour experimental psychotherapy sessions on average reported marked decreases in PTSD symptoms up to two months after the sessions (Mithoefer et al. 2011), as well as at long-term follow-up 17–74 months later (Mithoefer et al. 2013). A recent study exploring the role of adjunct MDMA on manualized therapy for PTSD suggested improvement in MDMA group over an inactive placebo; however, this work is limited by subjects largely having correctly identified treatment arm despite double blinding efforts and complex instructions for manualized therapy with uncertain application in clinical practice (Mitchell et al.

2021). Additional work is necessary to further understand the effectiveness of MDMA for PTSD and other disorders, especially given the abuse potential of this substance. Furthermore, the work with MDMA is farthest along of the psychoactive/psychedelic class, and these limitations, in particular the very difficult process to truly blind such studies, have yet to be overcome in demonstrating efficacy.

8 Cannabinoids

Many patients with PTSD report the use of cannabis to self-medicate due to its perceived effects on promoting relaxation and decreasing anxiety (Betthauser et al. 2015). The primary active compound within cannabis, Δ9-tetrahydrocannabinol (THC), is a partial agonist of the CB1 receptor, which allows it to produce wide-ranging effects in the central nervous system. Additionally, cannabis contains hundreds of bioactive compounds that modulate endogenous cannabinoid signaling within the brain (Volkow et al. 2016). Due to restrictive regulatory legislation, research on the use of cannabis to treat PTSD and anxiety disorders has been limited, with the small published literature yielding equivocal results (Black et al. 2019). Because of the heterogeneity of compounds within cannabis and the complexity of the biology of endogenous cannabinoid signaling, development of PTSD therapeutics that affect cannabinoid systems has focused upon molecules with more specific biological activity on endogenous cannabinoid signaling.

Extensive preclinical research implicates endogenous cannabinoid signaling in modulation of fear extinction learning. Anandamide (AEA) and 2-arachidonoylglycerol (2-AG), the two primary endogenous signaling cannabinoids in the brain, modulate activity of circuits within the basolateral amygdala (BLA) and ventromedial prefrontal cortex (vmPFC) in ways that facilitate fear extinction learning (Hill et al. 2018). Knockout of the CB1 receptor in mice leads to deficits in fear extinction learning, and fear extinction training is associated with increased AEA and 2-AG levels within the mouse BLA, but not the PFC (Marsicano et al. 2002). Systemic administration of CB1 receptor antagonists interferes with fear extinction learning (Bowers and Ressler 2015). Infusion of CB1 agonists directly into the rodent amygdala facilitates fear extinction (Lin et al. 2006). Levels of AEA and 2-AG can also be altered by blocking the enzymes responsible for their degradation, fatty acid amide hydrolase (FAAH) and monoacylglycerol lipase (MAGL), respectively. Interestingly, AEA and 2-AG may play opposing roles in fear extinction learning, as both systemic and local infusion of a FAAH inhibitor (increasing AEA levels) enhanced fear extinction learning, whereas systemic and local administration of a MAGL inhibitor (increasing 2-AG) impaired fear extinction learning (Hartley et al. 2016).

Clinical application of cannabinoid-targeting therapeutics for treatment of PTSD and other stress-related disorders remains at early stages but shows promise. A recent small-scale clinical trial of a FAAH inhibitor in healthy adults augmented fear extinction learning (Mayo et al. 2020a). A genetic loss of function in a common polymorphism within the FAAH gene, 385C→A, results in elevated baseline AEA, facilitated fear extinction learning and enhanced extinction recall (Mayo et al. 2020b; Dincheva et al. 2015). There are other ongoing clinical trials of FAAH inhibitors for fear extinction learning (Mayo et al. 2022). However, trials of using THC to stimulate CB1 receptors in PTSD treatment are ongoing but still inconclusive. MAGL inhibition, which raises 2-AG levels, is also being tested clinically; however, no results have yet been reported. In summary, multiple lines of preclinical evidence, naturalistic observations, and limited RCTs to date suggest that there may be important roles for cannabinoid pathways in extinction, however how to best target these pathways to specifically enhance extinction, without potential deleterious effects remains to be determined.

9 BDNF and Growth Factors

Because extinction of threat memories is an active process that requires new learning (Bouton et al. 2021; Maren 2015), modulation of neuronal growth factors that enhance plasticity has been considered as a potential mechanism to influence extinction learning and for the treatment of PTSD. Among the four growth factors expressed in the brain, brain-derived neurotrophic factor (BDNF) has been by far the most extensively studied (Notaras and van den Buuse 2020). BDNF signaling has been shown to be critical both for fear learning and for fear extinction processes, depending on the brain region and cell-types being studied (Bouton et al. 2021). Furthermore, direct agonists of the tyrosine receptor kinase B (TrkB) receptor, such as 7,8-Dihydroxyflavone (7,8-DHF), have been shown to enhance extinction and recover deficits in fear extinction found in stress models in mice (Andero et al. 2011). Additionally, because BDNF is thought to be a downstream final common pathway for the action of traditional selective serotonin reuptake inhibitors (SSRIs) (Zanos et al. 2018; Duman 1998), current treatments for PTSD likely already work in part through modification of BDNF signaling.

For a variety of reasons, including the cell-type specificity of BDNF signaling, challenging pharmacokinetic properties, and the poor blood-brain barrier penetrability of BDNF, BDNF is not a likely direct therapeutic candidate per se; although some TrkB-targeting small molecules have shown promise in preclinical studies (Andero et al. 2011; Stylianakis et al. 2022; Andero and Ressler 2012). Nonetheless, indirect means of increasing BDNF activity, whether through exercise or SSRI use, also may provide potential means of enhancing fear extinction learning.

10 HPA Modulation

The hypothalamic–pituitary–adrenal (HPA) axis is a major player linking stress cues to neurotransmitter/neuropeptide expression in the central nervous system, and hormone synthesis and release in the periphery. This robust mechanism allows for a quick homeostatic response to stress and consists of feedback loops between the hypothalamus, anterior pituitary gland, and the adrenal gland. Stress conditions stimulate the hypothalamus to release corticotropin-releasing hormone (CRH) and vasopressin which, in turn, stimulates release of adrenocorticotropic hormone (ACTH) from the pituitary gland, followed by corticosteroids (cortisol in humans, corticosterone in rodents) from the adrenal gland (Daskalakis et al. 2013). Corticosteroids exert their activity through the mineralocorticoid and glucocorticoid receptors (MR and GR, respectively). While MR are activated under basal conditions, GR are activated under stressful conditions when a large amount of corticosteroids are present. Through a series of feedback loops, action at the GR represses the ongoing stress response, returning the system to homeostasis.

Alterations in the reactivity of the HPA axis and its end product cortisol have been consistently reported in PTSD (Daskalakis et al. 2013; Mason et al. 1986; Ehlert et al. 2001). During times of heightened stress, corticosteroids are elevated in the amygdala. Stress also leads to a transient increase in noradrenaline from the sympathetic nervous system in the same region, which acts with corticosteroids to enhance memory formation, with insufficient elevation of corticosteroids hypothesized as a pathway to over-consolidation of traumatic memories from elevated sympathetic nervous system activation (Yehuda et al. 2015). Modulators of this system, including GR and CRH antagonists, have thus gained interest as potential modulators of fear-related circuits.

Mifepristone, a selective GR antagonist, is currently in use for termination of pregnancy and treatment of hyperglycemia in Cushing's syndrome. Notably, recent RCTs of oral mifepristone during traumatic memory reactivation did not demonstrate any difference from placebo (Wood et al. 2015). A randomized, controlled study of daily oral mifepristone administration in veterans also failed to demonstrate an improvement in symptoms of PTSD; however, results suggested in improvement in verbal learning, suggesting a possible cognitive enhancing effect in chronic mental illness (Golier et al. 2016). Work from animal studies suggests that a GR antagonist RU38486 enhances fear extinction in rats (Dadkhah et al. 2018). Another study found no effect of subcutaneous mifepristone on contextual fear extinction in rats (Ninomiya et al. 2010), while intra-amygdalar injection of mifepristone blocked extinction of conditioned fear (Yang et al. 2006). Thus, timing, route, and duration of drug administration and GR antagonism may play key roles in behavioral response, and further carefully controlled studies are needed.

The known dysregulation of the HPA axis in PTSD, a disorder largely of deficits in extinction, also lead to testing the hypothesis that suppressing or stabilizing cortisol levels through prior dexamethasone treatment might serve to enhance extinction processes. Participants were randomized to receive dexamethasone or placebo prior to fear conditioning and extinction, in a counterbalanced design, and it was found that extinction and discrimination deficits in subjects with PTSD were markedly reversed with dexamethasone (Michopoulos et al. 2017). These data suggest that dexamethasone may serve as a pharmacological agent with which to facilitate fear extinction and discrimination in individuals with PTSD. However, despite prior evidence that dexamethasone stabilization of HPA function enhanced extinction in both preclinical and human laboratory settings, it was not associated with improvements in a virtual reality-based exposure therapy trial for participants with PTSD (Maples-Keller et al. 2019). Thus, it remains unclear whether, as with many other agents discussed in this chapter, the failure to translate prior findings to clinical trials is due to overall lack of effect vs. not targeting the right patient population (e.g., in this case, pre-testing for those who had HPA dysregulation) with the right treatment.

Another potential treatment approach includes antagonists of corticotropin-releasing factor (CRF, same as CRH) receptors. Patients with PTSD have elevated CRH, making CRH modulation a potential target for PTSD medications. A recent study of an oral CRHR1 antagonist (GSK561679) failed to show efficacy in women subjects with PTSD (Dunlop et al. 2017), but further research into specific patient populations is needed.

11 Neurosteroids (Allopregnanolone)

Allopregnanolone (Allo), a metabolite of progesterone, is a neurosteroid hormone that is increased in the brain in response to stress. Allo is a modulator at extrasynaptic GABA-A receptors, leading to an anxiolytic and antidepressant effect, and is reduced in certain populations of patients with PTSD (Pineles et al. 2018). In rodents, decreased Allo in the mPFC, hippocampus, and amygdala is correlated with impaired fear extinction (Pibiri et al. 2008) and increased contextual fear responses. Recent work suggests that Allo in the mouse amygdala leads to emotional/affective regulation through modulation of electrical oscillatory states (Antonoudiou et al. 2022). Another study found that a synthetic analog of Allo (ganaxolone) enhanced fear extinction retention, making Allo a prime target for drug development for extinction modulation (Pinna and Rasmusson 2014).

Currently, Allo is used under the name Brexanolone to treat postpartum depression (Edinoff et al. 2021), however, the effects of Allo in patients with PTSD are only beginning to be elucidated. There is an ongoing clinical study to examine the effects of Allo on extinction consolidation in PTSD (clinicaltrials.gov identifier NCT04468360), as well as an ongoing study of PRAX-114, a potent allosteric regulator of GABA-A, in subjects with PTSD (clinicaltrials.gov identifier NCT05260541). Overall, allopregnanolone and other allosteric regulators of the GABAergic system may prove to be powerful tools for treating stress- and trauma-related disorders, though whether they will enhance fear extinction and related symptoms in patients remains to be seen.

12 Conclusions

In summary, significant advances have been made in recent years to identify molecular pathways underlying extinction learning. Work has also expanded to translate these findings to new pharmacological approaches to enhance the neuroplasticity underlying extinction as well as direct targeting of fear inhibition processes. A few of those pathways have been discussed here, including discoveries that were followed by some level of translational success with D-cycloserine, cannabinoid pathways, MDMA, angiotensin, ketamine, scopolamine, HPA-cortisol pathways, allopregnanolone, and neural plasticity factors such as the BDNF-TrkB pathway.

One early observation, that has yet to be overcome, is that with any reactivation of a fear memory trace, opposing reconsolidation and extinction learning mechanisms are both activated. Thus, generally enhancing plasticity combined with re-exposure to trauma memories may either strengthen the initial memory trace (if reconsolidation prevails) or strengthen fear inhibitory mechanisms (if extinction prevails). We note that many synaptic plasticity mechanisms underlying learning and memory are involved in both reconsolidation and extinction learning – thus general plasticity enhancers (e.g., NMDA activation via D-cycloserine or growth factors via BDNF-TrkB) may not differentiate enhancement of these effects, such that the therapeutic mechanisms will be required to favor extinction over reconsolidation for augmentation. Notably, expansion of our understanding of relevant neural circuits suggests that extinction vs. reconsolidation molecular pathways may differ at some circuit levels, thus targeting these "valence" or pathway-specific components may be particularly powerful for directly enhancing extinction.

Additionally, many of these pathways may be useful to target in some individuals with extinction deficits but not others. Biomarker, genetic, and physiological predictors of which patients should be targeted for which treatment approaches – a precision medicine approach in psychiatry – will likely be needed for optimization and maximizing efficacy.

Overall, there are many relatively specific pharmacological pathways that are now plausible targets for enhancing fear extinction – a key physiological process that has implications for a wide range of stress-, trauma-, and anxiety-related psychiatric disorders. Further understanding the specific nature of extinction vs. reconsolidation will be important for progress, as well as advances in biomarkers for precision medicine guidance of which patients may best respond to which pharmacological approaches for maximal response. The future is bright and targeting extinction may prove to be one of the early translational successes between neuroscience and psychiatry.

Acknowledgments

This work was supported by Veterans Administration (VA) Office of Academic Affairs (OAA) Psychiatric Research/Neurosciences Advanced Fellowship (OYP), NIH awards (P50-MH115874, R01-MH108665), and the Frazier Institute at McLean Hospital (KJR).

Disclosures Dr. Ressler has performed scientific consultation for BioXcel, Bionomics, Acer, Takeda, and Jazz Pharma; serves on Scientific Advisory Boards for Sage and the Brain Research Foundation, and he has received sponsored research support from Takeda, BrainsWay, and Alto Neuroscience. He receives research funding from the NIH.

References

Abdallah CG, Roache JD, Gueorguieva R et al (2022) Dose-related effects of ketamine for antidepressant-resistant symptoms of posttraumatic stress disorder in veterans and active duty military: a double-blind, randomized, placebo-controlled multi-center clinical trial. Neuropsychopharmacology 47(8):1574–1581. https://doi.org/10.1038/s41386-022-01266-9

Albott CS, Lim KO, Forbes MK et al (2018) Efficacy, safety, and durability of repeated ketamine infusions for comorbid posttraumatic stress disorder and treatment-resistant depression. J Clin Psychiatry 79(3):17m11634. https://doi.org/10.4088/JCP.17m11634

Amenta F, Mignini F, Rabbia F, Tomassoni D, Veglio F (2002) Protective effect of anti-hypertensive treatment on cognitive function in essential hypertension. J Neurol Sci 203-204: 147–151. https://doi.org/10.1016/S0022-510X(02)00281-2

Andero R, Ressler KJ (2012) Fear extinction and BDNF: translating animal models of PTSD to the clinic. Genes Brain Behav 11(5):503–512. https://doi.org/10.1111/j.1601-183X.2012.00801.x

Andero R, Heldt SA, Ye K, Liu X, Armario A, Ressler KJ (2011) Effect of 7,8-dihydroxyflavone, a small-molecule TrkB agonist, on emotional learning. Am J Psychiatry 168(2):163–172. https://doi.org/10.1176/appi.ajp.2010.10030326

Anderson C, Teo K, Gao P et al (2011) Renin-angiotensin system blockade and cognitive function in patients at high risk of cardiovascular disease: analysis of data from the ONTARGET and TRANSCEND studies. Lancet Neurol 10(1):43–53. https://doi.org/10.1016/S1474-4422(10)70250-7

Antonoudiou P, Colmers PLW, Walton NL et al (2022) Allopregnanolone mediates affective switching through modulation of oscillatory states in the basolateral amygdala. Biol Psychiatry 91(3):283–293. https://doi.org/10.1016/j.biopsych.2021.07.017

Armando I, Volpi S, Aguilera G, Saavedra JM (2007) Angiotensin II AT1 receptor blockade prevents the hypothalamic corticotropin-releasing factor response to isolation stress. Brain Res 1142:92–99. https://doi.org/10.1016/j.brainres.2007.01.037

Autry AE, Adachi M, Nosyreva E et al (2011) NMDA receptor blockade at rest triggers rapid behavioural antidepressant responses. Nature 475(7354):91–95. https://doi.org/10.1038/nature10130

Baker JF, Cates ME, Luthin DR (2017) D-cycloserine in the treatment of posttraumatic stress disorder. Ment Health Clin 7(2):88–94. https://doi.org/10.9740/mhc.2017.03.088

Baron H, Epstein IG, Mulinos MG, Nair KG (1955) Absorption, distribution, and excretion of cycloserine in man. Antibiot Annu 3:136–140

Berry MD, Gainetdinov RR, Hoener MC, Shahid M (2017) Pharmacology of human trace amine-associated receptors: therapeutic opportunities and challenges. Pharmacol Ther 180:161–180. https://doi.org/10.1016/j.pharmthera.2017.07.002

Betthauser K, Pilz J, Vollmer LE (2015) Use and effects of cannabinoids in military veterans with posttraumatic stress disorder. Am J Health-Syst Pharm 72(15):1279–1284. https://doi.org/10.2146/ajhp140523

Black N, Stockings E, Campbell G et al (2019) Cannabinoids for the treatment of mental disorders and symptoms of mental disorders: a systematic review and meta-analysis. Lancet Psychiatry 6(12):995–1010. https://doi.org/10.1016/S2215-0366(19)30401-8

Bouton ME, Maren S, McNally GP (2021) Behavioral and neurobiological mechanisms of Pavlovian and instrumental extinction learning. Physiol Rev 101(2):611–681. https://doi.org/10.1152/physrev.00016.2020

Bowers ME, Ressler KJ (2015) An overview of translationally informed treatments for posttraumatic stress disorder: animal models of Pavlovian fear conditioning to human clinical trials. Biol Psychiatry 78(5):E15–E27. https://doi.org/10.1016/j.biopsych.2015.06.008

Bregonzio C, Seltzer A, Armando I, Pavel J, Saavedra JM (2008) Angiotensin II AT $_1$ receptor blockade selectively enhances brain AT $_2$ receptor expression, and abolishes the cold-restraint stress-induced increase in tyrosine hydroxylase mRNA in the locus coeruleus of spontaneously hypertensive rats. Stress 11(6):457–466. https://doi.org/10.1080/10253890801892040

Browne CA, Lucki I (2013) Antidepressant effects of ketamine: mechanisms underlying fast-acting novel antidepressants. Front Pharmacol 4:161. https://doi.org/10.3389/fphar.2013.00161

Chang SD, Liang KC (2012) Roles of hippocampal GABAA and muscarinic receptors in consolidation of context memory and context–shock association in contextual fear conditioning: a double dissociation study. Neurobiol Learn Mem 98(1):17–24. https://doi.org/10.1016/j.nlm.2012.04.004

Craske MG, Hermans D, Vervliet B (2018) State-of-the-art and future directions for extinction as a translational model for fear and anxiety. Philos Trans R Soc B Biol Sci 373(1742):20170025. https://doi.org/10.1098/rstb.2017.0025

Craske MG, Fanselow M, Treanor M, Bystritksy A (2019) Cholinergic modulation of exposure disrupts hippocampal processes and augments extinction: proof-of-concept study with social anxiety disorder. Biol Psychiatry 86(9):703–711. https://doi.org/10.1016/j.biopsych.2019.04.012

Dadabayev AR, Joshi SA, Reda MH et al (2020) Low dose ketamine infusion for comorbid posttraumatic stress disorder and chronic pain: a randomized double-blind clinical trial. Chronic Stress (Thousand Oaks) 4:2470547020981670. https://doi.org/10.1177/2470547020981670

Dadkhah M, Abdullahi PR, Rashidy-Pour A, Sameni HR, Vafaei AA (2018) Infralimbic dopamine D2 receptors mediate glucocorticoid-induced facilitation of auditory fear memory extinction in rats. Brain Res 1682:84–92. https://doi.org/10.1016/j.brainres.2018.01.006

Daskalakis NP, Lehrner A, Yehuda R (2013) Endocrine aspects of post-traumatic stress disorder and implications for diagnosis and treatment. Endocrinol Metab Clin North Am 42(3):503–513. https://doi.org/10.1016/j.ecl.2013.05.004

Difede J, Cukor J, Wyka K et al (2014) D-cycloserine augmentation of exposure therapy for post-traumatic stress disorder: a pilot randomized clinical trial. Neuropsychopharmacology 39(5):1052–1058. https://doi.org/10.1038/npp.2013.317

Dincheva I, Drysdale AT, Hartley CA et al (2015) FAAH genetic variation enhances fronto-amygdala function in mouse and human. Nat Commun 6:6395. https://doi.org/10.1038/ncomms7395

Douchamps V, Jeewajee A, Blundell P, Burgess N, Lever C (2013) Evidence for encoding versus retrieval scheduling in the hippocampus by theta phase and acetylcholine. J Neurosci 33(20):8689–8704. https://doi.org/10.1523/JNEUROSCI.4483-12.2013

Dravid SM, Burger PB, Prakash A et al (2010) Structural determinants of D-cycloserine efficacy at the NR1/NR2C NMDA receptors. J Neurosci 30(7):2741–2754. https://doi.org/10.1523/JNEUROSCI.5390-09.2010

Duman RS (1998) Novel therapeutic approaches beyond the serotonin receptor. Biol Psychiatry 44(5):324–335. https://doi.org/10.1016/s0006-3223(98)00031-6

Dumont GJH, Sweep FCGJ, van der Steen R et al (2009) Increased oxytocin concentrations and prosocial feelings in humans after ecstasy (3,4-methylenedioxymethamphetamine) administration. Soc Neurosci 4(4):359–366. https://doi.org/10.1080/17470910802649470

Dunlop BW, Binder EB, Iosifescu D et al (2017) Corticotropin-releasing factor receptor 1 antagonism is ineffective for women with posttraumatic stress disorder. Biol Psychiatry 82(12):866–874. https://doi.org/10.1016/j.biopsych.2017.06.024

Edinoff AN, Odisho AS, Lewis K et al (2021) Brexanolone, a GABAA modulator, in the treatment of postpartum depression in adults: a comprehensive review. Front Psych 12:699740. https://doi.org/10.3389/fpsyt.2021.699740

Edmondson D (2014) An enduring somatic threat model of posttraumatic stress disorder due to acute life-threatening medical events: enduring somatic threat model. Soc Personal Psychol Compass 8(3):118–134. https://doi.org/10.1111/spc3.12089

Ehlert U, Gaab J, Heinrichs M (2001) Psychoneuroendocrinological contributions to the etiology of depression, posttraumatic stress disorder, and stress-related bodily disorders: the role of the hypothalamus-pituitary-adrenal axis. Biol Psychol 57(1–3):141–152. https://doi.org/10.1016/s0301-0511(01)00092-8

Feder A, Parides MK, Murrough JW et al (2014) Efficacy of intravenous ketamine for treatment of chronic posttraumatic stress disorder: a randomized clinical trial. JAMA Psychiatry 71(6):681–688. https://doi.org/10.1001/jamapsychiatry.2014.62

Feder A, Costi S, Rutter SB et al (2021) A randomized controlled trial of repeated ketamine administration for chronic posttraumatic stress disorder. Am J Psychiatry 178(2):193–202. https://doi.org/10.1176/appi.ajp.2020.20050596

Golier JA, Caramanica K, Michaelides AC et al (2016) A randomized, double-blind, placebo-controlled, crossover trial of mifepristone in gulf war veterans with chronic multisymptom illness. Psychoneuroendocrinology 64:22–30. https://doi.org/10.1016/j.psyneuen.2015.11.001

Harris DS, Baggott M, Mendelson JH, Mendelson JE, Jones RT (2002) Subjective and hormonal effects of 3,4-methylenedioxymethamphetamine (MDMA) in humans. Psychopharmacology (Berl) 162(4):396–405. https://doi.org/10.1007/s00213-002-1131-1

Hartley ND, Gunduz-Cinar O, Halladay L, Bukalo O, Holmes A, Patel S (2016) 2-arachidonoylglycerol signaling impairs short-term fear extinction. Transl Psychiatry 6:e749. https://doi.org/10.1038/tp.2016.26

Hill MN, Campolongo P, Yehuda R, Patel S (2018) Integrating endocannabinoid signaling and cannabinoids into the biology and treatment of posttraumatic stress disorder. Neuropsychopharmacology 43(1):80–102. https://doi.org/10.1038/npp.2017.162

Hirota K, Lambert DG (1996) Ketamine: its mechanism(s) of action and unusual clinical uses. Br J Anaesth 77(4):441–444. https://doi.org/10.1093/bja/77.4.441

Hurt RC, Garrett JC, Keifer OP et al (2015) Angiotensin type 1a receptors on corticotropin-releasing factor neurons contribute to the expression of conditioned fear. Genes Brain Behav 14(7):526–533. https://doi.org/10.1111/gbb.12235

Inslicht SS, Niles AN, Metzler TJ et al (2021) Randomized controlled experimental study of hydrocortisone and D-cycloserine effects on fear extinction in PTSD. Neuropsychopharmacology 19. https://doi.org/10.1038/s41386-021-01222-z

Khoury NM, Marvar PJ, Gillespie CF et al (2012) The renin-angiotensin pathway in posttraumatic stress disorder: angiotensin-converting enzyme inhibitors and angiotensin receptor blockers are associated with fewer traumatic stress symptoms. J Clin Psychiatry 73(6):849–855. https://doi.org/10.4088/JCP.11m07316

Krystal JH, Karper LP, Seibyl JP et al (1994) Subanesthetic effects of the noncompetitive NMDA antagonist, ketamine, in humans. Psychotomimetic, perceptual, cognitive, and neuroendocrine responses. Arch Gen Psychiatry 51(3):199–214. https://doi.org/10.1001/archpsyc.1994.03950030035004

Ledgerwood L, Richardson R, Cranney J (2003) Effects of D-cycloserine on extinction of conditioned freezing. Behav Neurosci 117(2):341–349. https://doi.org/10.1037/0735-7044.117.2.341

Lee DJ, Schnitzlein CW, Wolf JP, Vythilingam M, Rasmusson AM, Hoge CW (2016) Psychotherapy versus pharmacotherapy for posttraumatic stress disorder: systemic review and meta-analyses to determine first-line treatments. Depress Anxiety 33(9):792–806. https://doi.org/10.1002/da.22511

Lin HC, Mao SC, Gean PW (2006) Effects of intra-amygdala infusion of CB1 receptor agonists on the reconsolidation of fear-potentiated startle. Learn Mem 13(3):316–321. https://doi.org/10.1101/lm.217006

Maples-Keller JL, Jovanovic T, Dunlop BW et al (2019) When translational neuroscience fails in the clinic: dexamethasone prior to virtual reality exposure therapy increases drop-out rates. J Anxiety Disord 61:89–97. https://doi.org/10.1016/j.janxdis.2018.10.006

Maren S (2015) Out with the old and in with the new: synaptic mechanisms of extinction in the amygdala. Brain Res 1621:231–238. https://doi.org/10.1016/j.brainres.2014.10.010

Marsicano G, Wotjak CT, Azad SC et al (2002) The endogenous cannabinoid system controls extinction of aversive memories. Nature 418(6897):530–534. https://doi.org/10.1038/nature00839

Marvar PJ, Goodman J, Fuchs S, Choi DC, Banerjee S, Ressler KJ (2014) Angiotensin type 1 receptor inhibition enhances the extinction of fear memory. Biol Psychiatry 75(11): 864–872. https://doi.org/10.1016/j.biopsych.2013.08.024

Mas M, Farré M, de la Torre R et al (1999) Cardiovascular and neuroendocrine effects and pharmacokinetics of 3, 4-methylenedioxymethamphetamine in humans. J Pharmacol Exp Ther 290(1):136–145

Mason JW, Giller EL, Kosten TR, Ostroff RB, Podd L (1986) Urinary free-cortisol levels in posttraumatic stress disorder patients. J Nerv Ment Dis 174(3):145–149. https://doi.org/10.1097/00005053-198603000-00003

Mataix-Cols D, Fernández de la Cruz L, Monzani B et al (2017) D-cycloserine augmentation of exposure-based cognitive behavior therapy for anxiety, obsessive-compulsive, and posttraumatic stress disorders: a systematic review and meta-analysis of individual participant data. JAMA Psychiatry 74(5):501–510. https://doi.org/10.1001/jamapsychiatry.2016.3955

Mayo LM, Asratian A, Lindé J et al (2020a) Elevated anandamide, enhanced recall of fear extinction, and attenuated stress responses following inhibition of fatty acid amide hydrolase: a randomized, controlled experimental medicine trial. Biol Psychiatry 87(6):538–547. https://doi.org/10.1016/j.biopsych.2019.07.034

Mayo LM, Asratian A, Lindé J et al (2020b) Protective effects of elevated anandamide on stress and fear-related behaviors: translational evidence from humans and mice. Mol Psychiatry 25(5): 993–1005. https://doi.org/10.1038/s41380-018-0215-1

Mayo LM, Rabinak CA, Hill MN, Heilig M (2022) Targeting the endocannabinoid system in the treatment of posttraumatic stress disorder: a promising case of preclinical-clinical translation? Biol Psychiatry 91(3):262–272. https://doi.org/10.1016/j.biopsych.2021.07.019

McGhee LL, Maani CV, Garza TH, Gaylord KM, Black IH (2008) The correlation between ketamine and posttraumatic stress disorder in burned service members. J Trauma 64(2 Suppl): S195–S198.; Discussion S197–198. https://doi.org/10.1097/TA.0b013e318160ba1d

Michopoulos V, Norrholm SD, Stevens JS et al (2017) Dexamethasone facilitates fear extinction and safety discrimination in PTSD: a placebo-controlled, double-blind study. Psychoneuroendocrinology 83:65–71. https://doi.org/10.1016/j.psyneuen.2017.05.023

Mitchell JM, Bogenschutz M, Lilienstein A et al (2021) MDMA-assisted therapy for severe PTSD: a randomized, double-blind, placebo-controlled phase 3 study. Nat Med 27(6):1025–1033. https://doi.org/10.1038/s41591-021-01336-3

Mithoefer MC, Wagner MT, Mithoefer AT, Jerome L, Doblin R (2011) The safety and efficacy of {+/−}3,4-methylenedioxymethamphetamine-assisted psychotherapy in subjects with chronic, treatment-resistant posttraumatic stress disorder: the first randomized controlled pilot study. J Psychopharmacol Oxf Engl 25(4):439–452. https://doi.org/10.1177/0269881110378371

Mithoefer MC, Wagner MT, Mithoefer AT et al (2013) Durability of improvement in posttraumatic stress disorder symptoms and absence of harmful effects or drug dependency after 3,4-methylenedioxymethamphetamine-assisted psychotherapy: a prospective long-term follow-up study. J Psychopharmacol Oxf Engl 27(1):28–39. https://doi.org/10.1177/0269881112456611

Navarria A, Wohleb ES, Voleti B et al (2015) Rapid antidepressant actions of scopolamine: role of medial prefrontal cortex and M1-subtype muscarinic acetylcholine receptors. Neurobiol Dis 82: 254–261. https://doi.org/10.1016/j.nbd.2015.06.012

Newman EL, Venditto SJC, Climer JR, Petter EA, Gillet SN, Levy S (2017) Precise spike timing dynamics of hippocampal place cell activity sensitive to cholinergic disruption. Hippocampus 27(10):1069–1082. https://doi.org/10.1002/hipo.22753

Ninomiya EM, Martynhak BJ, Zanoveli JM, Correia D, da Cunha C, Andreatini R (2010) Spironolactone and low-dose dexamethasone enhance extinction of contextual fear conditioning. Prog Neuropsychopharmacol Biol Psychiatry 34(7):1229–1235. https://doi.org/10.1016/j.pnpbp.2010.06.025

Notaras M, van den Buuse M (2020) Neurobiology of BDNF in fear memory, sensitivity to stress, and stress-related disorders. Mol Psychiatry 25(10):2251–2274. https://doi.org/10.1038/s41380-019-0639-2

Orser BA, Pennefather PS, MacDonald JF (1997) Multiple mechanisms of ketamine blockade of N-methyl-D-aspartate receptors. Anesthesiology 86(4):903–917. https://doi.org/10.1097/00000542-199704000-00021

Peroutka SJ, Newman H, Harris H (1988) Subjective effects of 3,4-methylenedioxymethamphetamine in recreational users. Neuropsychopharmacology 1(4): 273–277

Pibiri F, Nelson M, Guidotti A, Costa E, Pinna G (2008) Decreased corticolimbic allopregnanolone expression during social isolation enhances contextual fear: a model relevant for posttraumatic stress disorder. Proc Natl Acad Sci 105(14):5567–5572. https://doi.org/10.1073/pnas.0801853105

Pineles SL, Nillni YI, Pinna G et al (2018) PTSD in women is associated with a block in conversion of progesterone to the GABAergic neurosteroids allopregnanolone and pregnanolone measured in plasma. Psychoneuroendocrinology 93:133–141. https://doi.org/10.1016/j.psyneuen.2018.04.024

Pinna G, Rasmusson AM (2014) Ganaxolone improves behavioral deficits in a mouse model of post-traumatic stress disorder. Front Cell Neurosci 8:256. https://doi.org/10.3389/fncel.2014.00256

Ressler KJ, Rothbaum BO, Tannenbaum L et al (2004) Cognitive enhancers as adjuncts to psychotherapy: use of D-cycloserine in phobic individuals to facilitate extinction of fear. Arch Gen Psychiatry 61(11):1136–1144. https://doi.org/10.1001/archpsyc.61.11.1136

Rodrigues H, Figueira I, Lopes A et al (2014) Does D-cycloserine enhance exposure therapy for anxiety disorders in humans? A meta-analysis. PloS One 9(7):e93519. https://doi.org/10.1371/journal.pone.0093519

Rosenfield D, Smits JAJ, Hofmann SG et al (2019) Changes in dosing and dose timing of D-cycloserine explain its apparent declining efficacy for augmenting exposure therapy for anxiety-related disorders: an individual participant-data meta-analysis. J Anxiety Disord 68: 102149. https://doi.org/10.1016/j.janxdis.2019.102149

Saavedra JM, Benicky J (2007) Brain and peripheral angiotensin II play a major role in stress. Stress Amst Neth 10(2):185–193. https://doi.org/10.1080/10253890701350735

Seligowski AV, Duffy LA, Merker JB et al (2021) The renin-angiotensin system in PTSD: a replication and extension. Neuropsychopharmacology 46(4):750–755. https://doi.org/10.1038/s41386-020-00923-1

Sewart A, McGlade A, Treanor M, Fanselow M, Craske M (2021) Pre-treatment hippocampal functioning impacts context renewal for cholinergic modulated exposure therapy. Biol Psychol 165:108167. https://doi.org/10.1016/j.biopsycho.2021.108167

Shiroma PR, Thuras P, Wels J et al (2020) Neurocognitive performance of repeated versus single intravenous subanesthetic ketamine in treatment resistant depression. J Affect Disord 277:470–477. https://doi.org/10.1016/j.jad.2020.08.058

Singewald N, Schmuckermair C, Whittle N, Holmes A, Ressler KJ (2015) Pharmacology of cognitive enhancers for exposure-based therapy of fear, anxiety and trauma-related disorders. Pharmacol Ther 149:150–190. https://doi.org/10.1016/j.pharmthera.2014.12.004

Stein MB, Jain S, Simon NM et al (2021) Randomized, placebo-controlled trial of the angiotensin receptor antagonist losartan for posttraumatic stress disorder. Biol Psychiatry 90(7):473–481. https://doi.org/10.1016/j.biopsych.2021.05.012

Stylianakis AA, Baker KD, Richardson R (2022) Pharmacological enhancement of extinction retention in non-stressed adolescent rats but not those exposed to chronic corticosterone. Front Neurosci 16:822709. https://doi.org/10.3389/fnins.2022.822709

Thompson MR, Callaghan PD, Hunt GE, Cornish JL, McGregor IS (2007) A role for oxytocin and 5-HT(1A) receptors in the prosocial effects of 3,4 methylenedioxymethamphetamine ("ecstasy"). Neuroscience 146(2):509–514. https://doi.org/10.1016/j.neuroscience.2007.02.032

Verrico CD, Miller GM, Madras BK (2007) MDMA (Ecstasy) and human dopamine, norepinephrine, and serotonin transporters: implications for MDMA-induced neurotoxicity and treatment. Psychopharmacology (Berl) 189(4):489–503. https://doi.org/10.1007/s00213-005-0174-5

Volkow ND, Swanson JM, Evins AE et al (2016) Effects of cannabis use on human behavior, including cognition, motivation, and psychosis: a review. JAMA Psychiatry 73(3):292–297. https://doi.org/10.1001/jamapsychiatry.2015.3278

Vollenweider FX, Gamma A, Liechti M, Huber T (1998) Psychological and cardiovascular effects and short-term sequelae of MDMA ("ecstasy") in MDMA-naïve healthy volunteers. Neuropsychopharmacology 19(4):241–251. https://doi.org/10.1016/S0893-133X(98)00013-X

Walker DL, Ressler KJ, Lu KT, Davis M (2002) Facilitation of conditioned fear extinction by systemic administration or intra-amygdala infusions of D-cycloserine as assessed with fear-potentiated startle in rats. J Neurosci 22(6):2343–2351

Wolff K, Tsapakis EM, Winstock AR et al (2006) Vasopressin and oxytocin secretion in response to the consumption of ecstasy in a clubbing population. J Psychopharmacol Oxf Engl 20(3): 400–410. https://doi.org/10.1177/0269881106061514

Wood NE, Rosasco ML, Suris AM et al (2015) Pharmacological blockade of memory reconsolidation in posttraumatic stress disorder: three negative psychophysiological studies. Psychiatry Res 225(1–2):31–39. https://doi.org/10.1016/j.psychres.2014.09.005

Yang YL, Chao PK, Lu KT (2006) Systemic and intra-amygdala administration of glucocorticoid agonist and antagonist modulate extinction of conditioned fear. Neuropsychopharmacology 31(5):912–924. https://doi.org/10.1038/sj.npp.1300899

Yehuda R, Bierer LM, Pratchett LC et al (2015) Cortisol augmentation of a psychological treatment for warfighters with posttraumatic stress disorder: randomized trial showing improved treatment retention and outcome. Psychoneuroendocrinology 51:589–597. https://doi.org/10.1016/j.psyneuen.2014.08.004

Young MB, Andero R, Ressler KJ, Howell LL (2015) 3,4-Methylenedioxymethamphetamine facilitates fear extinction learning. Transl Psychiatry 5:e634. https://doi.org/10.1038/tp.2015.138

Zanos P, Thompson SM, Duman RS, Zarate CA, Gould TD (2018) Convergent mechanisms underlying rapid antidepressant action. CNS Drugs 32(3):197–227. https://doi.org/10.1007/s40263-018-0492-x

Zelikowsky M, Hast TA, Bennett RZ et al (2013) Cholinergic blockade frees fear extinction from its contextual dependency. Biol Psychiatry 73(4):345–352. https://doi.org/10.1016/j.biopsych.2012.08.006

Zhu M, Nix DE, Adam RD, Childs JM, Peloquin CA (2001) Pharmacokinetics of cycloserine under fasting conditions and with high-fat meal, orange juice, and antacids. Pharmacotherapy 21(8): 891–897. https://doi.org/10.1592/phco.21.11.891.34524

Reconsolidation and Fear Extinction: An Update

Marissa Raskin and Marie-H. Monfils

Contents

Abstract Fear memories can be updated behaviorally by delivering extinction trials during the reconsolidation window, which results in a persistent attenuation of fear memories (Monfils et al., Science 324:951–955, 2009). This safe and non-invasive paradigm, termed retrieval-extinction (or post-retrieval extinction), has also been found to be successful at preventing the return of fear in healthy fear conditioned humans (Schiller et al., Nature 463:49–53, 2010), and in the time since its discovery, there has been an explosion of research on the use of retrieval-extinction in fear memories in humans and other animals, some of which have found a long-term

M. Raskin
Institute for Neuroscience, University of Texas at Austin, Austin, TX, USA

M.-H. Monfils (✉)
Department of Psychology, University of Texas at Austin, Austin, TX, USA
e-mail: marie.monfils@utexas.edu

Curr Topics Behav Neurosci (2023) 64: 307–334
https://doi.org/10.1007/7854_2023_438
Published Online: 11 August 2023

reduction in conditioned responding, and some who have not. These discrepant findings have raised concerns as to whether retrieval-extinction really results in updating of the original fear memory, or if it simply enhances extinction. We will first review the progress made on elucidating the cellular mechanisms underlying the fear attenuating effects of retrieval-extinction and how they differ from traditional extinction. Special attention will be paid to the molecular events necessary for retrieval-extinction to successfully occur and how these reconsolidated memories are represented in the brain. Next, we will examine the parameters that determine whether or not a memory will be updated via extinction during the reconsolidation window (also known as boundary conditions). These boundary conditions will also be discussed as possible explanations for discrepant findings of the retrieval-extinction effect. Then we will examine the factors that can determine whether an individual's fears will successfully be attenuated by retrieval-extinction. These individual differences include genetics, age, and psychopathology. Finally, we will discuss recent attempts to bring the retrieval-extinction paradigm from the bench to the bedside for the behavioral treatment of anxiety and trauma disorders.

Keywords Extinction · Fear · Memory · Reconsolidation · Retrieval · Retrieval-extinction

1 Introduction

During extinction learning, fear cues are repeatedly presented without an aversive outcome so that a new, safe association with the cue is learned. However, this does not modify the original memory but rather creates a second, competing memory that often loses out to the original for expression, often resulting in the return of fear (Craske and Mystkowski 2007; Myers and Davis 2002). Approaches which modify the original memory should generally result in persistent fear attenuation. When retrieved, memories are thought to be rendered temporarily labile before restabilizing in a process known as reconsolidation (Misanin et al. 1968; Sara 2000). Applying pharmacological agents that interfere with reconsolidation mechanisms during an opportunistic window can persistently modify memories (Bustos et al. 2006; Dębiec and Ledoux 2004; Nader et al. 2000). Reconsolidation-based approaches have garnered much interest as potential treatments for the maladaptive fear memories that underlie post-traumatic stress disorder, phobias, and anxiety; however, most amnesic agents are harmful to humans and those that are not, such as the beta-adrenergic antagonist propranolol, have only had limited success (Bos et al. 2014; Schroyens et al. 2017).

Fear memories can be updated behaviorally by delivering extinction trials during the reconsolidation window, which results in a persistent attenuation of fear memories (Monfils et al. 2009). This safe and non-invasive paradigm, termed retrieval-extinction (or post-retrieval extinction), has also been found to be successful at preventing the return of fear in healthy fear conditioned humans (Schiller et al. 2010), and in the time since its discovery, there has been an explosion of research on the use of retrieval-extinction in fear memories in humans and other animals, some of which have found a long-term reduction in conditioned responding, and some who have not. For a comprehensive review current at time of publication, see Monfils and Holmes (2018) and Table 1 for relevant studies. These discrepant findings have raised concerns as to whether retrieval-extinction really results in updating of the original fear memory, or if it simply enhances extinction.

We will first review the progress made on elucidating the cellular mechanisms underlying the fear attenuating effects of retrieval-extinction and how they differ from traditional extinction. Special attention will be paid to the molecular events necessary for retrieval-extinction to successfully occur and how these reconsolidated memories are represented in the brain. Next, we will examine the parameters that determine whether or not a memory will be updated via extinction during the reconsolidation window (also known as boundary conditions). These boundary conditions will also be discussed as possible explanations for discrepant findings of the retrieval-extinction effect. Then we will examine the factors that can determine whether an individual's fears will successfully be attenuated by retrieval-extinction. These individual differences include genetics, age, and psychopathology. Finally, we will discuss recent attempts to bring the retrieval-extinction paradigm from the bench to the bedside for the behavioral treatment of anxiety and trauma disorders.

2 Mechanisms

Though reconsolidation and extinction are both induced through retrieval of the CS after conditioning, reconsolidation-based interference does not simply represent a brief extinction session. The length of retrieval determines which process is initiated, as denoted by distinct cellular signatures. Shorter retrieval intervals induce reconsolidation which can be blocked pharmacologically through with protein synthesis inhibitors, antagonism of N-methyl-D-aspartate (NMDA) or β-adrenergic receptors, or GABA-A receptor modulators, and longer retrieval intervals induce extinction and can be blocked with antagonism of cannabinoid receptor 1 (CB1) or L-type voltage-gated calcium channel (LVGCC) and enhanced with NMDAR agonism (Ben Mamou et al. 2006; Bustos et al. 2006; Dębiec and Ledoux 2004; Lee et al. 2006; Suzuki et al. 2004). More recent data have further elucidated the distinct NMDAR subtypes necessary for each phase of memory reconsolidation, showing that GluN2B-containing NMDARs are necessary for destabilization and GluN2A-NMDARs are necessary for restabilization (Milton et al. 2013). They also found that AMPA receptors are necessary for memory retrieval, but not destabilization. These unique biochemical states can be challenged in order to elucidate which process dominates during a certain task or point in time. Using this approach, it has been determined that as the duration of CS retrieval increases, there are two null

Table 1 Behavioral studies on retrieval-extinction

Author, year	Species	Retrieval-extinction better at persistently attenuating fear than standard extinction?
Agren et al. (2012a)	Human	Yes
An et al. (2018)	Rat	Yes
Auchter et al. (2017a)	Rat	Yes
Björkstrand et al. (2015)	Human	Yes
Björkstrand et al. (2016)	Human	Yes
Björkstrand et al. (2017)	Human	Yes
Chalkia et al. (2020)	Human	No
Chan et al. (2010)	Rat	No
Chen et al. (2021a)	Human	Yes
Clem and Huganir (2010)	Mouse	Yes
Costanzi et al. (2011)	Mouse	No
Fernandez-Rey et al. (2018)	Human	Yes
Flavell et al. (2011)	Rat	Yes
Fricchione et al. (2016)	Human	No
Golkar et al. (2012)	Human	No
Goode et al. (2017)	Rat	No
Ishii et al. (2012)	Mouse	No
Ishii et al. (2015)	Mouse	No
Johnson and Casey (2015)	Human	Yes
Kindt and Soeter (2013)	Human	No
Klucken et al. (2016)	Human	No
Kredlow et al. (2018)	Human	No
Lancaster et al. (2020)	Human	Yes
H. J. Lee et al. (2016)	Rat	Yes
Luyten and Beckers (2017)	Rat	No
Monti et al. (2017)	Rat	Yes
Olshavsky et al. (2013)	Rat	Yes
Oyarzún et al. (2012)	Human	Yes
Piñeyro et al. (2014)	Mouse	Yes
Ponnusamy et al. (2016)	Rat	Yes
Rao-Ruiz et al. (2011)	Mouse	Yes
Schiller et al. (2010)	Human	Yes

(continued)

Table 1 (continued)

Author, year	Species	Retrieval-extinction better at persistently attenuating fear than standard extinction?
Shumake and Monfils (2015)	Rat	Yes
Soeter and Kindt (2011)	Human	No
Stafford et al. (2013)	Mouse	No
Steinfurth et al. (2014)	Human	Yes
Tedesco et al. (2014)	Rat	Yes
Telch et al. (2017)	Human	Yes
Thompson and Lipp (2017)	Human	Yes
Vermes et al. (2020)	Human	Yes
Warren et al. (2014)	Human	Yes
Zimmermann and Bach (2020)	Human	No

points during which neither reconsolidation nor extinction mechanisms are engaged: with brief retrievals before reconsolidation is engaged, and with intermediate retrievals during the transition between reconsolidation and extinction (Merlo et al. 2014). In addition, destabilization is necessary for memories to be updated using retrieval-extinction (Piñeyro et al. 2014). This tells us that in order for retrieval-extinction procedures to be successful, the initial retrieval must be of appropriate length to sufficiently recall a memory trace and engage reconsolidation mechanisms. Then, an interval of sufficient duration must be present to ensure the initial retrieval be distinct from those that follow, such that the retrieval does not simply engage extinction mechanisms. And finally, this isolated retrieval must be followed by an extinction session before the memory is restabilized (Fig. 1). Should any of these events fail to occur, the retrieval-extinction procedure will not result in a persistent attenuation of conditioned fear; however, since many of the necessary mechanistic events cannot be directly observed and can only be inferred through pharmacological challenge, it makes it difficult to determine whether unsuccessful studies of the retrieval-extinction effect are due to boundary conditions, failure to engage all of the necessary processes, or a true failure to replicate.

Nevertheless, a number of studies have provided evidence that retrieval- extinction can, in fact, be mechanistically distinct from extinction and results in weakening of the original memory. Key differences are visualized in Fig. 2. First, in the lateral amygdala (LA) the glutamate receptor 1 subunit (GluR1) of the AMPA receptor becomes phosphorylated after a single CS retrieval but is dephosphorylated by a second CS retrieval 1 h later (Monfils et al. 2009), leading to a decrease in AMPAR-mediated transmission in the LA, indicating that it does in fact lead to depotentiation the CS-US association (Clem and Huganir 2010). This is caused by the removal of calcium-permeable AMPA receptors, which in turn requires phosphorylation of GluR1 (Clem and Huganir 2010; Hong et al. 2013). For contextual fear memories,

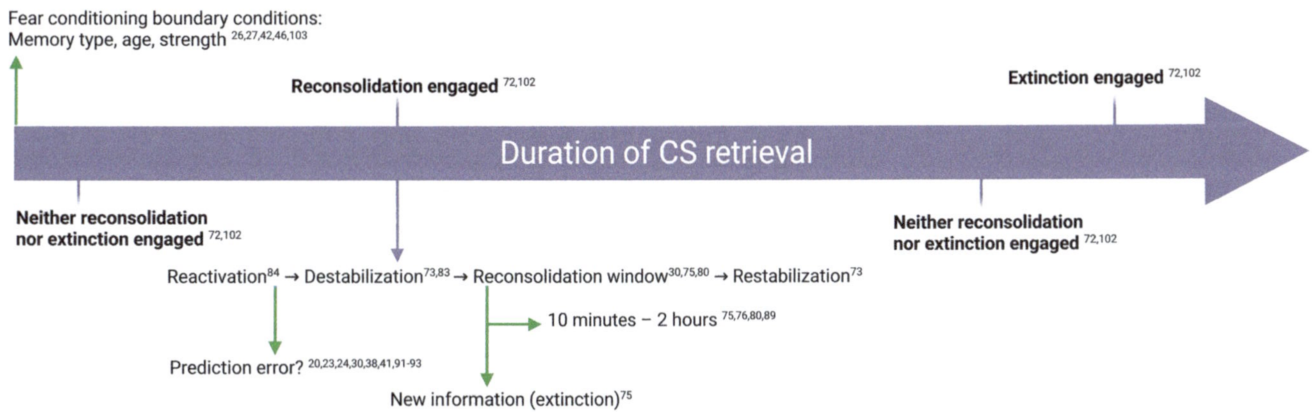

Fig. 1 Processes and boundary conditions of retrieval-extinction. There are boundary conditions on the types of fear memories that are susceptible to modification through reconsolidation processes such as memory type, age, and strength (Clem and Huganir 2010; Costanzi et al. 2011; Gräff et al. 2014; Ishii et al. 2012; Steinfurth et al. 2014). The duration of the CS retrieval determines whether reconsolidation, extinction, or neither process will be engaged (Merlo et al. 2014; Suzuki et al. 2004) A CS retrieval of the appropriate duration is necessary for reactivation of the memory, which allows it to become destabilized (Milton et al. 2013; Piñeyro et al. 2014). It remains an open question whether prediction error is a necessary condition for destabilization (Cahill et al. 2019; Chen et al. 2018, 2021a; Díaz-Mataix et al. 2013; Gershman et al. 2013, 2017; Sevenster et al. 2012, 2013, 2014) Before the memory is reconsolidated, there is a window of time (10 min–2 h (Monfils et al. 2009; Monfils and Holmes 2018; Nader et al. 2000; Schiller et al. 2013)) during which new information such as extinction can be incorporated (Monfils et al. 2009). Created with BioRender.com

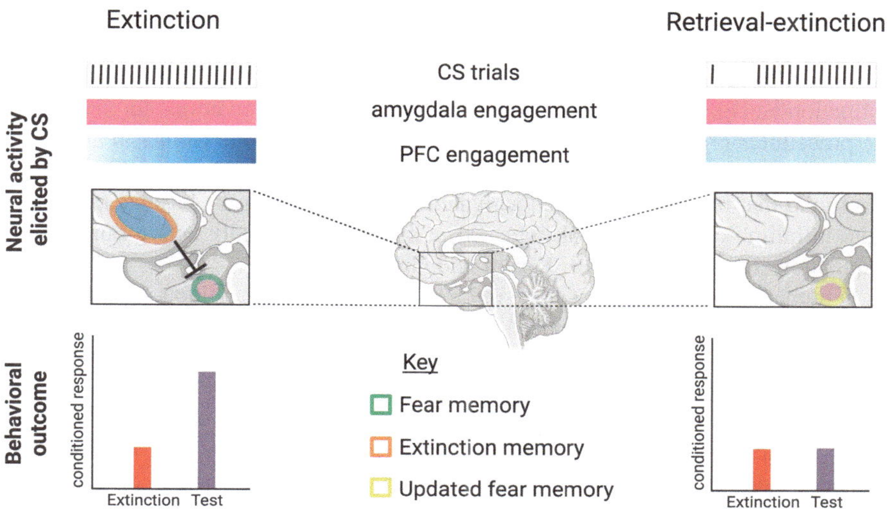

Fig. 2 Initial CS presentations during extinction and retrieval-extinction both activate the prefrontal cortex (PFC) and amygdala (Cahill and Milton 2019). However, as the procedures progress, their patterns of neural activation diverge. Extinction continues to engage the PFC while retrieval-extinction does not (Cahill and Milton 2019). While both paradigms continue to activate the amygdala, retrieval-extinction activates the same cells that were originally active during fear acquisition (Khalaf et al. 2018; Khalaf and Gräff 2019). Upon test, extinction relies upon the PFC to suppress the fear memory in the amygdala, resulting in a return of conditioned response (Björkstrand et al. 2015; Schiller et al. 2013). Retrieval-extinction updates the original fear memory in the amygdala, preventing the return of fear (Clem and Huganir 2010; Monfils et al. 2009). Created with BioRender.com

downregulation of AMPAR subunits is also necessary for memory updating: endocytosis of GluA2-AMPARs in the mouse dorsal hippocampus is necessary for retrieval-extinction (Rao-Ruiz et al. 2011). For a detailed review, see Cahill and Milton (2019).

A number of studies have used immunohistochemistry to compare the location, types, and number of neurons that are active during retrieval-extinction and extinction. One study in rats assessed zinc-finger protein 268 (Zif268), a marker of reconsolidation, and phosphorylated ribosomal protein S6 (rpS6P), which is associated with GluR1 activation (Tedesco et al. 2014). They found that retrieval-extinction increased Zif268 and rpS6P in prefrontal cortex (PL and IL) and LA but not hippocampal CA1. Retrieval alone led to the same increases in Zif268 but to a lesser extent, but neither retrieval alone nor extinction alone increased rpS6P. In another study in rats, *Arc* cellular compartment analysis of temporal activity using fluorescence in situ hybridization (catFISH) was performed in the brains of rats that had undergone either retrieval-extinction or extinction in order to compare the patterns of neural activation at the start and end of the two paradigms (Lee et al. 2016). While the two groups showed similar cytoplasmic *Arc* expression in the PL,

IL, and LA, higher nuclear and double nuclear-cytoplasmic expression were observed in the extinction group only. Since *Arc* moves from the nucleus into the cytoplasm as time passes after neural activation, these results indicate that retrieval-extinction and extinction initially show similar patterns of activation, but diverge as the procedures progress. Another study utilizing catFISH of the *Homer1a* and *cFos* genes in mice that underwent retrieval-extinction of remote contextual fear memories found that retrieval-extinction activated the same cells in the basolateral amygdala, infralimbic cortex, and dentate gyrus that were originally active during fear acquisition (Khalaf et al. 2018; Khalaf and Gräff 2019).

In humans, neural activity can be inferred by analysis of the blood-oxygen-level-dependent (BOLD) signal acquired through magnetic resonance imaging (MRI). The first study to apply this to retrieval-extinction examined how the amygdala and ventromedial prefrontal cortex (vmPFC) were differentially activated by the reminded and non-reminded CS+ during extinction (Schiller et al. 2013). They found comparable amygdala BOLD activation during extinction to both the reminded and non-reminded CS+. In the vmPFC, activation to the non-reminded CS+ decreased throughout the course of extinction, whereas activation to the reminded CS+ was similar to that of the CS− throughout. This suggests that unlike traditional extinction, retrieval-extinction does not depend on the vmPFC to suppress fear responding. However, it should be noted that a later study failed to replicate this finding (Klucken et al. 2016).

Another study assessed differences BOLD signal to the CSs the day after individuals that had already undergone either extinction or retrieval-extinction (Agren et al. 2012a). They found that the extinction group had greater bilateral amygdala activation than the retrieval-extinction group, and that this was positively correlated with their conditioned responding during reinstatement. Further, these differences persisted when tested 18 months later (Björkstrand et al. 2015). According to the authors, these results suggest that retrieval-extinction erased the fear memory trace in the amygdala that mediated fear responding to the CS.

3 Boundary Conditions

In the initial studies on reconsolidation blockade of fear memories using the protein synthesis inhibitor anisomycin, it was observed that memory updating would only occur under specific circumstances, referred to as boundary conditions (Duvarci and Nader 2004; Nader et al. 2000). They found that anisomycin had to be injected into the LA shortly after memory retrieval; a delay of 6 h, injection alone without CS retrieval, or injection of the vehicle after CS retrieval all left the fear memory intact. The effects of reconsolidation blockade could only be observed 24 h, but not 4 h later. Other work has also found that memories which are older or stronger are less prone to reconsolidation (Suzuki et al. 2004). These findings indicate that potential boundary conditions of reconsolidation processes include timing and memory type, which will be discussed in the context of retrieval-extinction, in addition to other potential boundary conditions of outcome measure and prediction error. For a more detailed discussion of possible boundary conditions of retrieval-extinction, see Auber et al. (2013); Kredlow et al. (2016).

3.1 Timing

Given that the only difference between retrieval-extinction and traditional extinction is the larger interval between the first and second CS presentations, timing is of the utmost importance in determining whether the procedure will be successful at preventing the return of fear. In the original experiment, an interval of either 10 min or 1 h resulted in persistent fear attenuation, whereas an interval of 6 or 24 h did not (Monfils et al. 2009). While our group has consistently replicated this effect with a 1 h interval (Auchter et al. 2017a, b; Jones et al. 2013; Lee et al. 2016; Olshavsky et al. 2013; Shumake and Monfils 2015; Tedesco et al. 2014), other studies of rodents that underwent cued fear conditioning found no benefit when compared to traditional extinction with retrieval-extinction intervals ranging from 10 to 90 min (Chan et al. 2010; Flavell et al. 2011; Goode et al. 2017; Ishii et al. 2012, 2015; Luyten and Beckers 2017). Procedural differences in the timing of other events, housing conditions, or precise details of rodent handling could play a role in some of the discrepancies; for example, higher variability in the inter-trial intervals during extinction results in reduced reinstatement when preceded by a retrieval cue (Auchter et al. 2017a), and housing conditions were found to be a significant moderator of outcome (Kredlow et al. 2016). Still, even instances in which attempts at replication included communication between study authors and procedures that were followed as closely as possible resulted in discrepant findings (Luyten and Beckers 2017), so procedural differences are unlikely to be responsible for all of the variance.

In rodents that have been contextually fear conditioned, retrieval-extinction has successfully prevented the return of fear when retrieval consisted of context exposure for 2–4 min but not 1 min followed by extinction between 10 min and 2 h later (An et al. 2018; Flavell et al. 2011; Gräff et al. 2014; Monti et al. 2017; Piñeyro et al. 2014; Rao-Ruiz et al. 2011), but see Stafford et al. (2013) for an exception.

In humans, an interval of 10 min but not 6 h prevents reinstatement (Schiller et al. 2010; Steinfurth et al. 2014), and intervals of 10 and 20 min both reduce spontaneous recovery (Fernandez-Rey et al. 2018), although others have not replicated the effect (Chalkia et al. 2020; Golkar et al. 2012).

Interestingly, one study found that in rats that underwent cued fear conditioning, a reminder of 3 CSs led to reduced spontaneous recovery when it was either 10 min before *or* after the extinction session (Ponnusamy et al 2016). However, the retrieval-extinction group showed higher freezing than the extinction and extinction-retrieval groups in the short-term memory test 3 h later. Since reconsolidation processes are protein synthesis dependent, evidence of memory updating is only apparently on long-term, not short-term memory tests (Duvarci and Nader 2004; Nader et al. 2000). This suggests that retrieval-extinction engaged reconsolidation-based mechanisms whereas extinction-retrieval likely enhanced extinction. Similarly in humans, a single CS retrieval 10 min before or after extinction resulted in reduced reinstatement 24 h later, but only retrieval-extinction resulted in increased recall when tested 3 or 12 h later the same day (Chen et al. 2021b). By testing fear recall before and after a night of sleep, these studies suggest that retrieval-extinction likely engages a different mechanism than extinction or extinction-retrieval that may require the completion of sleep-induced memory reconsolidation processes for expression.

In terms of the longevity of fear attenuation after retrieval-extinction, this effect has been reported to persist in rats for at least 1 month (Monfils et al. 2009) and in humans for at least a year (Björkstrand et al. 2015; Schiller et al. 2010; Telch et al. 2017). Taken together, these findings show that successful retrieval-extinction procedures have used 1–3 CSs for retrieval or context exposure for 2–4 min, followed by extinction 10 min to 2 h later. The memory impairing effects of this paradigm are only apparent when tested at least 24 h later and seem to be persistent. However, given that many studies were not able to replicate the effect using these parameters, it is possible that other boundary conditions may have played a role.

3.2 *Memory/Stimulus Type*

The original studies in rats and humans targeted memories that were recent (1 day old) and conditioned with simple, neutral stimuli. Recently, many have sought to extend the retrieval-extinction paradigm to the attenuation of fear memories of varying age, strength, and complexity that may more realistically model the types of fears that are often addressed in a therapeutic context.

Attempts to attenuate remote (7–29 day old) fear memories using retrieval-extinction have largely been unsuccessful. In rats, retrieval-extinction has not shown to be effective in attenuating remote cued or contextual fear memories (An et al. 2018; Clem and Huganir 2010; Costanzi et al. 2011; Gräff et al. 2014). However, it should be noted that this effect can be rescued with injection of a histone deacetylase 2-targeting inhibitor to induce neuroplasticity-related genes (Gräff et al. 2014). In humans, retrieval-extinction can prevent spontaneous recovery of 7-day-old fear memories (Steinfurth et al. 2014).

Given that maladaptive fears are more likely to occur in response to potentially dangerous stimuli than neutral cues, many have sought to apply the retrieval-extinction paradigm to reduce conditioned responses to fear-relevant stimuli such as snakes. In the first study of this kind, the authors compared the ability of extinction and the beta-adrenergic blocker propranolol to attenuate conditioned fear in humans when applied within the reconsolidation window (Soeter and Kindt 2011). Retrieval-extinction prevented spontaneous recovery, but not reinstatement, when measured by fear-potentiated startle (FPS) but not skin conductance response (SCR), subjective distress, or expectancy ratings. However, propranolol was able to prevent spontaneous recovery, reinstatement, and generalization as measured via FPS. In another study, humans were conditioned with both fear-relevant and fear irrelevant stimuli which were later retrieved with a US of reduced intensity 10 min before the extinction session (Thompson and Lipp 2017). Those that underwent retrieval-extinction showed reduced spontaneous recovery and reinstatement to both stimuli as compared to those that underwent extinction only. It should be noted that several subsequent studies have not found any added benefit of a CS retrieval prior to extinction in the persistent reduction of conditioned responding to fear-relevant stimuli (Fricchione et al. 2016; Golkar et al. 2012; Kindt and Soeter 2013; Marks and Zoellner 2014; Meir Drexler et al. 2014). Taken together, this suggests that stronger memories, such as those of fear-relevant stimuli, may require a stronger cue such as the US in order to trigger reconsolidation.

In one interesting study, individuals with a phobia of spiders had their fears reactivated while undergoing magnetic resonance imaging to assess amygdala activity (Björkstrand et al. 2016). Using several distinct images of spiders as cues, two cues were activated, only one of which was extinguished either 10 min or 6 h later, in addition to another cue which had not been activated. When tested the following day, the 10-min group showed reduced activity to the activated and exposed cue, while the 6-h group showed increased activity. In addition, amygdala activity to the other cues increased in the 6-h group but not 10-min group, which also showed less activity to a novel cue. The 10-min group showed greater approach behavior when offered money to voluntarily look at new spider cues. These differences in brain activity and behavior persisted 6 months later (Björkstrand et al. 2017). Amygdala activity inversely correlated with approach behavior in the 6-h group but not the 10-min group, suggesting that the procedure had successfully targeted the same brain region guiding this behavior (Björkstrand et al. 2016). Indeed, later studies found reduced avoidance in individuals with specific phobia that had their fear memories reactivated before exposure therapy, further supporting the idea that phobias can be persistently reduced with retrieval-extinction (Lancaster et al. 2020; Telch et al. 2017).

Another type of conditioned stimulus that is more resistant to extinction is a compound stimulus, wherein the CS consists of multiple cues presented simultaneously. In rats that were conditioned with a light and tone in compound, neither retrieval-extinction nor extinction of the compound cue was able to prevent spontaneous recovery (Jones et al. 2013). Interestingly, retrieval-extinction of either cue component prevented the return of conditioned fear to the light, whereas only the tone was able to reduce spontaneous recovery to both components and the compound. Even more effective was the sequential retrieval-extinction of the component cues within a 3-h window, but only if the tone was targeted first. In humans that were fear conditioned to a compound stimulus consisting of three visual cues, the number of component cues retrieved before compound extinction were compared for their ability to prevent the return of fear (Li et al. 2017). While those that had a retrieval trial with zero or one of the component cues experienced both spontaneous recovery and reinstatement, retrieval with two or three components prevented reinstatement, but only retrieval with two components additionally prevented spontaneous recovery. These studies indicate that while compound fear memories can successfully be attenuated by retrieval-extinction, retrieving the whole compound may not be the most effective method. On the other hand, in participants that were conditioned to an entire category of stimuli (i.e., fish or birds) using unique cues on each trial, retrieval-extinction was not able to prevent reinstatement (Kroes et al. 2017).

The strength of a fear memory can also depend on the type of US it was conditioned with. In rats, retrieval-extinction is still able to prevent spontaneous recovery regardless of whether conditioning occurred with a 0.7 or 1.0 mA shock (Olshavsky et al. 2013). In humans, fear responses that were conditioned with an aversive auditory US instead of electric shock were also more persistently attenuated when extinction was preceded by CS retrieval (Fernandez-Rey et al. 2018; Oyarzún et al. 2012). Fear memory strength can also be manipulated through the schedule of reinforcement, and one study has examined whether high or low reinforcement during conditioning could affect return of fear after retrieval-extinction (Kitamura et al. 2020). They found that humans who were fear conditioned with either a 40 or 80% reinforcement rate showed both reduced spontaneous recovery and reinstatement to the CS that was retrieved before extinction than the CS that was not retrieved. In contrast, the 80% group showed significantly greater spontaneous recovery of the non-retrieved stimulus than the 40% group, indicating that retrieval-extinction may be superior to standard extinction in preventing the return of conditioned responses to strong fear memories.

Taken together, these data suggest that while retrieval-extinction may not be able to update remote fear memories, it does seem to be able to update stronger or more complex memories, though it is difficult to draw firm conclusions for fear-relevant memories based on the current limited data.

3.3 *Outcome Measures*

Fear extinction studies typically index conditioned responding via freezing in rodents and skin conductance response (SCR) in humans and probe the return of fear via spontaneous recovery and reinstatement. However, the use of other outcome measures and return of fear assays can give a more complete picture of the extent and aspects of fear expression that have been modified.

In humans, other common measures of conditioned responding include fear-potentiated startle (FPS) and US expectancy ratings. While all of these outcomes are responsive to differential conditioning procedures, there do appear to be less reliable in their ability to detect subtle differences in conditioned responding after retrieval-extinction procedures. For example, some studies have found that retrieval-extinction results in less reinstatement than extinction when measured via SCR (Chen et al. 2021b) or FPS (Soeter and Kindt 2011), but not US expectancy. Others have not found any differences in reinstatement or spontaneous recovery in FPS, SCR, conditioned pupil size or US expectancy when collecting more than one of these measures simultaneously (Golkar et al. 2012; Kindt and Soeter 2013; Zimmermann and Bach 2020).

In order to determine whether the collection of US expectancy ratings could influence conditioned responding on psychophysiological measures in retrieval-extinction paradigms (Warren et al. 2014), fear conditioned participants while measuring FPS: half received a retrieval trial 10 min before extinction, and half were asked to rate their expectation that the US would occur on each trial. They found that the inclusion of US expectancy ratings strengthened FPS during acquisition and extinction recall. Interestingly, those that provided US expectancy also showed weakened FPS during re-extinction and, in the retrieval-extinction group, weakened reinstatement. Retrieval-extinction also decreased US expectancy during reinstatement as compared to extinction alone.

Other outcome measures have been used to assess the ability of retrieval-extinction to reduce other aspects of fear memories. In one study, a distressing video clip was used as the CS, and the number of self-reported intrusive thoughts was measured the day after it underwent retrieval-extinction. In this case, the participants that had the memory retrieved with a still from the video 10 min before extinction actually reported a higher number of intrusive thoughts than those that had the memory retrieved with a scrambled image or the same still but 10 min later (Marks and Zoellner 2014). However, another study utilizing distressing video clips found that retrieval with a video still followed by interference by playing the game Tetris reduced the number of intrusive thoughts compared to those who had only retrieval, only Tetris, or neither (James et al. 2015). Another study conditioned participants to a category of stimuli by pairing unique exemplars with a shock (Kroes et al. 2017). They found that retrieval with a prototypical exemplar before extinction did not prevent reinstatement via SCR, but it did improve episodic memory for item recognition memory, temporal context, and retrospective shock estimation. These studies suggest that the fear attenuating effects of retrieval-extinction may be limited to conditioned physiological responses and threat expectancy, while possibly strengthening episodic components and that reductions in intrusive thoughts may be better accomplished through interference with non-extinction stimuli.

In rodents, there are other behaviors beyond freezing which are affected by fear conditioning. One such behavior is conditioned suppression, wherein a conditioned behavior such as reward seeking is suppressed by the presentation of a fear CS. Retrieval-extinction was found to be superior to extinction alone in preventing the reinstatement of conditioned freezing, but there was no difference in conditioned suppression of reward seeking (Shumake and Monfils 2015). This indicates that retrieval-extinction may be able to reduce conditioned fear but not to the extent necessary to resume pleasurable activities in the presence of the CS. Another study assessed the non-associative components of remote contextual fear memories that often become maladaptive in individuals with PTSD: anxiety-like behavior, social behavior, and spatial learning (Costanzi et al. 2011). However, neither retrieval-extinction nor extinction was able to correct the increased freezing, anxiety-like behavior, and decreased social behavior and spatial learning to levels comparable to animals that had not been fear conditioned. Taken together, these data indicate that certain behaviors may be easily modified by fear conditioning but are less susceptible to modification through retrieval-extinction procedures. While this suggests that not all aspects of a fear memory can be erased, it may be useful to maintain some aspects of the memory, but not to the extent that it becomes maladaptive. In the case of those with anxiety disorders with excessive avoidance, this could mean that fear is reduced enough to resume engagement with anxiety-inducing situations, which would likely further reduce symptoms.

While fear may successfully be attenuated in the extinction context, it can often reemerge upon return to the conditioning context – a phenomenon known as renewal (Bouton 2002, 2004). In humans, context can be manipulated by presenting the CSs in one of two visually distinct frames (Meir Drexler et al. 2014). Using fear-relevant CS, neither retrieval-extinction nor extinction was able to prevent spontaneous recovery. Further, there was no evidence of a renewal effect – that is, conditioned responding was equally high in both the conditioning and extinction contexts at test. Another study utilized immersive virtual reality to contextually fear condition participants (Houtekamer et al. 2020). Neither retrieval-extinction nor extinction prevented spontaneous recovery or reinstatement via FPS, which were also equally high in both contexts in the early part of these phases. It is possible that attempts to model different contexts in human studies result in generalization of fear responses to the safe context and are also more difficult to extinguish.

These studies indicate that the fear attenuating effects of retrieval-extinction may be specific to certain types of memory expression, such as freezing in rodents and SCR in humans, and may not modify other aspects of fear memories such as US expectancy and conditioned suppression.

3.4 Prediction Error

Some theories of reconsolidation postulate that a memory will only be updated if there is new information to be learned. Prediction error, defined as the difference between the expected and actual outcome, is an indicator that there is new information to be learned. Several studies have been done in order to determine the specific conditions under which prediction error can be detected in order to trigger reconsolidation.

The first of these studies found that a single retrieval trial without expectation that the US could occur was not sufficient to trigger reconsolidation (Sevenster et al. 2012). US expectancy was manipulated by detaching participants from the shock electrodes during retrieval. Since propranolol was administered, failure to trigger reconsolidation was determined by the return of fear the following day. Using similar methodology, the same group found that retrieval trials that induce either positive or negative prediction error are able to trigger reconsolidation so long as the reinforcement schedule is different than it was during acquisition (Sevenster et al. 2013, 2014). Examples include a single unreinforced CS after 100% reinforcement, two unreinforced CSs after 50% reinforcement, and one reinforced CS after 33% reinforcement. Likewise, in rats, the CS-US pairing can be used to trigger reconsolidation so long as the temporal structure between the stimuli differs from that of acquisition, as evidenced by a reduction in freezing after anisomycin injection (Díaz-Mataix et al. 2013).

These findings have been extended from pharmacological to behavioral updating of reconsolidation. In humans that were fear conditioned with a 50% reinforcement schedule, retrieval that induced either positive and negative prediction error (two CS trials with or without the US) before extinction led to reduced spontaneous recovery and reinstatement (Chen et al. 2018). In humans that were fear conditioned with a predictable temporal association between the CS and US with 100% reinforcement, a single CS-US trial before extinction was a sufficient prediction error to prevent spontaneous recovery and reinstatement only if the temporal association was altered. But, if the temporal association during acquisition was unpredictable, then two CS-US trials was able to prevent spontaneous recovery and reinstatement (Chen et al. 2021a).

However, a study in rats casts doubt on the requirement of prediction error to trigger reconsolidation, since it found that both retrieval with the CS only (prediction error) or the CS-US (no prediction error) before extinction lead to reduced spontaneous recovery and reacquisition (Cahill et al. 2019) but see also Goode et al. (2017). Indeed, some theoretical accounts posit that new associative memories are formed when a new latent cause is inferred, such as when the CS is suddenly no longer reinforced during extinction (Gershman et al. 2017). As such, it is during conditions which are similar to that of the acquisition phase that the original memory is updated. This theory is supported by the finding that gradual extinction – where the reinforcement rate of the CS is gradually reduced during extinction – is superior to both traditional and reverse gradual extinction at preventing spontaneous recovery and reinstatement (Gershman et al. 2013). In line with this, several studies have found that fear memories can be attenuated using either retrieval with a US of reduced intensity followed by extinction, or CS retrieval followed by CS-US pairings of reduced intensity (Liu et al. 2014; Popik et al. 2020; Thompson and Lipp 2017). These studies suggest that inclusion of the US during retrieval-extinction procedures, whether of reduced frequency or intensity, may prevent the inference of a novel latent cause and result in memory updating as opposed to extinction (Gershman et al. 2013, 2017; Monfils and Holmes 2018).

4 Individual Differences

While there has been ample research on individual differences in fear extinction, there has been relatively little on such factors that may affect the response to retrieval-extinction. Individual differences that have been studied in this paradigm, focusing on genetics, age, and psychopathology, will be discussed here and are summarized in Table 2.

Table 2 Individual differences in the response to retrieval-extinction

Trait	Finding
Genetics	
5–HTTLPR	Short-allele carriers preferentially respond to retrieval-extinction (Agren et al. 2012b)
COMT	Val/val homozygotes preferentially respond to retrieval-extinction (Agren et al. 2012b)
COMT	No difference between alleles (Klucken et al. 2016)
BDNF	Met-allele carriers preferentially respond to retrieval-extinction (Asthana et al. 2016)
Mouse strain	No difference between 129S1/SvImJ (S1) and C57BL/6 J (B6) strains (MacPherson et al. 2013)
Age	
Adolescent humans	Retrieval-extinction prevented return of fear while extinction did not (Johnson and Casey 2015)
Adolescent rats	Less spontaneous recovery and renewal after retrieval-extinction than extinction (Baker et al. 2013)
Adolescent rats	Retrieval-extinction can attenuate remote memories acquired as a juvenile but not as an adult (Jones & Monfils 2016)
Juvenile mice	No persistent fear attenuation after retrieval-extinction (Ishii et al. 2015)
Psychopathology	
Spider phobia	Retrieval-extinction reduces amygdala activation and increases approach behavior to spider cues compared to extinction (Björkstrand et al. 2016)
Anxiety	No difference between anxious and healthy adults or effect of trait anxiety on reinstatement after retrieval-extinction or extinction (Kredlow et al. 2018)
Orienting and extinction phenotypes	
Orienting	Rats that display more appetitive cue-directed behavior show less spontaneous recovery after retrieval-extinction (Olshavsky et al. 2013)
Extinction	No difference between extinction and retrieval-extinction in distribution of freezing phenotypes during extinction and reinstatement (Shumake et al. 2018)

4.1 Genetics

The first study of this kind compared those that had a 10-min or 6-h interval between retrieval and extinction and assessed the influence of alleles in the serotonin-transporter gene-linked polymorphic region (5–HTTLPR) and functional polymorphism val158met (rs4680) catechol-O-methyltransferase (COMT) on reacquisition (Agren et al. 2012b). For 5–HTTLPR, short-allele carriers displayed significantly more reacquisition in the 6-h group than the 10-min group, whereas there was no difference in the long-allele homozygotes. For COMT, val/val homozygotes displayed significantly more reacquisition in the 6-h group than the 10-min group, whereas there was no difference in the met carriers. This suggests that 5–HTTLPR short-allele carriers and COMT val/val homozygotes preferentially respond to reconsolidation paradigms than to regular extinction. However, another study found no significant differences in reinstatement to reminded and un-reminded CSs in regard to the COMT val158met polymorphism (Klucken et al. 2016).

Given its roles in learning and memory and brain plasticity, genes related to brain derived neurotrophic factor (BDNF) have also been investigated as potential moderators of response to retrieval-extinction. The BDNF val66met polymorphism (rs6265) was genotyped in participants that underwent fear conditioning and then either extinction or retrieval-extinction (Asthana et al. 2016). Met-allele carriers showed significant spontaneous recovery in the extinction but not retrieval-extinction group whereas val66val homozygotes showed no difference. Similar to the findings of Agren et al., this suggests that met-allele carriers preferentially respond to reconsolidation-based approaches than to extinction. This is consistent with other studies demonstrating impaired extinction learning in BDNF met-allele carriers (Felmingham et al. 2013, 2018; Soliman et al. 2010). These studies point to a potential moderating factor of the polymorphisms in the 5–HTTLPR, COMT, and BDNF genes in the response to retrieval-extinction procedures, though additional work is needed to verify if these findings might extend to those with anxiety and trauma disorders.

Studies assessing these genetic differences in rodents are very limited. One study has compared two common inbred mouse strains, the 129S1/SvImJ (S1) and C57BL/6 J (B6) (MacPherson et al. 2013). In both strains, there was no difference between retrieval-extinction and extinction on return of fear the following day. This indicates that some rodent strains may be better suited for studying the behavioral and mechanistic differences between retrieval-extinction and traditional extinction, though further study is needed to confirm this.

4.2 Age

All of the aforementioned studies have been conducted in adult subjects; however, it is important to know whether retrieval-extinction can be applied to fears acquired in adolescence as well. In humans, retrieval-extinction prevented fear recovery tested 24 h later while extinction did not (Johnson and Casey 2015). In adolescent rats, those that received a retrieval trial before extinction had less spontaneous recovery and renewal than those that did not (Baker et al. 2013). This finding was robust: it held true whether the retrieval trial was in the conditioning or extinction context, or whether it occurred before or after extinction. In contrast, another study did not find any persistent fear attenuation in juvenile mice that underwent retrieval-extinction (Ishii et al. 2015).

The previous studies tested return of fear soon after extinction, but it is important to know whether remote fears can be attenuated this way in adolescence, and if the effect will persist throughout adulthood as well. One study sought to address this by fear conditioning juvenile rats at either p17 or p25 (Jones and Monfils 2016). At both ages, those that received retrieval-extinction in adolescence showed reduced spontaneous recovery when tested as adults, but intact reacquisition. Rats that were fear conditioned at p25 and had retrieval-extinction as adults showed reduced reacquisition as adults. This shows that retrieval-extinction can attenuate remote fear memories in adolescence, even though it is not able to modify memories of the same age acquired in adulthood.

4.3 Psychopathology

Most fear conditioning studies exclude individuals who self-report a psychiatric diagnosis even when the stated objective of research is to improve treatments for individuals with anxious psychopathology. To that end, few studies have assessed whether retrieval-extinction is able to attenuate fears in these participants. As previously discussed, subjects with a phobia of spiders showed reduced amygdala activation and increased approach behavior after extinction of spider cues within the reconsolidation window as compared to those who underwent extinction outside of the window (Björkstrand et al. 2016). But in another study with a mixed sample of healthy and anxious participants, there was no difference between retrieval-extinction and extinction on reinstatement (Kredlow et al. 2018). Further, there were no differences between healthy and anxious subjects nor a moderating effect of anxiety symptom severity on return of fear. These studies indicate that more work is needed to determine the efficacy of retrieval-extinction procedures in the reduction of fear responses in those with anxiety disorders.

4.4 Orienting and Extinction Phenotypes

Individual differences in both appetitive and aversive conditioned responses have been examined for their potential to predict retrieval-extinction phenotypes in rats. A sample that underwent appetitive conditioning several days prior to fear conditioning and extinction with or without retrieval was divided based on their orienting response to the appetitive CS (Olshavsky et al. 2013). When conditioned with a 1.0 mA but not a 0.7 mA shock, non-orienters showed significantly more spontaneous recovery than orienters both 1 and 21 days later.

Another study identified several different subtypes of fear responding based on freezing phenotypes during extinction and reinstatement using data from multiple studies comparing retrieval-extinction and extinction (Shumake et al. 2018). However, there were no significant differences in subgroup membership, indicating that individual fear recovery trajectories may not be differentially influenced by the two paradigms. Nevertheless, numerous studies have demonstrated heterogeneity in the response to extinction learning. For example, rats show an average distribution of responses during extinction recall, with the ends of this distribution showing distinct trajectories during extinction and recall (Bush et al. 2007; Galatzer-Levy et al. 2013). Additional work has assessed factors which might predict these distinct responses. Activation of orexin neurons in the hypothalamus during extinction recall predicts freezing levels (Sharko et al. 2017). These neurons can also be activated by exposure to a carbon dioxide challenge, which are predictive of behavioral reactivity during the challenge, which in turn is predictive of extinction recall (Monfils et al. 2019). Given that some individuals show a poorer response to extinction than others, it is possible that these would be better candidates for a reconsolidation-based approach. Indeed, it has been shown that fear attenuation after extinction is dependent upon genetics and extinction learning ability, but not for retrieval-extinction (Auchter et al. 2017b). Future studies should investigate factors that can predict which method will result in stronger, more persistent fear attenuation for a given individual.

5 Treatment Studies

Given the potential of retrieval-extinction principles to be used clinically for the non-invasive treatment of anxiety and trauma disorders, a number of studies have sought to extend the paradigm to update real-life maladaptive fear memories. These will be discussed here and are summarized in Table 3. Two studies incorporated a reactivation cue of either a neutral or phobic stimulus 10 min before exposure

Table 3 Clinical applications

Target	Protocol	Outcome
Specific phobia	Reactivation cue of a neutral or phobia stimulus 10 min before virtual reality exposure therapy	Comparable short- and long-term reduction in symptoms in both groups (Maples-Keller et al. 2017; Shiban et al. 2015)
Specific phobia	Reactivation cue 30 min before in vivo exposure therapy	Reduced self-reported fear at follow-up and number of sessions needed for symptom reduction compared to reactivation after exposure (Lancaster et al. 2020; Telch et al. 2017)
Traumatic memory	Retrieval of memory followed by interference with positive, neutral, or negative story	Interference with negative story reduced details recalled 1 week later (Kredlow and Otto 2015)
PTSD	Intrusive memory reactivated by writing of trauma script followed by interference with Tetris + treatment as usual	Decreased number of intrusions of targeted memory (Kessler et al. 2018)
PTSD	Reactivation 1 h before imaginal exposure	SCR to trauma script lower in reactivation group but no differences in subjective measures (Vermes et al. 2020)
PTSD	Reactivation before incorporation of cognitive restructuring	2/3 s of treatment completers achieved and maintained remission (Gray et al. 2019)

therapy in virtual reality in patients with specific phobia (Maples-Keller et al. 2017; Shiban et al. 2015). In both studies, patients in both groups responded equally well to therapy and maintained their treatment gains, with no differences in clinical measures at any time points. However, this may simply be a reflection of the high efficacy of the exposure therapy protocols used in these studies, since none of the groups experienced a return of fear after treatment. Two other studies have tested a 10-s reactivation 25–30 min before in vivo exposure therapy for specific phobia (Lancaster et al. 2020; Telch et al. 2017). Both studies found a benefit for adding reactivation before exposure: it reduced self-reported phobic fear at follow-up as compared to reactivation after exposure (Telch et al. 2017) and reduced by 21% the exposure dosage needed to achieve the same symptom reduction as deepened exposure or exposure alone (Lancaster et al. 2020). Taken together, these studies suggest that retrieval-extinction procedures may be better suited for phobias that can be treated with in vivo exposure therapy rather than virtual reality.

The first study to use retrieval-extinction principles to update traumatic memories did so in an analog sample of subjects that were exposed to the Boston Marathon bombing (Kredlow and Otto 2015). After retrieving their negative autobiographical memories of the day's events, they underwent interference with a story of either positive, neutral or negative valence. Those that had interference with a negative story recalled significantly fewer details than the no-story group when tested 1 week later, although the percentage of positive or negative words did not significantly differ between groups or change over time. In a sample of PTSD inpatients, an intrusive memory was reactivated by briefly writing a trauma script, after which they underwent interference with the game Tetris (Kessler et al. 2018). This was done once per week in addition to treatment as usual. The reactivation treatment was able to decrease the frequency of targeted intrusions by 64%, significantly more than the reduction of 11% for non-targeted intrusions. Since exposure therapy for PTSD typically involves imaginal exposure, and retrieval before imaginal extinction has also been shown to prevent reinstatement (Agren et al. 2017), these approaches were combined through a 5-min reactivation of either a traumatic or neutral event 1 h before 2 h-long imaginal exposure sessions (Vermes et al. 2020). Treatment group was a significant predictor of SCR to the trauma script after treatment, which was lower in those who had a traumatic memory reactivated than those that had a neutral memory reactivated, though the groups did not differ in subjective measures, and changes in symptom severity were not assessed. Another study incorporated elements of retrieval-extinction into a novel treatment protocol for PTSD (Gray et al. 2019). The treatment involved reactivating the traumatic memory by retelling it just until physiological responses were observed, and then incorporating cognitive distancing and restructuring techniques into the trauma script. Of those who completed the three 2-h long sessions, 67% achieved remission, and their symptoms reduced significantly more than those on the wait list. In addition, there was no return of fear after treatment, with post-treatment symptom severity scores remaining stable 2 and 6 weeks later. Though more research is needed, these studies show that behavioral reconsolidation is a promising technique for the treatment of intrusive thoughts, physiological reactivity, and symptom burden associated with PTSD.

6 Conclusion

The use of retrieval-extinction for the persistent, non-invasive attenuation of maladaptive fear memories has garnered much attention over the past decade. However, many have reported a return of fear after its use, leading some to question whether it actually results in the updating of the original fear memory, or simply enhances extinction. There are possible alternative explanations for these failures to replicate. First, there are boundary conditions on the types of fear memories that are susceptible to disruption through reconsolidation. Then, we know that in order for the procedure to work, the memory must successfully be reactivated and destabilized, after which there is a limited temporal window during which extinction must be incorporated before the memory restabilizes. While there are molecular markers for each of these processes, and pharmacological agents we can use to disrupt them in order to determine whether they occurred, this only offers a limited snapshot of the mechanisms that may be occurring and cannot be used in human studies. As a result, it is difficult to disentangle whether unsuccessful studies of the retrieval-extinction effect are due to boundary conditions, failure to engage all of the necessary processes, or a true failure to replicate. These issues will remain unresolved until reliable indices for each of these processes are identified. Another possibility is that not all subjects are well suited to treatment with retrieval-extinction; just as there are individual differences in the response to extinction. More research is needed to determine whether there may be factors that can dissociate those who will benefit most from each treatment.

In spite of these limitations, a number of studies have in fact discovered neural mechanisms that differentiate retrieval-extinction from extinction that supports the idea that it likely works through updating or partial updating of the original fear memory. There have also been a number of successful translations from the bench to the bedside in the treatment of phobias and PTSD. While there is still much basic and clinical work to be done in order to ensure optimal treatment protocols, retrieval-extinction remains a promising alternative to current treatments.

References

Agren T, Engman J, Frick A, Björkstrand J, Larsson EM, Furmark T, Fredrikson M (2012a) Disruption of reconsolidation erases a fear memory trace in the human amygdala. Science 337(6101):1550–1552. https://doi.org/10.1126/science.1223006

Agren T, Furmark T, Eriksson E, Fredrikson M (2012b) Human fear reconsolidation and allelic differences in serotonergic and dopaminergic genes. Transl Psychiatry 2:8–13. https://doi.org/10.1038/tp.2012.5

Agren T, Björkstrand J, Fredrikson M (2017) Disruption of human fear reconsolidation using imaginal and in vivo extinction. Behav Brain Res 319:9–15. https://doi.org/10.1016/j.bbr.2016.11.014

An X, Yang P, Chen S, Zhang F, Yu D (2018) An additional prior retrieval alters the effects of a retrieval-extinction procedure on recent and remote fear memory. Front Behav Neurosci 11:1–14. https://doi.org/10.3389/fnbeh.2017.00259

Asthana MK, Brunhuber B, Mühlberger A, Reif A, Schneider S, Herrmann MJ (2016) Preventing the return of fear using reconsolidation update mechanisms depends on the met-allele of the brain derived neurotrophic factor Val66Met polymorphism. Int J Neuropsychopharmacol 19(6): 1–9. https://doi.org/10.1093/ijnp/pyv137

Auber A, Tedesco V, Jones CE, Monfils MH, Chiamulera C (2013) Post-retrieval extinction as reconsolidation interference: methodological issues or boundary conditions? Psychopharmacology 226(4):631–647. https://doi.org/10.1007/s00213-013-3004-1

Auchter A, Cormack LK, Niv Y, Gonzalez-Lima F, Monfils MH (2017a) Reconsolidation-extinction interactions in fear memory attenuation: the role of inter-trial interval variability. Front Behav Neurosci 11:1–9. https://doi.org/10.3389/fnbeh.2017.00002

Auchter A, Shumake J, Gonzalez-Lima F, Monfils MH (2017b) Preventing the return of fear using reconsolidation updating and methylene blue is differentially dependent on extinction learning. Sci Rep 7:1–13. https://doi.org/10.1038/srep46071

Baker KD, McNally GP, Richardson R (2013) Memory retrieval before or after extinction reduces recovery of fear in adolescent rats. Learn Mem 20(9):467–473. https://doi.org/10.1101/lm.031989.113

Ben Mamou C, Gamache K, Nader K (2006) NMDA receptors are critical for unleashing consolidated auditory fear memories. Nat Neurosci 9(10):1237–1239. https://doi.org/10.1038/nn1778

Björkstrand J, Agren T, Frick A, Engman J, Larsson EM, Furmark T, Fredrikson M (2015) Disruption of memory reconsolidation erases a fear memory trace in the human amygdala: an 18-month follow-up. PLoS One 10(7):1–8. https://doi.org/10.1371/journal.pone.0129393

Björkstrand J, Agren T, Åhs F, Frick A, Larsson EM, Hjorth O, Furmark T, Fredrikson M (2016) Disrupting reconsolidation attenuates long-term fear memory in the human amygdala and facilitates approach behavior. Curr Biol 26(19):2690–2695. https://doi.org/10.1016/j.cub.2016.08.022

Björkstrand J, Agren T, Åhs F, Frick A, Larsson E-M, Hjorth O, Furmark T, Fredrikson M (2017) Think twice, it's all right: long lasting effects of disrupted reconsolidation on brain and behavior in human long-term fear. Behav Brain Res 324:125–129. https://doi.org/10.1016/j.bbr.2017.02.016

Bos MGN, Beckers T, Kindt M (2014) Noradrenergic blockade of memory reconsolidation: a failure to reduce conditioned fear responding. Front Behav Neurosci 8:1–8. https://doi.org/10.3389/fnbeh.2014.00412

Bouton ME (2002) Context, ambiguity, and unlearning: sources of relapse after behavioral extinction. Biol Psychiatry 52(10):976–986

Bouton ME (2004) Context and behavioral processes in extinction. In: Learning and memory, vol 11, pp 485–494. https://doi.org/10.1101/lm.78804

Bush D, Sotres-Bayon F, LeDoux JE (2007) Individual differences in fear: isolating fear reactivity and fear recovery phenotypes. J Trauma Stress 20(4):413–422. https://doi.org/10.1002/jts

Bustos SG, Maldonado H, Molina VA (2006) Midazolam disrupts fear memory reconsolidation. Neuroscience 139(3):831–842. https://doi.org/10.1016/j.neuroscience.2005.12.064

Cahill EN, Milton AL (2019) Neurochemical and molecular mechanisms underlying the retrieval-extinction effect. Psychopharmacology 236(1):111–132. https://doi.org/10.1007/s00213-018-5121-3

Cahill EN, Wood MA, Everitt BJ, Milton AL (2019) The role of prediction error and memory destabilization in extinction of cued-fear within the reconsolidation window. Neuropsychopharmacology 44(10):1762–1768. https://doi.org/10.1038/s41386-018-0299-y

Chalkia A, Schroyens N, Leng L, Vanhasbroeck N, Zenses AK, van Oudenhove L, Beckers T (2020) No persistent attenuation of fear memories in humans: a registered replication of the reactivation-extinction effect. Cortex 129:496–509. https://doi.org/10.1016/j.cortex.2020.04.017

Chan WYM, Leung HT, Westbrook RF, McNally GP (2010) Effects of recent exposure to a conditioned stimulus on extinction of Pavlovian fear conditioning. Learn Mem 17(10): 512–521. https://doi.org/10.1101/lm.1912510

Chen W, Li J, Xu L, Zhao S, Fan M, Zheng X (2018) Effects of prediction error on post-retrieval extinction of fear to compound stimuli. Acta Psychol Sin 50(7):739. https://doi.org/10.3724/sp.j.1041.2018.00739

Chen W, Li J, Xu L, Zhao S, Fan M, Zheng X (2021a) Destabilizing different strengths of fear memories requires different degrees of prediction error during retrieval. Front Behav Neurosci 14:1–18. https://doi.org/10.3389/fnbeh.2020.598924

Chen W, Li J, Zhang X, Dong Y, Shi P, Luo P, Zheng X (2021b) Retrieval-extinction as a reconsolidation-based treatment for emotional disorders: evidence from an extinction retention test shortly after intervention. Behav Res Ther 139:103831. https://doi.org/10.1016/j.brat.2021.103831

Clem RL, Huganir RL (2010) Calcium-permeable AMPA receptor dynamics mediate fear memory erasure. Science 330(6007):1108–1112. https://doi.org/10.1126/science.1195298

Costanzi M, Cannas S, Saraulli D, Rossi-Arnaud C, Cestari V (2011) Extinction after retrieval: effects on the associative and nonassociative components of remote contextual fear memory. Learn Mem 18(8):508–518. https://doi.org/10.1101/lm.2175811

Craske MG, Mystkowski JL (2007) Exposure therapy and extinction: clinical studies. In: Fear and learning: from basic processes to clinical implications. American Psychological Association, pp 217–233. https://doi.org/10.1037/11474-011

Dębiec J, Ledoux JE (2004) Disruption of reconsolidation but not consolidation of auditory fear conditioning by noradrenergic blockade in the amygdala. Neuroscience 129(2):267–272. https://doi.org/10.1016/j.neuroscience.2004.08.018

Díaz-Mataix L, Ruiz Martinez RC, Schafe GE, Ledoux JE, Doyère V (2013) Detection of a temporal error triggers reconsolidation of amygdala-dependent memories. Curr Biol 23(6): 467–472. https://doi.org/10.1016/j.cub.2013.01.053

Duvarci S, Nader K (2004) Characterization of fear memory reconsolidation. J Neurosci 24(42): 9269–9275. https://doi.org/10.1523/JNEUROSCI.2971-04.2004

Felmingham KL, Dobson-Stone C, Schofield PR, Quirk GJ, Bryant RA (2013) The brain-derived neurotrophic factor Val66Met polymorphism predicts response to exposure therapy in posttraumatic stress disorder. Biol Psychiatry 73(11):1059–1063. https://doi.org/10.1016/j.biopsych.2012.10.033

Felmingham KL, Zuj DV, Hsu KCM, Nicholson E, Palmer MA, Stuart K, Vickers JC, Malhi GS, Bryant RA (2018) The BDNF Val66Met polymorphism moderates the relationship between posttraumatic stress disorder and fear extinction learning. Psychoneuroendocrinology 91:142–148. https://doi.org/10.1016/j.psyneuen.2018.03.002

Fernandez-Rey J, Gonzalez-Gonzalez D, Redondo J (2018) Preventing the return of fear memories with postretrieval extinction: a human study using a burst of white noise as an aversive stimulus. Behav Neurosci 132(4):230–239. https://doi.org/10.1037/bne0000245

Flavell CR, Barber DJ, Lee JLC (2011) Behavioural memory reconsolidation of food and fear memories. Nat Commun 2(1). https://doi.org/10.1038/ncomms1515

Fricchione J, Greenberg MS, Spring J, Wood N, Mueller-Pfeiffer C, Milad MR, Pitman RK, Orr SP (2016) Delayed extinction fails to reduce skin conductance reactivity to fear-conditioned stimuli. Psychophysiology 53(9):1343–1351. https://doi.org/10.1111/psyp.12687

Galatzer-Levy IR, Bonanno GA, Bush DEA, LeDoux JE (2013) Heterogeneity in threat extinction learning: substantive and methodological considerations for identifying individual difference in response to stress. Front Behav Neurosci 7:1–7. https://doi.org/10.3389/fnbeh.2013.00055

Gershman SJ, Jones CE, Norman KA, Monfils MH, Niv Y (2013) Gradual extinction prevents the return of fear: implications for the discovery of state. Front Behav Neurosci 7:1–6. https://doi.org/10.3389/fnbeh.2013.00164

Gershman SJ, Monfils MH, Norman KA, Niv Y (2017) The computational nature of memory modification. elife 6:1–41. https://doi.org/10.7554/eLife.23763

Golkar A, Bellander M, Olsson A, Öhman A (2012) Are fear memories erasable? – Reconsolidation of learned fear with fear relevant and fear-irrelevant stimuli. Front Behav Neurosci 6:1–10. https://doi.org/10.3389/fnbeh.2012.00080

Goode TD, Holloway-Erickson CM, Maren S (2017) Extinction after fear memory reactivation fails to eliminate renewal in rats. Neurobiol Learn Mem 142:41–47. https://doi.org/10.1016/j.nlm.2017.03.001

Gräff J, Joseph NF, Horn ME, Samiei A, Meng J, Seo J, Rei D, Bero AW, Phan TX, Wagner F, Holson E, Xu J, Sun J, Neve RL, Mach RH, Haggarty SJ, Tsai LH (2014) Epigenetic priming of memory updating during reconsolidation to attenuate remote fear memories. Cell 156(1–2): 261–276. https://doi.org/10.1016/j.cell.2013.12.020

Gray R, Budden-Potts D, Bourke F (2019) Reconsolidation of traumatic memories for PTSD: a randomized controlled trial of 74 male veterans. Psychother Res 29(5):621–639. https://doi.org/10.1080/10503307.2017.1408973

Hong I, Kim J, Kim J, Lee S, Ko H-G, Nader K, Kaang B-K, Tsien RW, Choi S (2013) AMPA receptor exchange underlies transient memory destabilization on retrieval. Proc Natl Acad Sci 110(20):8218–8223. https://doi.org/10.1073/pnas.1322627111

Houtekamer MC, Henckens MJAG, Mackey WE, Dunsmoor JE, Homberg JR, Kroes MCW (2020) Investigating the efficacy of the reminder-extinction procedure to disrupt contextual threat memories in humans using immersive virtual reality. Sci Rep 10(1):1–18. https://doi.org/10.1038/s41598-020-73139-4

Ishii D, Matsuzawa D, Matsuda S, Tomizawa H, Sutoh C, Shimizu E (2012) No erasure effect of retrieval-extinction trial on fear memory in the hippocampus-independent and dependent paradigms. Neurosci Lett 523:76–81. https://doi.org/10.1016/j.neulet.2012.06.048

Ishii D, Matsuzawa D, Matsuda S, Tomizawa H, Sutoh C, Shimizu E (2015) An isolated retrieval trial before extinction session does not prevent the return of fear. Behav Brain Res 287:139–145. https://doi.org/10.1016/j.bbr.2015.03.052

James EL, Bonsall MB, Hoppitt L, Tunbridge EM, Geddes JR, Milton AL, Holmes EA (2015) Computer game play reduces intrusive memories of experimental trauma via reconsolidation-update mechanisms. Psychol Sci 26(8):1201–1215. https://doi.org/10.1177/0956797615583071

Johnson DC, Casey BJ (2015) Extinction during memory reconsolidation blocks recovery of fear in adolescents. Sci Rep 5(8863):1–5. https://doi.org/10.1038/srep08863

Jones CE, Monfils MH (2016) Post-retrieval extinction in adolescence prevents return of juvenile fear. Learn Mem 23(10):567–575. https://doi.org/10.1101/lm.043281.116

Jones CE, Ringuet S, Monfils MH (2013) Learned together, extinguished apart: reducing fear to complex stimuli. Learn Mem 20(12):674–685. https://doi.org/10.1101/lm.031740.113

Kessler H, Holmes EA, Blackwell SE, Schmidt AC, Schweer JM, Bücker A, Herpertz S, Axmacher N, Kehyayan A (2018) Reducing intrusive memories of trauma using a visuospatial interference intervention with inpatients with posttraumatic stress disorder (PTSD). J Consult Clin Psychol 86(12):1076–1090. https://doi.org/10.1037/ccp0000340

Khalaf O, Gräff J (2019) Reactivation of recall-induced neurons in the infralimbic cortex and the basolateral amygdala after remote fear memory attenuation. Front Mol Neurosci 12:1–8. https://doi.org/10.3389/fnmol.2019.00070

Khalaf O, Resch S, Dixsaut L, Gorden V, Glauser L, Gräff J (2018) Reactivation of recall-induced neurons contributes to remote fear memory attenuation. Science 1242:1239–1242

Kindt M, Soeter M (2013) Reconsolidation in a human fear conditioning study: a test of extinction as updating mechanism. Biol Psychol 92(1):43–50. https://doi.org/10.1016/j.biopsycho.2011.09.016

Kitamura H, Johnston P, Johnson L, Strodl E (2020) Boundary conditions of post-retrieval extinction: a direct comparison of low and high partial reinforcement. Neurobiol Learn Mem 174:107285. https://doi.org/10.1016/j.nlm.2020.107285

Klucken T, Kruse O, Schweckendiek J, Kuepper Y, Mueller EM, Hennig J, Stark R (2016) No evidence for blocking the return of fear by disrupting reconsolidation prior to extinction learning. Cortex 79:112–122. https://doi.org/10.1016/j.cortex.2016.03.015

Kredlow MA, Otto MW (2015) Interference with the reconsolidation of trauma-related memories in adults. Depress Anxiety 32(1):32–37. https://doi.org/10.1002/da.22343

Kredlow MA, Unger LD, Otto MW (2016) Harnessing reconsolidation to weaken fear and appetitive memories: a meta-analysis of post-retrieval extinction effects. Psychol Bull 142(3): 314–336. https://doi.org/10.1037/bul0000034

Kredlow MA, Orr SP, Otto MW (2018) Exploring the boundaries of post-retrieval extinction in healthy and anxious individuals. Behav Res Ther 108:45–57. https://doi.org/10.1016/j.brat.2018.06.010

Kroes MCW, Dunsmoor JE, Lin Q, Evans M, Phelps EA (2017) A reminder before extinction strengthens episodic memory via reconsolidation but fails to disrupt generalized threat responses. Sci Rep 7(1):1–14. https://doi.org/10.1038/s41598-017-10682-7

Lancaster CL, Monfils MH, Telch MJ (2020) Augmenting exposure therapy with pre-extinction fear memory reactivation and deepened extinction: a randomized controlled trial. Behav Res Ther 135:103730. https://doi.org/10.1016/j.brat.2020.103730

Lee JLC, Milton AL, Everitt BJ (2006) Reconsolidation and extinction of conditioned fear: inhibition and potentiation. J Neurosci 26(39):10051–10056. https://doi.org/10.1523/JNEUROSCI.2466-06.2006

Lee HJ, Haberman RP, Roquet RF, Monfils MH (2016) Extinction and retrieval + extinction of conditioned fear differentially activate medial prefrontal cortex and amygdala in rats. Front Behav Neurosci 9:1–10. https://doi.org/10.3389/fnbeh.2015.00369

Li J, Chen W, Caoyang J, Wu W, Jie J, Xu L, Zheng X (2017) Moderate partially reduplicated conditioned stimuli as retrieval cue can increase effect on preventing relapse of fear to compound stimuli. Front Hum Neurosci 11:1–12. https://doi.org/10.3389/fnhum.2017.00575

Liu J, Zhao L, Xue Y, Shi J, Suo L, Luo Y, Chai B, Yang C, Fang Q, Zhang Y, Bao Y, Pickens CL, Lu L (2014) An unconditioned stimulus retrieval extinction procedure to prevent the return of fear memory. Biol Psychiatry 76:895–901. https://doi.org/10.1016/j.biopsych.2014.03.027

Luyten L, Beckers T (2017) A preregistered, direct replication attempt of the retrieval-extinction effect in cued fear conditioning in rats. Neurobiol Learn Mem 144:208–215. https://doi.org/10.1016/j.nlm.2017.07.014

MacPherson K, Whittle N, Camp M, Gunduz-Cinar O, Singewald N, Holmes A (2013) Temporal factors in the extinction of fear in inbred mouse strains differing in extinction efficacy. Biol Mood Anxiety Disord 3(1):1–11. https://doi.org/10.1186/2045-5380-3-13

Maples-Keller JL, Price M, Jovanovic T, Norrholm SD, Odenat L, Post L, Zwiebach L, Breazeale K, Gross R, Kim SJ, Rothbaum BO (2017) Targeting memory reconsolidation to prevent the return of fear in patients with fear of flying. Depress Anxiety 34(7):610–620. https://doi.org/10.1002/da.22626

Marks EH, Zoellner LA (2014) Attenuating fearful memories: effect of cued extinction on intrusions. Emotion 14(6):1143–1154. https://doi.org/10.1037/a0037862

Meir Drexler S, Merz CJ, Hamacher-Dang TC, Marquardt V, Fritsch N, Otto T, Wolf OT (2014) Effects of postretrieval-extinction learning on return of contextually controlled cued fear. Behav Neurosci 128(4):474–481. https://doi.org/10.1037/a0036688

Merlo E, Milton AL, Goozée ZY, Theobald DE, Everitt BJ (2014) Reconsolidation and extinction are dissociable and mutually exclusive processes: behavioral and molecular evidence. J Neurosci 34(7):2422–2431. https://doi.org/10.1523/JNEUROSCI.4001-13.2014

Milton AL, Merlo E, Ratano P, Gregory BL, Dumbreck JK, Everitt BJ (2013) Double dissociation of the requirement for GluN2B-and GluN2A-containing NMDA receptors in the destabilization and restabilization of a reconsolidating memory. J Neurosci 33(3):1109–1115. https://doi.org/10.1523/JNEUROSCI.3273-12.2013

Misanin JR, Miller RR, Lewis DJ (1968) Retrograde amnesia produced by electroconvulsive shock after reactivation of a consolidated memory trace. Science 160(3827):554–555. https://doi.org/10.1126/science.160.3827.554

Monfils MH, Holmes EA (2018) Memory boundaries: opening a window inspired by reconsolidation to treat anxiety, trauma-related, and addiction disorders. Lancet Psychiatry 5(12):1032–1042. https://doi.org/10.1016/S2215-0366(18)30270-0

Monfils MH, Cowansage KK, Klann E, Ledoux JE (2009) Extinction-reconsolidation boundaries: key to persistent attenuation of fear memories. Science 324(5929):951–955. https://doi.org/10.1126/science.1167975

Monfils MH, Lee HJ, Keller NE, Roquet RF, Quevedo S, Agee L, Cofresi R, Shumake J (2019) Predicting extinction phenotype to optimize fear reduction. Psychopharmacology 236(1): 99–110. https://doi.org/10.1007/s00213-018-5005-6

Monti RIF, Alfei JM, Mugnaini M, Bueno AM, Beckers T, Urcelay GP, Molina VA (2017) A comparison of behavioral and pharmacological interventions to attenuate reactivated fear memories. Learn Mem 24(8):369–374. https://doi.org/10.1101/lm.045385.117

Myers KM, Davis M (2002) Behavioral and neural analysis of extinction. Neuron 36(4):567–584. https://doi.org/10.1016/S0896-6273(02)01064-4

Nader K, Schafe GE, Le Doux JE (2000) Fear memories require protein synthesis in the amygdala for reconsolidation after retrieval. Nature 406:722–726

Olshavsky ME, Jones CE, Lee HJ, Monfils MH (2013) Appetitive behavioral traits and stimulus intensity influence maintenance of conditioned fear. Front Behav Neurosci 7:1–7. https://doi.org/10.3389/fnbeh.2013.00179

Oyarzún JP, Lopez-Barroso D, Fuentemilla L, Cucurell D, Pedraza C, Rodriguez-Fornells A, de Diego-Balaguer R (2012) Updating fearful memories with extinction training during reconsolidation: a human study using auditory aversive stimuli. PLoS One 7(6). https://doi.org/10.1371/journal.pone.0038849

Piñeyro M, Monti RIF, Alfei JM, Bueno AM, Urcelay GP (2014) Memory destabilization is critical for the success of the reactivation-extinction procedure. Learn Mem 21(1):46–54. https://doi.org/10.1101/lm.032714.113

Ponnusamy R, Zhuravka I, Poulos AM, Shobe J, Merjanian M, Huang J, Wolvek D, O'Neill PK, Fanselow MS (2016) Retrieval and reconsolidation accounts of fear extinction. Front Behav Neurosci 10:1–11. https://doi.org/10.3389/fnbeh.2016.00089

Popik B, Amorim FE, Amaral OB, Alvares, L. de O. (2020) Shifting from fear to safety through deconditioning-update. elife 9:1–20. https://doi.org/10.7554/eLife.51207

Rao-Ruiz P, Rotaru DC, van der Loo RJ, Mansvelder HD, Stiedl O, Smit AB, Spijker S (2011) Retrieval-specific endocytosis of GluA2-AMPARs underlies adaptive reconsolidation of contextual fear. Nat Neurosci 14(10):1302–1308. https://doi.org/10.1038/nn.2907

Sara SJ (2000) Retrieval and reconsolidation: toward a neurobiology of remembering. Learn Mem 7(2):73–84. https://doi.org/10.1101/lm.7.2.73

Schiller D, Monfils MH, Raio CM, Johnson DC, Ledoux JE, Phelps EA (2010) Preventing the return of fear in humans using reconsolidation update mechanisms. Nature 463(7277):49–53. https://doi.org/10.1038/nature08637

Schiller D, Kanen JW, LeDoux JE, Monfils MH, Phelps EA (2013) Extinction during reconsolidation of threat memory diminishes prefrontal cortex involvement. Proc Natl Acad Sci U S A 110(50):20040–20045. https://doi.org/10.1073/pnas.1320322110

Schroyens N, Beckers T, Kindt M (2017) In search for boundary conditions of reconsolidation: a failure of fear memory interference. Front Behav Neurosci 11:1–13. https://doi.org/10.3389/fnbeh.2017.00065

Sevenster D, Beckers T, Kindt M (2012) Retrieval per se is not sufficient to trigger reconsolidation of human fear memory. Neurobiol Learn Mem 97(3):338–345. https://doi.org/10.1016/j.nlm.2012.01.009

Sevenster D, Beckers T, Kindt M (2013) Prediction error governs pharmacologically induced amnesia for learned fear. Science 339:830–834. https://doi.org/10.7551/mitpress/3629.003.0022

Sevenster D, Beckers T, Kindt M (2014) Prediction error demarcates the transition from retrieval, to reconsolidation, to new learning. Learn Mem 21(11):580–584. https://doi.org/10.1101/lm.035493.114

Sharko AC, Fadel JR, Kaigler KF, Wilson MA (2017) Activation of orexin/hypocretin neurons is associated with individual differences in cued fear extinction. Physiol Behav 178:93–102. https://doi.org/10.1016/j.physbeh.2016.10.008

Shiban Y, Brütting J, Pauli P, Mühlberger A (2015) Fear reactivation prior to exposure therapy: does it facilitate the effects of VR exposure in a randomized clinical sample? J Behav Ther Exp Psychiatry 46:133–140. https://doi.org/10.1016/j.jbtep.2014.09.009

Shumake J, Monfils MH (2015) Assessing fear following retrieval + extinction through suppression of baseline reward seeking vs. freezing. Front Behav Neurosci 9:1–9. https://doi.org/10.3389/fnbeh.2015.00355

Shumake J, Jones C, Auchter A, Monfils MH (2018) Data-driven criteria to assess fear remission and phenotypic variability of extinction in rats. Philos Trans R Soc B Biol Sci 373(1742). https://doi.org/10.1098/rstb.2017.0035

Soeter M, Kindt M (2011) Disrupting reconsolidation: pharmacological and behavioral manipulations. Learn Mem 18(6):357–366. https://doi.org/10.1101/lm.2148511

Soliman F, Glatt CE, Bath KG, Levita L, Jones RM, Pattwell SS, Jing D, Tottenham N, Amso D, Somerville LH, Voss HU, Glover G, Ballon DJ, Liston C, Teslovich T, Van Kempen T, Lee FS, Casey BJ (2010) A genetic variant BDNF polymorphism alters extinction learning in both mouse and human. Science 327:863–867. https://doi.org/10.7551/mitpress/8876.003.0036

Stafford JM, Maughan DAK, Ilioi EC, Lattal KM (2013) Exposure to a fearful context during periods of memory plasticity impairs extinction via hyperactivation of frontal-amygdalar circuits. Learn Mem 20(3):156–163. https://doi.org/10.1101/lm.029801.112

Steinfurth ECK, Kanen JW, Raio CM, Clem RL, Huganir RL, Phelps EA (2014) Young and old Pavlovian fear memories can be modified with extinction training during reconsolidation in humans. Learn Mem 21(7):338–341. https://doi.org/10.1101/lm.033589.113

Suzuki A, Josselyn SA, Frankland PW, Masushige S, Silva AJ, Kida S (2004) Memory reconsolidation and extinction have distinct temporal and biochemical signatures. J Neurosci 24(20):4787–4795. https://doi.org/10.1523/JNEUROSCI.5491-03.2004

Tedesco V, Roquet RF, DeMis J, Chiamulera C, Monfils MH (2014) Extinction, applied after retrieval of auditory fear memory, selectively increases zinc-finger protein 268 and phosphorylated ribosomal protein S6 expression in prefrontal cortex and lateral amygdala. Neurobiol Learn Mem 115:78–85. https://doi.org/10.1016/j.nlm.2014.08.015

Telch MJ, York J, Lancaster CL, Monfils MH (2017) Use of a brief fear memory reactivation procedure for enhancing exposure therapy. Clin Psychol Sci 5(2):367–378. https://doi.org/10.1177/2167702617690151

Thompson A, Lipp O v (2017) Extinction during reconsolidation eliminates recovery of fear conditioned to fear-irrelevant and fear-relevant stimuli. Behav Res Ther 92:1–10. https://doi.org/10.1016/j.brat.2017.01.017

Vermes JS, Ayres R, Goés AS, del Real N, Araújo ÁC, Schiller D, Neto FL, Corchs F (2020) Targeting the reconsolidation of traumatic memories with a brief 2-session imaginal exposure intervention in post-traumatic stress disorder. J Affect Disord 276:487–494. https://doi.org/10.1016/j.jad.2020.06.052

Warren VT, Anderson KM, Kwon C, Bosshardt L, Jovanovic T, Bradley B, Norrholm SD (2014) Human fear extinction and return of fear using reconsolidation update mechanisms: the contribution of on-line expectancy ratings. Neurobiol Learn Mem 113:165–173. https://doi.org/10.1016/j.nlm.2013.10.014

Zimmermann J, Bach DR (2020) Impact of a reminder/extinction procedure on threat-conditioned pupil size and skin conductance responses. Learn Mem 27(4):164–172. https://doi.org/10.1101/lm.050211.119

Extinction-Based Exposure Therapies Using Virtual Reality

Jessica L. Maples-Keller, Andrew Sherrill, Preethi Reddi, Seth D. Norrholm, and Barbara O. Rothbaum

Contents

Abstract The focus of this chapter is an overview of integrating virtual reality (VR) technology within the context of exposure therapy for anxiety disorders, a gold standard treatment, with a focus on how VR can help facilitate extinction learning processes integral to these interventions. The chapter will include an overview of advantages of incorporating VR within exposure therapy, and benefits specifically within an inhibitory learning approach for extinction training. A review of the empirical literature on the effectiveness of VR exposure therapy for specific phobia

---

J. L. Maples-Keller (✉), A. Sherrill, and B. O. Rothbaum
Department of Psychiatry and Behavioral Sciences, Emory University School of Medicine, Atlanta, GA, USA
e-mail: jmaple2@emory.edu

P. Reddi
Medical College of Georgia, Augusta University, Augusta, GA, USA

S. D. Norrholm
Department of Psychiatry and Behavioral Neurosciences, Wayne State University School of Medicine, Detroit, MI, USA

Curr Topics Behav Neurosci (2023) 64: 335–352
https://doi.org/10.1007/7854_2023_437
Published Online: 12 August 2023

and PTSD will be provided, as well as practical overview of how to effectively incorporate VR within exposure therapy.

Keywords Anxiety, Extinction learning, PTSD, Specific phobia, Translational science, Virtual reality

1 Introduction

Exposure therapy is a first-line treatment for anxiety disorders (e.g., Stewart and Chambless 2009). This chapter will provide an overview of the integration of virtual reality (VR) methods within exposure therapy, with a focus on how they may facilitate extinction learning processes. Advantages of VR for extinction training, including facilitating presence and precision, decreasing avoidance behaviors, and enhancing feasibility, will be reviewed. Specific benefits of using VR using an inhibitory learning approach will be provided. The extant empirical literature on VR exposure therapy will be reviewed, including for specific phobia and posttraumatic stress disorder (PTSD), and a brief summary of how to implement effective VR will be provided.

2 Virtual Reality Exposure Therapy

Meta-analytic results indicate that exposure therapy for anxiety disorders is an effective treatment approach (e.g., Norton and Price 2007; Hofmann and Smits 2008; Stewart and Chambless 2009) with basic, translational, and clinical empirical support (e.g., Craske et al. 2018; Milad et al. 2014). Exposure therapy is based on emotional processing theory, which posits that anxiety disorders involve pathological fear responses based on an underlying "fear network" in which neutral stimuli or responses become associated with exaggerated likelihood of threat or danger (Foa and Kozak 1986). As such, in this treatment approach, traditional methods involve systematically, gradually, and therapeutically exposing a patient to feared stimuli in order to facilitate a reduction in the fear response; the reduction of fear associated with repeated exposure trials is known as *extinction learning* and the maintenance of this learning is termed *extinction retention.* Starting in the 1990s, psychologists began to utilize VR methods within exposure therapy, with the first published study finding that VR-graded exposure treatment for fear of heights (i.e., acrophobia), which involved systematic, gradual exposures to virtual reality environments of increasing height facilitated by a therapist, resulted in significant reductions in symptoms compared to a waitlist condition (Rothbaum et al. 1995). Since this initial study was performed, the empirical literature for the effectiveness of virtual reality exposure therapy (VRE) for multiple anxiety disorders has grown substantially.

VR is a technological interface in which a user participates in a computer-generated three-dimensional virtual world while experiencing sensory input that can perceptually simulate real-world interactive experiences. Within the VR environment, the experience is tailored to be immersive and interactive to facilitate an experience of presence. A typical VR set-up will involve a head-mounted display (HMD) with a head tracker (to allow replication of in vivo movements within the VR environment), its synthesized sounds, and a vibrotactile platform to facilitate simulation of multiple senses, including auditory, tactile, and olfactory systems. Technological advances have led to newer innovations that can increase accessibility to VR methods such as cost-effective, smartphone-based VR head-mounted displays and mixed reality environments that allow synergy, interaction, and manipulation of virtual items and "real life" environments; the latter serving to further enhance ability to engage a feeling of presence and immersion.

3 Advantages of VR for Extinction Training

Conventional exposure modalities are composed of two primary elements: (1) in vivo exposure, which is "real life" confrontation with fear stimuli, and (2) imaginal exposure, which is the use of one's own memory, episodic recall, and imagination to confront feared stimuli and past events. For VR applications, patients instead confront VR-simulated anxiety and trauma-related stimuli. For an imaginal exposure VR approach, patients recount their traumatic memories with eyes open while the clinician matches multisensory VR stimuli to optimize engagement with the traumatic memory (Rothbaum et al. 2001; Sherrill et al. 2020). Anxiety-related disorders are conceptualized as the product of a conditioning process by which a neutral stimulus acquires the excitatory stimulus functions of a paired aversive unconditioned stimulus (US; Myers and Davis 2002, 2007; VanElzakker et al. 2014). The learned association between the US and the conditioned stimuli (CS) is never "unlearned" or "erased" (Myers and Davis 2002). However, fear-conditioned responses can be inhibited through therapeutic exposures, which are learning experiences during which the same previously neutral, now fear-evoking, stimulus is experienced directly and with less frequent and intense US association (Craske et al. 2014). This new learning, called extinction training, occurs when inhibitory responses to previously reinforced stimuli become more prominent than excitatory fear responses. Immersive VR simulations can facilitate extinction training by (a) providing opportunities for the patient to be present with the CS, (b) making exposure exercises more precise and individualized, (c) reducing opportunities for avoidance, and (d) making repeated and gradual exposure feasible.

3.1 VR-Facilitated Presence in Exposures

Critical to the execution of VR for extinction training is the ability of the technological equipment and therapist to create a sense of *presence* (Schubert et al. 2001; Wirth et al. 2007), or of "being there." This experience is widely viewed by therapists as an essential element to activating an emotional response to computer simulations and achieving meaningful long-term outcomes following exposure therapy (Rothbaum et al. 1995). Evidence indicates that presence and state anxiety are positively correlated (Ling et al. 2014). Providing further evidence for the importance of presence, the patient's tolerability of using VR equipment has been positively correlated to one's perceived presence (Weech et al. 2019). To evoke a sense of presence, effective VR environments contain realistic spatial relationships between objects, opportunities for agency, and a seamless coordination of sensory signals. The successful integration of these technological capabilities to create and deliver a convincing simulation is called *immersion* (Slater and Wilbur 1997).

Immersive VR technologies can potentially equip exposure therapists with the capacity to present patients with an exhaustive menu of stimuli. However, clinical guidance for VR exposure therapy suggests that immersion is a necessary but insufficient condition for presence (Sherrill et al. 2020). VR is simply a tool, not the therapy itself. The VR exposure therapist arranges the multisensory experience in a clinically informed manner that fits user characteristics and is reported by clients to feel "natural" or "real." The therapist deploys patient-relevant VR components to trigger patient-specific emotional reactions and behavioral responses that will, in turn, promote inhibitory learning. Additionally, to achieve a sense of naturalness of the experience, the therapist needs to deploy VR components consistently with the patient's expectations about how objects and people should interact in the real world. More realistic VR environments have been shown to evoke a greater sense of presence (e.g., Newman et al. 2022). The therapist aims to limit opportunities for the patient to revert to avoidant thinking such as, "this is not a real airplane" which would likely prevent inhibitory learning.

3.2 VR-Facilitated Precision in Exposures

Over the past several decades, there has been a proliferation of available VR environments useful for exposure therapy. The diversity of VR simulations equips the exposure therapist with the capability to precisely match stimulus content to specific patient needs and to select how and when it is delivered. At its core, exposure therapy is rooted in the idea that the patient must attend to, and engage with, stimuli that activate fear (Foa and Kozak 1986). The exposure therapist and patient collaborate to target specific stimuli at specific dosages (e.g., the size of a spider) and physical proximities (e.g., the distance between the patient and spider). They also collaborate on behavioral responses to inhibit (e.g., patient-specific avoidance behaviors) and select (e.g., awareness of and engagement with the stimuli). Importantly, the therapist can control outcomes within the simulations, setting up robust expectancy violations that are central to inhibitory learning (Craske et al. 2014). This degree of control over the parameters of stimulus presentation often results in far greater customizability than conventional in vivo exposure and allows the VR therapist to end exposures after achieving conditions that violate expectancies.

The precision of VR simulations allows these systems to provide graduated exposure tasks that are typical for a given fear. For example, one fear of flying system (Rothbaum et al. 1999b) included the following sequence of tasks: sitting near the gate to the jetway, walking through the jetway, walking through the airplane aisle, sitting down, taxiing from the gate to runway, taking off, flying, and then landing. Depending on what a specific patient needs to learn, each of these tasks can be repeated or the patient can progress through all tasks in whatever order is optimal to facilitate learning. However, therapists should keep in mind that logical order of events may be needed to foster increased engagement and immersion.

3.3 VR-Facilitated Avoidance Prevention in Exposures

Avoidance behavior is central to all anxiety-related disorders and prevents extinction learning from being acquired or consolidated (Craske et al. 2014; Rothbaum and Davis 2003). Expectedly, patients are hesitant to approach most exposure exercises. One advantage of VR exposures is that patients often find them more acceptable as an entry point than traditional exposures (Garcia-Palacios et al. 2007). Patients may be less likely than usual to engage in avoidance strategies when they detect an objectively safe and controlled environment, which allows them to engage with CSs and process their experiences with the therapist. Although patients do not completely habituate to the contrived nature of simulations, evidence suggests extinction learning in VRE generalized into in vivo approach behavior (Morina et al. 2015; Rothbaum et al. 1995). Once VRE patients are in a real-life feared situation (e.g., an actual airplane taking off), they recall becoming more comfortable the last time they were in this situation, even if it was in VR.

Even with awareness of the inherent safety of VR, avoidant strategies must be prevented to achieve extinction learning. Delivering VR exposures is an active process that requires mastery in exposure therapy, which requires expertise in stimulus presentation and clinical observation. The clinician provides verbal cues to encourage the patient to continue engaging with the environment. Additionally, the clinician detects and redirects avoidance strategies such as overly scanning the environment, looking away from aspects of the simulation, soothing through touch (e.g., rubbing knees) or movement (e.g., bouncing feet), cognitive avoidance (e.g., "It'll be over soon," "this isn't real,"), or distraction (e.g., counting). VR exposures often allows for the detection of avoidant strategies because the VR equipment allows the therapist to see, hear, and smell everything the patient experiences in real time. One avoidant behavior that may be difficult to detect is if the patient is closing their eyes.

3.4 VR-Facilitated Feasibility in Exposures

While extinction training requires precise stimuli presentation without avoidance, it also requires exposure tasks that are practical and achievable, which is a large reason why clinicians are interested in VRE (Segal et al. 2011). Using traditional methods, exposures can be conceptually sound but impossible to execute. For example, an exposure to giving three academic speeches in a single day might be clinically indicated but logistically infeasible. VR can reduce logistical barriers, diminish financial costs, and improve acceptability. Logistical preparation is minimal. Even practical reality can be temporarily suspended; for example, if the patient needs to only learn that standing in silence in front of a large group is tolerable, the therapist can present that patient with a full auditorium of people who will calmly and attentively look at the patient for however long as clinically indicated. An additional feasibility advantage of VR is the capability to provide exposures in varied contexts to assist in generalization of new learning. Given that the original associations between the CS and the US will always remain intact, it is not uncommon for fear responses to return after treatment, especially if the patient encounters CSs in a context that differs from exposure contexts (Vervliet et al. 2013). VR provides opportunities for variable exposures in multiple contexts without leaving the therapist's office. For example, to target a fear of heights, a VR therapist can present the following environments: walking up a staircase, walking across a low bridge, taking a glass elevator up to the tenth floor of a building, walking across a high bridge, looking over the wall of a tenth-floor balcony, taking a glass elevator to the rooftop of a skyscraper, and looking of the edge of the skyscraper's rooftop. These environments not only present varied contexts but also varying levels of stimulus intensity. To facilitate an inhibitory learning model of extinction, the VR therapist can present these varying levels of intensity in a random order to promote consolidation across contexts (Craske et al. 2014). A graded exposure model would dictate moving up the hierarchy from least to most difficult in that order and only proceeding to the next once fear has decreased to the current situation.

A feasible and cost-efficient entry point for clinicians looking to provide VR exposure therapy is using a smartphone-based HMD, which is simply a plastic or rubber casing in which to place a contemporary smartphone that will provide immersive visuals and sounds. This low-cost approach has preliminary evidence of effectiveness (Lindner et al. 2017). In addition to VR, smartphone-based HMDs can play 360° videos using online video players such as YouTube, which contain a repository of hundreds of videos for common exposure targets including flying, heights, driving, crowds, animals, and public speaking (Stevens and Sherrill 2021). These 360° videos are not "true" VR due to the passive user experience, though they can provide a sense of presence with CSs and can be used out of the therapist's office.

3.5 *Summary of VR Advantages for Extinction Learning*

The advantages of VR for extinction training previously described above map onto contemporary suggestions for maximizing effective exposure suggested by the inhibitory learning theory of extinction (Craske et al. 2014). Outlined in Table 1 are specific ways in which VR technologies can satisfy these recommendations, using fear of flying as an example. Importantly, all the advantages of VR for extinction training build upon conventional practices of exposure therapy. VR does not make exposure more effective but rather provides practical opportunities to ensure extinction training is achievable, implemented efficiently, and long-lasting.

4 Virtual Reality Exposure Therapy Research

Since the first published study using virtual reality in the context of therapy for mental health problems (Rothbaum et al. 1995), this area of clinical research has expanded significantly with studies conducted in several anxiety disorders as well as other mental health conditions such as schizophrenia, acute and chronic pain, addiction, eating pathology, and autism (Maples-Keller et al. 2017). The bulk of the clinical research in this area has focused on virtual reality extinction for specific phobia and PTSD, given their optimal match with VR methods. The wide body of translational and clinical research aimed at better understanding PTSD within a fear extinction model (e.g., Norrholm et al. 2011; Rothbaum and Davis 2003) provides strong fit with extinction VR methods. Specific phobia involves excessive fear related to a specific object, providing a strong opportunity to facilitate extinction learning via VR exposure to these specific stimuli. As such, we will briefly review the empirical literature surrounding virtual reality extinction for specific phobias and PTSD.

5 Virtual Reality Exposure for Specific Phobia

Specific phobia is defined by DSM-5 criteria as an anxiety disorder marked by persistent fear or anxiety about a specific object or situation that typically lasts for a period of 6 months or more and causes significant distress or impairment (APA 2013). In vivo exposure therapy is a first-line treatment for specific phobia due to strong evidence for its efficacy (Choy et al. 2007; Koch et al. 2004; Wolitzky-Taylor et al. 2008). There is some prior research indicating that patients prefer VRE over

Table 1 Advantages of VR for enhancing inhibitory learning and retrieving inhibitory learning

Inhibitory learning strategy	Specific benefits of using VR in exposures
Expectancy violation	• Therapist has 100% control of the frequency and intensity of aversive outcomes with simulations, thus guaranteeing expectancies violations (e.g., the expectation of the plane crashing will be violated) • For expectations regarding ability to tolerate emotion, patient and therapist can pre-determine the exact VR stimuli presentation and duration to test excitatory expectancies and rehearse inhibitory expectancies
Deepened extinction	• Therapist can deploy distinct VR elements separately (e.g., the sound of turbulence, then the visuals of bad weather) before deploying all elements together • Therapist can pair an extinguished VR element with a conditioned stimulus not yet targeted (e.g., after a fear response to sitting in plane is diminished, the therapist can make the plane take off)
Reinforced extinction	• Therapist can incorporate occasional CS-US pairings using novel US (e.g., while flying, patient can hear an avatar scream) • Given the complete control of VR, therapists can control ideal balance between CS-US pairings and CS-No US pairing to cultivate inhibitory expectances (e.g., "I can handle flying even if I hear jarring sounds")
Removal of safety signals	• During VR exposures using an HMD, patients cannot see therapist or office, which might function as a safety signal • VR systems include interaction with simulations (e.g., holding controller with a virtual boarding pass) that prevents patient from possessing safety objects
Stimulus variability	• VR systems often contain graduated exposures, and the therapist can choose an intensity level at random (e.g., presenting a flight with turbulence and then one without) • VR therapists can use distinct environments that target the same CS (e.g., after flying in simulated plane, patient can view a 360° video taken mid-flight within a fuselage)
Retrieval cues	• VR therapist can use the many neutral stimuli within rich environments to retrieve inhibitory learning (e.g., patient attends to engine sounds and later retrieves CS-No US relationship) • Given that VR can bring exposures into the therapist office (e.g., a plane), the therapist can immediately discuss what was learned and develop ways of remembering CS-No CS learning (e.g., "the view above the cloud was actually beautiful and peaceful")
Multiple contexts	• VR inherently brings patient into contexts that are experienced as distinct from the therapist office (e.g., patients walk in terminal as if the terminal were real) • VR simulations can create a sense the patient is alone or with people different from the therapist (e.g., patients interact with flight attendant as if she were real)

in vivo exposure therapy for this disorder (Garcia-Palacios et al. 2007). Several randomized trials for specific phobias have reported significant improvements following VRE compared to control conditions (Garcia-Palacios et al. 2002; Maltby et al. 2002; Michaliszyn et al. 2010; Rothbaum et al. 2000, 2006), with some studies showing similar effects between VRE therapy and in vivo exposure therapy

(Michaliszyn et al. 2010; Rothbaum et al. 2000, 2006). Yet, three non-randomized interventional studies showed inconsistent differences between efficacy of VRE and in vivo exposure therapy (Coelho et al. 2006; Emmelkamp et al. 2001; Michaliszyn et al. 2010).

For the treatment of flying phobia, VRE therapy has consistently been shown to reduce self-reported and physiological symptoms of fear of flying (Maltby et al. 2002; Mühlberger et al. 2001, 2003; Rothbaum et al. 2000, 2006). A randomized controlled study found a decrease in anxiety during an actual flight following VRE therapy compared to waitlist control (Rothbaum et al. 2000). In the treatment of acrophobia, or fear of heights, several interventional studies found decreased acrophobia symptoms post-VRE compared to pre-VRE as measured by behavioral avoidance, attitudes toward heights, and acrophobic distress (Coelho et al. 2006; Emmelkamp et al. 2002). In a randomized trial investigating a reconsolidation paradigm ($N = 89$), participants who received VRE for fear of flying demonstrate significant reductions in fear of flying, and gains were maintained across 12-month follow-up time period (Maples-Keller et al. 2017). Multiple randomized controlled and interventional studies have shown VRE therapy to be effective compared to waitlist controls and post-treatment compared to pre-treatment in treating arachnophobia, or fear of spiders, as measured by behavioral avoidance tests (Bouchard et al. 2006; Côté and Bouchard 2005; Garcia-Palacios et al. 2002; Michaliszyn et al. 2010), standardized questionnaires (Bouchard et al. 2006; Côté and Bouchard 2005; Garcia-Palacios et al. 2002; Michaliszyn et al. 2010), and psychophysiological data (Côté and Bouchard 2005). Preliminary research has supported the use of VRE therapy in treating driving phobia but the body of available literature is small (Beck et al. 2007; Wald and Taylor 2003), particularly because driving is a difficult virtual environment to create. The VR driving environment must include several environments – the interior of the vehicle and the multiple exterior roads and surroundings through which the user drives, all while including vehicle controls that feel natural. Additional research is needed to fully understand whether differences exist in VRE therapy efficacy based on subtypes of specific phobia (Choy et al. 2007; Wolitzky-Taylor et al. 2008).

For the treatment of flight phobia, randomized controlled trials have found that VRE therapy has similar efficacy as in vivo exposure therapy as measured by standardized questionnaires (Rothbaum et al. 2000, 2006), willingness of participants to fly on actual airplanes (Rothbaum et al. 2000, 2006), and intermittent anxiety ratings during flight (Rothbaum et al. 2000). Interventional studies show similar results following VRE therapy and in vivo exposure therapy in patients with acrophobia as indicated by acrophobia symptoms (Coelho et al. 2006; Emmelkamp et al. 2001, 2002), attitude toward heights (Coelho et al. 2006; Emmelkamp et al. 2002), behavioral avoidance (Coelho et al. 2006; Emmelkamp et al. 2001, 2002), and anxiety ratings during flight (Rothbaum et al. 2000). Another interventional study conducted by Michaliszyn et al. (2010) shows similar outcomes between VRE therapy and in vivo exposure therapy groups measured by a fear of spiders questionnaire and behavioral avoidance test in patients with arachnophobia (Michaliszyn et al. 2010).

The findings regarding the effectiveness of VR extinction for specific phobia compared to traditional in vivo exposure therapy have been inconsistent. One interventional study found significantly lower scores following VRE therapy compared to in vivo exposure therapy as measured by the attitude toward heights questionnaire (Emmelkamp et al. 2001). A randomized trial comparing virtual and in vivo exposure for fear of spiders found significant improvements in both groups, with comparable outcomes on four of five measures and the in vivo group demonstrating greater improvement on questionnaire related to spider beliefs (Michaliszyn et al. 2010).

While VRE therapy is effective in treating fear of flying, the impact on longer term maintenance of gains requires additional research. For instance, in a randomized controlled study of patients with a fear of flying, VRE was superior to attention placebo control at post-treatment but there was not a significant difference between VRE and attention placebo control at 6 month-follow up (Maltby et al. 2002). Yet, a later randomized controlled study showed superiority of VRE therapy compared to waitlist control on all questionnaire measures of fear of flying at 6 months and 12 months following treatment (Rothbaum et al. 2006). Randomized controlled studies examining the efficacy of VRE therapy in patients with arachnophobia also showed consistent improvements following exposure to VRE therapy compared to waitlist control groups post-treatment (Garcia-Palacios et al. 2002; Michaliszyn et al. 2010). A few studies have examined the effects of VR extinction in conjunction with other intervention methods for specific phobia. For example, in a study of one session of VRE with and without cognitive treatment, a reduction in fear of flying symptoms was found only in VRE exposure group (Mühlberger et al. 2003). An earlier randomized study by this group compared one VR test flight plus four VR exposure flights to treatment with one VR test flight followed by relaxation training. The authors found that four sessions of repeated VRE therapy reduced fear responses to a greater extent than one session of VRE therapy. Over time, symptoms improved in both treatment groups, but reduction was greater with repeated VR exposure (Mühlberger et al. 2001).

Given that exposure therapy is based on fear extinction principles, combining this treatment with pharmacological intervention that may enhance extinction learning shows promise. D-cycloserine (DCS) is a partial agonist of the glutamatergic (N-methyl-D-aspartate (NMDA) receptor that has been shown to enhance learning in pre-clinical and clinical paradigms. For example, a randomized controlled study in which participants were treated with two sessions of virtual reality exposure to heights preceded by DCS or placebo found that those who received DCS had significantly less fear as compared to subjects receiving placebo (Ressler et al. 2004). A subsequent study by a different group who investigated *post-session* administration of DCS following VRE for acrophobia did not find any between-group differences (Tart et al. 2013).

In sum, research indicates promise for the use of VRE therapy in patients with specific phobia. VRE therapy been found to be as efficacious as in vivo exposure therapy, the current first-line therapy for specific phobia and has been found to be preferred by patients (Garcia-Palacios et al. 2007). While studies show consistent improvements following VRE therapy in some specific phobia subtypes, including fear of flying, acrophobia, and arachnophobia, additional research is needed to indicate whether VRE therapy is as efficacious in treating other specific phobia subtypes. In addition, most current studies on VRE therapy include small sample sizes, indicating the need for larger studies on the use of VRE in treating specific phobia.

6 Virtual Reality Exposure for PTSD

Posttraumatic stress disorder (PTSD) is characterized by a heightened response to threatening stimuli following a traumatic event, which may include avoidance, intrusion, negative alterations in cognition and mood, or hyperarousal symptoms (APA 2013). Exposure therapy (Foa et al. 2019) is a gold standard treatment for PTSD (DiMauro 2014; Powers et al. 2010). There is an extensive body of literature in the form of case studies and series that have shown potential for the use of VRE therapy in treating PTSD following various traumatic events. These case studies have found decreased PTSD symptoms following VRE therapy in a former helicopter pilot in Vietnam (Rothbaum et al. 1999a, b), a survivor of the 9/11/2001 World Trade Center attacks (Difede and Hoffman 2004), a survivor of a deadly terrorist bulldozer crash (Freedman et al. 2010), a combat engineer (Gerardi et al. 2008), and a soldier of the Polish Military Contingent in Iraq, who narrowly escaped death three times (Tworus et al. 2010). Some of these case studies reported a significant decrease in Clinician Administered PTSD Scale (CAPS) symptoms immediately post-treatment compared to pre-treatment, with a maintenance in these gains at 6-month follow-up (Freedman et al. 2010; Rothbaum et al. 1999a, b). These case studies have also shown potential for use of VRE therapy to augment other treatment modalities for PTSD, including imaginal exposure (Freedman et al. 2010), desensitization (Tworus et al. 2010), and traditional exposure therapy (McLay et al. 2010).

Several trials have found a significant decrease in PTSD symptoms following VRE therapy (Ready et al. 2006; Rizzo et al. 2010; Rothbaum et al. 2014), with symptom reduction being maintained across long-term follow-ups (Ready et al. 2006; Rothbaum et al. 2014). One open clinical trial ($N = 20$) in active-duty servicemen previously engaged in PTSD treatment without benefit found that 16 participants no longer met DSM IV criteria for PTSD following VRE treatment (Rizzo et al. 2010). Research also indicates VRE treated groups experience greater reduction in PTSD symptoms as compared to waitlist control groups (Difede et al. 2007; Reger et al. 2016).

Most studies have shown similar outcomes following VRE therapy and in vivo exposure therapy in PTSD patients (Botella et al. 2015; Gonçalves et al. 2012; Reger et al. 2016). Randomized controlled studies have also found no significant difference in dropout rates between individuals treated with VRE therapy and those treated with standard or prolonged exposure therapy (Gonçalves et al. 2012; Reger et al. 2016). Additionally, uncontrolled interventional studies have found high satisfaction and acceptability among patients treated with VRE therapy (Beck et al. 2007; Botella et al. 2015). One open trial in active-duty soldiers in theater ($N = 20$) found a significantly greater improvement in PTSD symptoms in individuals who underwent VRE therapy compared to those who underwent treatment as usual (McLay et al. 2011).

While most studies have shown similar outcomes in PTSD patients following VRE and in vivo exposure therapy, a few have noted varying outcomes (DiMauro 2014). A randomized trial comparing PE, VRE, and a minimal attention waitlist group found no difference between PE and VRE with regard to dropout or treatment outcome; post-hoc analysis found PE was associated with greater reduction at 3- and 6-month follow-up (Reger et al. 2016). Novel uses of VRE for PTSD are being explored; for example, a recent feasibility study found military sexual trauma (MST) focused VRE was associated with PTSD symptom reduction across treatment and gains were maintained across follow-up (Loucks et al. 2019).

Literature accumulated to date indicates a vast potential for the use of VRE therapy to treat PTSD in civilian, active-duty service members, and veteran populations. However, it is important to recognize that there are few large randomized controlled studies [Reger et al. 2016 ($n = 162$); Rothbaum et al. 2014 ($n = 156$)] with most research involving case studies or studies with sample sizes under 30 individuals. Recently, one smaller study has shown DCS may be effective in further reducing PTSD symptoms as an adjunctive treatment to VRE therapy (Difede et al. 2014). A larger controlled trial did not find significant differences on PTSD treatment outcomes when comparing VRE with DCS augmentation to placebo control (Rothbaum et al. 2014), but the VRE treatment did result in significant PTSD symptom reduction from pre to post-treatment and gains were maintained at follow-up. This study did find advantages of DCS over placebo or alprazolam on psychobiological measures including salivary cortisol and acoustic startle response. Dexamethasone has been shown to facilitate fear extinction and was investigated in a randomized placebo-controlled trial as augmentation to VRE for PTSD, but dropout rates were significantly higher in the dexamethasone group (Maples-Keller et al. 2019). As the use of VRE therapy continues to increase, it is important to understand its relation to additional PTSD treatments and adjunctive treatments and the degree to which the combination of treatment modalities facilitates or hinders extinction learning. VR offers the advantage of precise methodological control in research, allowing the exact same stimuli to be presented across sessions, participants, and conditions.

7 Effective Approaches to VR Extinction

While the guidance provided in this chapter provides recommendations on how to maximize extinction training using VR methods (Table 1), brief general recommendations for effective VR extinction approaches are provided here (for a more in-depth discussion of clinical implementation of VR extinction, please see Sherrill et al. 2020). First, prior to attempting VR-based extinction training, clinical training and competency in exposure therapy should be obtained. This is foundational and a requisite. The successful use of VR technology itself is a clinical competency and, as such, it is important for clinicians pursuing VR extinction to obtain training and consultation in VR extinction methods. This training includes familiarity with the hardware and software, troubleshooting technical issues, and best practices for skillfully operating VR equipment while concurrently attending to patients in therapy sessions. Practicing VR exposure via role play or confederates is often helpful to practice being clinically attentive while facilitating VR extinction learning throughout the session. In addition to a standard and thorough clinical assessment at start of treatment consistent with standard clinical practice, VR extinction will also involve assessment at start of treatment of the specifics of the feared stimuli to tailor the VR experience during exposure sessions. For instance, this may include details related to the specific phobia, or specific details about an index trauma so that the clinician can practice and prepare the VR environment in advance of therapeutic exposure in the presence of a patient. When facilitating the VR exposure, it is important to avoid the inclusion of details that are inconsistent with the traumatic memory that may negatively impact the patient's ability to engage in the experience. During the VR exposure, a collaborative approach can be implemented through discussion with the patient to match various salient features of the trauma, including querying whether the time of day, weather, noises, or other specifics within the fear memory match the stimuli presented.

8 Future Research Directions

While the empirical literature regarding VR extinction training has expanded significantly in the past two decades, unfortunately, many clinical studies enroll small samples and lack long-term follow-up; well-powered randomized trials with long-term follow-up are necessary. The extant research typically indicates comparable outcome to traditional exposure therapy; future research could identify individual differences or variables that may select patients who specifically would benefit from VR extinction training as a medium for exposure compared to traditional in vivo or imaginal methods. This could include investigating level of avoidances, difficulties with imagination or visualization, or significant engagement in distraction-related avoidance; the latter being an integrative effort to inform personalized medicine and treatment matching approaches. For example, a recent study found that younger age, the absence of antidepressant pharmacotherapy, higher degree of hyperarousal symptoms, and greater than minimal suicide risk were associated with greater PTSD symptom reduction in active-duty soldiers with PTSD randomized to VR exposure as compared to Prolonged Exposure (PE) (Norr et al. 2018). Future research can directly test hypothesized advantages of VR versus traditional exposure methods and treatment mechanisms within VR exposure. We would like to highlight the promise of VR-based extinction for conducting clinical research, particularly related to extinction models. Fear extinction is a translational model, with many studies involving objective assessment such as psychophysiological reactivity to feared stimuli or trauma cues (e.g., Maples-Keller et al. 2017). Within VR-based extinction, psychophysiological responses can be assessed during extinction sessions or at pre- and post-treatment as outcome measures. Brain mechanisms of fear extinction have been previously characterized; for example, in a healthy human study, activation in ventromedial prefrontal cortex and hippocampus occurred in response to extinguished versus unextinguished stimulus, and level of activation was associated with the magnitude of extinction retention (Milad et al. 2007). As such, future research could also use VR methods during brain imaging extinction protocols in order to provide standardized and more immersive extinction training and retention techniques. VR extinction training provides the opportunity to standardize dose across participants, or to measure specifics of dose or exposure administration (e.g., time spent in exposure, specific stimuli used, repetition) in ways that could inform the continued research on extinction learning and impact on clinical outcomes.

References

American Psychiatric Association (2013) Diagnostic and statistical manual of mental disorders, 5th edn. American Psychiatric Publishing, Arlington

Beck JG, Palyo SA, Winer EH, Schwagler BE, Ang EJ (2007) Virtual reality exposure therapy for PTSD symptoms after a road accident: an uncontrolled case series. Behav Ther 38(1):39–48. https://doi.org/10.1016/J.BETH.2006.04.001

Botella C, Serrano B, Baños RM, Garcia-Palacios A (2015) Virtual reality exposure-based therapy for the treatment of post-traumatic stress disorder: a review of its efficacy, the adequacy of the treatment protocol, and its acceptability. Neuropsychiatr Dis Treat 11:2533. https://doi.org/10.2147/NDT.S89542

Bouchard S, Côté S, St-Jacques J, Robillard G, Renaud P (2006) Effectiveness of virtual reality exposure in the treatment of arachnophobia using 3D games. Technol Health Care 14(1):19–27. https://doi.org/10.3233/THC-2006-14103

Choy Y, Fyer A, Lipsitz J (2007) Treatment of specific phobia in adults. Clin Psychol Rev 27(3): 266–286. https://doi.org/10.1016/J.CPR.2006.10.002

Coelho CM, Santos JA, Silvério J, Silva CF (2006) Virtual reality and acrophobia: one-year follow-up and case study. Cyberpsychol Behav 9(3):336–341. https://doi.org/10.1089/CPB.2006.9.336

Côté S, Bouchard S (2005) Documenting the efficacy of virtual reality exposure with psychophysiological and information processing measures. Appl Psychophysiol Biofeedback 30(3): 217–232. https://doi.org/10.1007/S10484-005-6379-X

Craske MG, Treanor M, Conway CC, Zbozinek T, Vervliet B (2014) Maximizing exposure therapy: an inhibitory learning approach. Behav Res Ther 58:10–23

Craske MG, Hermans D, Vervliet B (2018) State-of-the-art and future directions for extinction as a translational model for fear and anxiety. Philos Trans Royal Soc B: Biol Sci 373:20170025

Difede J, Hoffman HG (2004) Virtual reality exposure therapy for world trade center post-traumatic stress disorder: a case report. Cyberpsychol Behav 5(6):529–535. https://doi.org/10.1089/109493102321018169

Difede J, Cukor J, Jayasinghe N, Patt I, Jedel S, Spielman L et al (2007) Virtual reality exposure therapy for the treatment of posttraumatic stress disorder following. J Clin Psychiatry 68:11

Difede J, Cukor J, Wyka K, Olden M, Hoffman H, Lee FS, Altemus M (2014) D-cycloserine augmentation of exposure therapy for post-traumatic stress disorder: a pilot randomized clinical trial. Neuropsychopharmacology 39(5):1052. https://doi.org/10.1038/NPP.2013.317

DiMauro J (2014) Exposure therapy for posttraumatic stress disorder: a meta-analysis. Mil Psychol 26(2):120–130. https://doi.org/10.1037/MIL0000038

Emmelkamp P, Bruynzeel M, Drost L, van der Mast C (2001) Virtual reality treatment in acrophobia: a comparison with exposure in vivo. Cyberpsychol Behav 4(3):335–339. https://doi.org/10.1089/109493101300210222

Emmelkamp P, Krijn M, Hulsbosch A, de Vries S, Schuemie M, van der Mast C (2002) Virtual reality treatment versus exposure in vivo: a comparative evaluation in acrophobia. Behav Res Ther 40(5):509–516. https://doi.org/10.1016/S0005-7967(01)00023-7

Foa EB, Kozak MJ (1986) Emotional processing of fear: exposure to corrective information. Psychol Bull 99:20–35

Foa E, Hembree E, Rothbaum BO, Rauch SAM (2019) Prolonged exposure therapy for PTSD: emotional processing of traumatic experiences, 2nd edn. Oxford University Press, New York

Freedman SA, Hoffman HG, Garcia-Palacios A, Weiss PL, Avitzour S, Josman N (2010) Prolonged exposure and virtual reality–enhanced imaginal exposure for PTSD following a terrorist bulldozer attack: a case study. Cyberpsychol Behav Soc Netw 13(1):95–101. https://doi.org/10.1089/CYBER.2009.0271

Garcia-Palacios A, Hoffman H, Carlin A, Furness T, Botella C (2002) Virtual reality in the treatment of spider phobia: a controlled study. Behav Res Ther 40(9):983–993. https://doi.org/10.1016/S0005-7967(01)00068-7

Garcia-Palacios A, Botella C, Hoffman H, Fabregat S (2007) Comparing acceptance and refusal rates of virtual reality exposure therapy vs in vivo exposure by patients with specific phobias. Cyberpsychol Behav 10:722–724

Gerardi M, Rothbaum BO, Ressler K, Heekin M, Rizzo A (2008) Virtual reality exposure therapy using a virtual Iraq: case report. J Trauma Stress 21(2):209. https //doi.org/10.1002/JTS.20331

Gonçalves R, Pedrozo AL, Coutinho ESF, Figueira I, Ventura P (2012) Efficacy of virtual reality exposure therapy in the treatment of PTSD: a systematic review. PLoS One 7(12):48469. https://doi.org/10.1371/JOURNAL.PONE.0048469

Hofmann SG, Smits JA (2008) Cognitive-behavioral therapy for adult anxiety disorders: a meta-analysis of randomized placebo-controlled trials. J Clin Psychiatry 69(4):0–0

Koch E, Spates C, Himle J (2004) Comparison of behavioral and cognitive-behavioral one-session exposure treatments for small animal phobias. Behav Res Ther 42(12):1483–1504. https://doi.org/10.1016/J.BRAT.2003.10.005

Lindner P, Miloff A, Hamilton W, Reuterskiöld L, Andersson G, Powers MB, Carlbring P (2017) Creating state of the art, next-generation virtual reality exposure therapies for anxiety disorders using consumer hardware platforms: design considerations and future directions. Cogn Behav Ther 46:404–420

Ling Y, Nefs HT, Morina N, Heynderickx I, Brinkman W-P (2014) A meta-analysis on the relationship between self-reported presence and anxiety in virtual reality exposure therapy for anxiety disorders. PLoS One 9(5):e96144

Loucks L, Yasinski C, Norrholm SD, Maples-Keller J, Post L, Zwiebach L et al (2019) You can do that?!: feasibility of virtual reality exposure therapy in the treatment of PTSD due to military sexual trauma. J Anxiety Disord 61:55–63

Maltby N, Kirsch I, Mayers M, Allen GJ (2002) Virtual reality exposure therapy for the treatment of fear of flying: a controlled investigation. J Consult Clin Psychol 70(5):1112–1118. https://doi.org/10.1037//0022-006X.70.5.1112

Maples-Keller JL, Bunnell BE, Kim SJ, Rothbaum BO (2017) The use of virtual reality technology in the treatment of anxiety and other psychiatric disorders. Harvard Rev Psychiatry 25(3):103

Maples-Keller JL, Jovanovic T, Dunlop BW, Rauch S, Yasinski C, Michopoulos V et al (2019) When translational neuroscience fails in the clinic: dexamethasone prior to virtual reality exposure therapy increases drop-out rates. J Anxiety Disord 61:89–97

McLay RN, McBrien C, Wiederhold MD, Wiederhold BK (2010) Exposure therapy with and without virtual reality to treat PTSD while in the combat theater: a parallel case series. Cyberpsychol Behav Soc Netw 13(1):37–42. https://doi.org/10.1089/CYBER.2009.0346

McLay RN, Wood DP, Webb-Murphy JA, Spira JL, Wiederhold MD, Pyne JM, Wiederhold BK (2011) A randomized, controlled trial of virtual reality-graded exposure therapy for post-traumatic stress disorder in active duty service members with combat-related post-traumatic stress disorder. Cyberpsychol Behav Soc Netw 14(4):223–229. https://doi.org/10.1089/CYBER.2011.0003

Michaliszyn D, Marchand A, Bouchard S, Martel MO, Poirier-Bisson J (2010) A randomized, controlled clinical trial of in virtuo and in vivo exposure for spider phobia. Cyberpsychol Behav Soc Netw 13(6):689–695. https://doi.org/10.1089/CYBER.2009.0277

Milad MR, Wright CI, Orr SP, Pitman RK, Quirk GJ, Rauch SL (2007) Recall of fear extinction in humans activates the ventromedial prefrontal cortex and hippocampus in concert. Biol Psychiatry 62(5):446–454

Milad MR, Rosenbaum BL, Simon NM (2014) Neuroscience of fear extinction: implications for assessment and treatment of fear-based and anxiety related disorders. Behav Res Ther 62:17–23

Morina N, Ijntema H, Meyerbroker K, Emmelkamp PMG (2015) Can virtual reality exposure therapy gains be generalized to real-life? A meta-analysis of studies applying behavioral assessments. Behav Res Ther 74:18–24

Mühlberger A, Herrmann M, Wiedemann G, Ellgring H, Pauli P (2001) Repeated exposure of flight phobics to flights in virtual reality. Behav Res Ther 39(9):1033–1050. https://doi.org/10.1016/S0005-7967(00)00076-0

Mühlberger A, Wiedemann G, Pauli P (2003) Efficacy of a one-session virtual reality exposure treatment for fear of flying. Psychother Res: J Soc Psychother Res 13(3):323–336. https://doi.org/10.1093/PTR/KPG030

Myers KM, Davis M (2002) Behavioral and neural analysis of extinction. Neuron 36(4):567–584

Myers KM, Davis M (2007) Mechanisms of fear extinction. Mol Psychiatry 12(2):120–150. https://doi.org/10.1038/sj.mp.4001939

Newman MARK, Gatersleben B, Wyles KJ, Ratcliffe E (2022) The use of virtual reality in environment experiences and the importance of realism. J Environ Psychol 79:101733

Norr AM, Smolenski DJ, Katz AC, Rizzo AA, Rothbaum BO, Difede J et al (2018) Virtual reality exposure versus prolonged exposure for PTSD: which treatment for whom? Depress Anxiety 35(6):523–529

Norrholm SD, Jovanovic T, Olin IW, Sands LA, Bradley B, Ressler KJ (2011) Fear extinction in traumatized civilians with posttraumatic stress disorder: relation to symptom severity. Biol Psychiatry 69(6):556–563

Norton PJ, Price EC (2007) A meta-analytic review of adult cognitive-behavioral treatment outcome across the anxiety disorders. J Nerv Ment Dis 195(6):521–531

Powers MB, Halpern JM, Ferenschak MP, Gillihan SJ, Foa EB (2010) A meta-analytic review of prolonged exposure for posttraumatic stress disorder. Clin Psychol Rev 30(6):635–641. https://doi.org/10.1016/J.CPR.2010.04.007

Ready DJ, Pollack S, Olasov Rothbaum B, Alarcon RD (2006) Virtual reality exposure for veterans with posttraumatic stress disorder. J Aggress Maltreat Trauma 12:1–2. https://doi.org/10.1300/J146v12n01_11

Reger GM, Koenen-Woods P, Zetocha K, Smolenski DJ, Holloway KM, Rothbaum BO et al (2016) Randomized controlled trial of prolonged exposure using imaginal exposure vs. virtual reality exposure in active duty soldiers with deployment-related posttraumatic stress disorder (PTSD). J Consult Clin Psychol 84(11):946–959. https://doi.org/10.1037/CCP0000134

Ressler KJ, Rothbaum BO, Tannenbaum L, Anderson P, Graap K, Zimand E et al (2004) Cognitive enhancers as adjuncts to psychotherapy: use of D-cycloserine in phobic individuals to facilitate extinction of fear. Arch Gen Psychiatry 61(11):1136–1144. https://doi.org/10.1001/ARCHPSYC.61.11.1136

Rizzo A, Difede J, Rothbaum BO, Reger G, Spitalnick J, Cukor J, Mclay R (2010) Development and early evaluation of the virtual Iraq/Afghanistan exposure therapy system for combat-related PTSD. Ann N Y Acad Sci 1208(1):114–125. https://doi.org/10.1111/J.1749-6632.2010.05755.X

Rothbaum BO, Davis M (2003) Applying learning principles to the treatment of post-trauma reactions. Ann N Y Acad Sci 1008(1):112–121

Rothbaum BO, Hodges LF, Kooper R, Opdyke D, Williford J, North M (1995) Effectiveness of computer-generated (virtual reality) graded exposure in the treatment of acrophobia. Am J Psychiatr 152:626–628

Rothbaum BO, Hodges L, Alarcon R, Ready D, Shahar F, Graap K et al (1999a) Virtual reality exposure therapy for PTSD Vietnam veterans: a case study. J Trauma Stress 12(2):263–271. https://doi.org/10.1023/A:1024772308758

Rothbaum BO, Hodges L, Smith S (1999b) Virtual reality exposure therapy abbreviated treatment manual: fear of flying application. Cogn Behav Pract 6:234–244

Rothbaum BO, Hodges L, Smith S, Lee S, Price L (2000) A controlled study of virtual reality exposure therapy for the fear of flying. J Consult Clin Psychol 68(6):1020–1026. https://doi.org/10.1037//0022-006X.68.6.1020

Rothbaum BO, Hodges L, Ready D, Graap K, Alarcon R (2001) Virtual reality exposure therapy for Vietnam veterans with posttraumatic stress disorder. J Clin Psychiatry 62:617–622

Rothbaum BO, Anderson P, Zimand E, Hodges L, Lang D, Wilson J (2006) Virtual reality exposure therapy and standard (in vivo) exposure therapy in the treatment of fear of flying. Behav Ther 37(1):80–90. https://doi.org/10.1016/J.BETH.2005.04.004

Rothbaum BO, Price M, Jovanovic T, Norrholm SD, Gerardi M, Dunlop B et al (2014) A randomized, double-blind evaluation of D-cycloserine or alprazolam combined with virtual reality exposure therapy for posttraumatic stress disorder (PTSD) in Iraq and Afghanistan war veterans. Am J Psychiatry 171(6):640. https://doi.org/10.1176/APPI.AJP.2014.13121625

Schubert T, Friedmann F, Regenbrecht H (2001) The experience of presence: factor analytic insights. Presence Teleop Virt Environ 10(3):266–281

Segal R, Bhatia M, Drapeau M (2011) Therapists' perception of benefits and costs of using virtual reality treatments. Cyberpsychol Behav Soc Netw 14:29–34

Sherrill AM, Goodnight JR, Burton MS, Rothbaum BO (2020) Use of virtual reality exposure therapy for trauma- and anxiety-related disorders. In: Reger GM (ed) Technology and mental health: a clinician's guide to improving outcomes. Routledge, New York

Slater M, Wilbur S (1997) A framework for immersive virtual environments (FIVE): speculations on the role of presence in virtual environments. Presence Teleop Virt 6(6):603–616. https://doi.org/10.1162/pres.1997.6.6.603

Stevens TM, Sherrill AM (2021) A clinician's introduction to 360° video for exposure therapy. Trans Issues Psychol Sci 7(3):261

Stewart RE, Chambless DL (2009) Cognitive–behavioral therapy for adult anxiety disorders in clinical practice: a meta-analysis of effectiveness studies. J Consult Clin Psychol 77(4):595

Tart C, Handelsman P, Deboer L, Rosenfield D, Pollack M, Hofmann S et al (2013) Augmentation of exposure therapy with post-session administration of D-cycloserine. J Psychiatr Res 47(2):168–174. https://doi.org/10.1016/J.JPSYCHIRES.2012.09.024

Tworus R, Szymanska S, Ilnicki S (2010) A soldier suffering from PTSD, treated by controlled stress exposition using virtual reality and behavioral training. Cyberpsychol Behav Soc Netw 13(1):103–107. https://doi.org/10.1089/CYBER.2009.0329

VanElzakker MB, Dahlgren MK, Davis FC, Dubois S, Shin LM (2014) From Pavlov to PTSD: the extinction of conditioned fear in rodents, humans, and anxiety disorders. Neurobiol Learn Mem 113:3–18

Vervliet B, Baeyens F, Van den Bergh O, Hermans D (2013) Extinction, generalization, and return of fear: a critical review of renewal research in humans. Biol Psychol 92(1):51–58

Wald J, Taylor S (2003) Preliminary research on the efficacy of virtual reality exposure therapy to treat driving phobia. Cyberpsychol Behav 6(5):459–465. https://doi.org/10.1089/109493103769710488

Weech S, Kenny S, Barnett-Cowan M (2019) Presence and cybersickness in virtual reality are negatively related: a review. Front Psychol 10:158

Wirth W, Hartmann T, Böcking S, Vorderer P, Klimmt C, Schramm H et al (2007) A process model of the formation of spatial presence experiences. Media Psychol 9(3):493–525

Wolitzky-Taylor K, Horowitz J, Powers M, Telch M (2008) Psychological approaches in the treatment of specific phobias: a meta-analysis. Clin Psychol Rev 28(6):1021–1037. https://doi.org/10.1016/J.CPR.2008.02.007

Contemporary Approaches Toward Neuromodulation of Fear Extinction and Its Underlying Neural Circuits

Claudia R. Becker and Mohammed R. Milad

Contents

Abstract Neuroscience and neuroimaging research have now identified brain nodes that are involved in the acquisition, storage, and expression of conditioned fear and its extinction. These brain regions include the ventromedial prefrontal cortex (vmPFC), dorsal anterior cingulate cortex (dACC), amygdala, insular cortex, and hippocampus. Psychiatric neuroimaging research shows that functional dysregulation of these brain regions might contribute to the etiology and symptomatology of various psychopathologies, including anxiety disorders and post traumatic stress disorder (PTSD) (Barad et al. Biol Psychiatry 60:322–328, 2006; Greco and Liberzon Neuropsychopharmacology 41:320–334, 2015; Milad et al. Biol Psychiatry 62:1191–1194, 2007a, Biol Psychiatry 62:446–454, b; Maren and Quirk Nat Rev Neurosci 5:844–852, 2004; Milad and Quirk Annu Rev Psychol 63:129, 2012; Phelps et al. Neuron 43:897–905, 2004; Shin and Liberzon Neuropsychopharmacology 35:169–191, 2009). Combined, these findings indicate that targeting the activation of these nodes and modulating their functional interactions might offer an opportunity to further our understanding of how fear and threat responses are formed and regulated in the human brain, which could lead to enhancing the efficacy of current treatments or creating novel treatments for PTSD

C. R. Becker and M. R. Milad (✉)
Department of Psychiatry, NYU Grossman School of Medicine, New York, NY, USA
e-mail: claudiabecker@tamu.edu; Mohammed.r.milad@uth.tmc.edu

Curr Topics Behav Neurosci (2023) 64: 353–388
https://doi.org/10.1007/7854_2023_442
Published Online: 2 September 2023

and other psychiatric disorders (Marin et al. Depress Anxiety 31:269–278, 2014; Milad et al. Behav Res Ther 62:17–23, 2014). Device-based neuromodulation techniques provide a promising means for directly changing or regulating activity in the fear extinction network by targeting functionally connected brain regions via stimulation patterns (Raij et al. Biol Psychiatry 84:129–137, 2018; Marković et al. Front Hum Neurosci 15:138, 2021). In the past ten years, notable advancements in the precision, safety, comfort, accessibility, and control of administration have been made to the established device-based neuromodulation techniques to improve their efficacy. In this chapter we discuss ten years of progress surrounding device-based neuromodulation techniques—Electroconvulsive Therapy (ECT), Transcranial Magnetic Stimulation (TMS), Magnetic Seizure Therapy (MST), Transcranial Focused Ultrasound (TUS), Deep Brain Stimulation (DBS), Vagus Nerve Stimulation (VNS), and Transcranial Electrical Stimulation (tES)—as research and clinical tools for enhancing fear extinction and treating PTSD symptoms. Additionally, we consider the emerging research, current limitations, and possible future directions for these techniques.

1 Introduction

Prior research has identified the underlying neural circuits of fear extinction and their relevance to psychopathology (Maren and Quirk 2004; Milad and Quirk 2012; Shin and Liberzon 2009). Through psychiatric neuroimaging studies, scientists have successfully pinpointed brain regions connected through activity into "networks" that are consistently implicated in fear extinction across studies (Milad et al. 2014). These studies have established that the fear extinction network includes the ventromedial prefrontal cortex (vmPFC), dorsal anterior cingulate cortex (dACC), amygdala, insular cortex, and hippocampus (Barad et al. 2006; Linnman et al. 2011; Milad et al. 2007a, b; Phelps et al. 2004). The same lines of research that identified the fear extinction network suggest patients with post-traumatic stress disorder (PTSD) commonly show dysfunctional activation of the established fear extinction network (Greco and Liberzon 2015; Milad et al. 2009; Milad and Quirk 2012; Wen et al. 2022). These findings in patients with PTSD have sparked myriad hypotheses supporting the notion that regulating function and connectivity within the fear extinction network could reduce psychiatric symptoms in patients with PTSD (Marin et al. 2014; Milad et al. 2014).

While traditionally psychopharmacology and exposure therapy have been considered the first-line, gold standard for modulating brain activity and reducing psychiatric symptoms in patients with PTSD, a significant subset of patients do not respond to these approaches or experience adverse side effects from these techniques (Cukor et al. 2009; Sippel et al. 2018). Recent research has demonstrated that administering precise stimulation patterns of electro-magnetic pulses or electric currents via neuromodulation devices can effectively regulate activity in the fear extinction network if targeted at the neural nodes implicated in fear extinction (Raij et al. 2018; Marković et al. 2021). Since the emergence of the first publications highlighting the efficacy of device-based neuromodulation for fear extinction, the stimulation tools used, patterns of pulses administered, and locations targeted have been increasingly optimized to improve device-based neuromodulation techniques for PTSD treatments (Lewis et al. 2016). These recent advancements suggest more targeted neuromodulation techniques increase treatment efficacy for PTSD while reducing the side effects and discomfort associated with less specific neuromodulation (van Rooij et al. 2021).

Not only may targeted device-based neuromodulation techniques provide a viable and effective treatment for PTSD, but these techniques also present a unique research opportunity. Scientists can study how changing activity in specific networks or brain regions implicated in fear extinction and PTSD influences overall symptomatology or behaviors within disorders (Fitzgerald et al. 2002; George and Aston-Jones 2009). In this chapter, we will discuss clinical developments and refinements in device-based brain stimulation in the past 10 years that may effectively enhance fear extinction for the treatment of PTSD and other anxiety-, trauma-, and stressor-related disorders. Furthermore, we will review recent research on device-based neuromodulation as a research tool to probe specific neural circuits and increase our understanding of underlying neurobiological mechanisms of fear extinction. We will review the data on these new neuromodulation approaches and discuss current understanding of how these techniques change neuroplastic activity in the fear extinction network to reduce psychiatric symptoms.

2 Electroconvulsive Therapy (ECT)

To administer ECT, clinicians place electrodes at specific locations – either unilaterally or bilaterally – on a patient's scalp (Ropper et al. 2022; Leaver et al. 2022). The electrodes induce electrical currents that penetrate the skull to reach the brain, triggering a short and controlled seizure in the patient (Ding and White 2002; Ropper et al. 2022; Lisanby 2007). This process is conducted under anesthesia or sedation with a muscle relaxant to ensure patient safety and comfort (Ding and White 2002; Ropper et al. 2022). While the underlying mechanisms of ECT have not yet been fully elucidated, many hypotheses aim to explain how ECT is successful at treating psychiatric disorders (including network regulation hypotheses) (Pang et al. 2022; Leaver et al. 2022). Several functional magnetic resonance imaging (fMRI) and electroencephalogram (EEG) studies suggest seizures induced via ECT change structural and functional connectivity between brain regions associated with psychiatric conditions such as the DLFC, amygdala, basal ganglia, or hippocampus – regions implicated in fear extinction (Abbott et al. 2014; Joshi et al. 2016; Krishnan 2016; Nordanskog et al. 2010; Perrin et al. 2012; Sartorius et al. 2016; Singh and Kar 2017; Takamiya et al. 2019). In the past 10 years, while many studies still investigate structural changes in the brain following ECT, many new studies whose aim is to explain ECT mechanisms have shifted from studying ECT's impact on brain *regions* to ECT's impact on brain *networks* (Cano et al. 2016; Van Waarde et al. 2015; Wei et al. 2018).

Scientists have established ECT's success at modulating neural activity in areas of the brain (Leaver et al. 2022; Pang et al. 2022). Since these areas – the DLFC, amygdala, basal ganglia, or hippocampus – are often involved in the fear extinction network and memory, some emerging research aims to directly test the impact of electroconvulsive seizure (ECS) – a rat analog to ECT – on fear extinction (Elahi et al. 2020; Joshi et al. 2016; Van Buel et al. 2017; Van Waarde et al. 2014). Rodent studies have provided helpful insight into the impact of ECT on fear extinction in a controlled, reliable setting (Misanin et al. 1968). In the past 10 years, studies using rats aimed to increase translatability to fear extinction in humans by comparing ECS and sham ECS, rats with and without trauma, and the influence of modeled comorbid conditions on the efficacy of ECS for improving fear extinction in rats (Van Buel et al. 2017; Elahi et al. 2020). These studies' results suggest ECS impairs fear learning memory when applied several times prior to fear conditioning, and possibly impairs extinction memory consolidation when administered after extinction phases, although the results are somewhat mixed (Van Buel et al. 2017; Elahi et al. 2020). Collectively, the results of these studies provide some increasing evidence that ECT might modulate the fear extinction network.

Due to the involved nature of ECT procedures and the adverse cognitive side effects associated with ECT, few studies directly explore the impact of ECT on fear extinction in humans. One study suggested ECT administered following memory reactivation disrupted reconsolidation of fear memories in patients with depression, although more studies are needed to determine if these results are replicable (Kroes et al. 2013). Most studies conducted are treatment-based clinical trials or retrospective studies examining the impact of ECT on PTSD symptoms. These trials suggest ECT shows promise as an effective treatment for PTSD and PTSD comorbid with depression (Ahmadi et al. 2016; Ahmadi et al. 2018; Tang et al. 2021c). A few early studies exemplified ECT's potential efficacy in reducing PTSD symptoms. For example, Watts (2007) conducted a retrospective chart review studying the impact of ECT on PTSD symptoms in individuals with comorbid depression (Watts 2007). Their results suggested ECT significantly decreased PTSD symptoms as measured via the PTSD symptom checklist (Watts 2007). Margoob and colleagues conducted a trial studying the impact of ECT on PTSD symptoms as assessed through the CAPS, and PTSD symptoms significantly decreased over the course of ECT (Margoob et al. 2010). Recent studies have also demonstrated ECT as a possibly effective treatment for reducing PTSD symptoms as assessed via changes on the Clinical Global Impressions Scale (CGI), Clinical Administered PTSD Scale for DSM-5 (CAPS-5), and the Modified PTSD Symptom Scale (MPSS-SR) (Ahmadi et al. 2016, 2018; Tang et al. 2021c). A case study suggested that ECT was effective in reducing PTSD symptoms when administered after reactivation of a traumatic memory (Gahr et al. 2014). Because of the relatively small sample sizes of the studies conducted on ECT for PTSD symptoms, large randomized, control studies may be warranted to fully discern the efficacy of ECT for PTSD.

While ECT is considered a highly effective treatment for many psychiatric disorders and shows promise in treating PTSD symptoms, the procedure requires anesthesia and is not without side effects or risk (Leaver et al. 2022; Ousdal et al. 2022). The utilization of anesthesia can prevent patients from comfortably resuming their daily activities after treatment, can intimidate many patients from undergoing the treatment, comes with additional safety risks, and can increase the cost associated with administering ECT (Ding and White 2002). Furthermore, ECT is a highly stigmatized procedure due to media portrayal which can intimidate many patients from considering ECT as a viable treatment option (McDonald and Walter 2001; Sienaert 2016). From an efficacy and safety standpoint, ECT modulation is less specific to individual brain regions than other device-based neuromodulation treatments because the skull opposes the electrical field and distributes the stimuli across the brain (Kirkcaldie et al. 1997)). Individual topographic skull variation results in unpredictable distribution of the electrical field which often produces unwanted side effects such as confusion and memory loss and can diminish effectiveness in some patients (Lisanby 2007; Sackeim et al. 2006). Since whole brain stimulation and a distributed electrical field means less specificity in targeting individual brain regions or networks, the technique potentially offers less utility for understanding how specific networks influence emotions and behaviors. Furthermore, ECT's lack of specificity to networks or brain regions makes it difficult to clearly distinguish the mechanisms of action behind symptom reduction from ECT (Kirkcaldie et al. 1997). While emerging research exemplifies ECT's efficacy in treating PTSD, ECT is also not yet cleared by the FDA for PTSD treatment, and more precise, lower-risk device-based neuromodulation techniques have emerged as viable options for changing network activity in symptomatic populations.

3 Transcranial Magnetic Stimulation (TMS)

TMS possibly provides a more precise method than ECT for modulating brain activity in brain networks implicated in fear learning and extinction, and, as such, PTSD etiology and treatment, because the stimulation is more selective in recruiting neurons (Mutz et al. 2019). Magnetic pulses administered via a magnetic coil easily penetrate a patient's skull with precision and without resistance (D. R. Kim et al. 2009). The coil is placed superficially on the patient's scalp and the treatment is non-invasive (Cusin and Dougherty 2012; George et al. 1999). TMS can be administered without anesthesia in an outpatient clinic (Cusin and Dougherty 2012). The magnetic pulses change neural activity by inducing an electric current at the targeted area of the brain, possibly more focalized than stimulation with ECT (Kobayashi and Pascual-Leone 2003; Mutz et al. 2019; Pascual-Leone and Meador 1998). These currents modulate activity in the targeted brain region and the brain regions functionally connected to the target region with the goal of regulating abnormal neural activity in patients with psychiatric conditions (Kobayashi and Pascual-Leone 2003; Mutz et al. 2019; Pascual-Leone and Meador 1998). Emerging research suggests many mental health disorders are characterized by irregular activity in brain regions and networks, and that TMS reduces symptoms for many patients by regulating the dysfunctional activity in networks implicated in specific conditions ((Kang et al. 2016; Liston et al. 2014; Richieri et al. 2017). Different stimulation parameters can achieve this effect via activating or inhibiting activity in specific brain regions (Chen et al. 1997; Lang et al. 2006; Speer et al. 2000). While initially primarily only regions of the brain were targeted via TMS, much like ECT research, neuroscience research regarding TMS has transitioned to a network-focused approach, utilizing neuroimaging techniques to guide targeting.

Like ECT research, some of the earliest studies examining the impact of targeted stimulation on fear extinction utilized rat models (Milad et al. 2004; Vidal-Gonzalez et al. 2006). These studies exemplified the efficacy of using targeted, brief electrical stimulation to modulate fear extinction learning (Milad et al. 2004; Vidal-Gonzalez et al. 2006). With advancements in the stimulation parameters of TMS, including the introduction of repetitive transcranial magnetic stimulation (rTMS) – the process of repeatedly administering a high-frequency TMS to excite neurons, creating long-term potentiation (LTP) and changes in network connectivity (Hoogendam et al. 2010) – new stimulation parameters have also been used to study fear extinction not only in rodents, but also in humans. The early rodent studies examining brief electrical stimulation for fear extinction targeted the infralimbic cortex – the human analog of the vmPFC – and the mPFC – a target shown to inhibit amygdala activation – with success (Milad et al. 2004; Vidal-Gonzalez et al. 2006). Thus, in the past 10 years studies have aimed to translate these findings to TMS applications for human participants. An early study examining the impact of rTMS on fear extinction was conducted by Baek and colleagues – who studied the impact of rTMS on fear extinction in rats (Baek et al. 2012). Baek and colleagues exemplified the efficacy of administering 10 Hz rTMS paired with the conditioned stimulus (CS) during extinction to improve fear extinction learning (Baek et al. 2012). Fear extinction learning was assessed via freezing behavior and rTMS was broadly applied to the scalp due to limitations in localization (Baek et al. 2012). Since TMS cannot target deep brain regions such as the vmPFC, human participant studies localizing targets for fear extinction use functional connectivity analysis to find superficial brain regions connected to the vmPFC (Pennington and Fanselow 2018; Raij et al. 2018). For example, one study targeted the vmPFC in human participants via an area on the left posterolateral PFC that was functionally connected to the vmPFC (Raij et al. 2018). Participants were administered 20 Hz rTMS during fear extinction 100 ms after a conditioned stimulus was presented (Raij et al. 2018). Fear extinction was enhanced (Raij et al. 2018). In another study, 10 Hz rTMS applied to the mPFC after fear acquisition and before extinction learning enhanced extinction retention in participants (Guhn et al. 2014). Other superficial targets have also been identified such as the DLPFC (Deng et al. 2021). A recent study testing low-frequency rTMS targeted at the left and right DLPFC resulted in significant enhancement of fear extinction when rTMS was administered after a reminder cue of the CS (Borgomaneri et al. 2020).

Some studies have also examined the impact of theta burst stimulation (TBS), an rTMS paradigm, to the left DLPFC on fear extinction (Deng et al. 2021). TBS was developed by exploring the neural firing pattern of rats during tasks and mimicking this observed pattern through an artificially stimulated pattern (Huang et al. 2005). TBS resulted in long-term potentiation (LTP) in the hippocampus in rodents and was soon adapted in humans to form new TMS protocols (Huang et al. 2005). TBS uses high-frequency pulses to suppress or increase cortical excitability (Chung et al. 2015, 2018). Intermittent theta burst stimulation (iTBS) sends high-frequency bursts for 2 s every 8 s for 192 s resulting in an increase in cortical excitability (Chung et al. 2015). Studies such as that by Deng and colleagues (2021) have investigated using iTBS targeted at the left DLPFC before and after extinction to enhance fear extinction and show promising results (Deng et al. 2021).

Clinically, TMS has been studied for PTSD since the late 1990s; however, TMS has not yet been FDA cleared for the treatment of PTSD, and thus, current studies aim to propose an optimal delivery of TMS as an effective and safe treatment for PTSD (Berlim and Van Den Eynde 2014; George and Aston-Jones 2009; Karsen et al. 2014; Cohen et al. 2004; Osuch et al. 2009). These current studies predominantly suggest that rTMS can help improve symptoms in patients with PTSD (although the results are somewhat mixed and there are inconsistent findings as to where and how to best stimulate) (Ahmadizadeh and Rezaei 2018; Belsher et al. 2021; Carpenter et al. 2018; Harris and Reece 2021; Madore et al. 2022; Philip et al. 2019b; Thierrée et al. 2022). These studies have aimed to find effective superficial targets, stimulation parameters, and frequencies for modulating the fear extinction circuit with TMS. However, heterogeneity in study design poses barriers for drawing concrete conclusions regarding optimal treatment parameters (Petrosino et al. 2021). The targets for PTSD treatment are translated from fear extinction research. Generally, the right or left DLPFC is targeted unilaterally or bilaterally to modulate activity in the vmPFC – there is still debate as to which amongst these targets are best – but a clinical treatment is typically administered over several sessions in the absence of a fear conditioning paradigm or as an adjunctive to exposure or traditional cognitive behavior therapy (Gouveia et al. 2020; Trevizol et al. 2016). Recent studies have also aimed to combine rTMS or deep TMS – a type of TMS that utilizes a different coil to penetrate broader and deeper into the brain – and exposure therapy (Fryml et al. 2019; Isserles et al. 2021) or other cognitive behavioral therapies (Kozel et al. 2018), which generally show promising, yet somewhat inconsistent results.

Another ongoing debate clinically is whether to utilize high- or low-frequency rTMS (Gouveia et al. 2020; Harris and Reece 2021). For example, a 2010 study from Boggio and colleagues indicated that 20 Hz daily rTMS treatments to the right DLPFC resulted in improvements in re-experiencing and avoidance symptoms; however, many patients experienced significant anxiety as a result (Boggio et al. 2010). Other studies reported PTSD symptom reduction from 20 Hz rTMS or 10 Hz (Isserles et al. 2013; Kozel et al. 2019). In 2016, an open-label trial studied daily treatments of 5 Hz rTMS to the left DLPFC to determine if a lower frequency stimulation could have similar efficacy for PTSD symptoms while reducing anxiety side effects (Philip et al. 2016). Although the sample size was small, participants experienced benefits and reduced anxiety (Philip et al. 2016). Other studies indicate 5 Hz as an effective frequency for PTSD symptom reduction (Carpenter et al. 2018). Even further complicating the picture, Watts and co-authors administered 1 Hz rTMS to the right DLPFC showing reduction in PTSD symptoms but not significant reductions in anxiety, and the Leong research group compared 1 Hz, 10 Hz, and sham – resulting in the most significant PTSD symptom reduction from 1 Hz, but little difference in anxiety symptoms (Watts et al. 2012; Leong et al. 2020). However, contrary studies suggest high-frequency rTMS may be more effective for PTSD treatment (Belsher et al. 2021) or that high- and low-frequency TMS may be more effective for differing symptom clusters (Yan et al. 2017).

Other studies have targeted different regions, for example the mPFC – the location currently FDA approved for treatment of OCD via TMS – using high frequencies (Isserles et al. 2013). Stimulating these targets with deep TMS has showed some efficacy for reducing PTSD symptoms in one study (Isserles et al. 2013). Furthermore, the intensity which stimulation is delivered at also varies widely between studies, making it difficult to fully distinguish the difference between treatment effects at various locations and frequencies (Clark et al. 2015). Additionally, scientists have also begun to use TBS as an option for treatment, thus far showing similar results (Philip et al. 2019b). Synchronized TMS (sTMS) – a more personalized TMS therapy – has also been studied as a treatment for PTSD (Philip et al. 2019a). Synchronized TMS exemplifies the field's overall trend toward personalized, optimized TMS treatments. Not only are researchers working to optimize current TMS administration, but in efforts to further personalize treatments, scientists are aiming to identify biomarkers that predict PTSD symptom reduction with TMS. Philip and colleagues identified differences in functional connectivity in the subgenual anterior cingulate (sgACC), default mode network (DMN), and salience network (SN) as biomarkers for TMS response in participants with comorbid PTSD and major depressive disorder (MDD) (Philip et al. 2018). A year later, Barredo and colleagues found fractional anisotropy (FS) in the anterior thalamic radiations (ATRs) predicted TMS response in MDD and PTSD patients, and Zandvakili and colleagues acquired EEG data and used machine learning techniques to accurately predict rTMS response in PTSD and MDD patients (Barredo et al. 2019; Zandvakili et al. 2019).

TMS can be administered in an outpatient clinic without anesthesia and is generally considered low-risk and tolerable; however, side effects – depending on the protocol administered and other patient risk factors – can include temporary headache, syncope, scalp pain, and temporary hearing changes (Najib and Horvath 2014; Rossi et al. 2021). Although rare, with a risk of less than 1%, seizures can occur with TMS (Najib and Horvath 2014; Stultz et al. 2020). Additionally, some patients experience negative changes in mood or behavior (Rossi et al. 2021). Furthermore, although new protocols such as the Stanford Accelerated Intelligent Neuromodulation Therapy (SAINT) protocol, which uses higher dose iTBS pulses administered via image-guided targeting multiple times a day, and other iTBS methods have shortened the time needed to administer TMS, TMS still requires a potentially significant time commitment for many patients (Cole et al. 2020, 2022).

However, despite these potential drawbacks, TMS has many strengths as a treatment and a research tool. Due to the low risks associated with TMS, the general tolerability of the procedure, and clinicians' ability to administer TMS in minutes in an outpatient clinic, TMS can be highly studied, providing a wealth of literature on the safety and efficacy of TMS. Furthermore, as previously mentioned, TMS can more precisely target brain regions or networks than more widespread brain stimulation like ECT thus making TMS a highly effective research tool (Mutz et al. 2019). However, since TMS only penetrates a few centimeters into the brain, and it can only target superficial brain regions, researchers and clinicians must target deep brain regions or networks via superficial regions functionally connected to the deeper brain regions implicated in disorders (Marin et al. 2014). Because of this, TMS provides less utility to directly study regions implicated in fear extinction such as the amygdala.

4 Magnetic Seizure Therapy (MST)

Since ECT is highly effective in treating a range of psychiatric conditions, scientists have long sought to improve the safety of ECT so that the procedure can help a wider range of patients with treatment-resistant mental health disorders (Rowny et al. 2009). MST uses high-intensity TMS targeted at a location on the cortex (typically at the vertex) to induce a seizure under anesthesia via more focalized, intense, and precise electrical stimulation (M. Chen et al. 2021; De Deng et al. 2011; Mutz et al. 2019; Rowny et al. 2008, 2009). Because the stimulation is significantly more targeted than ECT stimulation, MST is associated with a lower side effect profile and potentially more efficacy – although MST is a relatively early treatment and is not FDA cleared for any conditions yet (M. Chen et al. 2021; De Deng et al. 2011; McClintock et al. 2011).

Recent MST studies have examined current efficacy and possible optimization methods such as differing frequencies and protocols to treat a range of conditions, although MST is mostly used to treat depression (Tang et al. 2021a; Weissman et al. 2022). Studies also attempt to identify which patients might best respond to MST and to discern the impact of MST on cognition compared to traditional ECT (Chen et al. 2021; Kayser et al. 2019; Lisanby and Deng 2015). Thus far, most studies show MST appears to be almost equally effective as ECT in improving depressive symptoms, suicidal ideation, and schizophrenia symptoms, with comparatively fewer unwanted cognitive side effects (Li et al. 2022a, b; Tang et al. 2021a, b, c). Furthermore, currently high-frequency MST appears the most effective form of MST therapy in reducing depression symptoms, and cortical inhibition seems to be a viable method to determine who responds best to MST (Daskalakis et al. 2019). Future studies should aim to replicate these findings, compare ECT to MST in double blind, controlled trials, or use sham conditions to control for placebo effects in MST. Ongoing trials, such as Confirmatory Efficacy and Safety Trial of Magnetic Seizure Therapy for Depression (CREST-MST), aim to establish non-inferiority to

ECT and establish superior safety (Daskalakis et al. 2021b) or seek to identify novel biomarkers to further optimize MST and expand its applications (Daskalakis et al. 2021a).

Thus far, no studies to the best of our knowledge have investigated MST's impact on the fear extinction circuit. However, recent studies pairing MST with techniques such as EEG indicate that MST induces neural plasticity, likely a mechanism of change in the treatment, and fMRI studies show changes in resting state functional connectivity after MST (Ge et al. 2021; Hadas et al. 2020; Sun et al. 2018). These results suggest MST could be a viable tool for modulating activity in the fear extinction network. Additionally, one study suggested MST might decrease anxiety in some participants (Kayser et al. 2019). Another study suggested MST did not positively impact OCD symptoms (Tang et al. 2021b). While MST shows promise as an effective tool for modulating brain activity, with fewer side effects than ECT, more research must be done targeting the neural nodes of the fear extinction circuit to determine if MST could be an effective treatment for enhancing fear extinction and reducing PTSD symptoms (M. Chen et al. 2021). Furthermore, although MST promises fewer cognitive side effects than ECT, the procedure must still be conducted under anesthesia and a seizure is still induced during the treatment (Chen et al. 2021; Kallioniemi et al. 2019). Additionally, patients typically must undergo 12 treatments spread out to two to three times a week, which could provide another time barrier to receiving this treatment (Engel and Kayser 2016).

5 Transcranial Focused Ultrasound (tFUS)

Transcranial Focused Ultrasound (tFUS) is currently administered non-invasively by using a machine to create ultrasound waves, generally equal to or greater than 220 kHz, directed at regions or networks within the brain via a transducer (Krishna et al. 2018; Kubanek 2018). The waves created are high-frequency sound waves that use thermal heating and acoustics to inhibit or increase activity in the targeted brain region or network, and likely change local neural connectivity and brain networks responses to stimuli (Kuhn et al. 2023). A high-intensity ultrasound causes ablation; however, a low-intensity ultrasound may activate targeted neurons (Tufail et al. 2011). Although, the mechanisms of change of tFUS are not fully understood (Tufail et al. 2011). TFUS can penetrate deep into the brain, similar to the depth of penetration achieved with DBS and deeper than TMS can penetrate the brain – making the technique a good candidate for precise neuromodulation of deep structures with potential beneficial effects to enhance fear extinction (Folloni et al. 2019; Kubanek 2018; Kuhn et al. 2023; Spivak et al. 2022). Since there are several deep brain structures implicated in the fear extinction network, such as the amygdala, that cannot be reached by other superficial techniques such as TMS, tFUS could provide a helpful method of targeting these structures to modulate activity.

Since tFUS is a relatively new treatment and research tool for psychiatric conditions, and is not yet FDA cleared for psychiatric conditions, much of the research has focused on elucidating the underlying mechanisms of action of tFUS and establishing tFUS as a safe technique. Recent studies have also aimed to improve the safety and precision of Focused Ultrasound (FUS) techniques by employing image-guided techniques and comparing high-frequency and low-frequency FUS (Davidson et al. 2020; Folloni et al. 2019; Grohs-Metz et al. 2022). Sophisticated imaging techniques allow physicians and researchers to monitor the changes in neural waves and the structural components of individuals' brains in real-time, improving safety and accuracy (Spivak et al. 2022). For example, the use of magnetic-guided resonance ultrasound (MRgFUS) and structural MRI scans following FUS have allowed researchers to observe no long-term damage caused by FUS techniques (Davidson et al. 2020; Jung et al. 2014). Additionally, new techniques like acoustically targeted chemogenetics (ATC) – which open the blood-brain barrier to allow for more spatiotemporal specificity in neuromodulation – and ultrasound microbubble techniques have also started to emerge (Krishna et al. 2018; Szablowski et al. 2018; Menard and Russo 2018). Furthermore, some researchers are interested in combining neuromodulation therapies such as TMS and FUS to enhance treatment results (Spivak et al. 2022).

In the past 10 years, little research has been done specifically investigating the impact of tFUS directly on fear extinction translationally across humans, rats, and primates. One study targeted the amygdala (Chou et al. 2023). The study indicated that fear and anxiety ratings decreased after using tFUS to inhibit activity in the amygdala (Chou et al. 2023). Participants who underwent tFUS and exhibited increased fear extinction also showed decreased BOLD activation during the experimental paradigm in their left amygdala, hippocampus, insular, and dorsal anterior cingulate, pinpointing a potential mechanism for enhancing fear extinction learning (Chou et al. 2023).

Additionally, some studies also display FUS's impact on learning and memory, which provides momentum to the idea that FUS could possibly successfully enhance fear extinction learning and memory. For example, Zhao and colleagues demonstrated stimulation to sensory cortical regions improved learning and memory in mice by impacting brain activation and synaptic plasticity (Zhao et al. 2023). Furthermore, in the past decade, MRgFUS and FUS have shown efficacy thus far for treating OCD, although few studies have examined the influence of FUS on PTSD (Na et al. 2015; Cui et al. 2023; Davidson et al. 2020). Some of the studies examining OCD benefits from FUS also distinctly tested anxiety symptoms and reported decreased anxiety scores following MRgFUS anterior capsulotomy (Kim et al. 2018). In a case study conducted in 2021, Zielinksi and colleagues also showed FUS to the amygdala decreased anxiety in a participant (Zielinski et al. 2021).

A benefit of tFUS, as opposed to other techniques such as TMS, is that focused ultrasound can be used to target deep neural structures like the amygdala (Folloni et al. 2019). Additionally, tFUS has higher spatial resolution than ECT, MST, or TMS (Di Biase et al. 2019). However, tFUS's ability to target deep structures also means the angle of administration is not optimal for structures near the surface (Krishna et al. 2018); but unlike other deep stimulating neuromodulation techniques like DBS – which we will discuss next in this chapter – focused ultrasound is a non-invasive procedure administered without anesthesia and is a highly accessible tool for researchers and clinicians (Di Biase et al. 2019). Furthermore, methods of FUS can disrupt the blood-brain barrier (Hynynen et al. 2005; Leinenga et al. 2016). The blood-brain barrier divides blood flow from the brain's tissue by creating a cell blockade to protect the brain (Hynynen et al. 2005; Leinenga et al. 2016). When the barrier is disrupted successfully in a controlled setting pharmacological treatments can reach the brain with less resistance (Leinenga et al. 2016). FUS's ability to disrupt the blood-brain barrier means the technique could be helpful when used adjunctively with medication, but this ability also poses a risk during administration.

The full safety concerns of tFUS still need further consideration and study and variation in skull shape and size can influence the outcomes for patients. For example, random cavitation and microhemorrhages remain a risk, although new imaging techniques can help monitor this side effect (Pasquinelli et al. 2019). Additionally, some patients experience temporary headaches, scalp burns, gait disturbances, paresthesia, dizziness, nausea, sensory problems, or even deep vein thrombosis (Krishna et al. 2018); (Pasquinelli et al. 2019). Since FUS must typically be administered in direct contact with skin, hair also usually needs to be shaved at the stimulation site, another potential downside of the procedure (Spivak et al. 2022). Furthermore, from a research standpoint, tFUS is difficult to mimic, and thus sham techniques still must be improved for double-blinded clinical trials (Spivak et al. 2022).

6 Deep Brain Stimulation (DBS)

DBS involves implanting electrodes into regions of the brain via a surgical procedure (Fitzgerald and Segrave 2015; Hamani and Temel 2012). The electrodes are connected to a neurostimulator device – typically implanted into the chest by the collarbone – which is programmed to send electrical signals to the electrodes modulating neural activity in the implanted brain regions and the networks connected to them (Fitzgerald and Segrave 2015); (Hamani and Temel 2012). Like the other neuromodulation techniques mentioned in this chapter, this device-based technique likely regulates dysregulated function in networks implicated in conditions; however, the exact mechanisms of action have not yet been fully discerned (Ashkan et al. 2017; Deniau et al. 2010; Lozano et al. 2019; McIntyre and Hahn 2010; Sullivan et al. 2021). Since the invention of DBS, significant improvements in the treatment have been made –especially in the past 10 years. For example, more precise targeting, closed-loop DBS (a type of DBS which responds to input from the patient's body and environment), and wireless DBS have all revolutionized the procedure, increasing personalization and convenience (Hebb et al. 2014; Hosain et al. 2014; Krauss et al. 2020; Vedam-Mai et al. 2021). DBS has often been studied as a treatment for OCD and anxiety disorders before and after its FDA clearance for OCD in 2009 (Denys et al. 2010; Graat et al. 2017; Sullivan et al. 2021; Wu et al. 2020). The results of these studies exhibit DBS's potential utility in modulating the fear extinction circuit for PTSD treatment. Since 1998, DBS on the ventral capsule/ ventral striatum (VC/VS) region has been identified as a viable treatment for OCD and a potential modulator of fear (Graat et al. 2017; Rodriguez-Romaguera et al. 2015). Since the infralimbic (IL) cortex and prelimbic (PL) cortex are functionally connected to the VS some researchers believe that the same circuits that are stimulated for OCD could be stimulated to modulate fear extinction in rodents (Rodriguez-Romaguera et al. 2015).

Studies using rats (e.g., Do-Monte et al. 2013; Rodriguez-Romaguera et al. 2012, 2015) tested this hypothesis and showed stimulating the VS in rodents with a high-frequency stimulation during extinction at varying time points before and after the CS modulates fear extinction in both directions – enhances and decreases – based on the location within the VS that is stimulated. Additionally, DBS to the basolateral amygdala (BLA), hippocampus, lateral hypothalamus, and PFC in animals after conditioning, after extinction, or during extinction has also been shown to enhance fear extinction in rodents (Li et al. 2022a, b; Reznikov et al. 2016; Reznikov and Hamani 2017; Sui et al. 2014). This research shows promising implications for translational clinical psychiatry applications.

DBS is not yet cleared for PTSD, but rodent models have aided in the development of potentially viable treatment targets for PTSD treatment in humans (Koek et al. 2014; Langevin et al. 2010; Stidd et al. 2013). It has yet to be discovered how effective or long-lasting the positive benefits are of DBS for fear extinction and PTSD symptoms. Furthermore, few human studies have been conducted to test if the same effects of DBS on fear extinction might be seen in humans (Reznikov and Hamani 2017). Additionally, the existing studies including human participants are typically case studies and test different stimulation targets, making it difficult to determine efficacy of DBS for PTSD in humans (Langevin et al. 2016). For example, two studies examined high-frequency DBS on the BLA in combat veterans suffering from PTSD (Langevin et al. 2016; Md et al. 2019). Another study examined DBS to the mPFC (Hamani et al. 2020). These studies showed promising results; however, more studies must be conducted to determine efficacy.

DBS could possibly be more precise than TMS because clinicians can select specific deep brain regions to directly modulate (Lozano et al. 2019). Additionally, researchers and clinicians can easily adjust the stimulation administered and turn the device on and off (Hamani and Temel 2012). However, DBS is an invasive device-based brain stimulation technique and involves a surgical procedure which incurs additional risks and costs (Grant et al. 2014). The risks of the procedure include more serious side effects such as infection where the electrodes or neurostimulator are placed, bleeding at these sites in rare cases, speech or language problems, headache, nausea or dizziness, changes in mood, personality, and behavior including depression, anxiety, irritability (Fenoy and Simpson 2014). However, these side effects are typically uncommon with DBS. Additionally, since the procedure requires surgical implantation, the financial costs associated with studying DBS are higher than other techniques (Lozano et al. 2019).

7 Vagus Nerve Stimulation (VNS)

The vagus nerve is widely distributed, runs down the neck and back, and is implicated in vital functions such as heart rate, breathing, and digestion—to name a few (Groves and Brown 2005; Nemeroff et al. 2006). Vagus nerve stimulation sends electrical pulses to the vagus nerve either wired or wirelessly (George et al. 2000). Wired versions involve connection from electrodes wrapped around the nerve to a device implanted into the chest, like a pacemaker, although non-invasive, portable, and wireless VNS devices have been created relatively recently (Austelle et al. 2022; Clancy et al. 2014; Goadsby et al. n.d.; Habibagahi et al. 2022; Nemeroff et al. 2006). The vagus nerve is a mixed nerve that carries sensory and motor information to structures of the brain including the amygdala, hippocampus, and insular cortex (George et al. 2000; Groves and Brown 2005; Nemeroff et al. 2006). Scientists hypothesize this process might be critically involved in consolidating and storing fearful memories, and thus, is likely implicated in disorders like PTSD (George et al. 2008), meaning modulation of the nerve might also alter activity in these connected regions (Austelle et al. 2022; George et al. 2008; C. K. McIntyre 2018). In fact, several studies indicate VNS might promote neural plasticity and regulate neural activity in brain regions implicated in PTSD such as the BLA (Hays et al. 2013; Hassert et al. 2004; Peña et al. 2014).

Like the other device-based neuromodulation techniques we have discussed in this chapter, the stimulation parameters can be adjusted to better treat a specific condition or individual, or optimize the treatment safety, efficacy, or comfort (Nemeroff et al. 2006). While the FDA has only cleared a specific intensity range for clinical treatment of depression and epilepsy, many research studies have investigated adjusting this threshold, the location of stimulation, the pulse-width, voltage, pattern of pulses administered, and the timing that the pulses are administered (O'Reardon et al. 2006; Souza et al. 2020, 2021). Furthermore, improvements in VNS devices such as the further development of wireless solutions, improved non-invasive capabilities, smaller models, and longer battery lives may increase patient comfort and the accessibility to treatment (Austelle et al. 2022; Goggins et al. 2022). An additional notable advancement in VNS is the invention of closed-loop stimulation – meaning the neurostimulator adjusts the stimulation given to a patient according to physiological cues (heart rate) from the patient's body (Afra et al. 2021; Winston et al. 2021). In other words, the stimulator responds to the patient's body in real-time to personalize and optimize the stimulation administered at a given time (Afra et al. 2021; Sun and Morrell 2014; Winston et al. 2021). Since closed-loop stimulation is still relatively new and predominantly employed to treat epilepsy, it is not yet FDA cleared or used for psychiatric treatment but holds promise as an advancing precision neuromodulation technique (Afra et al. 2021; Datta et al. 2020; Spindler et al. 2021; Sun and Morrell 2014; Tzadok et al. 2019).

Vagus nerve stimulation has been extensively studied as a method of promoting fear extinction (Noble et al. 2017, 2019; Peña et al. 2013, 2014). Studies using rats have demonstrated that VNS improves fear extinction by modulating neural activity in the hippocampus, prefrontal cortex, and amygdala (Alvarez-Dieppa et al. 2016; Hassert et al. 2004; Peña et al. 2013; Ura et al. 2013). Recent studies have aimed to perfect the timing of the administration when paired with fear extinction learning paradigms. For example, Souza and colleagues studied presenting the CS at various points before and after administering VNS during extinction (Souza et al. 2022). The study exemplified that delivering VNS at the same time as the CS presentation during extinction or every 8 s throughout the entire extinction phase enhances fear extinction more robustly than when VNS is administered between the CS presentations or every 32 s throughout extinction (Souza et al. 2022). Other studies also suggest VNS paired with exposure to the CS during extinction enhanced extinction learning (Alvarez-Dieppa et al. 2016; Noble et al. 2019; Peña et al. 2013). These results suggest that VNS could be effective when paired with the CS at extinction for improving fear extinction.

Additionally, other methods of administering VNS and differing intensities for administering VNS have also been studied. Less invasive methods such as Transcutaneous Vagal Nerve Stimulation (tVNS) – which stimulates the vagus nerve via electrodes attached to the surface of the skin – administered during extinction have also been observed to improve fear extinction (Burger et al. 2016; Szeska et al. 2020; Yap et al. 2020). Although, the results of some studies examining the impact of tVNS on fear extinction are somewhat mixed (Burger et al. 2017, 2019).

VNS has been studied for subthreshold and threshold anxiety disorders in human participants with some success (George et al. 2008; Trottier-Duclos et al. 2018). However, new studies examining VNS for PTSD predominantly feature rat models of PTSD. Studies using rats have shown that VNS paired with extinction can effectively enhance extinction and reduce PTSD symptoms (Noble et al. 2017; Souza et al. 2020). Additionally, VNS shows efficacy in improving extinction even in rats with prolonged and repeated extreme stress (Souza et al. 2019). Translational studies with human participants aim to examine and optimize non-invasive VNS administration methods and understand the underlying mechanisms of action (Gurel et al. 2020; Lamb et al. 2017). A few studies found that tVNS lowers stress responses to traumatic memories or emotional stimuli in patients with PTSD (Gurel et al. 2020; Lamb et al. 2017). Another study found non-invasive VNS increased activation in the anterior cingulate and hippocampus – areas typically characterized by under activation in PTSD – during exposure to personalized trauma scripts, further bolstering the idea that tcVNS could be a viable PTSD treatment (Wittbrodt et al. 2020). Future aims for studies might also include pairing VNS with exposure therapy.

While VNS has a plethora of research exemplifying the tool's utility in enhancing fear extinction, and possibly treating PTSD and anxiety disorders, the procedure can be uncomfortable for some patients (O'Reardon et al. 2006; Sackeim et al. 2001). Side effects include infection, vocal changes or hoarseness, coughing or irritation of the throat, headache, dizziness, nausea, trouble swallowing, numbness in hands or arms, sweating, muscle twitches, pain, and rarely seizures (O'Reardon et al. 2006; Sackeim et al. 2001). Additionally, VNS often requires a surgical implant, although non-invasive devices have been created (O'Reardon et al. 2006; Wang et al. 2021).

However, VNS provides significant utility as a treatment and research tool because the stimulation can be extremely personalized to individual patients and can continue to stimulate patients outside the clinics as they engage in their lives – significantly reducing the time burden often associated with other device-based stimulation techniques (Gurel et al. 2020). Furthermore, as a research tool VNS can possibly provide more insight into the parasympathetic systems involved in PTSD and fear extinction.

8 Transcranial Electrical Stimulation (tES)

There are several types of common transcranial electrical stimulation (tES) techniques which include Transcranial Direct Current Stimulation (tDCS), Transcranial Alternating Current Stimulation (tACS), and Transcranial Random Noise Currents (tRNS; Fertonani and Miniussi 2017; Reed and Cohen Kadosh 2018; Yavari et al. 2018). TDCS is the type of tES most commonly used, especially in fear extinction studies, and as such, we will predominantly discuss tDCS in this chapter (Yavari et al. 2018). All three methods involve non-invasively administering electrical currents through the brain by placing electrodes on the scalp and using headgear to position the electrodes (Reed and Cohen Kadosh 2018). The electrodes deliver an electrical current through the brain via a current controlled stimulator with the aim of modulating neuronal activity (Reed and Cohen Kadosh 2018). In tDCS, an anode electrode is attached to the positive end of the stimulator and a cathode electrode is attached to the negative side of the stimulator (Reed and Cohen Kadosh 2018). A low-intensity electrical current passes from the stimulator through the anode electrode through the scalp through the brain to the cathode electrode (Reinhart et al. 2017). Like the other methods of device-based neuromodulation techniques, the location of stimulation, frequency, duration, and intensity can all be adjusted to personalize the treatment and increase efficacy, safety, or comfort (Truong and Bikson 2018). Furthermore, scientists are increasingly employing imaging methods such as fMRI and EEG to further individualize these treatments and learn more about the mechanisms of change in efforts to optimize them for a broader range of conditions or symptoms within conditions (Jog et al. 2016; Mosayebi-Samani et al. 2021).

In the past decade, a plethora of studies have examined the use of different methods and frequencies of tDCS to enhance fear extinction. While example studies examining tDCS as a method of enhancing fear extinction do so via direct stimulation at the vmPFC, there is a recent debate as to what the optimal timing of tDCS is paired with fear extinction paradigms. Some studies suggest that tDCS administered to the vmPFC prior to or during fear extinction training improves fear extinction (Dittert et al. 2018, Vicario et al. 2020). However, Abend and colleagues administered direct current stimulation, alternating current stimulation, and sham stimulation during a fear extinction paradigm in the extinction learning phase (Abend et al. 2016). Their results showed that tDCS did not enhance fear extinction, although this might have stemmed from dispersed distribution of the electrical field (Abend et al. 2016). In response to this research, Vicario et al. (2017) published a letter emphasizing the importance of timing and standardization of electrode placements – another largely debated and inconsistent application – in studies examining tCDS stimulation during encoding (Vicario et al. 2017). Recent trials have also supported this idea that the timing of stimulation impacts tDCS's ability to enhance fear extinction with some groups reporting that tDCS after extinction learning does not enhance fear extinction (Ney et al. 2021; van't Wout et al. 2017; Cybinski et al. 2020). Taken together, the extant literature seems to indicate administering tDCS prior to or during extinction learning might result in better fear extinction learning and results in changes in activity in the fear learning circuit. Some studies have also examined using anodal and cathodal tDCS to modulate neuronal activity in the DLPFC (Asthana et al. 2013; Ganho-Ávila et al. 2019; Van Schuerbeek et al. 2023; Guiomar et al. 2023; Lee et al. 2023; Mungee et al. 2016). The results suggest tDCS over the right DLPFC modulates the fear extinction network to some extent – although the results are somewhat mixed as to positive or negative benefit. Some meta-analysis studies have even reported tDCS as an ineffective method of modulating neural activity in any regions (Horvath et al. 2015), but an extensive body of research indicates otherwise, although some more standardization in the field could allow for clearer analysis on the efficacy of tDCS.

Currently, tES is still not FDA cleared for any conditions; however, research has been directed at using tES for PTSD, OCD, and GAD (Adams et al. 2022; Brunelin et al. 2018; de Lima et al. 2019; Marcolin et al. 2023). Many of these studies pair tDCS with exposure therapies, in vivo exposure, or novel exposure techniques such as virtual reality exposure (Cobb et al. 2021; van't Wout-Frank et al. 2019). One study tested implementing tDCS before exposure therapy (Cobb et al. 2021). The results indicated that patients who received tDCS prior to exposure therapy responded better to treatment than patients who received sham treatment before exposure therapy (Cobb et al. 2021). Other studies paired tDCS with virtual reality exposure, and the results suggested SCR declined during exposure in PTSD patients who received tDCS (van't Wout-Frank et al. 2019, 2023). Generally, the results are promising for tDCS as a PTSD treatment when paired with exposure. In addition to pairing the stimulation with exposure techniques, some studies aimed to examine the influence of tDCS on specific symptoms (Ahmadizadeh et al. 2019; Ahmadizadeh and Rezaei 2020; Hampstead et al. 2020; Smits et al. 2022). While these results were also mixed, the majority showed encouraging effects of tDCS on specific symptoms such as anxiety, re-imagining, and rumination (Ahmadizadeh et al. 2019; Ahmadizadeh and Rezaei 2020; Hampstead et al. 2020; Smits et al. 2022).

TES is generally safe, well tolerated, and non-invasive with few side effects (Brunoni et al. 2012; Priori et al. 2009). As previously mentioned, while the results of studies have been widely mixed, a large bulk of research supports tCDS as a neuromodulator, although further standardization could provide a clearer picture (Kekic et al. 2016). The procedure is considered easy to administer and cost effective (Priori et al. 2009; Sparing and Mottaghy 2008). Because of the ease of administration and access, an abundance of research has been conducted on tES. While tES tools are extremely portable and accessible, the electrical stimulation is less focalized and weaker than other device-based neuromodulation techniques like TMS (Sparing and Mottaghy 2008). Only around 10–50% of the current generated by tES passes through the skull to the brain, and thus, the stimulation can only reach superficial regions of the brain (Reed and Cohen Kadosh 2018 Reinhart et al. 2017). Furthermore, side effects include itching and tingling during administration, fatigue, and skin irritation (Brunoni et al. 2012). However, many of these side effects are very rare and tES has not shown long-term side effects or damage (Brunoni et al. 2012).

9 Conclusion

Neuroscience research has established device-based neuromodulation techniques as viable strategies for modulating the fear extinction circuit and reducing conditioned fear and threat responses. Not only have these treatments been shown to be effective, but strides have also been made to further improve the precision of device-based neuromodulation treatments and research tools by increasing accuracy, comfort, safety, predictions about who will respond to treatment, and most commonly: precision in administration. Neuroimaging techniques such as fMRI and EEG are now widely utilized to localize brain regions and circuits for personalized stimulation sites and to track the changes these neuromodulation devices can make to the fear extinction circuit. Furthermore, technological advancements improve targeting and controlled administration of electrical pulses to bolster safety, comfort, efficacy, and accessibility. Many researchers and clinicians also aim to combine therapeutic techniques, such as exposure therapy with device-based approaches, to bolster response rates even further for clinical treatments. While these advancements have improved the precision in administration of treatments, researchers have also shifted toward a precision medicine mentality by studying ways to modulate specific symptoms, behaviors, or clusters of symptoms within disorders. The advances in precision tools and attitude have led to an increase in the spectrum of patients who respond to device-based neuromodulation therapies, an increase in the indications of these techniques, and an influx of knowledge about how dysregulation of specific brain circuits can lead to behaviors, emotions, or disorders. Aims to increase the literature in device-based neuromodulation techniques for various symptoms and conditions include great strides in modulating the circuit implicated in fear extinction and PTSD – the vmPFC, amygdala, dAAC, hippocampus, and other regions. Because of these strides over the past 10 years, device-based neuromodulation techniques now show feasibility and efficacy not only as modulators of the fear extinction circuit, but also possibly as future PTSD treatments.

References

Abbott CC, Jones T, Lemke NT, Gallegos P, McClintock SM, Mayer AR, Bustillo J, Calhoun VD (2014) Hippocampal structural and functional changes associated with electroconvulsive therapy response. Transl Psychiatry 4(11):e483. https://doi.org/10.1038/TP.2014.124

Abend R, Jalon I, Gurevitch G, Sar-El R, Shechner T, Pine DS, Hendler T, Bar-Haim Y (2016) Modulation of fear extinction processes using transcranial electrical stimulation. Transl Psychiatry 6(10):e913–e913. https://doi.org/10.1038/tp.2016.197

Adams TG, Cisler JM, Kelmendi B, George JR, Kichuk SA, Averill CL, Anticevic A, Abdallah CG, Pittenger C (2022) Transcranial direct current stimulation targeting the medial prefrontal cortex modulates functional connectivity and enhances safety learning in obsessive-compulsive disorder: results from two pilot studies. Depress Anxiety 39(1):37–48. https://doi.org/10.1002/DA.23212

Afra P, Adamolekun B, Aydemir S, Watson GDR (2021) Evolution of the Vagus nerve stimulation (VNS) therapy system technology for drug-resistant epilepsy. Front Med Technol 3:31. https://doi.org/10.3389/FMEDT.2021.696543

Ahmadi N, Moss L, Simon E, Nemeroff CB, Atre-Vaidya N (2016) Efficacy and long-term clinical outcome of comorbid posttraumatic stress disorder and major depressive disorder after electroconvulsive therapy. Depress Anxiety 33(7):640–647. https://doi.org/10.1002/DA.22451

Ahmadi N, Moss L, Hauser P, Nemeroff C, Atre-Vaidya N (2018) Clinical outcome of maintenance electroconvulsive therapy in comorbid posttraumatic stress disorder and major depressive disorder. J Psychiatr Res 105:132–136. https://doi.org/10.1016/J.JPSYCHIRES.2018.08.023

Ahmadizadeh MJ, Rezaei M (2018) Unilateral right and bilateral dorsolateral prefrontal cortex transcranial magnetic stimulation in treatment post-traumatic stress disorder: a randomized controlled study. Brain Res Bull 140:334–340. https://doi.org/10.1016/J.BRAINRESBULL.2018.06.001

Ahmadizadeh M, Rezaei M (2020) Effectiveness of transcranial direct current stimulation (tDCS) on depression, anxiety and rumination of patients with post-traumatic stress disorder symptoms (PTSD). J Mil Med 22(3):264–272. https://doi.org/10.30491/JMM.22.3.264

Ahmadizadeh MJ, Rezaei M, Fitzgerald PB (2019) Transcranial direct current stimulation (tDCS) for post-traumatic stress disorder (PTSD): a randomized, double-blinded, controlled trial. Brain Res Bull 153:273–278. https://doi.org/10.1016/J.BRAINRESBULL.2019.09.011

Alvarez-Dieppa AC, Griffin K, Cavalier S, Mcintyre CK (2016) Vagus nerve stimulation enhances extinction of conditioned fear in rats and modulates arc protein, CaMKII, and GluN2B-containing NMDA receptors in the basolateral amygdala. Neural Plast. https://doi.org/10.1155/2016/4273280

Ashkan K, Rogers P, Bergman H, Ughratdar I (2017) Insights into the mechanisms of deep brain stimulation. Nat Rev Neurol 13(9):548–554. https://doi.org/10.1038/nrneurol.2017.105

Asthana M, Nueckel K, Mühlberger A, Neueder D, Polak T, Domschke K, Deckert J, Herrmann MJ (2013) Effects of transcranial direct current stimulation on consolidation of fear memory. Front Psych 4:107. https://doi.org/10.3389/FPSYT.2013.00107/BIBTEX

Austelle CW, O'Leary GH, Thompson S, Gruber E, Kahn A, Manett AJ, Short B, Badran BW (2022) A comprehensive review of vagus nerve stimulation for depression. Neuromodulation 25(3):309–315. https://doi.org/10.1111/NER.13528

Baek K, Chae JH, Jeong J (2012) The effect of repetitive transcranial magnetic stimulation on fear extinction in rats. Neuroscience 200:159–165. https://doi.org/10.1016/J.NEUROSCIENCE.2011.09.050

Barad M, Gean PW, Lutz B (2006) The role of the amygdala in the extinction of conditioned fear. Biol Psychiatry 60(4):322–328. https://doi.org/10.1016/J.BIOPSYCH.2006.05.029

Barredo J, Bellone JA, Edwards M, Carpenter LL, Correia S, Philip NS (2019) White matter integrity and functional predictors of response to repetitive transcranial magnetic stimulation for posttraumatic stress disorder and major depression. Depress Anxiety 36(11):1047–1057. https://doi.org/10.1002/DA.22952

Belsher BE, Beech EH, Reddy MK, Smolenski DJ, Rauch SAM, Kelber M, Issa F, Lewis C, Bisson JI (2021) Advances in repetitive transcranial magnetic stimulation for posttraumatic stress disorder: a systematic review. J Psychiatr Res 138:598–606. https://doi.org/10.1016/J.JPSYCHIRES.2021.05.011

Berlim MT, Van Den Eynde F (2014) Repetitive transcranial magnetic stimulation over the dorsolateral prefrontal cortex for treating posttraumatic stress disorder: an exploratory meta-analysis of randomized, double-blind and sham-controlled trials. Can J Psychiatry 59(9): 487–496. www.TheCJP.ca

Boggio PS, Rocha M, Oliveira MO, Fecteau S, Cohen RB, Campanhã C, Ferreira-Santos E, Meleiro A, Corchs F, Zaghi S, Pascual-Leone A, Fregni F (2010) Noninvasive brain stimulation with high-frequency and low-intensity repetitive transcranial magnetic stimulation treatment for posttraumatic stress disorder. J Clin Psychiatry 71(8):992. https://doi.org/10.4088/JCP.08M04638BLU

Borgomaneri S, Battaglia S, Garofalo S, Tortora F, Avenanti A, di Pellegrino G (2020) State-dependent TMS over prefrontal cortex disrupts fear-memory reconsolidation and prevents the return of fear. Curr Biol 30(18):3672–3679.e4. https://doi.org/10.1016/J.CUB.2020.06.091

Brunelin J, Mondino M, Bation R, Palm U, Saoud M, Poulet E (2018) Transcranial direct current stimulation for obsessive-compulsive disorder: a systematic review. Brain Sci 8(2):37. https://doi.org/10.3390/BRAINSCI8020037

Brunoni AR, Nitsche MA, Bolognini N, Bikson M, Wagner T, Merabet L, Edwards DJ, Valero-Cabre A, Rotenberg A, Pascual-Leone A, Ferrucci R, Priori A, Boggio PS, Fregni F (2012) Clinical research with transcranial direct current stimulation (tDCS): challenges and future directions. Brain Stimul 5(3):175–195. https://doi.org/10.1016/J.BRS.2011.03.002

Burger AM, Verkuil B, Van Diest I, Van der Does W, Thayer JF, Brosschot JF (2016) The effects of transcutaneous vagus nerve stimulation on conditioned fear extinction in humans. Neurobiol Learn Mem 132:49–56. https://doi.org/10.1016/J.NLM.2016.05.007

Burger AM, Verkuil B, Fenlon H, Thijs L, Cools L, Miller HC, Vervliet B, Van Diest I (2017) Mixed evidence for the potential of non-invasive transcutaneous vagal nerve stimulation to improve the extinction and retention of fear. Behav Res Ther 97:64–74. https://doi.org/10.1016/J.BRAT.2017.07.005

Burger AM, Van Diest I, Van der Does W, Korbee JN, Waziri N, Brosschot JF, Verkuil B (2019) The effect of transcutaneous vagus nerve stimulation on fear generalization and subsequent fear extinction. Neurobiol Learn Mem 161:192–201. https://doi.org/10.1016/J.NLM.2019.04.006

Cano M, Cardoner N, Urretavizcaya M, Martínez-Zalacaín I, Goldberg X, Via E, Contreras-Rodríguez O, Camprodon J, De Arriba-Arnau A, Hernández-Ribas R, Pujol J, Soriano-Mas C, Menchón JM (2016) Modulation of limbic and prefrontal connectivity by electroconvulsive therapy in treatment-resistant depression: a preliminary study. Brain Stimul 9(1):65–71. https://doi.org/10.1016/J.BRS.2015.08.016

Carpenter LL, Conelea C, Tyrka AR, Welch ES, Greenberg BD, Price LH, Niedzwiecki M, Yip AG, Barnes J, Philip NS (2018) 5 Hz repetitive transcranial magnetic stimulation for posttraumatic stress disorder comorbid with major depressive disorder. J Affect Disord 235: 414–420. https://doi.org/10.1016/J.JAD.2018.04.009

Chen R, Classen J, Gerloff C, Celnik P, Wassermann EM, Hallett M, Cohen LG (1997) Depression of motor cortex excitability by low-frequency transcranial magnetic stimulation. Neurology 48(5):1398–1403. https://doi.org/10.1212/WNL.48.5.1398

Chen M, Yang X, Liu C, Li J, Wang X, Yang C, Hu X, Li J, Zhao J, Li X, Xu Y, Liu S (2021) Comparative efficacy and cognitive function of magnetic seizure therapy vs. electroconvulsive therapy for major depressive disorder: a systematic review and meta-analysis. Transl Psychiatry 11(1):1–9. https://doi.org/10.1038/s41398-021-01560-y

Chou T, Long M, Hayden A, Wong KS, Borron B, Kanabar A, Deckersbach T, Dougherty D (2023) Transcranial focused ultrasound of the amygdala down-modulates fear neural circuit activation and facilitates fear extinction. Biol Psychiatry 93(9):S84–S85. https://doi.org/10.1016/J.BIOPSYCH.2023.02.221

Chung SW, Hoy KE, Fitzgerald PB (2015) Theta-burst stimulation: a new form of TMS treatment for depression? Depress Anxiety 32(3):182–192. https://doi.org/10.1002/DA.22335

Chung SW, Rogasch NC, Hoy KE, Sullivan CM, Cash RFH, Fitzgerald PB (2018) Impact of different intensities of intermittent theta burst stimulation on the cortical properties during TMS-EEG and working memory performance. Hum Brain Mapp 39(2):783–802. https://doi.org/10.1002/HBM.23882

Clancy JA, Mary DA, Witte KK, Greenwood JP, Deuchars SA, Deuchars J (2014) Non-invasive Vagus nerve stimulation in healthy humans reduces sympathetic nerve activity. Brain Stimul 7(6):871–877. https://doi.org/10.1016/j.brs.2014.07.031

Clark C, Cole J, Winter C, Williams K, Grammer G (2015) A review of transcranial magnetic stimulation as a treatment for post-traumatic stress disorder. Curr Psychiatry Rep 17(10):1–9. https://doi.org/10.1007/S11920-015-0621-X/TABLES/1

Cobb AR, O'Connor P, Zaizar E, Caulfield K, Gonzalez-Lima F, Telch MJ (2021) tDCS-augmented in vivo exposure therapy for specific fears: a randomized clinical trial. J Anxiety Disord 78:102344. https://doi.org/10.1016/J.JANXDIS.2020.102344

Cohen H, Kaplan Z, Kotler M, Kouperman I, Moisa R, Grisaru N (2004) Repetitive transcranial magnetic stimulation of the right dorsolateral prefrontal cortex in posttraumatic stress disorder: a double-blind, placebo-controlled study. Am J Psychiatry 161(3):515–524. https://doi.org/10.1176/APPI.AJP.161.3.515/ASSET/IMAGES/LARGE/M618T4.JPEG

Cole EJ, Stimpson KH, Bentzley BS, Gulser M, Cherian K, Tischler C, Nejad R, Pankow H, Choi E, Aaron H, Espil FM, Pannu J, Xiao X, Duvio D, Solvason HB, Hawkins J, Guerra A, Jo B, Raj KS et al (2020) Stanford accelerated intelligent neuromodulation therapy for treatment-resistant depression. Am J Psychiatry 177(8):716–726. https://doi.org/10.1176/APPI.AJP.2019.19070720/ASSET/IMAGES/LARGE/APPI.AJP.2019.19070720F3.JPEG

Cole EJ, Phillips AL, Bentzley BS, Stimpson KH, Nejad R, Barmak F, Veerapal C, Khan N, Cherian K, Felber E, Brown R, Choi E, King S, Pankow H, Bishop JH, Azeez A, Coetzee J, Rapier R, Odenwald N et al (2022) Stanford neuromodulation therapy (SNT): a double-blind randomized controlled trial. Am J Psychiatry 179(2):132–141. https://doi.org/10.1176/APPI.AJP.2021.20101429/ASSET/IMAGES/LARGE/APPI.AJP.2021.20101429F2.JPEG

Cui H, Zhang Y, Zhao Y, Zhao Y, Ding Q, Chen R, Manssuer L, Zhang C, Liu W, Li D, Sun B, Voon V (2023) Mechanisms underlying capsulotomy for refractory obsessive-compulsive disorder: neural correlates of negative affect processing overlap with deep brain stimulation targets. Mol Psychiatry 2023:1–12. https://doi.org/10.1038/s41380-023-01989-1

Cukor J, Spitalnick J, Difede JA, Rizzo A, Rothbaum BO (2009) Emerging treatments for PTSD. Clin Psychol Rev 29(8):715–726. https://doi.org/10.1016/J.CPR.2009.09.001

Cusin C, Dougherty DD (2012) Somatic therapies for treatment-resistant depression: ECT, TMS, VNS, DBS. Biol Mood Anxiety Disord 2(1):1–9. https://doi.org/10.1186/2045-5380-2-14

Cybinski L, Boehme S, Mühlberger A, Polak T, Herrmann M (2020) Non-invasive brain stimulation for modulating a consolidated fear memory. Pharmacopsychiatry 53(02):P5.11. https://doi.org/10.1055/S-0039-3403030

Daskalakis ZJ, Dimitrova J, McClintock SM, Sun Y, Voineskos D, Rajji TK, Goldbloom DS, Wong AHC, Knyahnytska Y, Mulsant BH, Downar J, Fitzgerald PB, Blumberger DM (2019) Magnetic seizure therapy (MST) for major depressive disorder. Neuropsychopharmacology 45(2):276–282. https://doi.org/10.1038/s41386-019-0515-4

Daskalakis ZJ, McClintock SM, Hadas I, Kallioniemi E, Zomorrodi R, Throop A, Palmer L, Farzan F, Thorpe KE, Tamminga C, Blumberger DM (2021a) Confirmatory efficacy and safety trial of magnetic seizure therapy for depression (CREST-MST): protocol for identification of novel biomarkers via neurophysiology. Trials 22(1):1–8. https://doi.org/10.1186/S13063-021-05873-7/PEER-REVIEW

Daskalakis ZJ, Tamminga C, Throop A, Palmer L, Dimitrova J, Farzan F, Thorpe KE, McClintock SM, Blumberger DM (2021b) Confirmatory efficacy and safety trial of magnetic seizure therapy for depression (CREST-MST): study protocol for a randomized non-inferiority trial of magnetic seizure therapy versus electroconvulsive therapy. Trials 22(1):1–15. https://doi.org/10.1186/S13063-021-05730-7/TABLES/1

Datta P, Galla KM, Sajja K, Wichman C, Wang H, Madhavan D (2020) Vagus nerve stimulation with tachycardia detection provides additional seizure reduction compared to traditional vagus nerve stimulation. Epilepsy Behav 111 107280. https://doi.org/10.1016/J.YEBEH.2020.107280

Davidson B, Hamani C, Rabin JS, Goubran M, Meng Y, Huang Y, Baskaran A, Sharma S, Ozzoude M, Richter MA, Levitt A, Giacobbe P, Hynynen K, Lipsman N (2020) Magnetic resonance-guided focused ultrasound capsulotomy for refractory obsessive compulsive disorder and major depressive disorder: clinical and imaging results from two phase I trials. Mol Psychiatry 25(9):1946–1957. https://doi.org/10.1038/s41380-020-0737-1

De Deng Z, Lisanby SH, Peterchev AV (2011) Electric field strength and focality in electroconvulsive therapy and magnetic seizure therapy: a finite element simulation study. J Neural Eng 8(1):016007. https://doi.org/10.1088/1741-2560/8/1/016007

de Lima AL, Braga FMA, da Costa RMM, Gomes EP, Brunoni AR, Pegado R (2019) Transcranial direct current stimulation for the treatment of generalized anxiety disorder: a randomized clinical trial. J Affect Disord 259:31–37. https://doi.org/10.1016/J.JAD.2019.08.020

Deng J, Fang W, Gong Y, Bao Y, Li H, Su S, Sun J, Shi J, Lu L, Shi L, Sun H (2021) Augmentation of fear extinction by theta-burst transcranial magnetic stimulation of the prefrontal cortex in humans. J Psychiatry Neurosci 46(2):E292. https://doi.org/10.1503/JPN.200053

Deniau JM, Degos B, Bosch C, Maurice N (2010) Deep brain stimulation mechanisms: beyond the concept of local functional inhibition. Eur J Neurosci 32(7):1080–1091. https://doi.org/10.1111/J.1460-9568.2010.07413.X

Denys D, Mantione M, Figee M, Van Den Munckhof P, Koerselman F, Westenberg H, Bosch A, Schuurman R (2010) Deep brain stimulation of the nucleus accumbens for treatment-refractory obsessive-compulsive disorder. Arch Gen Psychiatry 67(10):1061–1068. https://doi.org/10.1001/ARCHGENPSYCHIATRY.2010.122

Di Biase L, Falato E, Di Lazzaro V (2019) Transcranial focused ultrasound (tFUS) and transcranial unfocused ultrasound (tUS) neuromodulation: from theoretical principles to stimulation practices. Front Neurol 10:549. https://doi.org/10.3389/FNEUR.2019.00549/BIBTEX

Ding Z, White PF (2002) Anesthesia for electroconvulsive therapy. Anesth Analg 94(5):1351–1364. https://doi.org/10.1097/00000539-200205000-00057

Dittert N, Hüttner S, Polak T, Herrmann MJ (2018) Augmentation of fear extinction by transcranial direct current stimulation (tDCS). Front Behav Neurosci 12:76. https://doi.org/10.3389/FNBEH.2018.00076/BIBTEX

Do-Monte FH, Rodriguez-Romaguera J, Rosas-Vidal LE, Quirk GJ (2013) Deep brain stimulation of the ventral striatum increases BDNF in the fear extinction circuit. Front Behav Neurosci 7:102. https://doi.org/10.3389/FNBEH.2013.00102/BIBTEX

Elahi H, Hong V, Ploski JE (2020) Electroconvulsive shock does not impair the reconsolidation of cued and contextual Pavlovian threat memory. Int J Mol Sci 21(19):7072. https://doi.org/10.3390/IJMS21197072

Engel A, Kayser S (2016) An overview on clinical aspects in magnetic seizure therapy. J Neural Transm 123(10):1139–1146. https://doi.org/10.1007/S00702-016-1583-9/TABLES/1

Fenoy AJ, Simpson RK (2014) Risks of common complications in deep brain stimulation surgery: management and avoidance: clinical article. J Neurosurg 120(1):132–139. https://doi.org/10.3171/2013.10.JNS131225

Fertonani A, Miniussi C (2017) Transcranial electrical stimulation: what we know and do not know about mechanisms. Neuroscientist 23(2):109. https://doi.org/10.1177/1073858416631966

Fitzgerald PB, Segrave RA (2015) Deep brain stimulation in mental health: review of evidence for clinical efficacy. Aust N Z J Psychiatry 49(11):979–993. https://doi.org/10.1177/0004867415598011

Fitzgerald PB, Brown TL, Daskalakis ZJ (2002) The application of transcranial magnetic stimulation in psychiatry and neurosciences research. Acta Psychiatr Scand 105(5):324–340. https://doi.org/10.1034/J.1600-0447.2002.1R179.X

Folloni D, Verhagen L, Mars RB, Fouragnan E, Constans C, Aubry JF, Rushworth MFS, Sallet J (2019) Manipulation of subcortical and deep cortical activity in the primate brain using transcranial focused ultrasound stimulation. Neuron 101(6):1109–1116.e5. https://doi.org/10.1016/J.NEURON.2019.01.019

Fryml LD, Pelic CG, Acierno R, Tuerk P, Yoder M, Borckardt JJ, Juneja N, Schmidt M, Beaver KL, George MS (2019) Exposure therapy and simultaneous repetitive transcranial magnetic stimulation: a controlled pilot trial for the treatment of posttraumatic stress disorder. J ECT 35(1):53–60. https://doi.org/10.1097/YCT.0000000000000505

Gahr M, Schönfeldt-Lecuona C, Spitzer M, Graf H (2014) Electroconvulsive therapy and posttraumatic stress disorder: first experience with conversation-based reactivation of traumatic memory contents and subsequent ECT-mediated impairment of reconsolidation. J Neuropsychiatry Clin Neurosci 26(3). https://doi.org/10.1176/appi.neuropsych.13070159

Ganho-Ávila A, Gonçalves ÓF, Guiomar R, Boggio PS, Asthana MK, Krypotos AM, Almeida J (2019) The effect of cathodal tDCS on fear extinction: a cross-measures study. PloS One 14(9): e0221282. https://doi.org/10.1371/JOURNAL.PONE.0221282

Ge R, Gregory E, Wang J, Ainsworth N, Jian W, Yang C, Wang G, Vila-Rodriguez F (2021) Magnetic seizure therapy is associated with functional and structural brain changes in MDD: therapeutic versus side effect correlates. J Affect Disord 286:40–48. https://doi.org/10.1016/J.JAD.2021.02.051

George MS, Aston-Jones G (2009) Noninvasive techniques for probing neurocircuitry and treating illness: vagus nerve stimulation (VNS), transcranial magnetic stimulation (TMS) and transcranial direct current stimulation (tDCS). Neuropsychopharmacology 35(1):301–316. https://doi.org/10.1038/npp.2009.87

George MS, Lisanby SH, Sackeim HA (1999) Transcranial magnetic stimulation: applications in neuropsychiatry. Arch Gen Psychiatry 56(4):300–311. https://doi.org/10.1001/ARCHPSYC.56.4.300

George MS, Sackeim HA, Rush AJ, Marangell LB, Nahas Z, Husain MM, Lisanby S, Burt T, Goldman J, Ballenger JC (2000) Vagus nerve stimulation: a new tool for brain research and therapy. Biol Psychiatry 47(4):287–295. https://doi.org/10.1016/S0006-3223(99)00308-X

George MS, Ward HE, Ninan PT, Pollack M, Nahas Z, Anderson B, Kose S, Howland RH, Goodman WK, Ballenger JC (2008) A pilot study of vagus nerve stimulation (VNS) for treatment-resistant anxiety disorders. Brain Stimul 1(2):112–121. https://doi.org/10.1016/J.BRS.2008.02.001

Goadsby PJ, Grosberg BM, Mauskop A, Cady R, Simmons KA (n.d.) Effect of noninvasive vagus nerve stimulation on acute migraine: an open-label pilot study. Cephalalgia 34(12):986–993. https://doi.org/10.1177/0333102414524494

Goggins E, Mitani S, Tanaka S (2022) Clinical perspectives on vagus nerve stimulation: present and future. Clin Sci 136(9):695–709. https://doi.org/10.1042/CS20210507

Gouveia FV, Davidson B, Meng Y, Gidyk DC, Rabin JS, Ng E, Abrahao A, Lipsman N, Giacobbe P, Hamani C (2020) Treating post-traumatic stress disorder with neuromodulation therapies: transcranial magnetic stimulation, transcranial direct current stimulation, and deep brain stimulation. Neurotherapeutics 17(4):1747–1756. https://doi.org/10.1007/S13311-020-00871-0/TABLES/3

Graat I, Figee M, Denys D (2017) The application of deep brain stimulation in the treatment of psychiatric disorders. Int Rev Psychiatry 29(2):178–190. https://doi.org/10.1080/09540261.2017.1282439

Grant RA, Halpern CH, Baltuch GH, O'Reardon JP, Caplan A (2014) Ethical considerations in deep brain stimulation for psychiatric illness. J Clin Neurosci 21(1):1–5. https://doi.org/10.1016/J.JOCN.2013.04.004

Greco JA, Liberzon I (2015) Neuroimaging of fear-associated learning. Neuropsychopharmacology 41(1):320–334. https://doi.org/10.1038/npp.2015.255

Grohs-Metz G, Smausz R, Gigg J, Boeckers T, Hengerer B (2022) Functional ultrasound imaging of recent and remote memory recall in the associative fear neural network in mice. Behav Brain Res 428:113862. https://doi.org/10.1016/J.BBR.2022.113862

Groves DA, Brown VJ (2005) Vagal nerve stimulation: a review of its applications and potential mechanisms that mediate its clinical effects. Neurosci Biobehav Rev 29(3):493–500. https://doi.org/10.1016/J.NEUBIOREV.2005.01.004

Guhn A, Dresler T, Andreatta M, Müller LD, Hahn T, Tupak SV, Polak T, Deckert J, Herrmann MJ (2014) Medial prefrontal cortex stimulation modulates the processing of conditioned fear. Front Behav Neurosci 8:44. https://doi.org/10.3389/FNBEH.2014.00044/ABSTRACT

Guiomar R, Saraiva M, Branco B, Pocinho F, Melo A, Brunoni A, Ganho-Ávila A (2023) Mechanism of action of tDCS for fear conditioning in a clinical sample: methodological considerations. Brain Stimul 16(1):360. https://doi.org/10.1016/j.brs.2023.01.700

Gurel NZ, Wittbrodt MT, Jung H, Shandhi MMH, Driggers EG, Ladd SL, Huang M, Ko YA, Shallenberger L, Beckwith J, Nye JA, Pearce BD, Vaccarino V, Shah AJ, Inan OT, Bremner JD (2020) Transcutaneous cervical vagal nerve stimulation reduces sympathetic responses to stress in posttraumatic stress disorder: a double-blind, randomized, sham controlled trial. Neurobiol Stress 13:100264. https://doi.org/10.1016/J.YNSTR.2020.100264

Habibagahi I, Omidbeigi M, Hadaya J, Lyu H, Jang J, Ardell JL, Bari AA, Babakhani A (2022) Vagus nerve stimulation using a miniaturized wirelessly powered stimulator in pigs. Sci Rep. https://doi.org/10.1038/s41598-022-11850-0

Hadas I, Zomorrodi R, Hill AT, Sun Y, Fitzgerald PB, Blumberger DM, Daskalakis ZJ (2020) Subgenual cingulate connectivity and hippocampal activation are related to MST therapeutic and adverse effects. Transl Psychiatry 10(1):1–8. https://doi.org/10.1038/s41398-020-01042-7

Hamani C, Temel Y (2012) Deep brain stimulation for psychiatric disease: contributions and validity of animal models. Sci Transl Med 4(142). https://doi.org/10.1126/SCITRANSLMED.3003722/SUPPL_FILE/4-142RV8_SM.PDF

Hamani C, Davidson B, Levitt A, Meng Y, Corchs F, Abrahao A, Rabin JS, Giacobbe P, Lipsman N (2020) Patient with posttraumatic stress disorder successfully treated with deep brain stimulation of the medial prefrontal cortex and uncinate fasciculus. Biol Psychiatry 88(11):e57–e59. https://doi.org/10.1016/j.biopsych.2020.05.018

Hampstead BM, Mascaro N, Schlaefflin S, Bhaumik A, Laing J, Peltier S, Martis B (2020) Variable symptomatic and neurophysiologic response to HD-tDCS in a case series with posttraumatic stress disorder. Int J Psychophysiol 154:93–100. https://doi.org/10.1016/J.IJPSYCHO.2019.10.017

Harris A, Reece J (2021) Transcranial magnetic stimulation as a treatment for posttraumatic stress disorder: a meta-analysis. J Affect Disord 289:55–65. https://doi.org/10.1016/J.JAD.2021.04.003

Hassert DL, Miyashita T, Williams CL (2004) The effects of peripheral vagal nerve stimulation at a memory-modulating intensity on norepinephrine output in the basolateral amygdala. Behav Neurosci 118(1):79–88. https://doi.org/10.1037/0735-7044.118.1.79

Hays SA, Rennaker RL, Kilgard MP (2013) Targeting plasticity with Vagus nerve stimulation to treat neurological disease. Prog Brain Res 207:275–299. https://doi.org/10.1016/B978-0-444-63327-9.00010-2

Hebb AO, Zhang JJ, Mahoor MH, Tsiokos C, Matlack C, Chizeck HJ, Pouratian N (2014) Creating the feedback loop: closed-loop neurostimulation. Neurosurg Clin 25(1):187–204. https://doi.org/10.1016/J.NEC.2013.08.006

Hoogendam JM, Ramakers GMJ, Di Lazzaro V (2010) Physiology of repetitive transcranial magnetic stimulation of the human brain. Brain Stimul 3(2):95–118. https://doi.org/10.1016/J.BRS.2009.10.005

Horvath JC, Forte JD, Carter O (2015) Quantitative review finds no evidence of cognitive effects in healthy populations from single-session transcranial direct current stimulation (tDCS). Brain Stimul 8(3):535–550. https://doi.org/10.1016/J.BRS.2015.01.400

Hosain MK, Kouzani A, Tye S (2014) Closed loop deep brain stimulation: an evolving technology. Australas Phys Eng Sci Med 37(4):619–634. https://doi.org/10.1007/S13246-014-0297-2/TABLES/2

Huang YZ, Edwards MJ, Rounis E, Bhatia KP, Rothwell JC (2005) Theta burst stimulation of the human motor cortex. Neuron 45(2):201–206. https://doi.org/10.1016/J.NEURON.2004.12.033

Hynynen K, McDannold N, Sheikov NA, Jolesz FA, Vykhodtseva N (2005) Local and reversible blood–brain barrier disruption by noninvasive focused ultrasound at frequencies suitable for trans-skull sonications. Neuroimage 24(1):12–20. https://doi.org/10.1016/J.NEUROIMAGE.2004.06.046

Isserles M, Shalev AY, Roth Y, Peri T, Kutz I, Zlotnick E, Zangen A (2013) Effectiveness of deep transcranial magnetic stimulation combined with a brief exposure procedure in post-traumatic stress disorder – a pilot study. Brain Stimul 6(3):377–383. https://doi.org/10.1016/J.BRS.2012.07.008

Isserles M, Tendler A, Roth Y, Bystritsky A, Blumberger DM, Ward H, Feifel D, Viner L, Duffy W, Zohar J, Keller CJ, Bhati MT, Etkin A, George MS, Filipcic I, Lapidus K, Casuto L, Vaishnavi S, Stein A et al (2021) Deep transcranial magnetic stimulation combined with brief exposure for posttraumatic stress disorder: a prospective multisite randomized trial. Biol Psychiatry 90(10):721–728. https://doi.org/10.1016/J.BIOPSYCH.2021.04.019

Jog MV, Smith RX, Jann K, Dunn W, Lafon B, Truong D, Wu A, Parra L, Bikson M, Wang DJJ (2016) In-vivo imaging of magnetic fields induced by transcranial direct current stimulation (tDCS) in human brain using MRI. Sci Rep 6(1):1–10. https://doi.org/10.1038/srep34385

Joshi SH, Espinoza RT, Pirnia T, Shi J, Wang Y, Ayers B, Leaver A, Woods RP, Narr KL (2016) Structural plasticity of the hippocampus and amygdala induced by electroconvulsive therapy in major depression. Biol Psychiatry 79(4):282–292. https://doi.org/10.1016/J.BIOPSYCH.2015.02.029

Jung HH, Kim SJ, Roh D, Chang JG, Chang WS, Kweon EJ, Kim CH, Chang JW (2014) Bilateral thermal capsulotomy with MR-guided focused ultrasound for patients with treatment-refractory obsessive-compulsive disorder: a proof-of-concept study. Mol Psychiatry 20(10):1205–1211. https://doi.org/10.1038/mp.2014.154

Kallioniemi E, McClintock SM, De Deng Z, Husain MM, Lisanby SH (2019) Magnetic seizure therapy: towards personalized seizure therapy for major depression. Pers Med Psychiatry:37–42. https://doi.org/10.1016/J.PMIP.2019.04.003

Kang JI, Lee H, Jhung K, Kim KR, An SK, Yoon KJ, Kim SI, Namkoong K, Lee E (2016) Frontostriatal connectivity changes in major depressive disorder after repetitive transcranial magnetic stimulation: a randomized sham-controlled study. J Clin Psychiatry 77(9):7478. https://doi.org/10.4088/JCP.15M10110

Karsen EF, Watts BV, Holtzheimer PE (2014) Review of the effectiveness of transcranial magnetic stimulation for post-traumatic stress disorder. Brain Stimul 7(2):151–157. https://doi.org/10.1016/j.brs.2013.10.006

Kayser S, Bewernick BH, Wagner S, Schlaepfer TE (2019) Clinical predictors of response to magnetic seizure therapy in depression: a preliminary report. J ECT 35(1):48–52. https://doi.org/10.1097/YCT.0000000000000495

Kekic M, Boysen E, Campbell IC, Schmidt U (2016) A systematic review of the clinical efficacy of transcranial direct current stimulation (tDCS) in psychiatric disorders. J Psychiatr Res 74:70–86. https://doi.org/10.1016/J.JPSYCHIRES.2015.12.018

Kim DR, Pesiridou A, O'Reardon JP (2009) Transcranial magnetic stimulation in the treatment of psychiatric disorders. Curr Psychiatry Rep 11(6):447–452. https://doi.org/10.1007/S11920-009-0068-Z/METRICS

Kim J, Roh D, Hyun JH, Won CS, Kim C-H, Jin CW (2018) A study of novel bilateral thermal capsulotomy with focused ultrasound for treatment-refractory obsessive-compulsive disorder: 2-year follow-up. J Psychiatry Neurosci 43(5):327. https://doi.org/10.1503/jpn.170188

Kirkcaldie M, Pridmore S, Reid P (1997) Bridging the skull: electroconvulsive therapy (ECT) and repetitive transcranial magnetic stimulation (rTMS) in psychiatry. J ECT 13(2):83–19. https://journals.lww.com/ectjournal/Abstract/1997/06000/Bridging_the_Skull__Electroconvulsive_Therapy.4.aspx

Kobayashi M, Pascual-Leone A (2003) Transcranial magnetic stimulation in neurology. Lancet Neurol 2(3):145–156. https://doi.org/10.1016/S1474-4422(03)00321-1

Koek RJ, Langevin JP, Krahl SE, Kosoyan HJ, Schwartz HN, Chen JWY, Melrose R, Mandelkern MJ, Sultzer D (2014) Deep brain stimulation of the basolateral amygdala for treatment-refractory combat post-traumatic stress disorder (PTSD): study protocol for a pilot randomized controlled trial with blinded, staggered onset of stimulation. Trials 15(1):1–25. https://doi.org/10.1186/1745-6215-15-356

Kozel FA, Motes MA, Didehbani N, DeLaRosa B, Bass C, Schraufnagel CD, Jones P, Morgan CR, Spence JS, Kraut MA, Hart J (2018) Repetitive TMS to augment cognitive processing therapy in combat veterans of recent conflicts with PTSD: a randomized clinical trial. J Affect Disord 229:506–514. https://doi.org/10.1016/J.JAD.2017.12.046

Kozel FA, Van Trees K, Larson V, Phillips S, Hashimie J, Gadbois B, Johnson S, Gallinati J, Barrett B, Toyinbo P, Weisman M, Centorino M, Gibson CA, Catalano G (2019) One hertz versus ten hertz repetitive TMS treatment of PTSD: a randomized clinical trial. Psychiatry Res 273:153–162. https://doi.org/10.1016/J.PSYCHRES.2019.01.004

Krauss JK, Lipsman N, Aziz T, Boutet A, Brown P, Chang JW, Davidson B, Grill WM, Hariz MI, Horn A, Schulder M, Mammis A, Tass PA, Volkmann J, Lozano AM (2020) Technology of deep brain stimulation: current status and future directions. Nat Rev Neurol 17(2):75–87. https://doi.org/10.1038/s41582-020-00426-z

Krishna V, Sammartino F, Rezai A (2018) A review of the current therapies, challenges, and future directions of transcranial focused ultrasound technology: advances in diagnosis and treatment. JAMA Neurol 75(2):246–254. https://doi.org/10.1001/JAMANEUROL.2017.3129

Krishnan KRR (2016) How does electroconvulsive therapy work? Biol Psychiatry 79(4):264–265. https://doi.org/10.1016/J.BIOPSYCH.2015.10.015

Kroes MCW, Tendolkar I, Van Wingen GA, Van Waarde JA, Strange BA, Fernández G (2013) An electroconvulsive therapy procedure impairs reconsolidation of episodic memories in humans. Nat Neurosci 17(2):204–206. https://doi.org/10.1038/nn.3609

Kubanek J (2018) Neuromodulation with transcranial focused ultrasound. Neurosurg Focus 44(2):E14. https://doi.org/10.3171/2017.11.FOCUS17621

Kuhn T, Spivak NM, Dang BH, Becerra S, Halavi SE, Rotstein N, Rosenberg BM, Hiller S, Swenson A, Cvijanovic L, Dang N, Sun M, Kronemyer D, Berlow R, Revett MR, Suthana N, Monti MM, Bookheimer S (2023) Transcranial focused ultrasound selectively increases perfusion and modulates functional connectivity of deep brain regions in humans. Front Neural Circuits 17:1120410. https://doi.org/10.3389/FNCIR.2023.1120410

Lamb DG, Porges EC, Lewis GF, Williamson JB (2017) Non-invasive vagal nerve stimulation effects on hyperarousal and autonomic state in patients with posttraumatic stress disorder and history of mild traumatic brain injury: preliminary evidence. Front Med 4:124. https://doi.org/10.3389/FMED.2017.00124/BIBTEX

Lang N, Harms J, Weyh T, Lemon RN, Paulus W, Rothwell JC, Siebner HR (2006) Stimulus intensity and coil characteristics influence the efficacy of rTMS to suppress cortical excitability. Clin Neurophysiol 117(10):2292–2301. https://doi.org/10.1016/J.CLINPH.2006.05.030

Langevin JP, De Salles AAF, Kosoyan HP, Krahl SE (2010) Deep brain stimulation of the amygdala alleviates post-traumatic stress disorder symptoms in a rat model. J Psychiatr Res 44(16):1241–1245. https://doi.org/10.1016/J.JPSYCHIRES.2010.04.022

Langevin JP, Koek RJ, Schwartz HN, Chen JWY, Sultzer DL, Mandelkern MA, Kulick AD, Krahl SE (2016) Deep brain stimulation of the basolateral amygdala for treatment-refractory posttraumatic stress disorder. Biol Psychiatry 79(10):e82–e84. https://doi.org/10.1016/j.biopsych.2015.09.003

Leaver AM, Espinoza R, Wade B, Narr KL (2022) Parsing the network mechanisms of electroconvulsive therapy. Biol Psychiatry 92(3):193–203. https://doi.org/10.1016/J.BIOPSYCH.2021.11.016

Lee D, Guiomar R, Gonçalves ÓF, Almeida J, Ganho-Ávila A (2023) Effects of transcranial direct current stimulation on neural activity and functional connectivity during fear extinction. Int J Clin Health Psychol 23(1):100342. https://doi.org/10.1016/J.IJCHP.2022.100342

Leinenga G, Langton C, Nisbet R, Götz J (2016) Ultrasound treatment of neurological diseases – current and emerging applications. Nat Rev Neurol 12(3):161–174. https://doi.org/10.1038/nrneurol.2016.13

Leong K, Chan P, Ong L, Zwicker A, Willan S, Lam RW, McGirr A (2020) A randomized Sham-controlled trial of 1-Hz and 10-Hz repetitive transcranial magnetic stimulation (rTMS) of the right dorsolateral prefrontal cortex in civilian post-traumatic stress disorder: Un essai randomisé contrôlé simulé de stimulation magnétique transcrânienne repetitive (SMTr) de 1 Hz et 10 Hz du cortex préfrontal dorsolatéral droit dans le trouble de stress post-traumatique chez des civils. Can J Psychiatry 65(11):770–778. https://doi.org/10.1177/0706743720923064/FORMAT/EPUB

Lewis PM, Thomson RH, Rosenfeld JV, Fitzgerald PB (2016) Brain neuromodulation techniques: a review. Neuroscientist 22(4):406–421. https://doi.org/10.1177/1073858416646707/ASSET/IMAGES/LARGE/10.1177_1073858416646707-FIG10.JPEG

Li HT, Donegan DC, Peleg-Raibstein D, Burdakov D (2022a) Hypothalamic deep brain stimulation as a strategy to manage anxiety disorders. Proc Natl Acad Sci U S A 119(16). https://doi.org/10.1073/PNAS.2113518119/-/DCSUPPLEMENTAL

Li J, Zhang X, Jiang J, Zhang B, Tang Y, Zhang T, Jia Y, Li Q, Xia M, Sheng J, Li C, Wang J (2022b) Comparison of electroconvulsive therapy and magnetic seizure therapy in schizophrenia: structural changes/neuroplasticity. Psychiatry Res 312:114523. https://doi.org/10.1016/J.PSYCHRES.2022.114523

Linnman C, Rougemont-Bücking A, Beucke JC, Zeffiro TA, Milad MR (2011) Unconditioned responses and functional fear networks in human classical conditioning. Behav Brain Res 221(1):237–245. https://doi.org/10.1016/J.BBR.2011.02.045

Lisanby SH (2007) Electroconvulsive therapy for depression. N Engl J Med 357(19):1939–1945. https://doi.org/10.1056/NEJMCT075234

Lisanby SH, Deng ZD (2015) Magnetic seizure therapy for the treatment of depression. In: Brain stimulation: methodologies and interventions, pp 123–148. https://doi.org/10.1002/9781118568323.CH8

Liston C, Chen AC, Zebley BD, Drysdale AT, Gordon R, Leuchter B, Voss HU, Casey BJ, Etkin A, Dubin MJ (2014) Default mode network mechanisms of transcranial magnetic stimulation in depression. Biol Psychiatry 76(7):517–526. https://doi.org/10.1016/J.BIOPSYCH.2014.01.023

Lozano AM, Lipsman N, Bergman H, Brown P, Chabardes S, Chang JW, Matthews K, McIntyre CC, Schlaepfer TE, Schulder M, Temel Y, Volkmann J, Krauss JK (2019) Deep brain stimulation: current challenges and future directions. Nat Rev Neurol 15(3):148–160. https://doi.org/10.1038/s41582-018-0128-2

Madore MR, Kozel FA, Williams LM, Green LC, George MS, Holtzheimer PE, Yesavage JA, Philip NS (2022) Prefrontal transcranial magnetic stimulation for depression in US military veterans – a naturalistic cohort study in the veterans health administration. J Affect Disord 297:671–678. https://doi.org/10.1016/J.JAD.2021.10.025

Marcolin KADS, da Cunha ÂBM, Yoneyama BC, Ribeiro TA (2023) Effects of transcranial direct current stimulation (tDCS) in "Kiss nightclub fire" patients with post-traumatic stress disorder (PTSD): a phase II clinical trial. SAGE Open Med 11. https://doi.org/10.1177/20503121231160953/ASSET/IMAGES/LARGE/10.1177_20503121231160953-FIG1.JPEG

Maren S, Quirk GJ (2004) Neuronal signalling of fear memory. Nat Rev Neurosci 5(11):844–852. https://doi.org/10.1038/nrn1535

Margoob MA, Ali Z, Andrade C (2010) Efficacy of ECT in chronic, severe, antidepressant- and CBT-refractory PTSD: an open, prospective study. Brain Stimul 3(1):28–35. https://doi.org/10.1016/J.BRS.2009.04.005

Marin MF, Camprodon JA, Dougherty DD, Milad MR (2014) Device-based brain stimulation to augment fear extinction: implications for PTSD treatment and beyond. Depress Anxiety 31(4):269–278. https://doi.org/10.1002/DA.22252

Marković V, Vicario CM, Yavari F, Salehinejad MA, Nitsche MA (2021) A systematic review on the effect of transcranial direct current and magnetic stimulation on fear memory and extinction. Front Hum Neurosci 15:138. https://doi.org/10.3389/FNHUM.2021.655947/BIBTEX

McClintock SM, Tirmizi O, Chansard M, Husain MM (2011) A systematic review of the neurocognitive effects of magnetic seizure therapy. Int Rev Psychiatry 23(5):413–423. https://doi.org/10.3109/09540261.2011.623687

McDonald A, Walter G (2001) The portrayal of ECT in American movies. J ECT 17(4):264–274. https://journals.lww.com/ectjournal/fulltext/2001/12000/the_portrayal_of_ect_in_american_movies.6.aspx?casa_token=KK6mMawkuYcAAAAA:oNWK_7WG7RzfA3QULX1WWtWCKyuK4YuVXCqM4ogBOCFqVtM4Y8E1fMWAdgGVfnNH98l0EpN719Cv7j-QeiRkovPLFA

McIntyre CK (2018) Is there a role for vagus nerve stimulation in the treatment of posttraumatic stress disorder? Bioelectron Med 1(2):95–99. https://doi.org/10.2217/BEM-2018-0002

McIntyre CC, Hahn PJ (2010) Network perspectives on the mechanisms of deep brain stimulation. Neurobiol Dis 38(3):329–337. https://doi.org/10.1016/J.NBD.2009.09.022

Md RK, Langevin J, Krahl S, Chen J, Sultzer D, Mandelkern M, Kulick A (2019) Basolateral amygdala deep brain stimulation for treatment refractory combat PTSD: data from the first two cases. Brain Stimul 12(2):429–430. https://doi.org/10.1016/j.brs.2018.12.391

Menard C, Russo SJ (2018) Non-invasive chemogenetics. Nature Biomedical Engineering 2(7):467–468. https://doi.org/10.1038/s41551-018-0269-z

Milad MR, Quirk GJ (2012) Fear extinction as a model for translational neuroscience: ten years of progress. Annu Rev Psychol 63:129. https://doi.org/10.1146/ANNUREV.PSYCH.121208.131631

Milad MR, Vidal-Gonzalez I, Quirk GJ (2004) Electrical stimulation of medial prefrontal cortex reduces conditioned fear in a temporally specific manner. Behav Neurosci 118(2):389. https://doi.org/10.1037/0735-7044.118.2.389

Milad MR, Quirk GJ, Pitman RK, Orr SP, Fischl B, Rauch SL (2007a) A role for the human dorsal anterior cingulate cortex in fear expression. Biol Psychiatry 62(10):1191–1194. https://doi.org/10.1016/J.BIOPSYCH.2007.04.032

Milad MR, Wright CI, Orr SP, Pitman RK, Quirk GJ, Rauch SL (2007b) Recall of fear extinction in humans activates the ventromedial prefrontal cortex and hippocampus in concert. Biol Psychiatry 62(5):446–454. https://doi.org/10.1016/J.BIOPSYCH.2006.10.011

Milad MR, Pitman RK, Ellis CB, Gold AL, Shin LM, Lasko NB, Zeidan MA, Handwerger K, Orr SP, Rauch SL (2009) Neurobiological basis of failure to recall extinction memory in posttraumatic stress disorder. Biol Psychiatry 66(12):1075–1082. https://doi.org/10.1016/J.BIOPSYCH.2009.06.026

Milad MR, Rosenbaum BL, Simon NM (2014) Neuroscience of fear extinction: implications for assessment and treatment of fear-based and anxiety related disorders. Behav Res Ther 62:17–23. https://doi.org/10.1016/J.BRAT.2014.08.006

Misanin JR, Miller RR, Lewis DJ (1968) Retrograde amnesia produced by electroconvulsive shock after reactivation of a consolidated memory trace. Science 160(3827):554–555. https://doi.org/10.1126/SCIENCE.160.3827.554

Mosayebi-Samani M, Jamil A, Salvador R, Ruffini G, Haueisen J, Nitsche MA (2021) The impact of individual electrical fields and anatomical factors on the neurophysiological outcomes of tDCS: a TMS-MEP and MRI study. Brain Stimul 14(2):316–326. https://doi.org/10.1016/J.BRS.2021.01.016

Mungee A, Burger M, Bajbouj M (2016) No effect of cathodal transcranial direct current stimulation on fear memory in healthy human subjects. Brain Sci 6(4):55. https://doi.org/10.3390/BRAINSCI6040055

Mutz J, Vipulananthan V, Carter B, Hurlemann R, Fu CHY, Young AH (2019) Comparative efficacy and acceptability of non-surgical brain stimulation for the acute treatment of major depressive episodes in adults: systematic review and network meta-analysis. BMJ 364:1079. https://doi.org/10.1136/BMJ.L1079

Na YC, Jung HH, Chang JW (2015) Focused ultrasound for the treatment of obsessive-compulsive disorder. Neurosurgical treatments for psychiatric disorders, pp 125–141. https://doi.org/10.1007/978-94-017-9576-0_12/FIGURES/3

Najib U, Horvath JC (2014) Transcranial magnetic stimulation (TMS) safety considerations and recommendations. NeuroMethods 89:15–30. https://doi.org/10.1007/978-1-4939-0879-0_2/TABLES/6

Nemeroff CB, Mayberg HS, Krahl SE, McNamara J, Frazer A, Henry TR, George MS, Charney DS, Brannan SK (2006) VNS therapy in treatment-resistant depression: clinical evidence and putative neurobiological mechanisms. Neuropsychopharmacology 31(7):1345–1355. https://doi.org/10.1038/sj.npp.1301082

Ney LJ, Vicario CM, Nitsche MA, Felmingham KL (2021) Timing matters: transcranial direct current stimulation after extinction learning impairs subsequent fear extinction retention. Neurobiol Learn Mem 177:107356. https://doi.org/10.1016/J.NLM.2020.107356

Noble LJ, Gonzalez IJ, Meruva VB, Callahan KA, Belfort BD, Ramanathan KR, Meyers E, Kilgard MP, Rennaker RL, McIntyre CK (2017) Effects of vagus nerve stimulation on extinction of conditioned fear and post-traumatic stress disorder symptoms in rats. Transl Psychiatry 7(8):e1217. https://doi.org/10.1038/TP.2017.191

Noble LJ, Meruva VB, Hays SA, Rennaker RL, Kilgard MP, McIntyre CK (2019) Vagus nerve stimulation promotes generalization of conditioned fear extinction and reduces anxiety in rats. Brain Stimul 12(1):9–18. https://doi.org/10.1016/J.BRS.2018.09.013

Nordanskog P, Dahlstrand U, Larsson MR, Larsson EM, Knutsson L, Johanson A (2010) Increase in hippocampal volume after electroconvulsive therapy in patients with depression: a volumetric magnetic resonance imaging study. J ECT 26(1):62–67. https://doi.org/10.1097/YCT.0B013E3181A95DA8

O'Reardon JP, Cristancho P, Peshek AD (2006) Vagus nerve stimulation (VNS) and treatment of depression: to the brainstem and beyond. Psychiatry (Edgmont) 3(5):54–63

Osuch EA, Benson BE, Luckenbaugh DA, Geraci M, Post RM, McCann U (2009) Repetitive TMS combined with exposure therapy for PTSD: a preliminary study. J Anxiety Disord 23(1):54–59. https://doi.org/10.1016/J.JANXDIS.2008.03.015

Ousdal OT, Brancati GE, Kessler U, Erchinger V, Dale AM, Abbott C, Oltedal L (2022) The neurobiological effects of electroconvulsive therapy studied through magnetic resonance: what have we learned, and where do we go? Biol Psychiatry 91(6):540–549. https://doi.org/10.1016/J.BIOPSYCH.2021.05.023

Pang Y, Wei Q, Zhao S, Li N, Li Z, Lu F, Pang J, Zhang R, Wang K, Chu C, Tian Y, Wang J (2022) Enhanced default mode network functional connectivity links with electroconvulsive therapy response in major depressive disorder. J Affect Disord 306:47–54. https://doi.org/10.1016/J.JAD.2022.03.035

Pascual-Leone A, Meador K (1998) Is transcranial magnetic stimulation coming of age? J Clin Neurophysiol 15(4):285–287. https://journals.lww.com/clinicalneurophys/Fulltext/1998/07000/Is_Transcranial_Magnetic_Stimulation_Coming_of.1.aspx?casa_token=PTMKc_wUQCsAAAAA:QLfssjQ6gMrkPr8Vxq1Kwf7_tsF6xy3CEWgTAwYaUG3XVscV9Dqr6T3TtD5ifwpb_slIHhgUUgZ1-sh5ADV0bZGKRw

Pasquinelli C, Hanson LG, Siebner HR, Lee HJ, Thielscher A (2019) Safety of transcranial focused ultrasound stimulation: a systematic review of the state of knowledge from both human and animal studies. Brain Stimul 12(6):1367–1380. https://doi.org/10.1016/J.BRS.2019.07.024

Peña DF, Engineer ND, McIntyre CK (2013) Rapid remission of conditioned fear expression with extinction training paired with Vagus nerve stimulation. Biol Psychiatry 73(11):1071–1077. https://doi.org/10.1016/J.BIOPSYCH.2012.10.021

PeñA DF, Childs JE, Willett S, Vital A, McIntyre CK, Kroener S (2014) Vagus nerve stimulation enhances extinction of conditioned fear and modulates plasticity in the pathway from the ventromedial prefrontal cortex to the Amygdala. Front Behav Neurosci 8:327. https://doi.org/10.3389/FNBEH.2014.00327/BIBTEX

Pennington ZT, Fanselow MS (2018) Indirect targeting of subsuperficial brain structures with transcranial magnetic stimulation reveals a promising way forward in the treatment of fear. Biol Psychiatry 84:80–81. https://doi.org/10.1016/j.biopsych.2018.05.003

Perrin JS, Merz S, Bennett DM, Currie J, Steele DJ, Reid IC, Schwarzbauer C (2012) Electroconvulsive therapy reduces frontal cortical connectivity in severe depressive disorder. Proc Natl Acad Sci U S A 109(14):5464–5468. https://doi.org/10.1073/PNAS.1117206109/SUPPL_FILE/PNAS.201117206SI.PDF

Petrosino NJ, Cosmo C, Berlow YA, Zandvakili A, van't Wout-Frank M, Philip NS (2021) Transcranial magnetic stimulation for post-traumatic stress disorder. Ther Adv Psychopharmacol 11:204512532110499. https://doi.org/10.1177/20451253211049921

Phelps EA, Delgado MR, Nearing KI, Ledoux JE (2004) Extinction learning in humans: role of the amygdala and vmPFC. Neuron 43(6):897–905. https://doi.org/10.1016/J.NEURON.2004.08.042

Philip NS, Ridout SJ, Albright SE, Sanchez G, Carpenter LL (2016) 5-Hz transcranial magnetic stimulation for comorbid posttraumatic stress disorder and major depression. J Trauma Stress 29(1):93–96. https://doi.org/10.1002/JTS.22065

Philip NS, Barredo J, van't Wout-Frank M, Tyrka AR, Price LH, Carpenter LL (2018) Network mechanisms of clinical response to transcranial magnetic stimulation in posttraumatic stress disorder and major depressive disorder. Biol Psychiatry 83(3):263–272. https://doi.org/10.1016/J.BIOPSYCH.2017.07.021

Philip NS, Aiken EE, Kelley ME, Burch W, Waterman L, Holtzheimer PE (2019a) Synchronized transcranial magnetic stimulation for posttraumatic stress disorder and comorbid major depression. Brain Stimul 12(5):1335–1337. https://doi.org/10.1016/j.brs.2019.06.010

Philip NS, Barredo J, Aiken E, Larson V, Jones RN, Tracie Shea M, Greenberg BD, Van'T Wout-Frank M (2019b) Theta burst transcranial magnetic stimulation for posttraumatic stress disorder. Am J Psychiatry 176(11):939. https://doi.org/10.1176/APPI.AJP.2019.18101160

Priori A, Hallett M, Rothwell JC (2009) Repetitive transcranial magnetic stimulation or transcranial direct current stimulation? Brain Stimul 2(4):241–245. https://doi.org/10.1016/J.BRS.2009.02.004

Raij T, Nummenmaa A, Marin MF, Porter D, Furtak S, Setsompop K, Milad MR (2018) Prefrontal cortex stimulation enhances fear extinction memory in humans. Biol Psychiatry 84(2):129–137. https://doi.org/10.1016/J.BIOPSYCH.2017.10.022

Reed T, Cohen Kadosh R (2018) Transcranial electrical stimulation (tES) mechanisms and its effects on cortical excitability and connectivity. J Inherit Metab Dis 41(6):1123. https://doi.org/10.1007/S10545-018-0181-4

Reinhart RMG, Cosman JD, Fukuda K, Woodman GF (2017) Using transcranial direct-current stimulation (tDCS) to understand cognitive processing. Atten Percept Psychophys 79(1):3–23. https://doi.org/10.3758/S13414-016-1224-2/FIGURES/9

Reznikov R, Hamani C (2017) Posttraumatic stress disorder: perspectives for the use of deep brain stimulation. Neuromodulation 20(1):7–14. https://doi.org/10.1111/NER.12551

Reznikov R, Binko M, Nobrega JN, Hamani C (2016) Deep brain stimulation in animal models of fear, anxiety, and posttraumatic stress disorder. Neuropsychopharmacology 41(12):2810–2817. https://doi.org/10.1038/npp.2016.34

Richieri R, Jouvenoz D, Verger A, Fiat P, Boyer L, Lançon C, Guedj E (2017) Changes in dorsolateral prefrontal connectivity after rTMS in treatment-resistant depression: a brain perfusion SPECT study. Eur J Nucl Med Mol Imaging 44(6):1051–1055. https://doi.org/10.1007/S00259-017-3640-5/FIGURES/2

Rodriguez-Romaguera J, Do Monte FHM, Quirk GJ (2012) Deep brain stimulation of the ventral striatum enhances extinction of conditioned fear. Proc Natl Acad Sci U S A 109(22):8764–8769. https://doi.org/10.1073/PNAS.1200782109/SUPPL_FILE/SM02.AVI

Rodriguez-Romaguera J, Do-Monte FH, Tanimura Y, Quirk GJ, Haber SN (2015) Enhancement of fear extinction with deep brain stimulation: evidence for medial orbitofrontal involvement. Neuropsychopharmacology 40(7):1726–1733. https://doi.org/10.1038/npp.2015.20

Ropper AH, Espinoza RT, Kellner CH (2022) Electroconvulsive Therapy. N Engl J Med 386(7): 667–672. https://doi.org/10.1056/NEJMRA2034954

Rossi S, Antal A, Bestmann S, Bikson M, Brewer C, Brockmöller J, Carpenter LL, Cincotta M, Chen R, Daskalakis JD, Di Lazzaro V, Fox MD, George MS, Gilbert D, Kimiskidis VK, Koch G, Ilmoniemi RJ, Pascal Lefaucheur J, Leocani L et al (2021) Safety and recommendations for TMS use in healthy subjects and patient populations, with updates on training, ethical and regulatory issues: expert guidelines. Clin Neurophysiol 132(1):269–306. https://doi.org/10.1016/J.CLINPH.2020.10.003

Rowny S, Lisanby SH, Roose SP (2008) Brain stimulation in psychiatry. Psychiatry 2:2354–2371. https://doi.org/10.1002/9780470515167.CH109

Rowny SB, Benzl K, Lisanby SH (2009) Translational development strategy for magnetic seizure therapy. Exp Neurol 219(1):27–35. https://doi.org/10.1016/J.EXPNEUROL.2009.03.029

Sackeim HA, Rush AJ, George MS, Marangell LB, Husain MM, Nahas Z, Johnson CR, Seidman S, Giller C, Haines S, Simpson RK, Goodman RR (2001) Vagus nerve stimulation (VNSTM) for treatment-resistant depression: efficacy, side effects, and predictors of outcome. Neuropsychopharmacology 25(5):713–728. https://doi.org/10.1016/S0893-133X(01)00271-8

Sackeim HA, Prudic J, Fuller R, Keilp J, Lavori PW, Olfson M (2006) The cognitive effects of electroconvulsive therapy in community settings. Neuropsychopharmacology 32(1):244–254. https://doi.org/10.1038/sj.npp.1301180

Sartorius A, Demirakca T, Böhringer A, Clemm von Hohenberg C, Aksay SS, Bumb JM, Kranaster L, Ende G (2016) Electroconvulsive therapy increases temporal gray matter volume and cortical thickness. Eur Neuropsychopharmacol 26(3):506–517. https://doi.org/10.1016/J.EURONEURO.2015.12.036

Shin LM, Liberzon I (2009) The neurocircuitry of fear, stress, and anxiety disorders. Neuropsychopharmacology 35(1):169–191. https://doi.org/10.1038/npp.2009.83

Sienaert P (2016) Based on a true story? The portrayal of ECT in international movies and television programs. Brain Stimul 9(6):882–891. https://doi.org/10.1016/J.BRS.2016.07.005

Singh A, Kar SK (2017) How electroconvulsive therapy works?: Understanding the neurobiological mechanisms. Clin Psychopharmacol Neurosci 15(3):210–221. https://doi.org/10.9758/CPN.2017.15.3.210

Sippel LM, Holtzheimer PE, Friedman MJ, Schnurr PP (2018) Defining treatment-resistant posttraumatic stress disorder: a framework for future research. Biol Psychiatry 84(5):e37–e41. https://doi.org/10.1016/j.biopsych.2018.03.011

Smits FM, Geuze E, Schutter DJLG, van Honk J, Gladwin TE (2022) Effects of tDCS during inhibitory control training on performance and PTSD, aggression and anxiety symptoms: a randomized-controlled trial in a military sample. Psychol Med 52(16):3964–3974. https://doi.org/10.1017/S0033291721000817

Souza RR, Robertson NM, Pruitt DT, Gonzales PA, Hays SA, Rennaker RL, Kilgard MP, McIntyre CK (2019) Vagus nerve stimulation reverses the extinction impairments in a model of PTSD with prolonged and repeated trauma. Stress 22(4):509–520. https://doi.org/10.1080/10253890.2019.1602604

Souza RR, Robertson NM, Mathew E, Tabet MN, Bucksot JE, Pruitt DT, Rennaker RL, Hays SA, McIntyre CK, Kilgard MP (2020) Efficient parameters of vagus nerve stimulation to enhance extinction learning in an extinction-resistant rat model of PTSD. Prog Neuropsychopharmacol Biol Psychiatry 99:109848. https://doi.org/10.1016/J.PNPBP.2019.109848

Souza RR, Robertson NM, McIntyre CK, Rennaker RL, Hays SA, Kilgard MP (2021) Vagus nerve stimulation enhances fear extinction as an inverted-U function of stimulation intensity. Exp Neurol 341:113718. https://doi.org/10.1016/J.EXPNEUROL.2021.113718

Souza RR, Powers MB, Rennaker RL, McIntyre CK, Hays SA, Kilgard MP (2022) Timing of vagus nerve stimulation during fear extinction determines efficacy in a rat model of PTSD. Sci Rep 12(1):1–11. https://doi.org/10.1038/s41598-022-20301-9

Sparing R, Mottaghy FM (2008) Noninvasive brain stimulation with transcranial magnetic or direct current stimulation (TMS/tDCS) – from insights into human memory to therapy of its dysfunction. Methods 44(4):329–337. https://doi.org/10.1016/J.YMETH.2007.02.001

Speer AM, Kimbrell TA, Wassermann EM, Repella JD, Willis MW, Herscovitch P, Post RM (2000) Opposite effects of high and low frequency rTMS on regional brain activity in depressed patients. Biol Psychiatry 48(12):1133–1141. https://doi.org/10.1016/S0006-3223(00)01065-9

Spindler P, Vajkoczy P, Schneider UC (2021) Closed-loop vagus nerve stimulation. Patient-tailored therapy or undirected treatment? Interdiscip Neurosurg 23:101003. https://doi.org/10.1016/J.INAT.2020.101003

Spivak NM, Sanguinetti JL, Monti MM (2022) Focusing in on the future of focused ultrasound as a translational tool. Brain Sci 12(2). https://doi.org/10.3390/BRAINSCI12020158

Stidd DA, Vogelsang K, Krahl SE, Langevin JP, Fellous JM (2013) Amygdala deep brain stimulation is superior to paroxetine treatment in a rat model of posttraumatic stress disorder. Brain Stimul 6(6):837–844. https://doi.org/10.1016/J.BRS.2013.05.008

Stultz DJ, Osburn S, Burns T, Pawlowska-Wajswol S, Walton R (2020) Transcranial magnetic stimulation (Tms) safety with respect to seizures: a literature review. Neuropsychiatr Dis Treat 16:2989–3000. https://doi.org/10.2147/NDT.S276635

Sui L, Huang SJ, Peng BB, Ren J, Tian FY, Wang Y (2014) Deep brain stimulation of the amygdala alleviates fear conditioning-induced alterations in synaptic plasticity in the cortical-amygdala pathway and fear memory. J Neural Transm 121(7):773–782. https://doi.org/10.1007/S00702-014-1183-5/FIGURES/4

Sullivan CRP, Olsen S, Widge AS (2021) Deep brain stimulation for psychiatric disorders: from focal brain targets to cognitive networks. Neuroimage 225:117515. https://doi.org/10.1016/J.NEUROIMAGE.2020.117515

Sun FT, Morrell MJ (2014) Closed-loop neurostimulation: the clinical experience. Neurotherapeutics 11(3):553–563. https://doi.org/10.1007/S13311-014-0280-3/FIGURES/3

Sun Y, Blumberger DM, Mulsant BH, Rajji TK, Fitzgerald PB, Barr MS, Downar J, Wong W, Farzan F, Daskalakis ZJ (2018) Magnetic seizure therapy reduces suicidal ideation and produces neuroplasticity in treatment-resistant depression. Transl Psychiatry 8(1):1–11. https://doi.org/10.1038/s41398-018-0302-8

Szablowski JO, Lee-Gosselin A, Lue B, Malounda D, Shapiro MG (2018) Acoustically targeted chemogenetics for the non-invasive control of neural circuits. Nat Biomed Eng 2(7):475–484. https://doi.org/10.1038/s41551-018-0258-2

Szeska C, Richter J, Wendt J, Weymar M, Hamm AO (2020) Promoting long-term inhibition of human fear responses by non-invasive transcutaneous vagus nerve stimulation during extinction training. Sci Rep 10(1):1–16. https://doi.org/10.1038/s41598-020-58412-w

Takamiya A, Hirano J, Yamagata B, Takei S, Kishimoto T, Mimura M (2019) Electroconvulsive therapy modulates resting-state EEG oscillatory pattern and phase synchronization in nodes of the default mode network in patients with depressive disorder. Front Hum Neurosci 13:1. https://doi.org/10.3389/FNHUM.2019.00001/BIBTEX

Tang VM, Blumberger DM, Throop A, McClintock SM, Voineskos D, Downar J, Knyahnytska Y, Mulsant BH, Fitzgerald PB, Daskalakis ZJ (2021a) Continuation magnetic seizure therapy for treatment-resistant unipolar or bipolar depression. J Clin Psychiatry 82(6):37501. https://doi.org/10.4088/JCP.20M13677

Tang VM, Blumberger DM, Weissman CR, Dimitrova J, Throop A, McClintock SM, Voineskos D, Rajji TK, Downar J, Knyahnytska Y, Mulsant BH, Fitzgerald PB, Daskalakis ZJ (2021b) A pilot study of magnetic seizure therapy for treatment-resistant obsessive–compulsive disorder. Depress Anxiety 38(2):161–171. https://doi.org/10.1002/DA.23097

Tang VM, Trought K, Gicas KM, Kozak M, Josselyn SA, Daskalakis ZJ, Blumberger DM, Voineskos D, Knyahnytska Y, Pasricha S, Chung Y, Zhou Y, Isserles M, Wong AHC (2021c) Electroconvulsive therapy with a memory reactivation intervention for post-traumatic stress disorder: a randomized controlled trial. Brain Stimul 14(3):635–642. https://doi.org/10.1016/J.BRS.2021.03.015

Thierrée S, Raulin-Briot M, Legrand M, Le Gouge A, Vancappel A, Tudorache AC, Brizard B, Clarys D, Caille A, El-Hage W (2022) Combining trauma script exposure with rTMS to reduce symptoms of post-traumatic stress disorder: randomized controlled trial. Neuromodulation 25(4):549–557. https://doi.org/10.1111/NER.13505

Trevizol AP, Duarte Barros M, Silva PO, Osuch E, Cordeiro Q, Shiozawa P (2016) Brief communication rends in psychiatry and psychotherapy T transcranial magnetic stimulation for posttraumatic stress disorder: an updated systematic review and meta-analysis Estimulação magnética transcraniana para transtorno de estresse pós-traumático: revisão sistemática de literatura e metanálise. Trends Psychiatry Psychother 38(1). https://doi.org/10.1590/2237-6089-2015-0072

Trottier-Duclos F, Desbeaumes Jodoin V, Fournier-Gosselin MP, Richer F, Desjardins N, Tieu S, Lespérance P (2018) A 6-year follow-up study of Vagus nerve stimulation effect on quality of life in treatment-resistant depression: a pilot study. J ECT 34(4):E58–E60. https://doi.org/10.1097/YCT.0000000000000485

Truong DQ, Bikson M (2018) Physics of transcranial direct current stimulation devices and their history. J ECT 34(3):137–143. https://doi.org/10.1097/YCT.0000000000000531

Tufail Y, Yoshihiro A, Pati S, Li MM, Tyler WJ (2011) Ultrasonic neuromodulation by brain stimulation with transcranial ultrasound. Nat Protoc 6(9):1453–1470. https://doi.org/10.1038/nprot.2011.371

Tzadok M, Harush A, Nissenkorn A, Zauberman Y, Feldman Z, Ben-zeev B (2019) Clinical outcomes of closed-loop vagal nerve stimulation in patients with refractory epilepsy. Seizure 71:140–144. https://doi.org/10.1016/J.SEIZURE.2019.07.006

Ura H, Sugaya Y, Ohata H, Takumi I, Sadamoto K, Shibasaki T, Maru E (2013) Vagus nerve stimulation induced long-lasting enhancement of synaptic transmission and decreased granule cell discharge in the hippocampal dentate gyrus of urethane-anesthetized rats. Brain Res 1492:63–71. https://doi.org/10.1016/J.BRAINRES.2012.11.024

Van Buel EM, Sigrist H, Seifritz E, Fikse L, Bosker FJ, Schoevers RA, Klein HC, Pryce CR, Eisel ULM (2017) Mouse repeated electroconvulsive seizure (ECS) does not reverse social stress effects but does induce behavioral and hippocampal changes relevant to electroconvulsive therapy (ECT) side-effects in the treatment of depression. PloS One 12(9). https://doi.org/10.1371/JOURNAL.PONE.0184603

van Rooij SJH, Sippel LM, McDonald WM, Holtzheimer PE (2021) Defining focal brain stimulation targets for PTSD using neuroimaging. Depress Anxiety 38(7):768–785. https://doi.org/10.1002/DA.23159

Van Schuerbeek A, Vanderhasselt M-A, Baeken C, Pierre A, Smolders I, Van Waes V, De Bundel D (2023) Effects of repeated tDCS on fear extinction in mice. Brain Stimul 16(1):120. https://doi.org/10.1016/j.brs.2023.01.025

Van Waarde JA, Scholte HS, Van Oudheusden LJB, Verwey B, Denys D, Van Wingen GA (2014) A functional MRI marker may predict the outcome of electroconvulsive therapy in severe and treatment-resistant depression. Mol Psychiatry 20(5):609–614. https://doi.org/10.1038/mp.2014.78

Van Waarde JA, Scholte HS, Van Oudheusden LJB, Verwey B, Denys D, Van Wingen GA (2015) A functional MRI marker may predict the outcome of electroconvulsive therapy in severe and treatment-resistant depression. Mol Psychiatry 20(5):609–614. https://doi.org/10.1038/MP.2014.78

van't Wout M, Longo SM, Reddy MK, Philip NS, Bowker MT, Greenberg BD (2017) Transcranial direct current stimulation may modulate extinction memory in posttraumatic stress disorder. Brain Behav 7(5):e00681. https://doi.org/10.1002/BRB3.681

van't Wout-Frank M, Shea MT, Larson VC, Greenberg BD, Philip NS (2019) Combined transcranial direct current stimulation with virtual reality exposure for posttraumatic stress disorder: feasibility and pilot results. Brain Stimul 12(1):41–43. https://doi.org/10.1016/J.BRS.2018.09.011

van't Wout-Frank M, Brigido S, Cole J, Philip N (2023) Utilizing tDCS to augment the formation of safety signals for fear inhibition in posttraumatic stress disorder. Brain Stimul 16(1):120–121. https://doi.org/10.1016/j.brs.2023.01.026

Vedam-Mai V, Deisseroth K, Giordano J, Lazaro-Munoz G, Chiong W, Suthana N, Langevin JP, Gill J, Goodman W, Provenza NR, Halpern CH, Shivacharan RS, Cunningham TN, Sheth SA, Pouratian N, Scangos KW, Mayberg HS, Horn A, Johnson KA et al (2021) Proceedings of the eighth annual deep brain stimulation think tank: advances in optogenetics, ethical issues affecting DBS research, neuromodulatory approaches for depression, adaptive neurostimulation, and emerging DBS technologies. Front Hum Neurosci 15:169. https://doi.org/10.3389/FNHUM.2021.644593/BIBTEX

Vicario CM, Nitsche MA, Felmingham K (2017) Forgetting fear associations through tES: which memory process might be critical? Transl Psychiatry 7(2):–e1046. https://doi.org/10.1038/tp.2017.26

Vicario CM, Nitsche MA, Hoysted I, Yavari F, Avenanti A, Salehinejad MA, Felmingham KL (2020) Anodal transcranial direct current stimulation over the ventromedial prefrontal cortex enhances fear extinction in healthy humans: a single blind sham-controlled study. Brain Stimul 13(2):489–491. https://doi.org/10.1016/j.brs.2019.12.022

Vidal-Gonzalez I, Vidal-Gonzalez B, Rauch SL, Quirk GJ (2006) Microstimulation reveals opposing influences of prelimbic and infralimbic cortex on the expression of conditioned fear. Learn Mem 13(6):728. https://doi.org/10.1101/LM.306106

Wang Y, Zhan G, Cai Z, Jiao B, Zhao Y, Li S, Luo A (2021) Vagus nerve stimulation in brain diseases: therapeutic applications and biological mechanisms. Neurosci Biobehav Rev 127:37–53. https://doi.org/10.1016/J.NEUBIOREV.2021.04.018

Watts BV (2007) Electroconvulsive therapy for comorbid major depressive disorder and posttraumatic stress disorder. J ECT 23(2):93–95. https://doi.org/10.1097/01.YCT.0000264369.38033.34

Watts BV, Landon B, Groft A, Young-Xu Y (2012) A sham controlled study of repetitive transcranial magnetic stimulation for posttraumatic stress disorder. Brain Stimul 5(1):38–43. https://doi.org/10.1016/J.BRS.2011.02.002

Wei Q, Bai T, Chen Y, Ji G, Hu X, Xie W, Xiong Z, Zhu D, Wei L, Hu P, Yu Y, Wang K, Tian Y (2018) The changes of functional connectivity strength in electroconvulsive therapy for depression: a longitudinal study. Front Neurosci 12. https://doi.org/10.3389/FNINS.2018.00661

Weissman C, Blumberger D, Daskalakis Z (2022) MST for treatment-resistant depression. Managing treatment-resistant depression: road to novel therapeutics, pp 309–317. https://doi.org/10.1016/B978-0-12-824067-0.00020-7

Wen Z, Seo J, Pace-Schott EF, Milad MR (2022) Abnormal dynamic functional connectivity during fear extinction learning in PTSD and anxiety disorders. Mol Psychiatry 27(4):2216–2224. https://doi.org/10.1038/s41380-022-01462-5

Winston GM, Guadix S, Lavieri MT, Uribe-Cardenas R, Kocharian G, Williams N, Sholle E, Grinspan Z, Hoffman CE (2021) Closed-loop vagal nerve stimulation for intractable epilepsy: a single-center experience. Seizure 88:95–101. https://doi.org/10.1016/J.SEIZURE.2021.03.030

Wittbrodt MT, Gurel NZ, Nye JA, Ladd S, Shandhi MMH, Huang M, Shah AJ, Pearce BD, Alam ZS, Rapaport MH, Murrah N, Ko YA, Haffer AA, Shallenberger LH, Vaccarino V, Inan OT, Bremner JD (2020) Non-invasive vagal nerve stimulation decreases brain activity during trauma scripts. Brain Stimul 13(5):1333–1348. https://doi.org/10.1016/J.BRS.2020.07.002

Wu H, Hariz M, Visser-Vandewalle V, Zrinzo L, Coenen VA, Sheth SA, Bervoets C, Naesström M, Blomstedt P, Coyne T, Hamani C, Slavin K, Krauss JK, Kahl KG, Taira T, Zhang C, Sun B, Toda H, Schlaepfer T et al (2020) Deep brain stimulation for refractory obsessive-compulsive disorder (OCD): emerging or established therapy? Mol Psychiatry 26(1):60–65. https://doi.org/10.1038/s41380-020-00933-x

Yan T, Xie Q, Zheng Z, Zou K, Wang L (2017) Different frequency repetitive transcranial magnetic stimulation (rTMS) for posttraumatic stress disorder (PTSD): a systematic review and meta-analysis. J Psychiatr Res 89:125–135. https://doi.org/10.1016/J.JPSYCHIRES.2017.02.021

Yap JYY, Keatch C, Lambert E, Woods W, Stoddart PR, Kameneva T (2020) Critical review of transcutaneous Vagus nerve stimulation: challenges for translation to clinical practice. Front Neurosci 14:284. https://doi.org/10.3389/FNINS.2020.00284/BIBTEX

Yavari F, Jamil A, Mosayebi Samani M, Vidor LP, Nitsche MA (2018) Basic and functional effects of transcranial electrical stimulation (tES) – an introduction. Neurosci Biobehav Rev 85:81–92. https://doi.org/10.1016/J.NEUBIOREV.2017.06.015

Zandvakili A, Philip NS, Jones SR, Tyrka AR, Greenberg BD, Carpenter LL (2019) Use of machine learning in predicting clinical response to transcranial magnetic stimulation in comorbid posttraumatic stress disorder and major depression: a resting state electroencephalography study. J Affect Disord 252:47–54. https://doi.org/10.1016/J.JAD.2019.03.077

Zhao Z, Ji H, Zhang C, Pei J, Zhang X, Yuan Y (2023) Modulation effects of low-intensity transcranial ultrasound stimulation on the neuronal firing activity and synaptic plasticity of mice. Neuroimage 270:119952. https://doi.org/10.1016/J.NEUROIMAGE.2023.119952

Zielinski M, Mahdavi K, Jordan S, Haroon J, Habelhah B, Beccera S, Spivak N, Kuhn T, Bystrisky A (2021) A case study of low-intensity focused ultrasound for treatment-resistant generalized anxiety disorder and major depressive disorder. Brain Stimul 14(6):1667. https://doi.org/10.1016/j.brs.2021.10.252